London Mathematical Society Student Texts

T0269083

Managing editor: Professor J.W. Bruce, Department of Mathe[...]
University of Liverpool, UK

London Mathematical Society Student Texts 65

Elements of the Representation Theory of Associative Algebras
Volume 1 Techniques of Representation Theory

IBRAHIM ASSEM
Université de Sherbrooke

DANIEL SIMSON
Nicolaus Copernicus University

ANDRZEJ SKOWROŃSKI
Nicolaus Copernicus University

CAMBRIDGE
UNIVERSITY PRESS

CAMBRIDGE UNIVERSITY PRESS
Cambridge, New York, Melbourne, Madrid, Cape Town, Singapore,
São Paulo, Delhi, Dubai, Tokyo

Cambridge University Press
The Edinburgh Building, Cambridge CB2 8RU, UK

Published in the United States of America by Cambridge University Press, New York

www.cambridge.org
Information on this title: www.cambridge.org/9780521586313

First published 2006

A catalogue record for this publication is available from the British Library

Library of Congress Cataloguing in Publication data

Assem, Ibrahim.
Elements of the representation theory of associative algebras / Ibrahim Assem,
Daniel Simson, Andrzej Skowroński.
p. cm. – (London Mathematical Society student texts ; 65)
Includes bibliographical references and index.
Contents: v. 1. Techniques of representation theory
ISBN 0-521-58423-X (v. 1 : hardback) – ISBN 0-521-58631-3 (v. 1 : pbk.)
1. Associative algebras. 2. Representations of algebras. I. Simson, Daniel.
II. Skowroński, Andrzej. III. Title. IV. Series.
QA251.5.A767 2005
512′.46–dc22 2004045885

ISBN 978-0-521-58423-4 Hardback
ISBN 978-0-521-58631-3 Paperback

Transferred to digital printing 2010

Contents

Introduction

The idea of representing a complex mathematical object by a simpler one is as old as mathematics itself. It is particularly useful in classification problems. For instance, a single linear transformation on a finite dimensional vector space is very adequately characterised by its reduction to its rational or its Jordan canonical form. It is now generally accepted that the representation theory of associative algebras traces its origin to Hamilton's description of the complex numbers by pairs of real numbers. During the 1930s, E. Noether gave to the theory its modern setting by interpreting representations as modules. That allowed the arsenal of techniques developed for the study of semisimple algebras as well as the language and machinery of homological algebra and category theory to be applied to representation theory. Using these, the theory grew rapidly over the past thirty years.

Nowadays, studying the representations of an algebra (which we always assume to be finite dimensional over an algebraically closed field, associative, and with an identity) is understood as involving the classification of the (finitely generated) indecomposable modules over that algebra and the homomorphisms between them. The rapid growth of the theory and the extent of the published original literature became major obstacles for the beginners seeking to make their way into this area.

We are writing this textbook with these considerations in mind: It is therefore primarily addressed to graduate students starting research in the representation theory of algebras. It should also, we hope, be of interest to mathematicians working in other fields.

At the origin of the present developments of the theory is the almost simultaneous introduction and use on the one hand of quiver-theoretical techniques by P. Gabriel and his school and, on the other hand, of the theory of almost split sequences by M. Auslander, I. Reiten, and their students. An essential rôle in the theory is also played by integral quadratic forms. Our approach in this book consists in developing these theories on an equal footing, using their interplay to obtain our main results. Our strong belief is that this combination is best at yielding both concrete illustrations of the concepts and the theorems and an easier computation of actual examples. We have thus taken particular care in introducing in the text as many as possible of the latter and have included a large number of workable exercises.

With these purposes in mind, we divide our material into two parts.

The first volume serves as a general introduction to some of the techniques most commonly used in representation theory. We start by showing in Chapters II and III how one can represent an algebra by a bound quiver and a module by a linear representation of the bound quiver. We then turn in Chapter IV to the Auslander–Reiten theory of almost split sequences, giving various characterisations of these, showing their existence in module categories, and introducing one of our main working tools, the so-called Auslander–Reiten quiver. As a first and easy application of these concepts, we show in Chapter V how one can obtain a complete description of the representation theory of the Nakayama (or generalised uniserial) algebras. We return to theory in Chapter VI, giving an outline of tilting theory, another of our main working tools. A first application of tilting theory is the classification in Chapter VII of those hereditary algebras that are representation–finite (that is, admit only finitely many isomorphism classes of indecomposable modules) by means of the Dynkin diagrams, a result now known as Gabriel's theorem. We then study in Chapter VIII a class of algebras whose representation theory is as "close" as possible to that of hereditary algebras, the class of tilted algebras introduced by D. Happel and C. M. Ringel. Besides the general properties of tilted algebras, we give a very handy criterion, due to S. Liu and A. Skowroński, allowing verification of whether a given algebra is tilted or not. The last chapter in this volume deals with indecomposable modules not lying on an oriented cycle of nonzero nonisomorphisms between indecomposable modules.

Throughout this volume, we essentially use integral quadratic form techniques. We present them here in the spirit of Ringel [144].

The first volume ends with an appendix collecting, for the convenience of the reader, the notations and terminology on categories, functors, and homology and recalling some of the basic facts from category theory and homological algebra needed in the book. In Chapter I, we introduce the notation and terminology we use on algebras and modules, and we briefly recall some of the basic facts from module theory. We introduce the notions of the radical of an algebra and of a module; the notions of semisimple module, projective cover, injective envelope, the socle, and the top of a module, local algebra, primitive idempotent. We also collect basic facts from the module theory of finite dimensional K-algebras.

The reader interested mainly in linear representations of quivers and path algebras or familiar with elementary facts on rings and modules can skip Chapter I.

It is our experience that the contents of the first volume of this book can be covered during one (eight-month) course.

The main aim of the second volume, "Representation–Infinite Tilted Algebras," is to study some interesting classes of representation-infinite algebras A and, in particular, to give a fairly complete description of the representation theory of representation-infinite tilted algebras. If the algebra A is tame hereditary, that is, if the underlying graph of its quiver is a Euclidean diagram, we show explicitly how to compute the regular indecomposable modules over A, and then over any tame concealed algebra.

It was not possible to be encyclopedic in this work. Therefore many important topics from the theory have been left out. Among the most notable omissions are covering techniques, the use of derived categories and partially ordered sets. Some other aspects of the theory presented here are discussed in the books [21], [31], [76], [98], [84], [151], and especially [144].

Throughout this book, the symbols $\mathbb{N}$, $\mathbb{Z}$, $\mathbb{Q}$, $\mathbb{R}$, and $\mathbb{C}$ mean the sets of natural numbers, integers, rational, real, and complex numbers, and $\mathbb{M}_n(K)$ means the set of all square $n \times n$ matrices over K. The cardinality of a set X is denoted by $|X|$.

We take pleasure in thanking all our colleagues and students who helped us with their comments and suggestions. We wish particularly to express our appreciation to Sheila Brenner, Otto Kerner, and Kunio Yamagata for their helpful discussions and suggestions. Particular thanks are due to François Huard and Jessica Lévesque, and to Mrs. Jolanta Szelatyńska for her help in preparing a print-ready copy of the manuscript.

Chapter I

Algebras and modules

We introduce here the notations and terminology we use on algebras and modules, and we briefly recall some of the basic facts from module theory. Examples of algebras, modules, and functors are presented. We introduce the notions of the (Jacobson) radical of an algebra and of a module; the notions of semisimple module, projective cover, injective envelope, the socle and the top of a module, local algebra, and primitive idempotent. We also collect basic facts from the module theory of finite dimensional K-algebras. In this chapter we present complete proofs of most of the results, except for a few classical theorems. In these cases the reader is referred to the following textbooks on this subject [2], [6], [49], [61], [131], and [165].

Throughout, we freely use the basic notation and facts on categories and functors introduced in the Appendix.

The reader interested mainly in linear representations of quivers and path algebras or familiar with elementary facts on rings and modules can skip this chapter and begin with Chapter II.

For the sake of simplicity of presentation, we always suppose that K is an algebraically closed field and that an algebra means a finite dimensional K-algebra, unless otherwise specified.

I. 1 Algebras

By a **ring**, we mean a triple $(A, +, \cdot)$ consisting of a set A, two binary operations: addition $+ : A \times A \rightarrow A$, $(a, b) \mapsto a + b$; multiplication $\cdot : A \times A \rightarrow A$, $(a, b) \mapsto ab$, such that $(A, +)$ is an abelian group, with zero element $0 \in A$, and the following conditions are satisfied:

(i) $(ab)c = a(bc)$,

(ii) $a(b + c) = ab + ac$ and $(b + c)a = ba + ca$

for all $a, b, c \in A$. In other words, the multiplication is associative and both left and right distributive over the addition. A ring A is **commutative** if $ab = ba$ for all $a, b \in A$.

We only consider rings such that there is an element $1 \in A$ where $1 \neq 0$ and $1a = a1 = a$ for all $a \in A$. Such an element is unique with respect to this property; we call it the **identity** of the ring A. In this case the ring is a quadruple $(A, +, \cdot, 1)$. Throughout, we identify the ring $(A, +, \cdot, 1)$ with its underlying set A.

A ring K is a **skew field** (or division ring) if every nonzero element a in K is invertible, that is, there exists $b \in K$ such that $ab = 1$ and $ba = 1$. A skew field K is said to be a **field** if K is commutative.

A field K is **algebraically closed** if any nonconstant polynomial $h(t)$ in one indeterminate t with coefficients in K has a root in K.

If A and B are rings, a map $f : A \to B$ is called a **ring homomorphism** if $f(a + b) = f(a) + f(b)$ and $f(ab) = f(a)f(b)$ for all $a, b \in A$. If, in addition, A and B are rings with identity elements we assume that the ring homomorphism f preserves the identities, that is, that $f(1) = 1$.

Let K be a field. A K-**algebra** is a ring A with an identity element (denoted by 1) such that A has a K-vector space structure compatible with the multiplication of the ring, that is, such that

$$\lambda(ab) = (a\lambda)b = a(\lambda b) = (ab)\lambda$$

for all $\lambda \in K$ and all $a, b \in A$. A K-algebra A is said to be **finite dimensional** if the dimension $\dim_K A$ of the K-vector space A is finite.

A K-vector subspace B of a K-algebra A is a K-**subalgebra** of A if the identity of A belongs to B and $bb' \in B$ for all $b, b' \in B$. A K-vector subspace I of a K-algebra A is a **right ideal** of A (or **left ideal** of A) if $xa \in I$ (or $ax \in I$, respectively) for all $x \in I$ and $a \in A$. A two-sided ideal of A (or simply an ideal of A) is a K-vector subspace I of A that is both a left ideal and a right ideal of A.

It is easy to see that if I is a two-sided ideal of a K-algebra A, then the quotient K-vector space A/I has a unique K-algebra structure such that the canonical surjective linear map $\pi : A \to A/I$, $a \mapsto \bar{a} = a + I$, becomes a K-algebra homomorphism.

If I is a two-sided ideal of A and $m \geq 1$ is an integer, we denote by I^m the two-sided ideal of A generated by all elements $x_1 x_2 \ldots x_m$, where $x_1, x_2, \ldots, x_m \in I$, that is, I^m consists of all finite sums of elements of the form $x_1 x_2 \ldots x_m$, where $x_1, x_2, \ldots, x_m \in I$. We set $I^0 = A$. The ideal I is said to be **nilpotent** if $I^m = 0$ for some $m \geq 1$.

If A and B are K-algebras, then a ring homomorphism $f : A \to B$ is called a K-**algebra homomorphism** if f is a K-linear map. Two K-algebras A and B are called isomorphic if there is a K-algebra isomorphism $f : A \to B$, that is, a bijective K-algebra homomorphism. In this case we write $A \cong B$.

Throughout this book, K denotes an algebraically closed field.

1.1. Examples. (a) The ring $K[t]$ of all polynomials in the indeterminate t with coefficients in K and the ring $K[t_1, \ldots, t_n]$ of all polynomials in commuting indeterminates $t_1, \ldots, t_n$ with coefficients in K are infinite dimensional K-algebras.

(b) If A is a K-algebra and $n \in \mathbb{N}$, then the set $\mathbb{M}_n(A)$ of all $n \times n$ square matrices with coefficients in A is a K-algebra with respect to the usual matrix addition and multiplication. The identity of $\mathbb{M}_n(A)$ is the matrix $E = \mathrm{diag}(1, \ldots, 1) \in \mathbb{M}_n(A)$ with 1 on the main diagonal and zeros elsewhere. In particular $\mathbb{M}_n(K)$ is a K-algebra of dimension n^2. A K-basis of $\mathbb{M}_n(K)$ is the set of matrices e_{ij}, $1 \leq i, j \leq n$, where e_{ij} has the coefficient 1 in the position (i, j) and the coefficient 0 elsewhere.

(c) The subset
$$
\mathbb{T}_n(K) = \begin{bmatrix} K & 0 & \cdots & 0 \\ K & K & \cdots & 0 \\ \vdots & \vdots & \ddots & \vdots \\ K & K & \cdots & K \end{bmatrix}
$$
of $\mathbb{M}_n(K)$ consisting of all triangular matrices $[a_{ij}]$ in $\mathbb{M}_n(K)$ with zeros over the main diagonal is a K-subalgebra of $\mathbb{M}_n(K)$. If $n = 3$ then the subset
$$
A = \begin{bmatrix} K & 0 & 0 \\ 0 & K & 0 \\ K & K & K \end{bmatrix}
$$
of $\mathbb{M}_3(K)$ consisting of all lower triangular matrices $\lambda = [\lambda_{ij}] \in \mathbb{T}_3(K)$ with $\lambda_{21} = 0$ is a K-subalgebra of $\mathbb{M}_3(K)$, and also of $\mathbb{T}_3(K)$.

(d) Suppose that $(I; \preceq)$ is a finite **poset** (partially ordered set), where $I = \{a_1, \ldots, a_n\}$ and $\preceq$ is a partial order relation on I. The subset
$$
KI = \left\{ \lambda = [\lambda_{ij}] \in \mathbb{M}_n(K); \ \lambda_{st} = 0 \text{ if } a_s \npreceq a_t \right\}
$$
of $\mathbb{M}_n(K)$ consisting of all matrices $\lambda = [\lambda_{ij}]$ such that $\lambda_{ij} = 0$ if the relation $a_i \preceq a_j$ does not hold in I is a K-subalgebra of $\mathbb{M}_n(K)$. We call KI the **incidence algebra** of the poset $(I; \preceq)$ with coefficients in K. The matrices $\{e_{ij}\}$ with $a_i \preceq a_j$ form a basis of the K-vector space KI.

Without loss of generality, we may suppose that $I = \{1, \ldots, n\}$ and that $i \preceq j$ implies that $i \geq j$ in the natural order. This can easily be done by a suitable renumbering of the elements in I. In this case, KI takes the form of the lower triangular matrix algebra
$$
KI = \begin{bmatrix} K & 0 & \cdots & 0 \\ K_{21} & K & \cdots & 0 \\ \vdots & \vdots & \ddots & \vdots \\ K_{n1} & K_{n2} & \cdots & K \end{bmatrix},
$$
where $K_{ij} = K$ if $i \preceq j$ and $K_{ij} = 0$ otherwise. For example, if $(I; \preceq)$ is the poset $\{1 \succ 2 \succ 3 \succ \cdots \succ n\}$ then the algebra KI is isomorphic to the algebra $\mathbb{T}_n(K)$ in Example 1.1 (c). If $(I; \preceq)$ is the poset $\{1 \succ 3 \prec 2\}$ then

the incidence algebra KI is isomorphic to the five-dimensional algebra A in Example 1.1 (c). If the poset $(I; \preceq)$ is given by $I = \{1, 2, 3, 4\}$ and the relations $\{3 \succ 4 \prec 2 \prec 1 \succ 3\}$ then

$$KI = \begin{bmatrix} K & 0 & 0 & 0 \\ K & K & 0 & 0 \\ K & 0 & K & 0 \\ K & K & K & K \end{bmatrix}.$$

(e) The associative ring $K\langle t_1, t_2 \rangle$ of all polynomials in two noncommuting indeterminates t_1 and t_2 with coefficients in K is an infinite dimensional K-algebra. Note that, if I is the two-sided ideal in $K\langle t_1, t_2 \rangle$ generated by the element $t_1 t_2 - t_2 t_1$, then the K-algebra $K\langle t_1, t_2 \rangle / I$ is isomorphic to $K[t_1, t_2]$.

(f) Let $(G, \cdot)$ be a finite group with identity element e and let A be a K-algebra. The **group algebra** of G with coefficients in A is the K-vector space AG consisting of all the formal sums $\sum_{g \in G} g \lambda_g$, where $\lambda_g \in A$ and $g \in G$, with the multiplication defined by the formula

$$\left(\sum_{g \in G} g \lambda_g \right) \cdot \left(\sum_{h \in G} h \mu_h \right) = \sum_{f = gh \in G} f \lambda_g \mu_h.$$

Then AG is a K-algebra of dimension $|G| \cdot \dim_K A$ (here $|G|$ denotes the order of G) and the element $e = e1$ is the identity of AG. If $A = K$, then the elements $g \in G$ form a basis of KG over K.

For example, if G is a cyclic group of order m, then $KG \cong K[t]/(t^m - 1)$.

(g) Assume that A_1 and A_2 are K-algebras. The **product of the algebras** A_1 and A_2 is the algebra $A = A_1 \times A_2$ with the addition and the multiplication given by the formulas $(a_1, a_2) + (b_1, b_2) = (a_1 + b_1, a_2 + b_2)$ and $(a_1, a_2)(b_1, b_2) = (a_1 b_1, a_2 b_2)$, where $a_1, b_1 \in A_1$ and $a_2, b_2 \in A_2$. The identity of A is the element $1 = (1, 1) = e_1 + e_2 \in A_1 \times A_2$, where $e_1 = (1, 0)$ and $e_2 = (0, 1)$.

(h) For any K-algebra A we define the **opposite algebra** A^{op} of A to be the K-algebra whose underlying set and vector space structure are just those of A, but the multiplication $*$ in A^{op} is defined by formula $a * b = ba$.

1.2. Definition. The (Jacobson) **radical** $\operatorname{rad} A$ of a K-algebra A is the intersection of all the maximal right ideals in A.

It follows from (1.3) that $\operatorname{rad} A$ is the intersection of all the maximal left ideals in A. In particular, $\operatorname{rad} A$ is a two-sided ideal.

1.3. Lemma. *Let A be a K-algebra and let $a \in A$. The following conditions are equivalent:*

(a) $a \in \operatorname{rad} A$;

(a′) *a belongs to the intersection of all maximal left ideals of A;*
(b) *for any $b \in A$, the element $1 - ab$ has a two-sided inverse;*
(b′) *for any $b \in A$, the element $1 - ab$ has a right inverse;*
(c) *for any $b \in A$, the element $1 - ba$ has a two-sided inverse;*
(c′) *for any $b \in A$, the element $1 - ba$ has a left inverse.*

Proof. (a) implies (b′). Let $b \in A$ and assume to the contrary that $1 - ab$ has no right inverse. Then there exists a maximal right ideal I of A such that $1 - ab \in I$. Because $a \in \operatorname{rad} A \subseteq I$, $ab \in I$ and $1 \in I$; this is a contradiction. This shows that $1 - ab$ has a right inverse.

(b′) implies (a). Assume to the contrary that $a \notin \operatorname{rad} A$ and let I be a maximal right ideal of A such that $a \notin I$. Then $A = I + aA$ and therefore there exist $x \in I$ and $b \in A$ such that $1 = x + ab$. It follows that $x = 1 - ab \in I$ has no right inverse, contrary to our assumption. The equivalence of (a′) and (c′) can be proved in a similar way.

The equivalence of (b) and (c) is a consequence of the following two simple implications:
(i) If $(1 - cd)x = 1$, then $(1 - dc)(1 + dxc) = 1$.
(ii) If $y(1 - cd) = 1$, then $(1 + dyc)(1 - dc) = 1$.

(b′) implies (b). Fix an element $b \in A$. By (b′), there exists an element $c \in A$ such that $(1 - ab)c = 1$. Hence $c = 1 - a(-bc)$ and, according to (b′), there exists $d \in A$ such that $1 = cd = d + abcd = d + ab$. It follows that $d = 1 - ab$, c is the left inverse of $1 - ab$ and (b) follows. That (c′) implies (c) follows in a similar way. Because (b) implies (b′) and (c) implies (c′) obviously, the lemma is proved. □

1.4. Corollary. *Let $\operatorname{rad} A$ be the radical of an algebra A.*
(a) $\operatorname{rad} A$ *is the intersection of all the maximal left ideals of A.*
(b) $\operatorname{rad} A$ *is a two-sided ideal and $\operatorname{rad}(A/\operatorname{rad} A) = 0$.*
(c) *If I is a two-sided nilpotent ideal of A, then $I \subseteq \operatorname{rad} A$. If, in addition, the algebra A/I is isomorphic to a product $K \times \cdots \times K$ of copies of K, then $I = \operatorname{rad} A$.*

Proof. The statements (a) and (b) easily follow from (1.3).
(c) Assume that $I^m = 0$ for some $m > 0$. Let $x \in I$ and let a be an element of A. Then $ax \in I$ and therefore $(ax)^r = 0$ for some $r > 0$. It follows that the equality $(1 + ax + (ax)^2 + \cdots + (ax)^{r-1})(1 - ax) = 1$ holds for any element $a \in A$, and, according to (1.3), the element x belongs to $\operatorname{rad} A$. Consequently, $I \subseteq \operatorname{rad} A$. To prove the reverse inclusion, assume that the algebra A/I is isomorphic to a product of copies of K. It follows that $\operatorname{rad}(A/I) = 0$. Next, the canonical surjective algebra homomorphism $\pi : A \to A/I$ carries $\operatorname{rad} A$ to $\operatorname{rad}(A/I) = 0$. Indeed, if $a \in \operatorname{rad} A$ and $\pi(b) = b + I$, with $b \in A$, is any element of A/I then, by (1.3), $1 - ba$ is

invertible in A and therefore the element $\pi(1-ba) = 1-\pi(b)\pi(a)$ is invertible in A/I; thus $\pi(a) \in \operatorname{rad} A/I = 0$, by (1.3). This yields $\operatorname{rad} A \subseteq \operatorname{Ker} \pi = I$ and finishes the proof. $\qquad\qquad\qquad\qquad\qquad\qquad\qquad\qquad\qquad\qquad\qquad\qquad$ $\square$

1.5. Examples. (a) Let $s_1, \ldots, s_n$ be positive integers and let $A = K[t_1, \ldots, t_n]/(t_1^{s_1}, \ldots, t_n^{s_n})$. Because the ideal $I = (\bar{t}_1, \ldots, \bar{t}_n)$ of A generated by the cosets $\bar{t}_1, \ldots, \bar{t}_n$ of the indeterminates $t_1, \ldots, t_n$ modulo the ideal $(t_1^{s_1}, \ldots, t_n^{s_n})$ is nilpotent, then (1.4) yields $I \subseteq \operatorname{rad} A$. On the other hand, there is a K-algebra isomorphism $A/I \cong K$. It follows that I is a maximal ideal and therefore $\operatorname{rad} A = I$.

(b) Let I be a finite poset and $A = KI$ be its incidence K-algebra viewed, as in (1.1)(d), as a subalgebra of the full matrix algebra $\mathbb{M}_n(K)$. Then $\operatorname{rad} A$ is the set U of all matrices $\lambda = [\lambda_{ij}] \in KI$ with $\lambda_{ii} = 0$ for $i = 1, 2, \ldots, n$, and the algebra $A/\operatorname{rad} A$ is isomorphic to the product $K \times \cdots \times K$ of n copies of K. Indeed, we note that the set U is clearly a two-sided ideal of KI, it is easily seen that $U^n = 0$ and finally the algebra A/U is isomorphic to the product of n copies of K, thus (1.4)(c) applies.

(c) By applying the preceding arguments, one also shows that the radical $\operatorname{rad} A$ of the lower triangular matrix algebra $A = \mathbb{T}_n(K)$ of (1.1)(c) consists of all matrices in A with zeros on the main diagonal. It follows that $(\operatorname{rad} A)^n = 0$.

In the study of modules over finite dimensional K-algebras over an algebraically closed field K an important rôle is played by the following theorem, known as the Wedderburn–Malcev theorem.

1.6. Theorem. *Let A be a finite dimensional K-algebra. If the field K is algebraically closed, then there exists a K-subalgebra B of A such that there is a K-vector space decomposition $A = B \oplus \operatorname{rad} A$ and the restriction of the canonical surjective algebra homomorphism $\pi : A \to A/\operatorname{rad} A$ to B is a K-algebra isomorphism.*

Proof. See [61, section VI.2] and [131, section 11.6]. $\qquad\qquad\qquad$ $\square$

I.2   Modules

2.1. Definition. Let A be a K-algebra. A **right A-module** (or a right module over A) is a pair $(M, \cdot)$, where M is a K-vector space and $\cdot :$ $M \times A \to M$, $(m, a) \mapsto ma$, is a binary operation satisfying the following conditions:

(a) $(x + y)a = xa + ya$;
(b) $x(a + b) = xa + xb$;
(c) $x(ab) = (xa)b$;
(d) $x1 = x$;

(e) $(x\lambda)a = x(a\lambda) = (xa)\lambda$
for all $x, y \in M$, $a, b \in A$ and $\lambda \in K$.

The definition of a left A-module is analogous. Throughout, we write M or M_A instead of $(M, \cdot)$. We write A_A and $_AA$ whenever we view the algebra A as a right or left A-module, respectively.

A module M is said to be **finite dimensional** if the dimension $\dim_K M$ of the underlying K-vector space of M is finite.

A K-subspace M' of a right A-module M is said to be an A-**submodule** of M if $ma \in M'$ for all $m \in M'$ and all $a \in A$. In this case the K-vector space M/M' has a natural A-module structure such that the canonical epimorphism $\pi : M \to M/M'$ is an A-module homomorphism.

Let M be a right A-module and let I be a right ideal of A. It is easy to see that the set MI consisting of all sums $m_1a_1 + \ldots + m_sa_s$, where $s \geq 1$, $m_1, \ldots, m_s \in M$ and $a_1, \ldots, a_s \in I$, is a submodule of M.

A right A-module M is said to be **generated** by the elements $m_1, \ldots, m_s$ of M if any element $m \in M$ has the form $m = m_1a_1 + \cdots + m_sa_s$ for some $a_1, \ldots, a_s \in A$. In this case, we write $M = m_1A + \ldots + m_sA$. A module M is said to be **finitely generated** if it is generated by a finite subset of elements of M.

Let $M_1, \ldots, M_s$ be submodules of a right A-module M. We define $M_1 + \ldots + M_s$ to be the submodule of M consisting of all sums $m_1 + \cdots + m_s$, where $m_1 \in M_1, \cdots, m_s \in M_s$, and we call it the submodule generated by $M_1, \ldots, M_s$, or the sum of $M_1, \ldots, M_s$.

Note that a right module M over a finite dimensional K-algebra A is finitely generated if and only if M is finite dimensional. Indeed, if $x_1, \ldots, x_m$ is a K-basis of M, then it is obviously a set of A-generators of M. Conversely, if the A-module M is generated by the elements $m_1, \ldots, m_n$ over A and $\xi_1, \ldots \xi_s$ is a K-basis of A then the set $\{m_j\xi_i; j = 1, \ldots, n, i = 1, \ldots, s\}$ generates the K-vector space M.

Throughout, we frequently use the following lemma, known as Nakayama's lemma.

2.2. Lemma. *Let A be a K-algebra, M be a finitely generated right A-module, and $I \subseteq \operatorname{rad} A$ be a two-sided ideal of A. If $MI = M$, then $M = 0$.*

Proof. Suppose that $M = MI$ and $M = m_1A + \cdots + m_sA$, that is, M is generated by the elements $m_1, \ldots, m_s$. We proceed by induction on s. If $s = 1$, then the equality $m_1A = m_1I$ implies that $m_1 = m_1x_1$ for some $x_1 \in I$. Hence $m_1(1 - x_1) = 0$ and therefore $m_1 = 0$, because $1 - x_1$ is invertible. Consequently $M = 0$, as required.

Assume that $s \geq 2$. The equality $M = MI$ implies that there are elements $x_1, \ldots, x_s \in I$ such that $m_1 = m_1 x_1 + m_2 x_2 + \cdots + m_s x_s$. Hence $m_1(1 - x_1) = m_2 x_2 + \cdots + m_s x_s$ and therefore $m_1 \in m_2 A + \cdots + m_s A$ because $1 - x_1$ is invertible. This shows that $M = m_2 A + \cdots + m_s A$ and the inductive hypothesis yields $M = 0$. $\qquad\square$

2.3. Corollary. *If A is a finite dimensional K-algebra, then* $\operatorname{rad} A$ *is nilpotent.*

Proof. Because $\dim_K A < \infty$, the chain

$$A \supseteq \operatorname{rad} A \supseteq (\operatorname{rad} A)^2 \supseteq \cdots \supseteq (\operatorname{rad} A)^m \supseteq (\operatorname{rad} A)^{m+1} \supseteq \cdots$$

becomes stationary. It follows that $(\operatorname{rad} A)^m = (\operatorname{rad} A)^m \operatorname{rad} A$ for some m, and Nakayama's lemma (2.2) yields $(\operatorname{rad} A)^m = 0$. $\qquad\square$

Let M and N be right A-modules. A K-linear map $h : M \to N$ is said to be an **A-module homomorphism** (or simply an A-homomorphism) if $h(ma) = h(m)a$ for all $m \in M$ and $a \in A$. An A-module homomorphism $h : M \to N$ is said to be a **monomorphism** (or an **epimorphism**) if it is injective (or surjective, respectively). A bijective A-module homomorphism is called an **isomorphism**. The right A-modules M and N are said to be **isomorphic** if there exists an A-module isomorphism $h : M \to N$. In this case, we write $M \cong N$. An A-module homomorphism $h : M \to M$ is said to be an **endomorphism** of M.

The set $\operatorname{Hom}_A(M, N)$ of all A-module homomorphisms from M to N is a K-vector space with respect to the scalar multiplication $(f, \lambda) \mapsto f\lambda$ given by $(f\lambda)(m) = f(m\lambda)$ for $f \in \operatorname{Hom}_A(M, N)$, $\lambda \in K$ and $m \in M$. If the modules M and N are finite dimensional, then the K-vector space $\operatorname{Hom}_A(M, N)$ is finite dimensional. The K-vector space

$$\operatorname{End} M = \operatorname{Hom}_A(M, M)$$

of all A-module endomorphisms of any right A-module M is an associative K-algebra with respect to the composition of maps. The identity map 1_M on M is the identity of $\operatorname{End} M$.

It is easy to check that for any triple L, M, N of right A-modules the composition mapping $\cdot : \operatorname{Hom}_A(M, N) \times \operatorname{Hom}_A(L, M) \longrightarrow \operatorname{Hom}_A(L, N)$, $(h, g) \mapsto hg$, is K-bilinear.

It is clear that the **kernel** $\operatorname{Ker} h = \{m \in M \mid h(m) = 0\}$, the **image** $\operatorname{Im} h = \{h(m) \mid m \in M\}$, and the **cokernel** $\operatorname{Coker} h = N / \operatorname{Im} h$ of an A-module homomorphism $h : M \to N$ have natural A-module structures.

The **direct sum** of the right A-modules $M_1, \ldots, M_s$ is defined to be the K-vector space direct sum $M_1 \oplus \cdots \oplus M_s$ equipped with an A-module structure defined by $(m_1, \ldots, m_s)a = (m_1 a, \ldots, m_s a)$ for $m_1 \in M_1, \ldots, m_s \in M_s$

and $a \in A$. We set

$$M^s = M \oplus \cdots \oplus M, \quad (s \text{ copies}).$$

A right A-module M is said to be **indecomposable** if M is nonzero and M has no direct sum decomposition $M \cong N \oplus L$, where L and N are nonzero A-modules.

We denote by $\operatorname{Mod} A$ the abelian category of all right A-modules, that is, the category whose objects are right A-modules, the morphisms are A-module homomorphisms, and the composition of morphisms is the usual composition of maps. The reader is referred to Sections 1 and 2 of the Appendix for basic facts on categories and functors. Throughout, we freely use the notation introduced there.

We note that any left A-module can be viewed as a right A^{op}-module and conversely. Therefore, throughout the text, the category $\operatorname{Mod} A^{\mathrm{op}}$ is identified with the category of all left A-modules.

We denote by $\operatorname{mod} A$ the full subcategory of $\operatorname{Mod} A$ whose objects are the finitely generated modules. It follows that if A is a finite dimensional K-algebra, then all modules in $\operatorname{mod} A$ are finite dimensional.

An important idea in the study of A-modules is to view them as sets of K-vector spaces connected by K-linear maps. This is illustrated by the following three examples.

2.4. Example. Let A be the lower triangular matrix K-subalgebra

$$A = \begin{bmatrix} K & 0 \\ K & K \end{bmatrix}$$

of the matrix algebra $\mathbb{M}_2(K)$. We note that the matrices $e_1 = \left(\begin{smallmatrix} 1 & 0 \\ 0 & 0 \end{smallmatrix}\right)$, $e_2 = \left(\begin{smallmatrix} 0 & 0 \\ 0 & 1 \end{smallmatrix}\right)$, $e_{21} = \left(\begin{smallmatrix} 0 & 0 \\ 1 & 0 \end{smallmatrix}\right)$ form a K-basis of A over K, $1_R = e_1 + e_2$, and $e_1 e_2 = e_2 e_1 = 0$.

It follows that every module X in $\operatorname{mod} A$, viewed as a K-vector space, has a direct sum decomposition $X = X_1 \oplus X_2$, where X_1, X_2 are the vector spaces Xe_1, Xe_2 over K. Note that given $a = \left(\begin{smallmatrix} a_{11} & 0 \\ a_{21} & a_{22} \end{smallmatrix}\right) \in A$ and $x = (x_1, x_2) \in X$ with $x_1 \in X_1$ and $x_2 \in X_2$ we have

$$xa = (x_1 a_{11} + x_2 a_{21}, x_2 a_{22}) = (x_1 a_{11} + f_X(x_2) a_{21}, x_2 a_{22}),$$

where $f_X : X_2 \to X_1$ is the K-linear map given by the formula $f_X(x_2) = x_2 e_{21} = x_2 e_{21} e_{11}$. It follows that X, viewed as a right A-module, can be identified with the triple $(X_1 \xleftarrow{f_X} X_2)$. Moreover, any A-module homomorphism $h : X \to Y$ can be identified with the pair (h_1, h_2) of K-linear maps $h_1 : X_1 \to Y_1$, $h_2 : X_2 \to Y_2$ that are the restrictions of h to, respectively, X_1 and X_2. These satisfy the equation $h_1 f_X = f_Y h_2$.

The converse correspondence to $X \mapsto (X_1 \xleftarrow{f_X} X_2)$ is defined by associating to any triple $(X_1 \xleftarrow{f} X_2)$ with K-vector spaces X_1, X_2 and

$f \in \mathrm{Hom}_K(X_2, X_1)$, the K-vector space $X = X_1 \oplus X_2$ endowed with the right action $\cdot : X \times A \to X$ of A on X defined by the formula $(x_1, x_2)\begin{pmatrix} a_{11} & 0 \\ a_{21} & a_{22} \end{pmatrix} = (x_1 a_{11} + f(x_2) a_{21}, x_2 a_{22})$, where $x_1 \in X_1$, $x_2 \in X_2$, and $\begin{pmatrix} a_{11} & 0 \\ a_{21} & a_{22} \end{pmatrix} \in A$.

2.5. Example. Let A be the **Kronecker algebra**

$$A = \begin{bmatrix} K & 0 \\ K^2 & K \end{bmatrix}$$

whose elements are 2×2 matrices of the form $\begin{pmatrix} \lambda & 0 \\ (u_1, u_2) & \mu \end{pmatrix}$ with $\lambda, \mu \in K$, $(u_1, u_2) \in K^2$, and the multiplication in A is defined by the formula

$$\begin{pmatrix} d & 0 \\ (u_1, u_2) & c \end{pmatrix} \begin{pmatrix} f & 0 \\ (v_1, v_2) & e \end{pmatrix} = \begin{pmatrix} df & 0 \\ (u_1 f + v_1 c, u_2 f + v_2 c) & ce \end{pmatrix}.$$

Finite dimensional right A-modules are called **Kronecker modules**. Every such A-module X can be identified with a quadruple

$$\left(X_1 \underset{\varphi_2}{\overset{\varphi_1}{\longleftarrow}} X_2 \right),$$

where X_1, X_2 are the K-vector spaces Xe_1, Xe_2, respectively, $e_1 = \begin{pmatrix} 1 & 0 \\ 0 & 0 \end{pmatrix}$, $e_2 = \begin{pmatrix} 0 & 0 \\ 0 & 1 \end{pmatrix}$, φ_1, φ_2 are the K-linear maps defined by the formulas

$$\varphi_1(x) = x \cdot \begin{pmatrix} 0 & 0 \\ \xi_1 & 0 \end{pmatrix} = x \cdot \begin{pmatrix} 0 & 0 \\ \xi_1 & 0 \end{pmatrix} \cdot e_1, \quad \varphi_2(x) = x \cdot \begin{pmatrix} 0 & 0 \\ \xi_2 & 0 \end{pmatrix} = x \cdot \begin{pmatrix} 0 & 0 \\ \xi_2 & 0 \end{pmatrix} \cdot e_1,$$

for $x \in X_2$, where $\xi_1 = (1, 0)$ and $\xi_2 = (0, 1)$ are the standard basis vectors of K^2. Any A-module homomorphism $c : X' \to X$ can be identified with a pair (c_1, c_2) of K-linear maps $c_1 : X_1' \to X_1$ and $c_2 : X_2' \to X_2$ such that $c_1 \varphi_1' = \varphi_1 c_2$ and $c_1 \varphi_2' = \varphi_2 c_2$.

The converse correspondence to $X \mapsto \left(X_1 \underset{\varphi_2}{\overset{\varphi_1}{\longleftarrow}} X_2 \right)$ is defined by associating to any quadruple $\left(X_1 \underset{\varphi_2}{\overset{\varphi_1}{\longleftarrow}} X_2 \right)$ with finite dimensional K-vector spaces X_1, X_2 and $\varphi_1, \varphi_2 \in \mathrm{Hom}_K(X_2, X_1)$, the K-vector space $X = X_1 \oplus X_2$ endowed with the right action $\cdot : X \times A \to X$ of A on X defined by the formula

$$(x_1, x_2)\begin{pmatrix} \lambda & 0 \\ (u_1, u_2) & \mu \end{pmatrix} = (x_1 \lambda + \varphi_1(x_2) u_1 + \varphi_1(x_2) u_2, x_2 \mu),$$

where $x_1 \in X_1$, $x_2 \in X_2$ and $\begin{pmatrix} \lambda & 0 \\ (u_1, u_2) & \mu \end{pmatrix} \in A$.

It follows that the category of Kronecker modules is equivalent to the category of pairs $[\Phi_1, \Phi_2]$ of matrices Φ_1, Φ_2 over K of the same size, where the map from $[\Phi_1', \Phi_2']$ to $[\Phi_1, \Phi_2]$ is a pair (C_1, C_2) of matrices with coefficients in K such that $C_1 \Phi_1' = \Phi_1 C_2$ and $C_1 \Phi_2' = \Phi_2 C_2$.

2.6. Example. Let $K[t]$ be the K-algebra of all polynomials in the indeterminate t with coefficients in K. Note that every module V in $\mathrm{Mod}\, K[t]$

may be viewed as a pair (V, h), where V is the underlying K-vector space and $h : V \to V$ is the K-linear endomorphism $v \mapsto vt$. Every $K[t]$-module homomorphism $f : V \to V'$ may be viewed as a K-linear map such that $fh = h'f$.

The converse correspondence to $V \mapsto (V, h)$ is given by attaching to any pair (V, h), with a K-vector space V and $h \in \mathrm{End}_K V$, the K-vector space V endowed with the right action $\cdot : V \times K[t] \longrightarrow V$ of $K[t]$ on V given by the formula

$$v \cdot (\lambda_0 + t\lambda_1 + \cdots + t^m \lambda_m) = v\lambda_0 + h(v)\lambda_1 + \cdots + h^m(v)\lambda_m,$$

where $v \in V$ and $\lambda_0, \ldots, \lambda_m \in K$. The reader is referred to [49] for details.

2.7. Example. Assume that $A = A_1 \times A_2$ is the product of two K-algebras A_1 and A_2. The identity of A is the element $1 = (1, 1) = e_1 + e_2 \in A_1 \times A_2$, where $e_1 = (1, 0)$ and $e_2 = (0, 1)$. Note that $e_1 e_2 = e_2 e_1 = 0$. If X_A is a right A-module, then Xe_1 is a right A_1-module, Xe_2 is a right A_2-module and there is an A-module direct sum decomposition $X = Xe_1 \oplus Xe_2$, where Xe_j is viewed as a right A-module via the algebra projection $A \to A_j$ for $j = 1, 2$. Then the same type of arguments as in the previous examples shows that the correspondence $X_A \mapsto (Xe_1, Xe_2)$ defines an equivalence of categories $\mathrm{Mod}(A_1 \times A_2) \cong \mathrm{Mod}\, A_1 \times \mathrm{Mod}\, A_2$, which we use throughout as an identification.

2.8. A matrix notation. In presenting homomorphisms between direct sums of A-modules, we use the following matrix notation. Given a set of A-module homomorphisms $f_1 : X_1 \to Y, \ldots, f_n : X_n \to Y$ and $g_1 : Y \to Z_1, \ldots, g_m : Y \to Z_m$ in $\mathrm{Mod}\, A$ we define two A-module homomorphisms

$$f = [f_1 \; \cdots \; f_n] : X_1 \oplus \cdots \oplus X_n \longrightarrow Y, \quad g = \begin{bmatrix} g_1 \\ \vdots \\ g_m \end{bmatrix} : Y \longrightarrow Z_1 \oplus \cdots \oplus Z_m$$

by the following formulas $f(x_1, \ldots, x_n) = f_1(x_1) + \ldots + f_n(x_n)$ and $g(y) = (g_1(y), \ldots, g_m(y))$ for $x_j \in X_j$ and $y \in Y$. It is easy to see that f and g are the unique A-module homomorphisms in $\mathrm{Mod}\, A$ such that $fu_j = f_j$ for $j = 1, \ldots, n$ and $p_i g = g_i$ for $i = 1, \ldots, m$, where $u_j : X_j \to X_1 \oplus \cdots \oplus X_n$ is the jth summand embedding $x_j \mapsto (0, \ldots, 0, x_j, 0, \ldots, 0)$ and $p_i : Z_1 \oplus \cdots \oplus Z_m \to Z_i$ is the ith summand projection $(z_1, \ldots, z_m) \mapsto z_i$. If $X = X_1 \oplus \cdots \oplus X_n$ and $Z = Z_1 \oplus \cdots \oplus Z_m$, then any A-module homomorphism $h : X \to Z$ in $\mathrm{Mod}\, A$ can be written in the form of an $m \times n$ matrix

$$h = [h_{ij}] = \begin{bmatrix} h_{11} & h_{12} & \cdots & h_{1n} \\ h_{21} & h_{22} & \cdots & h_{2n} \\ \vdots & \vdots & \ddots & \vdots \\ h_{m1} & h_{m2} & \cdots & h_{mn} \end{bmatrix},$$

where $h_{ij} = p_i h u_j \in \mathrm{Hom}_A(X_j, Z_i)$.

2.9. Standard dualities. Let A be a finite dimensional K-algebra. We define the functor

$$D : \operatorname{mod} A \longrightarrow \operatorname{mod} A^{\operatorname{op}}$$

by assigning to each right module M in $\operatorname{mod} A$ the dual K-vector space

$$M^* = \operatorname{Hom}_K(M, K)$$

endowed with the left A-module structure given by the formula $(a\varphi)(m) = \varphi(ma)$ for $\varphi \in \operatorname{Hom}_K(M, K)$, $a \in A$ and $m \in M$, and to each A-module homomorphism $h : M \to N$ the dual K-homomorphism $D(h) = \operatorname{Hom}_K(h, K) :$ $D(N) \longrightarrow D(M)$, $\varphi \mapsto \varphi h$, of left A-modules. One shows that D is a duality of categories, called the **standard K-duality**. The quasi-inverse to the duality D is also denoted by

$$D : \operatorname{mod} A^{\operatorname{op}} \longrightarrow \operatorname{mod} A$$

and is defined by attaching to each left A-module Y the dual K-vector space $D(Y) = Y^* = \operatorname{Hom}_K(Y, K)$ endowed with the right A-module structure given by the formula $(\varphi a)(y) = \varphi(ay)$ for $\varphi \in \operatorname{Hom}_K(Y, K)$, $a \in A$ and $y \in Y$. A straightforward calculation shows that the evaluation K-linear map $\operatorname{ev} : M \to M^{**}$ given by the formula $\operatorname{ev}(m)(f) = f(m)$, where $m \in M$ and $f \in D(M)$, defines natural equivalences of functors $1_{\operatorname{mod} A} \cong D \circ D$ and $1_{\operatorname{mod} A^{\operatorname{op}}} \cong D \circ D$.

Any right A-module M is a left module over the algebra $\operatorname{End} M$ with respect to the left multiplication $(\operatorname{End} M) \times M \to M$, $(\varphi, m) \mapsto \varphi m = \varphi(m)$. It is easy to check that M is an $(\operatorname{End} M)$–A-bimodule in the following sense.

2.10. Definition. Let A and B be two K-algebras. An A-B-**bimodule** is a triple $_AM_B = (M, *, \cdot)$, where $_AM = (M, *)$ is a left A-module, $M_B = (M, \cdot)$ is a right B-module, and $(a * m) \cdot b = a * (m \cdot b)$ for all $m \in M$, $a \in A$, $b \in B$. Throughout, we write simply am and mb instead of $a * m$ and $m \cdot b$, respectively.

For any A-B-bimodule $_AM_B$ and for any right B-module X_B, the K-vector space $\operatorname{Hom}_B(_AM_B, X_B)$ of all B-module homomorphisms from M_B to X_B is a right A-module with respect to the A-scalar multiplication $(f, a) \mapsto fa$ given by $(fa)(m) = f(am)$ for $f \in \operatorname{Hom}_B(M_B, X_B)$, $a \in A$ and $m \in M$. If M and X are finite dimensional over K, then so is $\operatorname{Hom}_B(_AM_B, X_B)$.

Important examples of functors are the Hom-functors $\operatorname{Hom}_B(_AM_B, -)$ and $\operatorname{Hom}_B(-, _AM_B)$. We define the covariant Hom-functor

$$\operatorname{Hom}_B(_AM_B, -) : \operatorname{Mod} B \longrightarrow \operatorname{Mod} A$$

by associating to X_B in $\operatorname{Mod} B$ the K-vector space $\operatorname{Hom}_B({}_A M_B, X_B)$ endowed with the right A-module structure defined earlier. If $\varphi : X_B \to Y_B$ is a homomorphism of B-modules, we define the induced homomorphism $\operatorname{Hom}_B({}_A M_B, \varphi) : \operatorname{Hom}_B({}_A M_B, X_B) \to \operatorname{Hom}_B({}_A M_B, Y_B)$ of right A-modules by the formula $f \mapsto \varphi f$. The contravariant Hom-functor

$$\operatorname{Hom}_B(-, {}_A M_B) : \operatorname{Mod} B \longrightarrow \operatorname{Mod} A^{\mathrm{op}}$$

is defined by $X_B \mapsto \operatorname{Hom}_B(X_B, {}_A M_B)$ and by assigning to any homomorphism $\psi : X_B \longrightarrow Y_B$ of right B-modules the induced homomorphism $\operatorname{Hom}_B(\psi, {}_A M_B) : \operatorname{Hom}_B(Y_B, {}_A M_B) \to \operatorname{Hom}_B(X_B, {}_A M_B)$, $f \mapsto f\psi$, of left A-modules.

We recall also that, given an A-B-bimodule ${}_A M_B$, the covariant **tensor product functors**

$$(-) \otimes_A M_B : \operatorname{Mod} A \longrightarrow \operatorname{Mod} B, \quad {}_A M \otimes_B (-) : \operatorname{Mod} B^{\mathrm{op}} \longrightarrow \operatorname{Mod} A^{\mathrm{op}}$$

are defined by associating to any right A-module X_A and to any left B-module ${}_B Y$ the tensor products $X \otimes_A M_B$ and ${}_A M \otimes_B Y$ endowed with the natural right B-module and left A-module structure, respectively. It is well known that there exists an **adjunction isomorphism**

$$\operatorname{Hom}_B(X \otimes_A M_B, Z_B) \cong \operatorname{Hom}_A(X_A, \operatorname{Hom}_B({}_A M_B, Z_B)) \qquad \textbf{(2.11)}$$

given by attaching to a B-module homomorphism $\varphi : X \otimes_A M_B \longrightarrow Z_B$ the A-module homomorphism

$$\overline{\varphi} : X_A \longrightarrow \operatorname{Hom}_B({}_A M_B, Z_B)$$

adjoint to φ defined by the formula $\overline{\varphi}(x)(m) = \varphi(x \otimes m)$, where $x \in X$ and $m \in M$. A straightforward calculation shows that the inverse to $\varphi \mapsto \overline{\varphi}$ is defined by $\psi \mapsto (x \otimes m \mapsto \psi(x)(m))$, where $x \in X$ and $m \in M$.

Formula (2.11) shows that the functor $(-) \otimes_A M_B$ is left adjoint to the functor $\operatorname{Hom}_B(-, {}_A M_B)$, and that $\operatorname{Hom}_B(-, {}_A M_B)$ is right adjoint to $(-) \otimes_A M_B$ (see (A.2.3) of the Appendix).

I.3   Semisimple modules and the radical of a module

Throughout, we assume that K is an algebraically closed field and that A is a finite dimensional K-algebra. A right A-module S is **simple** if S is nonzero and any submodule of S is either zero or S. A module M is **semisimple** if M is a direct sum of simple modules.

3.1. Schur's lemma. *Let S and S' be right A-modules, and $f : S \to S'$ be a nonzero A-homomorphism.*

(a) *If S is simple, then f is a monomorphism.*

(b) *If S' is simple, then f is an epimorphism.*

(c) *If S and S' are simple, then f is an isomorphism.*

Proof. Because $f : S \to S'$ is an A-module homomorphism, $\operatorname{Ker} h$ and $\operatorname{Im} h$ are A-submodules of S and S', respectively. Then $f \neq 0$ yields $\operatorname{Ker} h = 0$ if S is simple, and $\operatorname{Im} h = S'$ if S' is simple. The lemma follows.
$\square$

3.2. Corollary. *If S is a simple A-module, then there is a K-algebra isomorphism* $\operatorname{End} S \cong K$.

Proof. It follows from Schur's lemma that any nonzero element in $\operatorname{End} S$ is invertible and therefore $\operatorname{End} S$ is a skew field. Because S is simple, S is a cyclic A-module and therefore $\dim_K S$ is finite. It follows that $\dim_K \operatorname{End} S$ is finite and, for any nonzero element $\varphi \in \operatorname{End} S$, the elements $1_S, \varphi, \varphi^2, \ldots, \varphi^m, \ldots$ are linearly dependent over K. Consequently, there exists an irreducible nonzero polynomial $f(t) \in K[t]$ such that $f(\varphi) = 0$. Because the field K is algebraically closed, f is of degree 1 and therefore φ acts on S as the multiplication by a scalar $\lambda_\varphi \in K$. The correspondence $\varphi \mapsto \lambda_\varphi$ establishes a K-algebra isomorphism $\operatorname{End} S \cong K$. $\square$

3.3. Lemma. (a) *A finite dimensional right A-module M is semisimple if and only if for any A-submodule N of M there exists a submodule L of M such that $L \oplus N = M$.*

(b) *A submodule of a semisimple module is semisimple.*

Proof. (a) Assume that $M = S_1 \oplus \cdots \oplus S_m$, where $S_1, \ldots . S_m$ are simple modules. Let N be a nonzero A-submodule of M and let $\{S_{j_1}, \ldots, S_{j_t}\}$ be a maximal family of modules in the set $\{S_1, \ldots, S_m\}$ such that the intersection of N with the module $L = S_{j_1} \oplus \cdots \oplus S_{j_t}$ is zero. It follows that $N \cap (L + S_t) \neq 0$, for all $t \notin \{j_1, \ldots, j_m\}$. This implies that $(L + N) \cap S_t \neq 0$ and hence we conclude that $S_t \subseteq L + N$, for all $t \notin \{j_1, \ldots, j_m\}$, because S_t is simple. Consequently, we get $M = L + N$ and therefore $M = L \oplus N$. The converse implication follows easily by induction on $\dim_K M$.

Because (b) is an immediate consequence of (a), the lemma is proved.$\square$

For any right A-module M, the submodule soc M of M generated by all simple submodules of M is a semisimple module (see [2], [131]); it is called the **socle** of M. The main properties of the socle are listed in Exercise I.17.

Throughout, we frequently use the following well-known result.

3.4. Wedderburn–Artin theorem. *For any finite dimensional algebra A over an algebraically closed field K the following conditions are equivalent:*

(a) *The right A-module A_A is semisimple.*
(b) *Every right A-module is semisimple.*
(a′) *The left A-module $_AA$ is semisimple.*
(b′) *Every left A-module is semisimple.*
(c) $\operatorname{rad} A = 0$.
(d) *There exist positive integers $m_1, \ldots, m_s$ and a K-algebra isomorphism*
$$A \cong \mathbb{M}_{m_1}(K) \times \cdots \times \mathbb{M}_{m_s}(K).$$
Proof. See [2], [49], [61], [131], and [164]. $\qquad\square$

A finite dimensional K-algebra A is called **semisimple** if one of the equivalent conditions in the Wedderburn–Artin theorem (3.4) is satisfied.

By (3.4), the commutative algebra $A = K[X_1, \ldots, X_n]/(X_1^{s_1}, \ldots, X_n^{s_n})$ of Example 1.5(a), where $s_1, \ldots, s_n$ are positive integers and $n \geq 1$, is semisimple if and only if $s_1 = \ldots = s_n = 1$.

In view of Example 1.5(b), the incidence K-algebra KI of a poset I is semisimple if and only if $a_i \not\leq a_j$ for every pair of elements $a_i \neq a_j$ of I.

The semisimple group algebras KG are characterised as follows.

3.5. Maschke's theorem. *Let G be a finite group and let K be a field. Then the group algebra KG is semisimple if and only if the characteristic of K does not divide the order of G.*

Proof. See [61], [131], [164] and Section 5 of Chapter V. $\qquad\square$

We now define the radical of a module.

3.6. Definition. Let M be a right A-module. The (Jacobson) **radical** $\operatorname{rad} M$ of M is the intersection of all the maximal submodules of M.

It follows from (1.2) that the radical $\operatorname{rad} A_A$ of the right A-module A_A is the radical $\operatorname{rad} A$ of the algebra A.

The main properties of the radical are collected in the following proposition.

3.7. Proposition. *Suppose that L, M, and N are modules in $\operatorname{mod} A$.*
(a) *An element $m \in M$ belongs to $\operatorname{rad} M$ if and only if $f(m) = 0$ for any $f \in \operatorname{Hom}_A(M, S)$ and any simple right A-module S.*
(b) $\operatorname{rad}(M \oplus N) = \operatorname{rad} M \oplus \operatorname{rad} N$.
(c) *If $f \in \operatorname{Hom}_A(M, N)$, then $f(\operatorname{rad} M) \subseteq \operatorname{rad} N$.*
(d) $M \operatorname{rad} A = \operatorname{rad} M$.
(e) *Assume that L and M are A-submodules of N. If $L \subseteq \operatorname{rad} N$ and $L + M = N$, then $M = N$.*

Proof. The statement (a) follows immediately from the definition, because $L \subseteq M$ is a maximal submodule if and only if M/L is simple. The statements (b) and (c) follow immediately from (a). We leave them as an exercise.

(d) Take $m \in M$ and define a homomorphism $f_m : A \to M$ of right A-modules by the formula $f_m(a) = ma$ for $a \in A$. It follows from (c) that for $a \in \operatorname{rad} A$ we get $ma = f_m(a) \in f_m(\operatorname{rad} A) \subseteq \operatorname{rad} M$ and therefore $M \operatorname{rad} A \subseteq \operatorname{rad} M$. To prove the inclusion $\operatorname{rad} M \subseteq M \operatorname{rad} A$ we note that $(M/M \operatorname{rad} A) \operatorname{rad} A = 0$ and therefore the A-module $M/M \operatorname{rad} A$ is a module over the algebra $A/\operatorname{rad} A$ with respect to the action $(m + M \operatorname{rad} A) \cdot (a + \operatorname{rad} A) = ma + M \operatorname{rad} A$. By the Wedderburn–Artin theorem (3.4), the algebra $A/\operatorname{rad} A$ is semisimple and the finite dimensional $A/\operatorname{rad} A$-module $M/M \operatorname{rad} A$ is a direct sum of simple modules. Because the radical of any simple module is zero, (b) yields $\operatorname{rad}(M/M \operatorname{rad} A) = 0$. By (c), the canonical A-module epimorphism $\pi : M \to M/M \operatorname{rad} A$ carries $\operatorname{rad} M$ to zero, that is, $\operatorname{rad} M \subseteq \operatorname{Ker} \pi = M \operatorname{rad} A$ and we are done.

(e) Assume that $L \subseteq \operatorname{rad} N$ and $L + M = N$, and suppose to the contrary that $M \neq N$. Because N is finite dimensional, M is a submodule of a maximal submodule $X \neq N$ of N. It follows that $L \subseteq \operatorname{rad} N \subseteq X$ and we get $N = L + M \subseteq X + M = X$, contrary to our assumption. $\square$

3.8. Corollary. *Suppose that M is a module in* $\operatorname{mod} A$.

(a) *The A-module $M/\operatorname{rad} M$ is semisimple and it is a module over the K-algebra $A/\operatorname{rad} A$.*

(b) *If L is a submodule of M such that M/L is semisimple, then* $\operatorname{rad} M \subseteq L$.

Proof. (a) We recall from (3.7)(d) that $\operatorname{rad} M = M \operatorname{rad} A$. It follows that $(M/\operatorname{rad} M) \operatorname{rad} A = 0$ and therefore the A-module $M/\operatorname{rad} M$ is a module over $A/\operatorname{rad} A$ with respect to the action $(m + M \operatorname{rad} A) \cdot (a + \operatorname{rad} A) = ma + M \operatorname{rad} A$. Now, by (3.4), the algebra $A/\operatorname{rad} A$ is semisimple, and the module $M/\operatorname{rad} M$ is semisimple.

(b) Assume that L is a submodule of M such that M/L is semisimple. Consider the canonical epimorphism $\varepsilon : M \to M/L$. Because (3.7)(c) yields $\varepsilon(\operatorname{rad} M) \subseteq \operatorname{rad}(M/L) = 0$, $\operatorname{rad} M \subseteq \operatorname{Ker} \varepsilon = L$, and (b) follows. $\square$

It follows from (3.7)(d) that $(M/\operatorname{rad} M) \operatorname{rad} A = 0$ and therefore the module

$$\operatorname{top} M = M/\operatorname{rad} M,$$

called the **top** of M, is a right $A/\operatorname{rad} A$-module with respect to the action of $A/\operatorname{rad} A$ defined by the formula $(m + \operatorname{rad} M) \cdot (a + \operatorname{rad} A) = ma + \operatorname{rad} M$.

We remark that if $f : M \to N$ is an A-homomorphism, then $f(\operatorname{rad} M) \subseteq \operatorname{rad} N$ and therefore f induces a homomorphism $\operatorname{top} f : \operatorname{top} M \longrightarrow \operatorname{top} N$

of $A/\operatorname{rad}A$-modules defined by the formula $(\operatorname{top}f)(m+\operatorname{rad}M)=f(m)+\operatorname{rad}N$.

3.9. Corollary. (a) *A homomorphism $f:M\to N$ in $\operatorname{mod}A$ is surjective if and only if the homomorphism $\operatorname{top}f:\operatorname{top}M\longrightarrow\operatorname{top}N$ is surjective.*

(b) *If S is a simple A-module, then $S\operatorname{rad}A=0$ and S is a simple $A/\operatorname{rad}A$-module.*

(c) *An A-module M is semisimple if and only if $\operatorname{rad}M=0$.*

Proof. (a) Assume that $\operatorname{top}f$ is surjective. Then $\operatorname{Im}f+\operatorname{rad}N=N$ and therefore f is surjective, because (3.7)(e) yields $\operatorname{Im}f=N$. Because the converse implication is easy, (a) follows.

(b) Because $S\neq0$ and S is simple, S is cyclic and, by Nakayama's lemma (2.2), $S\neq S\operatorname{rad}A$. Hence $S\operatorname{rad}A=0$ and (b) follows.

(c) If M is semisimple, then (b) yields $\operatorname{rad}M=0$. The converse implication is a consequence of (3.7)(d) and (3.8)(a). □

Suppose that A is a finite dimensional K-algebra. If M is a module in $\operatorname{mod}A$, then there exists a chain $0=M_0\subset M_1\subset M_2\subset\ldots\subset M_m=M$ of submodules of M such that the module M_{j+1}/M_j is simple for $j=0,1,\ldots,m-1$ (see [2], [61], and [131]). This chain is called a **composition series** of M and the simple modules $M_1/M_0,\ldots,M_m/M_{m-1}$ are called the **composition factors** of M.

3.10. Jordan–Hölder theorem. *If A is a finite dimensional K-algebra and*
$$0=M_0\subset M_1\subset M_2\subset\ldots\subset M_m=M,$$
$$0=N_0\subset N_1\subset N_2\subset\ldots\subset N_n=M$$
are two composition series of a module M in $\operatorname{mod}A$, then $m=n$, and there exists a permutation σ of $\{1,\ldots,m\}$ such that, for any $j\in\{0,1,\ldots,m-1\}$, there is an A-isomorphism $M_{j+1}/M_j\cong N_{\sigma(j+1)}/N_{\sigma(j)}$.

Proof. See [2], [61], [131], and [164]. □

It follows from (3.10) that the number m of modules in a composition series $0=M_0\subset M_1\subset M_2\subset\cdots\subset M_m=M$ of M depends only on M; it is called the **length** of M and is denoted by $\ell(M)$.

As an immediate consequence of (3.10) we get the following.

3.11. Corollary. (a) *If N is an A-submodule of M in $\operatorname{mod}A$, then $\ell(M)=\ell(N)+\ell(M/N)$.*

(b) *If L and N are A-submodules of M in $\operatorname{mod}A$, then $\ell(L+N)+\ell(L\cap N)=\ell(L)+\ell(N)$.* □

I.4 Direct sum decompositions

In the study of indecomposable modules over a K-algebra A, an important rôle is played by idempotent elements of A. An element $e \in A$ is called an **idempotent** if $e^2 = e$. The idempotent e is said to be **central** if $ae = ea$ for all $a \in A$. The idempotents $e_1, e_2 \in A$ are called **orthogonal** if $e_1 e_2 = e_2 e_1 = 0$. The idempotent e is said to be **primitive** if e cannot be written as a sum $e = e_1 + e_2$, where e_1 and e_2 are nonzero orthogonal idempotents of A.

Every algebra A has two trivial idempotents 0 and 1. If the idempotent e of A is nontrivial, then $1 - e$ is also a nontrivial idempotent, the idempotents e and $1 - e$ are orthogonal, and there is a nontrivial right A-module decomposition $A_A = eA \oplus (1 - e)A$. Conversely, if $A_A = M_1 \oplus M_2$ is a nontrivial A-module decomposition and $1 = e_1 + e_2$, $e_i \in M_i$, then e_1, e_2 is a pair of orthogonal idempotents of A, and $M_i = e_i A$ is indecomposable if and only if e_i is primitive.

If e is a central idempotent, then so is $1 - e$, and hence eA and $(1-e)A$ are two-sided ideals and they are easily shown to be K-algebras with identity elements $e \in eA$ and $1 - e \in (1 - e)A$, respectively. In this case the decomposition $A_A = eA \oplus (1-e)A$ is a direct product decomposition of the algebra A.

Because the algebra A is finite dimensional, the module A_A admits a direct sum decomposition $A_A = P_1 \oplus \cdots \oplus P_n$, where $P_1, \ldots, P_n$ are indecomposable right ideals of A. It follows from the preceding discussion that $P_1 = e_1 A, \ldots, P_n = e_n A$, where $e_1, \ldots, e_n$ are primitive pairwise orthogonal idempotents of A such that $1 = e_1 + \cdots + e_n$. Conversely, every set of idempotents with the preceding properties induces a decomposition $A_A = P_1 \oplus \cdots \oplus P_n$ with indecomposable right ideals $P_1 = e_1 A, \ldots, P_n = e_n A$.

Such a decomposition is called an **indecomposable decomposition** of A and such a set $\{e_1, \cdots, e_n\}$ is called a **complete set of primitive orthogonal idempotents** of A.

We say that an algebra A is **connected** (or indecomposable) if A is not a direct product of two algebras, or equivalently, if 0 and 1 are the only central idempotents of A.

4.1. Example. The K-subalgebra $A = \begin{bmatrix} K & 0 & 0 \\ 0 & K & 0 \\ K & K & K \end{bmatrix}$ of $\mathbb{M}_3(K)$ defined in $(1.1)(c)$ is connected, $\dim_K A = 5$, and A_A has an indecomposable decomposition $A_A = e_1 A \oplus e_2 A \oplus e_3 A$, where $e_1 = \begin{bmatrix} 1 & 0 & 0 \\ 0 & 0 & 0 \\ 0 & 0 & 0 \end{bmatrix}$, $e_2 = \begin{bmatrix} 0 & 0 & 0 \\ 0 & 1 & 0 \\ 0 & 0 & 0 \end{bmatrix}$, $e_3 = \begin{bmatrix} 0 & 0 & 0 \\ 0 & 0 & 0 \\ 0 & 0 & 1 \end{bmatrix}$ are primitive orthogonal idempotents of A such that $1_A =$

$e_1 + e_2 + e_3$. The right ideal $e_j A$ consists of all matrices $\lambda = [\lambda_{st}]$ in A with $\lambda_{st} = 0$ for $s \neq j$, that is, $\lambda_{st} = 0$ outside the jth row. The right A-modules $e_1 A$ and $e_2 A$ are one-dimensional; hence they are simple. We also note that the right A-module $M = e_3 A$ is of length 3. Indeed, the subspace M_1 of M consisting of the matrices $\lambda \in M$ such that $\lambda_{33} = \lambda_{32} = 0$ is a one-dimensional submodule of M (isomorphic to the simple ideal $e_{11}A$), the subspace M_2 consisting of the matrices $\lambda \in M$ such that $\lambda_{33} = 0$ is a two-dimensional submodule of M containing M_1, $\dim_K M_2/M_1 = 1$ and $\dim_K M/M_2 = 1$; hence $0 \subset M_1 \subset M_2 \subset M$ is a composition series of M and therefore $\ell(M) = 3$. $\square$

Assume that $e \in A$ is an idempotent and that M is a right A-module. It is easy to check that the K-vector subspace eAe of A is a K-algebra and that e is the identity element of eAe. Note that eAe is a subalgebra of A if and only if $e = 1$. The K-vector subspace Me of M is a right eAe-module if we set $(me) \cdot (eae) = meae$ for all $m \in M$ and $a \in A$. In particular, Ae is a right eAe-module and eA is a left eAe-module. It follows that the K-vector space $\mathrm{Hom}_A(eA, M)$ is a right eAe-module with respect to the action $(\varphi \cdot eae)(x) = \varphi(eaex)$ for $x \in eA$, $a \in A$, $\varphi \in \mathrm{Hom}_A(eA, M)$.

The following useful fact is frequently used.

4.2. Lemma. *Let A be a K-algebra, $e \in A$ be an idempotent, and M be a right A-module.*

(a) *The K-linear map*

$$\theta_M : \mathrm{Hom}_A(eA, M) \longrightarrow Me, \qquad (4.3)$$

defined by the formula $\varphi \mapsto \varphi(e) = \varphi(e)e$ for $\varphi \in \mathrm{Hom}_A(eA, M)$, is an isomorphism of right eAe-modules, and it is functorial in M.

(b) *The isomorphism $\theta_{eA} : \mathrm{End}\, eA \xrightarrow{\approx} eAe$ of right eAe-modules induces an isomorphism of K-algebras.*

Proof. It is easy to see that the map θ_M is a homomorphism of right eAe-modules and it is functorial at the variable M. We define a K-linear map $\theta'_M : Me \to \mathrm{Hom}_A(eA, M)$ by the formula $\theta'_M(me)(ea) = mea$ for $a \in A$ and $m \in M$. A straightforward calculation shows that, given $m \in M$, the map $\theta'_M(me) : eA \to M$ is well-defined (does not depend of the choice of a in the presentation ea), it is a homomorphism of A-modules, moreover θ'_M is a homomorphism of eAe-modules and θ'_M is an inverse of θ_M. This proves (a). The statement (b) easily follows from (a). $\square$

We also need the following technical but useful result.

4.4. Lemma (lifting idempotents). *For any K-algebra A the idempotents of the algebra $B = A/\mathrm{rad}\, A$ can be lifted modulo $\mathrm{rad}\, A$, that is, for* *any idempotent* $\eta = g + \operatorname{rad} A \in B$, $g \in A$, *there exists an idempotent* e *of* A *such that* $g - e \in \operatorname{rad} A$.

Proof. It follows from (2.3) that $(\operatorname{rad} A)^m = 0$ for some $m > 1$. Because $\eta^2 = \eta$, $g - g^2 \in \operatorname{rad} A$ and therefore $(g - g^2)^m = 0$. Hence, by Newton's binomial formula, we get $0 = (g - g^2)^m = g^m - g^{m+1}t$, where $t = \sum_{j=1}^{m} (-1)^{j-1} \binom{m}{j} g^{j-1}$. It follows that

　(i) $g^m = g^{m+1}t$;

　(ii) $gt = tg$.

We claim that the element $e = (gt)^m$ is the idempotent lifting η. First, we note that $e = g^m t^m = g^{m+1} t^{m+1} = \cdots = g^{2m} t^{2m} = ((gt)^m)^2 = e^2$ and therefore e is an idempotent. Next, we note that

　(iii) $g - g^m \in \operatorname{rad} A$,

because the relation $g - g^2 \in \operatorname{rad} A$ yields the equalities $g - g^m = g(1 - g^{m-1}) = g(1 - g)(1 + g + \cdots + g^{m-2}) = (g - g^2)(1 + g + \cdots + g^{m-2}) \in \operatorname{rad} A$. Moreover, we have

　(iv) $g - gt \in \operatorname{rad} A$,

because equalities (i)–(iii) yield

$$g + \operatorname{rad} A = g^m + \operatorname{rad} A = g^{m+1}t + \operatorname{rad} A = (g^{m+1} + \operatorname{rad} A)(t + \operatorname{rad} A) =$$
$$= (g^m + \operatorname{rad} A)(g + \operatorname{rad} A)(t + \operatorname{rad} A) = (g + \operatorname{rad} A)(g + \operatorname{rad} A)(t + \operatorname{rad} A) =$$
$$= (g^2 + \operatorname{rad} A)(t + \operatorname{rad} A) = (g + \operatorname{rad} A)(t + \operatorname{rad} A) = gt + \operatorname{rad} A.$$

Consequently, we get $e + \operatorname{rad} A = (gt)^m + \operatorname{rad} A = (gt + \operatorname{rad} A)^m = (g + \operatorname{rad} A)^m = g^m + \operatorname{rad} A = g + \operatorname{rad} A$ and our claim follows. □

4.5. Proposition. *Let* $B = A/\operatorname{rad} A$. *The following statements hold.*

　(a) *Every right ideal* I *of* B *is a direct sum of simple right ideals of the form* eB, *where* e *is a primitive idempotent of* B. *In particular, the right* B-*module* B_B *is semisimple.*

　(b) *Any module* N *in* $\operatorname{mod} B$ *is isomorphic to a direct sum of simple right ideals of the form* eB, *where* e *is a primitive idempotent of* B.

　(c) *If* $e \in A$ *is a primitive idempotent of* A, *then the* B-*module* $\operatorname{top} eA$ *is simple and* $\operatorname{rad} eA = e \operatorname{rad} A \subset eA$ *is the unique maximal proper submodule of* eA.

Proof. (a) Let S be a nonzero right ideal of B contained in I that is of minimal dimension. Then S is a simple B-module and $S^2 \neq 0$, because otherwise, in view of (1.4)(c), $0 \neq S \subseteq \operatorname{rad} B = 0$ and we get a contradiction. Hence $S^2 = S$ and there exists $x \in S$ such that $xS \neq 0$, $S = xS$ and $x = xe$ for some nonzero $e \in S$. Then, according to Schur's lemma, the B-homomorphism $\varphi : S \to S$ given by the formula $\varphi(y) = xy$ is bijective. Because $\varphi(e^2 - e) = x(e^2 - e) = xee - xe = xe - xe = 0$, $e^2 - e = 0$, the element $e \in S$ is a nonzero idempotent, and $S = eB$. It follows that $B = eB \oplus (1-e)B$ and $I = S \oplus (1-e)I$. Because $\dim_K (1-e)I < \dim_K I$, we can assume by induction that (a) is satisfied for $(1-e)I$ and therefore (a) follows.

(b) Let N be a B-module generated by the elements $n_1, \ldots, n_s$ and consider the B-module epimorphism $h : B^s \to N$ defined by the formula $h(\xi_i) = n_i$, where $\xi_1, \ldots, \xi_s$ is the standard basis of the B-module B^s. If N is simple, then $s = 1$ and (a) together with (3.3)(a) yields $N \cong eB$, where e is a primitive idempotent of B. Now suppose that N is arbitrary. Then, by (a), B^s is a direct sum of simple right ideals of the form eB, where e is a primitive idempotent of B, and it follows from (3.3)(a) that $B^s = \operatorname{Ker} h \oplus L$ for some B-submodule L of B^s. Then h induces an isomorphism $L \cong N$ and (b) follows from (3.3)(b).

(c) The element $\bar{e} = e + \operatorname{rad} A$ is an idempotent of B and $\operatorname{top} eA \cong \bar{e}B$. Assume to the contrary that $\bar{e}B$ is not simple. It follows from (a) that $\bar{e}B = \bar{e}_1 B \oplus \bar{e}_2 B$, where $\bar{e}_1, \bar{e}_2$ are nonzero idempotents of B such that $\bar{e} = \bar{e}_1 + \bar{e}_2$ and $\bar{e}_1 \bar{e}_2 = \bar{e}_2 \bar{e}_1 = 0$. Because $\bar{e}_1 = \bar{e}_1^2 = (\bar{e} - \bar{e}_2)\bar{e}_1 = \bar{e}\bar{e}_1$, $\bar{e}_1 = g_1 + \operatorname{rad} A$ for some $g_1 \in eA$. By (4.4), there exist $t \in A$ and $m \in \mathbb{N}$ such that the element $e_1 = (g_1 t)^m$ is an idempotent of A and $\bar{e}_1 = e_1 + \operatorname{rad} A$. It follows that $\operatorname{top} eA = \bar{e}B = \bar{e}_1 B \oplus \bar{e}_2 B$. Because $g_1 \in eA$, $e_1 \in eA$ and $e_1 A \subseteq eA$. Then the decomposition $A_A = e_1 A \oplus (1 - e_1)A$ induces the decomposition $eA = e_1 A \oplus \{(1 - e_1)A \cap eA\}$. It follows that $eA = e_1 A$, because the primitivity of e implies that eA is indecomposable. Hence $\bar{e}B = \operatorname{top} eA = \operatorname{top} e_1 A = \bar{e}_1 B$ and therefore $\bar{e}_2 B = 0$, contrary to our assumption. Consequently, the module $\operatorname{top} eA$ is simple and therefore $\operatorname{rad} eA = (eA)\operatorname{rad} A$ is a maximal proper A-submodule of eA. Now, if L is a proper A-submodule of eA that is not in $\operatorname{rad} eA$, then $L + \operatorname{rad} eA = eA$ and (3.7)(e) yields $L = eA$, a contradiction. This shows that $\operatorname{rad} eA$ contains all proper A-submodules of eA and finishes the proof. $\qquad \square$

An algebra A is said to be **local** if A has a unique maximal right ideal, or equivalently, if A has a unique maximal left ideal, see (4.6).

An example of a local algebra is the commutative algebra

$$A = K[X_1, \ldots, X_n]/(X_1^{s_1}, \ldots, X_n^{s_n}),$$

where $s_1, \ldots, s_n$ are nonzero natural numbers and $n \geq 1$. Indeed, it was shown in Example 1.5(a) that the radical $\operatorname{rad} A$ of A is a maximal ideal. It follows that $\operatorname{rad} A$ is the unique maximal ideal of A, that is, the algebra A is local.

Note that, in view of Example 1.5(b), the incidence K-algebra KI of a finite poset I is not local if $|I| \geq 2$.

Now we give a characterisation of algebras having only trivial idempotents.

4.6. Lemma. *Let A be a finite dimensional K-algebra. The following conditions are equivalent:*

(a) *A is a local algebra.*

(a$'$) *A has a unique maximal left ideal.*

(b) *The set of all noninvertible elements of A is a two-sided ideal.*

(c) *For any $a \in A$, one of the elements a or $1 - a$ is invertible.*

(d) *A has only two idempotents, 0 and 1.*

(e) *The K-algebra $A/\mathrm{rad}\,A$ is isomorphic to K.*

Proof. (a) implies (b). Because A is local, $\mathrm{rad}\,A$ is a unique proper maximal right ideal of A. It follows that $x \in \mathrm{rad}\,A$ if and only if x has no right inverse. Hence we conclude that any right invertible element $x \in A$ is invertible. Indeed, if $xy = 1$ then $(1 - yx)y = 0$. It follows that y has a right inverse and $1 - yx = 0$, because otherwise $y \in \mathrm{rad}\,A$, in view of (1.3), the element $1 - yx$ is invertible and we get $y = 0$, which is a contradiction.

This shows that $x \in \mathrm{rad}\,A$ if and only if x has no right inverse, or equivalently, if and only if x is not invertible. Then (b) follows.

That (a$'$) implies (b) follows in a similar way, and it is easy to see that (b) implies (c).

(c) implies (d). If $e \in A$ is an idempotent, then so is $1 - e$ and we have $e(1 - e) = 0$. It follows from (c) that $e = 0$ or $e = 1$.

(d) implies (e). Because, by (4.4), the idempotents of $A/\mathrm{rad}\,A$ can be lifted modulo $\mathrm{rad}\,A$, the semisimple algebra $B = A/\mathrm{rad}\,A$ has only two idempotents 0 and 1. By (4.5)(a), the right B-module B_B is simple and, in view of (3.2), there is a K-algebra isomorphism $\mathrm{End}\,B_B \cong K$. Hence we get K-algebra isomorphisms $B \cong \mathrm{Hom}_B(B_B, B_B) \cong K$ and (e) follows.

In view of (1.4), the statement (e) implies that $\mathrm{rad}\,A$ is the unique proper maximal right ideal and the unique proper maximal left ideal of A. Hence it follows that (e) implies (a) and that (e) implies (a$'$). The proof is complete. $\square$

We note that infinite dimensional algebras with only two idempotents 0 and 1 are not necessarily local. An example of such an algebra is the polynomial algebra $K[t]$, which is not local and has only two idempotents 0 and 1.

4.7. Corollary. *An idempotent $e \in A$ is primitive if and only if the algebra $eAe \cong \mathrm{End}\,eA$ has only two idempotents 0 and e, that is, the algebra eAe is local.* $\square$

4.8. Corollary. *Let A be an arbitrary K-algebra and M a right A-module.*

(a) *If the algebra $\mathrm{End}\,M$ is local, then M is indecomposable.*

(b) *If M is finite dimensional and indecomposable, then the algebra* $\operatorname{End} M$ *is local and any A-module endomorphism of M is nilpotent or is an isomorphism.*

Proof. (a) If M decomposes as $M = X_1 \oplus X_2$ with both X_1 and X_2 nonzero, then there exist projections $p_i : M \to X_i$ and injections $u_i : X_i \to M$ (for $i = 1, 2$) such that $u_1 p_1 + u_2 p_2 = 1_M$. Because $u_1 p_1$ and $u_2 p_2$ are nonzero idempotents in $\operatorname{End} M$, the algebra $\operatorname{End} M$ is not local, because otherwise 1_M belongs to the unique proper maximal ideal of $\operatorname{End} M$, a contradiction.

(b) Assume that M is finite dimensional and indecomposable. If $\operatorname{End} M$ is not local then, according to (4.6), the algebra $\operatorname{End} M$ has a pair of nonzero idempotents e_1, $e_2 = 1 - e_1$ and therefore $M \cong \operatorname{Im} e_1 \oplus \operatorname{Im} e_2$ is a nontrivial direct sum decomposition. Consequently, the algebra $\operatorname{End} M$ is local. By (4.6), every noninvertible A-module endomorphism $f : M \to M$ belongs to the radical of $\operatorname{End} M$ and therefore f is nilpotent, because $\operatorname{End} M$ is finite dimensional, and it follows from (2.3) that the radical of $\operatorname{End} M$ is nilpotent. $\qquad\square$

We note that infinite dimensional indecomposable modules over finite dimensional algebras do not necessarily have local endomorphism rings. An example of such a module over the Kronecker algebra (2.5) is presented in Exercise 4.15 of Chapter III.

4.9. Example. Let $A = \mathbb{T}_3(K) = \begin{bmatrix} K & 0 & 0 \\ K & K & 0 \\ K & K & K \end{bmatrix}$ be the K-subalgebra of $\mathbb{M}_3(K)$ defined in (1.1)(c), and let B be the subalgebra of A consisting of all matrices $\lambda = \begin{bmatrix} \lambda_{11} & 0 & 0 \\ \lambda_{21} & \lambda_{22} & 0 \\ \lambda_{31} & \lambda_{32} & \lambda_{33} \end{bmatrix}$ in A such that $\lambda_{11} = \lambda_{22} = \lambda_{33}$. The algebra B is noncommutative and local; because $\operatorname{rad} B$ consists of all matrices $\begin{bmatrix} 0 & 0 & 0 \\ \lambda_{21} & 0 & 0 \\ \lambda_{31} & \lambda_{32} & 0 \end{bmatrix}$ in B, there is an algebra isomorphism $B/\operatorname{rad} B \cong K$ and (4.6) applies (compare with (1.5)(c)).

The following result is fundamental for the representation theory of finite dimensional algebras.

4.10. Unique decomposition theorem. *Let A be a finite dimensional K-algebra.*

(a) *Every module M in* $\operatorname{mod} A$ *has a decomposition $M \cong M_1 \oplus \cdots \oplus M_m$, where $M_1, \ldots, M_m$ are indecomposable modules and the endomorphism K-algebra $\operatorname{End} M_j$ is local for each $j = 1, \ldots, m$.*

(b) *If* $M \cong \displaystyle\bigoplus_{i=1}^{m} M_i \cong \bigoplus_{j=1}^{n} N_j$, *where M_i and N_j are indecomposable,* *then $m = n$ and there exists a permutation σ of $\{1, \ldots, n\}$ such that $M_i \cong N_{\sigma(i)}$ for each $i = 1, \ldots, n$.*

Proof. (a) Because $\dim_K M$ is finite, M has an indecomposable decomposition, that is, a decomposition into a direct sum of indecomposable modules. In view of (4.8), the endomorphism algebra of every indecomposable direct summand of M is local. Then M has a decomposition as required.

(b) Without loss of generality, we may suppose that $M = \bigoplus_{i=1}^{m} M_i = \bigoplus_{j=1}^{n} N_j$. We proceed by induction on m. If $m = 1$, then M is indecomposable and there is nothing to show. Assume that $m > 1$ and put $M_1' = \bigoplus_{i>1} M_i$. Denote the injections and projections associated to the direct sum decomposition $M = M_1 \oplus M_1'$ by u, u', p, p' and those associated to the direct sum decomposition $M = \bigoplus_{j=1}^{n} N_j$ by u_j, p_j (with $1 \leq j \leq n$). We have $1_{M_1} = pu = p\left(\sum_{j=1}^{n} u_j p_j\right) u = \sum_{j=1}^{n} p u_j p_j u$. Because $\operatorname{End} M_1$ is local, by (4.6)(c), there exists j with $1 \leq j \leq n$ such that $v = p u_j p_j u$ is invertible. Rearranging the indices if necessary, we may suppose that $j = 1$. Then $w = v^{-1} p u_1 : N_1 \to M_1$ satisfies $w p_1 u = 1_{M_1}$ so that $p_1 u w \in \operatorname{End} N_1$ is an idempotent. Because $\operatorname{End} N_1$ is local, it must equal 1_{N_1} or 0, because of (4.6)(d). If $p_1 u w = 0$, then $p_1 u = 0$ (because w is an epimorphism), a contradiction, because $v = p u_1 p_1 u$ is invertible. Thus $p_1 u w = 1_{N_1}$ and $f_{11} = p_1 u \in \operatorname{Hom}_A(M_1, N_1)$ is an isomorphism. Setting $N_1' = \bigoplus_{j>1} N_j$, we can put the identity homomorphism $1_M : M_1 \oplus M_1' \xrightarrow{\cong} N_1 \oplus N_1'$ in the matrix form $f = \begin{bmatrix} f_{11} & f_{12} \\ f_{21} & f_{22} \end{bmatrix}$. The wanted result would then follow from the induction hypothesis if we could show that $M_1' \cong N_1'$. Because the composite A-module homomorphism $g = \begin{bmatrix} 1 & 0 \\ -f_{21}f_{11}^{-1} & 1 \end{bmatrix} f = \begin{bmatrix} f_{11} & f_{12} \\ 0 & f_{22}' \end{bmatrix}$, where $f_{22}' = -f_{21}f_{11}^{-1}f_{12} + f_{22}$, is an isomorphism $M_1 \oplus M_1' \xrightarrow{\cong} N_1 \oplus N_1'$, $f_{22}' : M_1' \xrightarrow{\cong} N_1'$ is also an isomorphism and the proof is complete. $\qquad\square$

It follows that if $A_A = P_1 \oplus \cdots \oplus P_n$ is an indecomposable decomposition, then it is unique in the sense of the unique decomposition theorem.

We end this section by defining representation-finite algebras, a class we study in detail in the following chapters.

4.11. Definition. A finite dimensional K-algebra A is defined to be **representation–finite** (or an **algebra of finite representation type**) if the number of the isomorphism classes of indecomposable finite dimensional right A-modules is finite. A K-algebra A is called **representation–infinite** (or an **algebra of infinite representation type**) if A is not representation–finite.

It follows from the standard duality D : $\operatorname{mod} A \longrightarrow \operatorname{mod} A^{\mathrm{op}}$ that this definition is right-left symmetric. One can prove that if A is representation–finite then the number of the isomorphism classes of all indecomposable left A-modules is finite, or equivalently, that every indecomposable right (and left) A-module is finite dimensional (see [12], [13], [69], [147], and [151]).

I.5. Projective and injective modules

We start with some definitions. Let $h : M \to N$ and $u : L \to M$ be homomorphisms of right A-modules. We call an A-homomorphism $s : N \to M$ a **section** of h if $hs = 1_N$, and we call an A-homomorphism $r : M \to L$ a **retraction** of u if $ru = 1_L$. If s is a section of h, then h is surjective, s is injective, there are direct sum decompositions $M = \operatorname{Im} s \oplus \operatorname{Ker} h \cong N \oplus \operatorname{Ker} h$, and h is a retraction of s. Similarly, if r is a retraction of u, then r is surjective, u is injective, u is a section of r, and there exist direct sum decompositions $M = \operatorname{Im} u \oplus \operatorname{Ker} r \cong L \oplus \operatorname{Ker} r$.

An A-homomorphism $h : M \to N$ is called a **section** (or a **retraction**) if h admits a retraction (or a section, respectively).

A sequence $\cdots \longrightarrow X_{n-1} \xrightarrow{h_{n-1}} X_n \xrightarrow{h_n} X_{n+1} \xrightarrow{h_{n+1}} X_{n+2} \longrightarrow \cdots$ (infinite or finite) of right A-modules connected by A-homomorphisms is called **exact** if $\operatorname{Ker} h_n = \operatorname{Im} h_{n-1}$ for any n. In particular

$$0 \longrightarrow L \xrightarrow{u} M \xrightarrow{r} N \longrightarrow 0$$

is called a **short exact sequence** if u is a monomorphism, r is an epimorphism and $\operatorname{Ker} r = \operatorname{Im} u$. Note that the homomorphism u admits a retraction $p : M \to L$ if and only if r admits a section $v : N \to M$. In this case there are direct sum decompositions $M = \operatorname{Im} u \oplus \operatorname{Ker} p = \operatorname{Im} v \oplus \operatorname{Ker} r$ of M, and we say that the short exact sequence **splits**.

The following lemma is frequently used.

5.1. Snake lemma. *Assume that the following diagram*

$$
\begin{array}{ccccccccc}
0 & \longrightarrow & L & \xrightarrow{u} & M & \xrightarrow{v} & N & \longrightarrow & 0 \\
& & \downarrow{\scriptstyle f} & & \downarrow{\scriptstyle g} & & \downarrow{\scriptstyle h} & & \\
0 & \longrightarrow & L' & \xrightarrow{u'} & M' & \xrightarrow{v'} & N' & \longrightarrow & 0
\end{array}
$$

in $\operatorname{mod} A$ has exact rows and is commutative. Then there exists a connecting A-homomorphism $\delta : \operatorname{Ker} h \to \operatorname{Coker} f$ such that the induced sequence

$$0 \longrightarrow \operatorname{Ker} f \xrightarrow{u} \operatorname{Ker} g \xrightarrow{v} \operatorname{Ker} h$$
$$\xrightarrow{\delta} \operatorname{Coker} f \xrightarrow{u'} \operatorname{Coker} g \xrightarrow{v'} \operatorname{Coker} h \longrightarrow 0$$

is exact.

Proof. See [49] , [112], [131], and [149].　　　　　　　　□

5.2. Definition. (a) A right A-module F is **free** if F is isomorphic to a direct sum of copies of the module A_A.

(b) A right A-module P is **projective** if, for any epimorphism $h : M \to N$, the induced map $\operatorname{Hom}_A(P, h) : \operatorname{Hom}_A(P, M) \longrightarrow \operatorname{Hom}_A(P, N)$ is surjective, that is, for any epimorphism $h : M \to N$ and any $f \in \operatorname{Hom}_A(P, N)$, there is an $f' \in \operatorname{Hom}_A(P, M)$ such that the following diagram is commutative

$$
\begin{array}{ccc}
 & P & \\
 f' \swarrow & \downarrow f & \\
M \xrightarrow{\ h\ } N & \longrightarrow 0 &
\end{array}
$$

(c) A right A-module E is **injective** if, for any monomorphism $u : L \to M$, the induced map $\operatorname{Hom}_A(u, E) : \operatorname{Hom}_A(M, E) \longrightarrow \operatorname{Hom}_A(L, E)$ is surjective, that is, for any monomorphism $u : L \to M$ and any $g \in \operatorname{Hom}_A(L, E)$, there is a $g' \in \operatorname{Hom}_A(M, E)$ such that the following diagram is commutative

$$
\begin{array}{ccc}
0 \longrightarrow L & \xrightarrow{\ u\ } & M \\
g \downarrow & \swarrow g' & \\
E & &
\end{array}
$$

5.3. Lemma. (a) *A right A-module P is projective if and only if there exist a free A-module F and a right A-module P' such that $P \oplus P' \cong F$.*

(b) *Suppose that $A_A = e_1 A \oplus \cdots \oplus e_n A$ is a decomposition of A_A into indecomposable submodules. If a right A-module P is projective, then $P = P_1 \oplus \cdots \oplus P_m$, where every summand P_j is indecomposable and isomorphic to some $e_s A$.*

(c) *Let M be an arbitrary right A-module. Then there exists an exact sequence*

$$\cdots \to P_m \xrightarrow{h_m} P_{m-1} \to \cdots \to P_1 \xrightarrow{h_1} P_0 \xrightarrow{h_0} M \to 0 \qquad (5.4)$$

in $\operatorname{Mod} A$, *where P_j is a projective right A-module for any $j \geq 0$. If, in addition, M is in* $\operatorname{mod} A$, *then there exists an exact sequence* (5.4), *where P_j is a projective module in* $\operatorname{mod} A$ *for any $j \geq 0$.*

Proof. (a) It is easy to check that any free module is projective and that a direct summand of a free module is a projective module. Conversely, suppose that P is a projective module generated by elements $\{m_j; j \in J\}$.

If $F = \bigoplus_{j \in J} x_j A$ is a free module with the set $\{x_j, \, j \in J\}$ of free generators and $f : F \to P$ is the epimorphism defined by $f(x_j) = m_j$, then, by the projectivity of P, there exists a section $s : P \to F$ of f and therefore $F \cong P \oplus \operatorname{Ker} f$.

(b) Let P be a projective module. By (a), there exist a free A-module F and a right A-module P' such that $P \oplus P' \cong F$. By our assumption, F is a direct sum of copies of the indecomposable modules $e_1 A, \dots, e_n A$. Because by (4.8) the algebra $\operatorname{End} e_j A$ is local for each $j = 1, \dots, n$, (b) is a consequence of the unique decomposition theorem (4.10).

(c) It was shown in (a) that, for any module M (or M in $\operatorname{mod} A$), there is an epimorphism $f : F \to M$, where F is a free module in $\operatorname{Mod} A$ (or in $\operatorname{mod} A$, respectively). We set $P_0 = F$ and $h_0 = f$. Let $f_1 : F_1 \to \operatorname{Ker} h_0$ be an epimorphism with a free module F_1 in $\operatorname{Mod} A$. We set $P_1 = F_1$ and we take for h_1 the composition of f_1 with the embedding $\operatorname{Ker} h_0 \subseteq P_0$. If M is in $\operatorname{mod} A$, then the free module F_1 can be chosen in $\operatorname{mod} A$, because A is finite dimensional, hence $\dim_K M$ and $\dim_K F_0$ are finite, and therefore $\operatorname{Ker} h_0$ is in $\operatorname{mod} A$. Continuing this procedure, we construct by induction the required exact sequence (5.4). $\qquad\square$

We define a **projective resolution** of a right A-module M to be a complex
$$P_\bullet : \quad \cdots \to P_m \xrightarrow{h_m} P_{m-1} \to \cdots \to P_1 \xrightarrow{h_1} P_0 \to 0$$
of projective A-modules together with an epimorphism $h_0 : P_0 \xrightarrow{h_0} M$ of right A-modules such that the sequence (5.4) is exact. For the sake of simplicity, we call the sequence (5.4) a projective resolution of the A-module M. By (5.3), any module M in $\operatorname{mod} A$ has a projective resolution in $\operatorname{mod} A$.

We define an **injective resolution** of M to be a complex
$$I^\bullet : \quad 0 \to I^0 \xrightarrow{d^1} I^1 \to \cdots \to I^m \xrightarrow{d^{m+1}} I^{m+1} \to \cdots$$
of injective A-modules together with a monomorphism $d^0 : M \to I^0$ of right A-modules such that the sequence
$$0 \to M \xrightarrow{d^0} I^0 \xrightarrow{d^1} I^1 \to \cdots \to I^m \xrightarrow{d^{m+1}} I^{m+1} \to \cdots$$
is exact. For the sake of simplicity, we call this sequence an injective resolution of the A-module M. We show later that any module M in $\operatorname{mod} A$ has an injective resolution in $\operatorname{mod} A$.

First, we show that if A is a finite dimensional K-algebra, then any module M in $\operatorname{mod} A$ admits an exact sequence (5.4) in $\operatorname{mod} A$, where the epimorphisms $h_j : P_j \to \operatorname{Im} h_j$ are minimal for all $j \geq 0$ in the following sense.

5.5. Definition. (a) An A-submodule L of M is **superfluous** if for every submodule X of M the equality $L + X = M$ implies $X = M$.

(b) An A-epimorphism $h : M \rightarrow N$ in $\operatorname{mod} A$ is **minimal** if $\operatorname{Ker} h$ is superfluous in M. An epimorphism $h : P \rightarrow M$ in $\operatorname{mod} A$ is called a **projective cover** of M if P is a projective module and h is a minimal epimorphism.

It follows from (3.7)(e) that the submodule $\operatorname{rad} M$ of M is superfluous if M is a finitely generated module over a finite dimensional algebra.

Now we give a useful characterisation of projective covers.

5.6. Lemma. *An epimorphism $h : P \rightarrow M$ is a projective cover of an A-module M if and only if P is projective and for any A-homomorphism $g : N \rightarrow P$ the surjectivity of hg implies the surjectivity of g.*

Proof. Assume that $h : P \rightarrow M$ is a projective cover of M and let $g : N \rightarrow P$ be a homomorphism such that hg is surjective. It follows that $\operatorname{Im} g + \operatorname{Ker} h = P$ and therefore g is surjective, because by assumption $\operatorname{Ker} h$ is superfluous in P. This shows the sufficiency.

Conversely, assume that $h : P \rightarrow M$ has the stated property. Let N be a submodule of P such that $N + \operatorname{Ker} h = P$. If $g : N \hookrightarrow P$ is the natural inclusion, then $hg : N \rightarrow M$ is surjective. Hence, by hypothesis, g is surjective. This shows that $\operatorname{Ker} h$ is superfluous and finishes the proof. $\square$

5.7. Definition. (a) An exact sequence

$$P_1 \xrightarrow{p_1} P_0 \xrightarrow{p_0} M \longrightarrow 0$$

in $\operatorname{mod} A$ is called a **minimal projective presentation** of an A-module M if the A-module homomorphisms $P_0 \xrightarrow{p_0} M$ and $P_1 \xrightarrow{p_1} \operatorname{Ker} p_0$ are projective covers.

(b) An exact sequence (5.4) in $\operatorname{mod} A$ is called a **minimal projective resolution** of M if $h_j : P_j \rightarrow \operatorname{Im} h_j$ is a projective cover for all $j \geq 1$ and $P_0 \xrightarrow{h_0} M$ is a projective cover.

It follows from the next result that any module M in $\operatorname{mod} A$ admits a minimal projective presentation and a minimal projective resolution in $\operatorname{mod} A$.

5.8. Theorem. *Let A be a finite dimensional K-algebra and let $A_A = e_1 A \oplus \cdots \oplus e_n A$, where $\{e_1, \ldots, e_n\}$ is a complete set of primitive orthogonal idempotents of A.*

(a) *For any A-module M in $\operatorname{mod} A$ there exists a projective cover*

$$P(M) \xrightarrow{h} M \longrightarrow 0$$

where $P(M) \cong (e_1 A)^{s_1} \oplus \cdots \oplus (e_n A)^{s_n}$ and $s_1 \geq 0, \ldots, s_n \geq 0$. The homomorphism h induces an isomorphism $P(M)/\operatorname{rad} P(M) \cong M/\operatorname{rad} M$.

(b) *The projective cover $P(M)$ of a module M in $\operatorname{mod} A$ is unique in the sense that if $h' : P' \to M$ is another projective cover of M, then there exists a commutative diagram*

$$
\begin{array}{ccc}
& & 0 \\
& & \uparrow \\
P(M) & \xrightarrow{\ h\ } M & \longrightarrow 0 \\
& {}_{g}\nwarrow \quad \uparrow{}_{h'} & \\
& P' &
\end{array}
$$

where g is an isomorphism.

Proof. We set $B = A/\operatorname{rad} A$, $\bar{e}_j = e_j + \operatorname{rad} A \in B$ and let $p : A \to B$ be the residual class K-algebra epimorphism. Because $\{e_1, \ldots, e_n\}$ is a complete set of primitive orthogonal idempotents of A, $\{\bar{e}_1, \ldots, \bar{e}_n\}$ is a complete set of primitive orthogonal idempotents of B and $B_B = \bar{e}_1 B \oplus \cdots \oplus \bar{e}_n B$ is an indecomposable decomposition. It follows from (4.5)(c) that $\operatorname{rad} e_j A \subset e_j A$ is the unique maximal A-submodule of $e_j A$, then $\operatorname{top} e_j A \cong \bar{e}_j B$ is a simple B-module and the epimorphism $p_j : e_j A \to \operatorname{top} e_j A$ induced by p is a projective cover of $\operatorname{top} e_j A$.

Let M be a module in $\operatorname{mod} A$. Then $\operatorname{top} M = M/\operatorname{rad} M$ is a module in $\operatorname{mod} B$ and, according to (3.8) and (4.5), there exist B-module isomorphisms

$$
\operatorname{top} M \cong (\bar{e}_1 B)^{s_1} \oplus \cdots \oplus (\bar{e}_n B)^{s_n} \cong (\operatorname{top} e_1 A)^{s_1} \oplus \cdots \oplus (\operatorname{top} e_n A)^{s_n},
$$

for some $s_1 \geq 0, \ldots, s_n \geq 0$. We set $P(M) = (e_1 A)^{s_1} \oplus \cdots \oplus (e_n A)^{s_n}$. By the projectivity of the module $P(M)$, there exists an A-module homomorphism $h : P(M) \to M$ making the diagram

$$
\begin{array}{ccc}
P(M) & \xrightarrow{\ h\ } & M \\
\downarrow{}_{t} & & \downarrow{}_{t'} \\
\operatorname{top} P(M) & \xrightarrow{\ \operatorname{top} h\ } & \operatorname{top} M
\end{array}
$$

commutative, where t and t' are the canonical epimorphisms. It follows that $\operatorname{top} h$ is an isomorphism and, from (3.9)(a), we infer that h is an epimorphism. Moreover, the commutativity of the diagram yields

$$
\operatorname{Ker} h \subseteq \operatorname{Ker} t = (\operatorname{rad} e_1 A)^{s_1} \oplus \cdots \oplus (\operatorname{rad} e_n A)^{s_n} = \operatorname{rad} P(M).
$$

Because, according to (3.7)(e), the module $\operatorname{rad} P(M)$ is superfluous in $P(M)$, $\operatorname{Ker} h$ is also superfluous in $P(M)$. Therefore the epimorphism h is a projective cover of M.

(b) The existence of a homomorphism $g : P' \to P(M)$ making the diagram shown in (b) commutative follows from the projectivity of P'. Because

$hg = h'$ is surjective, $\operatorname{Im} g + \operatorname{Ker} h = P(M)$ and therefore g is surjective, because $\operatorname{Ker} h$ is superfluous in $P(M)$. It follows that $\ell(P') \geq \ell(P(M))$. The preceding argument with $P(M)$ and P' interchanged shows that $\ell(P(M)) \geq \ell(P')$. Hence g is an isomorphism and the proof is complete. $\square$

Remark. The proof of (5.8) gives us a recipe for constructing the projective cover $P(M) \to M$ of any module in $\operatorname{mod} A$. We also refer simply to the module $P(M)$ as being a projective cover of M.

5.9. Corollary. *If P is a projective module in $\operatorname{mod} A$, then the canonical epimorphism $t : P \to \operatorname{top} P$ is a projective cover of $\operatorname{top} P$ and there exists an A-isomorphism $P \cong (e_1 A)^{s_1} \oplus \cdots \oplus (e_n A)^{s_n}$ for some $s_1 \geq 0, \ldots, s_n \geq 0$.*
$\square$

5.10. Corollary. *Let A be a K-algebra. Any module M in $\operatorname{mod} A$ admits a minimal projective presentation and a minimal projective resolution in $\operatorname{mod} A$.*

Proof. Let M be a module in $\operatorname{mod} A$. By (5.8), there is a projective cover $p_0 : P_0 \to M$ in $\operatorname{mod} A$. Then $\operatorname{Ker} p_0$ is finite dimensional and, according to (5.8), there is a projective cover $p_1 : P_0 \to \operatorname{Ker} p_0$. This yields a minimal projective presentation $P_1 \xrightarrow{p_1} P_0 \xrightarrow{p_0} M \longrightarrow 0$ of M. Continuing this procedure, we get by induction a minimal projective resolution of M in $\operatorname{mod} A$. $\square$

Now we shift our attention from projective to injective modules. For this purpose we recall from (2.9) that the functor $D(-) = \operatorname{Hom}_K(-, K)$ defines two dualities

$$\operatorname{mod} A \xrightarrow{D} \operatorname{mod} A^{\mathrm{op}} \xrightarrow{D} \operatorname{mod} A$$

such that there are natural equivalences of functors $D \circ D \cong 1_{\operatorname{mod} A}$ and $D \circ D \cong 1_{\operatorname{mod} A^{\mathrm{op}}}$. This allows us to study the injective modules in $\operatorname{mod} A$ by means of the projective modules in $\operatorname{mod} A^{\mathrm{op}}$.

We start by recalling the following important result.

5.11. Baer's criterion. *A right A-module E is injective if for any right ideal I of A and any A-homomorphism $f : I \to E$ there exists an A-homomorphism $f' : A_A \to E$ such that $f = f'u$, where u is the inclusion $u : I \hookrightarrow A$.*

Proof. See [2], [48], and [149]. $\square$

The notions dual to minimal epimorphism and to projective cover are defined as follows.

5.12. Definition. An A-module monomorphism $u : L \to M$ in $\operatorname{mod} A$ is **minimal** if every nonzero submodule X of M has a nonzero intersection with Im u. A monomorphism $u : L \to E$ in mod A is called an **injective envelope** of L if E is an injective module and u is a minimal monomorphism.

Now we are able to state the main transfer theorem via the standard duality.

5.13. Theorem. *Let A be a finite dimensional K-algebra and let $D : \operatorname{mod} A \longrightarrow \operatorname{mod} A^{\mathrm{op}}$ be the standard duality $D(-) = \operatorname{Hom}_K(-, K)$ (2.9). Then the following hold.*

(a) *A sequence $0 \longrightarrow L \xrightarrow{u} N \xrightarrow{h} M \longrightarrow 0$ in mod A is exact if and only if the induced sequence $0 \longrightarrow D(M) \xrightarrow{D(h)} D(N) \xrightarrow{D(u)} D(L) \longrightarrow 0$ is exact in mod A^{op}.*

(b) *A module E in mod A is injective if and only if the module $D(E)$ is projective in mod A^{op}. A module P in mod A is projective if and only if the module $D(P)$ is injective in mod A^{op}.*

(c) *A module S in mod A is simple if and only if the module $D(S)$ is simple in mod A^{op}.*

(d) *A monomorphism $u : M \to E$ in mod A is an injective envelope if and only if the epimorphism $D(u) : D(E) \to D(M)$ is a projective cover in mod A^{op}. An epimorphism $h : P \to M$ in mod A is a projective cover if and only if the $D(h) : D(M) \to D(P)$ is an injective envelope in mod A^{op}.*

Proof. This is straightforward and left to the reader (see [61]). □

5.14. Corollary. *Every module M in mod A has an injective envelope $u : M \to E(M)$ and the module $E(M)$ is uniquely determined by M, up to isomorphism.*

Proof. Let M be a module in mod A. By (5.8), the left A-module $D(M)$ has a projective cover $h : P \xrightarrow{} D(M)$. It follows from (5.13)(d) that the monomorphism $M \cong DD(M) \xrightarrow{D(h)} D(P)$ is an injective envelope of M in mod A. We set $E(M) = D(P)$. By (5.8), the left A-module P is uniquely determined by $D(M)$, up to isomorphism. It follows that the right module $E(M) = D(P)$ is uniquely determined by M, up to isomorphism. □

We refer simply to the module $E(M)$ as being an injective envelope of M.

5.15. Definition. (a) An exact sequence $0 \longrightarrow N \xrightarrow{u^0} I^0 \xrightarrow{u^1} I^1$ is a **minimal injective presentation** of an A-module N if the monomorphisms $u^0 : N \to I^0$ and Im $u^1 \hookrightarrow I^1$ are injective envelopes.

(b) An injective resolution $0 \to M \xrightarrow{d^0} I^0 \xrightarrow{d^1} I^1 \to \cdots \to I^m \xrightarrow{d^{m+1}} I^{m+1} \to \cdots$ of a module M in mod A is said to be **minimal** if Im $d^m \to I^m$ is an injective envelope for all $m \geq 1$ and $d^0 : M \to I^0$ is an injective envelope.

5.16. Corollary. *Every module M in* $\mathrm{mod}\,A$ *has a minimal injective presentation and a minimal injective resolution in* $\mathrm{mod}\,A$.

Proof. Let M be a module in $\mathrm{mod}\,A$. By (5.8), the left A-module $D(M)$ has a minimal projective presentation and a minimal projective resolution in $\mathrm{mod}\,A^{\mathrm{op}}$. It follows from (5.13) that the standard duality $D : \mathrm{mod}\,A^{\mathrm{op}} \longrightarrow \mathrm{mod}\,A$ carries a minimal projective presentation and a minimal projective resolution of $D(M)$ to a minimal injective presentation and a minimal injective resolution of the module $M \cong DD(M)$, respectively. $\square$

5.17. Corollary. *Suppose that $A_A = e_1 A \oplus \cdots \oplus e_n A$ is a decomposition of A into indecomposable submodules.*

(a) *Every simple right A-module is isomorphic to one of the modules*

$$S(1) = \mathrm{top}\,e_1 A, \ldots, S(n) = \mathrm{top}\,e_n A.$$

(b) *Every indecomposable projective right A-module is isomorphic to one of the modules*

$$P(1) = e_1 A, \ P(2) = e_2 A, \ldots, P(n) = e_n A.$$

Moreover, $e_i A \cong e_j A$ if and only if $S(i) \cong S(j)$.

(c) *Every indecomposable injective right A-module is isomorphic to one of the modules*

$$I(1) = D(Ae_1) \cong E(S(1)), \ldots, I(n) = D(Ae_n) \cong E(S(n)),$$

where $E(S(j))$ is an injective envelope of the simple module $S(j)$.

Proof. Apply (4.5), (4.7), (4.10), (5.9), and (5.13). $\square$

5.18. Example. Let $A = \mathbb{M}_2(K)$ and let $e_1 = \left(\begin{smallmatrix} 1 & 0 \\ 0 & 0 \end{smallmatrix}\right)$, $e_2 = \left(\begin{smallmatrix} 0 & 0 \\ 0 & 1 \end{smallmatrix}\right)$. Then e_1, e_2 are primitive orthogonal idempotents of A such that $1_A = e_1 + e_2$ and $A_A = e_1 A \oplus e_2 A$. The algebra A is semisimple, $S(1) = P(1) = I(1) \cong S(2) = P(2) = I(2)$ and $\dim_K S(1) = \dim_K S(2) = 2$. $\square$

I.6 Basic algebras and embeddings of module categories

Throughout, we need essentially the following class of algebras (see [73], [125], and [131] for historical notes).

6.1. Definition. Assume that A is a K-algebra with a complete set $\{e_1, \ldots, e_n\}$ of primitive orthogonal idempotents. The algebra A is called **basic** if $e_i A \not\cong e_j A$, for all $i \neq j$.

It is clear that every local finite dimensional algebra is basic. It follows from the following proposition that the algebras of Examples (1.1)(c) and (1.1)(d) are basic.

6.2. Proposition. (a) *A finite dimensional K-algebra A is basic if and only if the algebra $B = A/\mathrm{rad}\,A$ is isomorphic to a product $K \times K \times \cdots \times K$ of copies of K.*

(b) *Every simple module over a basic K-algebra is one-dimensional.*

Proof. (a) Let $A_A = e_1 A \oplus \cdots \oplus e_n A$ be an indecomposable decomposition of A. Then $\{e_1, \ldots, e_n\}$ is a complete set of primitive orthogonal idempotents of A, the element $\overline{e}_j = e_j + \mathrm{rad}\,A$ is an idempotent of $B = A/\mathrm{rad}\,A$, and in view of (4.5)(c) $\overline{e}_j B = \mathrm{top}\,e_j A$ is a simple B-module. Hence $B_B = \overline{e}_1 B \oplus \cdots \oplus \overline{e}_n B$ is an indecomposable decomposition of B_B. By (5.9), $e_j A \cong P(\overline{e}_j B)$ and therefore $e_j A \cong e_i A$ if and only if $\overline{e}_j B \cong \overline{e}_i B$.

It follows that if A is basic, then B is basic. Moreover, Schur's lemma (3.1) yields $\mathrm{Hom}_B(\overline{e}_i B, \overline{e}_j B) = 0$ for $i \neq j$, and (3.2) yields $\mathrm{End}\,\overline{e}_j B \cong K$ for $j = 1, \ldots, n$. Hence, given an element $b \in B$ and $j \leq n$, the multiplication map $b_j : \overline{e}_j B \to B_B$ defined by the formula $b_j(y) = \overline{e}_j by$, for $y \in \overline{e}_j B$, induces a homomorphism $b'_j : \overline{e}_j B \to \overline{e}_j B$ of right B-modules and the K-algebra homomorphism $\sigma_j : B \to \mathrm{End}\,\overline{e}_j B \cong K$ defined by the formula $\sigma_j(b) = b'_j$. Hence we get the K-algebra homomorphism

$$\sigma : B \longrightarrow \mathrm{End}(\overline{e}_1 B) \times \cdots \times \mathrm{End}(\overline{e}_n B) \cong K \times \cdots \times K$$

defined by $\sigma(b) = (\sigma_1(b), \ldots, \sigma_n(b))$, for $b \in B$. Because σ is obviously injective, by comparing the dimensions, we see that it is bijective. The sufficiency part of (a) follows.

Assume now that B is a product $K \times \cdots \times K$. Then B is commutative and $\overline{e}_1, \ldots, \overline{e}_n$ are central primitive pairwise orthogonal idempotents of B. It follows that $\overline{e}_i B \not\cong \overline{e}_j B$ for $i \neq j$ and (5.8) yields $e_i A \cong P(\overline{e}_i B) \not\cong P(\overline{e}_j B) \cong e_j A$. Consequently A is basic and (a) follows.

The statement (b) follows from (a) because, by (3.9)(b), any simple A-module S is a module over the quotient algebra $B = A/\mathrm{rad}\,A$ and, by (a), B is isomorphic to a product $K \times \cdots \times K$ if A is basic. Hence $\dim_K S = 1$ and the proof is complete. □

6.3. Definition. Assume that A is a K-algebra with a complete set $\{e_1, \ldots, e_n\}$ of primitive orthogonal idempotents. A **basic algebra** associated to A is the algebra

$$A^b = e_A A e_A,$$

where $e_A = e_{j_1} + \cdots + e_{j_a}$, and $e_{j_1}, \ldots, e_{j_a}$ are chosen such that $e_{j_i} A \not\cong e_{j_t} A$ for $i \neq t$ and each module $e_s A$ is isomorphic to one of the modules $e_{j_1} A, \ldots, e_{j_a} A$.

Example 6.4. Let $A = \mathbb{M}_n(K)$ and $\{e_1, \ldots, e_n\}$ be the standard set of matrix orthogonal idempotents of A. Then $e_i A \cong e_j A$ for all i, j, $e_A = e_1$ and $A^b \cong K$.

6.5. Lemma. *Let $A^b = e_A A e_A$ be a basic algebra associated to A.*

(a) *The idempotent $e_A \in A^b$ is the identity element of A^b and there is a K-algebra isomorphism $A^b \cong \mathrm{End}(e_{j_1} A \oplus \cdots \oplus e_{j_a} A)$.*

(b) *The algebra A^b does not depend on the choice of the sets $e_1, \ldots, e_n$ and $e_{j_1}, \ldots, e_{j_a}$, up to a K-algebra isomorphism.*

Proof. (a) By (4.2) applied to the A-module $M = e_A A$, there is a K-algebra isomorphism $\mathrm{End}\, e_A A \cong e_A A e_A$. Because there exists an A-module isomorphism $e_A A \cong e_{j_1} A \oplus \cdots \oplus e_{j_a} A$, we derive K-algebra isomorphisms

$$A^b = e_A A e_A \cong \mathrm{Hom}_A(e_A A, e_A A) \cong \mathrm{End}(e_{j_1} A \oplus \cdots \oplus e_{j_a} A).$$

(b) It follows from the unique decomposition theorem (4.10) that the A-module $e_A A$ depends only on A and not on the choice of the sets $\{e_1, \ldots, e_n\}$ and $\{e_{j_1}, \ldots, e_{j_a}\}$, up to isomorphism of A-modules. Then the statement (b) is a consequence of the K-algebra isomorphisms $A^b \cong \mathrm{End}\, e_A A \cong \mathrm{End}(e_{j_1} A \oplus \cdots \oplus e_{j_a} A)$. $\square$

We will show in (6.10) that the algebra A^b is basic and that there is an equivalence of categories $\mathrm{mod}\, A \cong \mathrm{mod}\, A^b$.

In the study of $\mathrm{mod}\, A$ we frequently use two embeddings of module categories induced by an algebra idempotent defined as follows.

Suppose that $e \in A$ is an idempotent in a finite dimensional K-algebra A and consider the algebra $B = eAe \cong \mathrm{End}\, eA$ with the identity element $e \in B$. We define three additive K-linear covariant functors

$$\mathrm{mod}\, B \xleftarrow[\mathrm{res}_e]{T_e, L_e} \mathrm{mod}\, A \qquad (6.6)$$

by the formulas

$$\mathrm{res}_e(-) = (-)e, \quad T_e(-) = - \otimes_B eA, \quad L_e(-) = \mathrm{Hom}_B(Ae, -).$$

If $f : X \to X'$ is a homomorphism of A-modules, we define a homomorphism of B-modules $\mathrm{res}_e(f) : \mathrm{res}_e(X) \to \mathrm{res}_e(X')$ by the formula $xe \mapsto f(x)e$, that is, $\mathrm{res}_e(f)$ is the restriction of f to the subspace Xe of X. We call res_e the **restriction functor**. The K-linear functors T_e, L_e are called **idempotent embedding functors**.

Example 6.7. Suppose that $A = KI \subseteq \mathbb{M}_n(K)$ is the incidence algebra of a poset $(I, \preceq)$, where $I = \{1, \ldots, n\}$ (see (1.1)(d)). Let J be a subposet of I and take for e the idempotent $e_J = \sum_{j \in J} e_j \in KI$, where $e_1, \ldots, e_n \in KI$

are the standard matrix idempotents. A simple calculation shows that if $\lambda' = [\lambda'_{pq}] \in KI$ and $\lambda = e_J\lambda'e_J$, then λ has an $n \times n$ matrix form $\lambda = [\lambda_{pq}] \in KI$, where $\lambda_{pq} = 0$ whenever $p \in I \setminus J$ or $q \in I \setminus J$. This shows that $e_J(KI)e_J$ is the K-vector subspace of KI consisting of all matrices $\lambda = [\lambda_{pq}] \in KI$ with $\lambda_{pq} = 0$ whenever $p \in I \setminus J$ or $q \in I \setminus J$. Therefore there is a K-algebra isomorphism $e_J(KI)e_J \cong KJ$.

The following result is very useful in applications.

Theorem 6.8. *Suppose that A is a finite dimensional K-algebra and that $e \in A$ is an idempotent, and let $B = eAe$. The functors T_e, L_e (6.6) associated to $e \in A$ satisfy the following conditions.*

(a) *T_e and L_e are full and faithful K-linear functors such that $\mathrm{res}_e T_e \cong 1_{\mathrm{mod}\, B} \cong \mathrm{res}_e L_e$, the functor L_e is right adjoint to res_e and T_e is left adjoint to res_e, that is, there are functorial isomorphisms*

$$
\begin{aligned}
\mathrm{Hom}_A(X_A, L_e(Y_B)) &\cong \mathrm{Hom}_B(\mathrm{res}_e(X_A), Y_B) \\
\mathrm{Hom}_A(T_e(Y_B), X_A) &\cong \mathrm{Hom}_B(Y_B, \mathrm{res}_e(X_A))
\end{aligned}
$$

for every A-module X_A and every B-module Y_B.

(b) *The restriction functor res_e is exact, T_e is right exact, and L_e is left exact.*

(c) *The functors T_e and L_e preserve indecomposability, T_e carries projectives to projectives, and L_e carries injectives to injectives.*

(d) *A module X_A is in the category $\mathrm{Im}\, T_e$ if and only if there is an exact sequence $P_1 \xrightarrow{h} P_0 \longrightarrow X_A \longrightarrow 0$, where P_1 and P_0 are direct sums of summands of eA.*

Proof. (a) By (4.2), the map θ_X, $f \mapsto f(e) = f(e)e$, is a functorial B-module isomorphism $\mathrm{Hom}_A(eA, X_A) \xrightarrow{\cong} Xe$. Hence, in view of the adjoint formula (2.11), we get

$$
\begin{aligned}
\mathrm{Hom}_A(T_e(Y_B), X_A) &= \mathrm{Hom}_A(Y \otimes_B eA, X_A) \\
&\cong \mathrm{Hom}_B(Y, \mathrm{Hom}_A(eA, X_A)) \\
&\cong \mathrm{Hom}_B(Y, Xe) \cong \mathrm{Hom}_B(Y_B, \mathrm{res}_e(X_A)),
\end{aligned}
$$

and similarly we get the first isomorphism required in (a). Moreover, there are isomorphisms $\mathrm{res}_e T_e(Y_B) = (Y \otimes_B eA)e \cong Y \otimes_B (eAe) = Y \otimes_B B \cong Y_B$ and $\mathrm{res}_e L_e(Y_B) \cong Y_B$. As a consequence, we get functorial isomorphisms

$$
\begin{aligned}
\mathrm{Hom}_B(Y_B, Y'_B) &\cong \mathrm{Hom}_B(Y_B, \mathrm{res}_e T_e(Y'_B)) \\
&\cong \mathrm{Hom}_A(T_e(Y_B), T_e(Y'_B))
\end{aligned}
$$

and $\operatorname{Hom}_B(Y_B, Y'_B) \cong \operatorname{Hom}_A(L_e(Y_B), L_e(Y'_B))$ such that $f \mapsto T_e(f)$ and $f \mapsto L_e(f)$, respectively. This proves that T_e and L_e are full and faithful and (a) follows.

(b) The exactness of the functor res_e is obvious. The functor T_e is right exact, because the tensor product functor is right exact. Because the functor $\operatorname{Hom}_A(M, -)$ is left exact, the functor L_e is left exact and (b) follows.

(c) It follows from (a) that L_e and T_e induce the algebra isomorphisms $\operatorname{End} X \cong \operatorname{End} L_e X$ and $\operatorname{End} X \cong \operatorname{End} T_e X$. Hence they preserve indecomposability, because of (4.8).

Now assume that P is a projective module in $\operatorname{mod} B$ and let $h : M \to N$ be an epimorphism in $\operatorname{mod} A$. In view of the natural isomorphism in (6.8)(a) for the functor T_e, there is a commutative diagram

$$
\begin{array}{ccc}
\operatorname{Hom}_A(T_e(P), M) & \xrightarrow{\operatorname{Hom}_A(T_e(P), h)} & \operatorname{Hom}_A(T_e(P), N) \\
\cong \downarrow & & \cong \downarrow \\
\operatorname{Hom}_B(P, \operatorname{res}_e(M)) & \xrightarrow{\operatorname{Hom}_B(P, \operatorname{res}_e(h))} & \operatorname{Hom}_B(P, \operatorname{res}_e(N)).
\end{array}
$$

Because P is projective, the homomorphism $\operatorname{Hom}_B(P, \operatorname{res}_e(h))$ is surjective. It follows that $\operatorname{Hom}_A(T_e(P), h)$ is also surjective and therefore the A-module $T_e(P)$ is projective. If E is injective, then we show that $L_e(E)$ is injective.

(d) Assume that $e = e_{j_1} + \ldots + e_{j_s}$ and $e_{j_1}, \ldots, e_{j_s}$ are primitive orthogonal idempotents. It follows that $B = e_{j_1} B \oplus \ldots \oplus e_{j_s} B$ and the modules $e_{j_1} B, \ldots, e_{j_s} B$ are indecomposable.

First, we show that the multiplication map

$$ m_{j_i} : e_{j_i} B \otimes_B eA \to e_{j_i} A, \tag{6.9} $$

$e_{j_i} x \otimes ea \mapsto e_{j_i} xea$, is an A-module isomorphism for $i = 1, \ldots, s$. It is clear that m_{j_i} is well-defined and an A-module epimorphism. Because m_{j_i} is the restriction of the A-module isomorphism $m : B \otimes_B eA \to eA$, $x \otimes ea \mapsto xea$, to the direct summand $e_{j_i} B \otimes_B eA$ of $B \otimes_B eA \cong eA$, m_{j_i} is injective and we are done.

To prove (d), assume that $\overline{P}_1 \to \overline{P}_0 \to Y_B \to 0$ is an exact sequence in $\operatorname{mod} B$, where $\overline{P}_0, \overline{P}_1$ are projective. Then the induced sequence

$$ \overline{P}_1 \otimes_B eA \to \overline{P}_0 \otimes_B eA \to Y \otimes_B eA \to 0 $$

in $\operatorname{mod} A$ is exact and the modules $P_1 = \overline{P}_1 \otimes_B eA$, $P_0 = \overline{P}_0 \otimes_B eA$ satisfy the conditions required in (d) because, according to (5.3), the modules $\overline{P}_1$ and $\overline{P}_0$ are direct sums of indecomposable modules isomorphic to some of the modules $e_{j_1} B, \ldots, e_{j_s} B$, and the preceding observation applies.

Conversely, assume there is an exact sequence $P_1 \xrightarrow{h} P_0 \to X_A \to 0$, in mod A with P_0, P_1 direct sums of summands of eA. Then P_0e and P_1e are obviously finite dimensional projective B-modules and by the observation, there are A-module isomorphisms $T_e(P_0e) = P_0e \otimes_B eA \cong P_0$, $T_e(P_1e) = P_1e \otimes_B eA \cong P_1$. If Y_B denotes the cokernel of the restriction $he : P_1e \to P_0e$ of h to $\mathrm{res}_e(P_1) = P_1e$, then we derive a commutative diagram

$$
\begin{array}{ccccccc}
P_1 & \longrightarrow & P_0 & \longrightarrow & X_A & \longrightarrow & 0 \\
f_1 \downarrow \cong & & f_0 \downarrow \cong & & & & \\
T_e(P_1e) & \longrightarrow & T_e(P_0e) & \longrightarrow & T_e(Y_B) & \longrightarrow & 0
\end{array}
$$

with exact rows and bijective vertical maps f_1, f_0. Hence we get an isomorphism $X_A \cong T_e(Y_B)$ induced by f_0 and the proof is complete. $\square$

6.10. Corollary. *Let $A^b = e_A A e_A$ be a basic K-algebra associated with A (see (6.3)). The algebra A^b is basic and the functors*

$$
\mathrm{mod}\, A^b \xleftarrow[\mathrm{res}_{e_A}]{T_{e_A}} \mathrm{mod}\, A
$$

are K-linear equivalences of categories quasi-inverse to each other.

Proof. Assume that $\{e_1, \ldots, e_n\}$ is a complete set of primitive orthogonal idempotents of A, $e_A = e_{j_1} + \cdots + e_{j_a}$ and $e_{j_1}, \ldots, e_{j_a}$ are chosen as in (6.3). Then $e_{j_1}, \ldots, e_{j_a}$ are orthogonal idempotents of A^b,

$$
A^b = e_A A^b = e_{j_1} A^b \oplus \ldots \oplus e_{j_a} A^b,
$$

and $e_{j_t} A^b e_{j_t} = e_{j_t} e_A A e_A e_{j_t} = e_{j_t} A e_{j_t}$ for all t. It follows from (4.7) that the algebra $\mathrm{End}\, e_{j_t} A^b \cong e_{j_t} A^b e_{j_t}$ is local, because $e_{j_t} A$ is indecomposable in mod A. Hence e_{j_t} is a primitive idempotent of A^b. To show that the algebra A^b is basic, assume that $e_{j_t} A^b \cong e_{j_r} A^b$. Because we have shown in (6.9) that the multiplication map $m_{j_i} : e_{j_i} A^b \otimes_{A^b} e_A A \to e_{j_i} A$, $e_{j_i} x \otimes e_A a \mapsto e_{j_i} x e_A a$, is an A-module isomorphism for $i = 1, \ldots, a$, we get A-module isomorphisms

$$
e_{j_t} A \cong e_{j_t} A^b \otimes_{A^b} e_A A \cong e_{j_r} A^b \otimes_{A^b} e_A A \cong e_{j_r} A
$$

and therefore $t = r$ by the choice of $e_{j_1}, \ldots, e_{j_a}$ in (6.3).

By (6.8), the functor T_{e_A} is full and faithful. Because

$$
e_A A \cong e_{j_1} A \oplus \cdots \oplus e_{j_a} A,
$$

each $e_{j_t} A$ is isomorphic to a summand of $e_A A$. This, together with (6.3) and (6.8), shows that every module X in mod A admits an exact sequence

$P' \to P \to X \to 0$, where P and P' are direct sums of summands of $e_A A$. It then follows from (6.8)(d) that any module X_A belongs to the image of the functor T_{e_A}. Consequently, T_{e_A} is dense, and according to (A.2.5) of the Appendix, the full and faithful K-linear functor T_{e_A} is an equivalence of categories. Therefore res_{e_A} is a quasi-inverse of T_{e_A}. $\square$

6.11. Corollary. *Let A be a K-algebra. For each $n \geq 1$, there exists a K-linear equivalence of categories* $\mathrm{mod}\, A \cong \mathrm{mod}\, \mathbb{M}_n(A)$.

Proof. Let $B = \mathbb{M}_n(A)$ and let $\xi_1, \ldots, \xi_n \in B$ be the standard set of matrix idempotents in B, that is, ξ_j is the matrix with 1 on the position (j, j) and zeros elsewhere. Because $B = \xi_1 B \oplus \cdots \oplus \xi_n B$, $\xi_1 B \cong \xi_2 B \cong \cdots \cong \xi_n B$ and $\xi_1 B \xi_1 \cong A$, applying (6.8) to $e = \xi_1 \in B$, we conclude as in the proof of (6.10) that the composite functor $\mathrm{mod}\, A \cong \mathrm{mod}\, \xi_1 B \xi_1 \xrightarrow{T_{\xi_1}} \mathrm{mod}\, \mathbb{M}_n(A)$ is an equivalence of categories. $\square$

I.7. Exercises

1. Let $f : A \to B$ be a homomorphism of K-algebras. Prove that $f(\mathrm{rad}\, A) \subseteq \mathrm{rad}\, B$.

2. Let A be the polynomial K-algebra $K[t_1, t_2]$. Prove that
(a) the algebra A is not local,
(b) the elements 0 and 1 are the only idempotents of A, and
(c) the radical of A is zero.

3. Prove that a homomorphism $u : L \to M$ of right A-modules admits a retraction $p : M \to L$ if and only if u is injective and $M = \mathrm{Im}\, u \oplus N$, where N is a submodule of M.

4. Prove that a homomorphism $r : M \to N$ of right A-modules admits a section $v : N \to M$ if and only if r is surjective and $M = L \oplus \mathrm{Ker}\, r$, where L is a submodule of M.

5. Suppose that the sequence $0 \longrightarrow L \xrightarrow{u} M \xrightarrow{r} N \longrightarrow 0$ of right A-modules is exact. Prove that the homomorphism u admits a retraction $p : M \to L$ if and only if r admits a section $v : N \to M$.

6. Let N be a submodule of a right A-module M. Prove that
(a) $\mathrm{rad}(M/N) \supseteq (N + \mathrm{rad}\, M)/N$, and
(b) if $N \subseteq \mathrm{rad}\, M$, then $\mathrm{rad}(M/N) = (\mathrm{rad}\, M)/N$.

7. Let $A = K[t]$. Prove that the cyclic A-module $M = K[t]/(t^3)$ has no projective cover in $\mathrm{Mod}\, A$.

8. Let A be a K-algebra and let $Z(A)$ be the **centre** of A, that is, the subalgebra of A consisting of all elements $a \in A$ such that $ay = ya$ for all $y \in A$. Show that the following three conditions are equivalent:

(a) The algebra A is connected.

(b) The algebra $Z(A)$ is connected.

(c) The elements 0 and 1 are the only central idempotents of A.

9. Assume that A is a K-algebra, $e \in A$ is an idempotent of A, and M is a right A-module. Prove the following statements:

(a) The K-subspace eAe of A is a K-algebra with respect to the multiplication of A, and e is the identity element of eAe.

(b) The K-vector space Me is a right eAe-module, and the K-vector space $\mathrm{Hom}_A(eA, M)$ is a right eAe-module with respect to the multiplication $(f, a) \mapsto fa$ for $f \in \mathrm{Hom}_A(eA, M)$ and $a \in A$, where we set $(fa)(x) = f(xa)$ for all $x \in eA$.

(c) The K-linear map $\theta_M : \mathrm{Hom}_A(eA, M) \longrightarrow Me$, $f \mapsto f(e)$, is an isomorphism of right eAe-modules, and it is functorial in M.

(d) The map $\theta_{eA} : \mathrm{Hom}_A(eA, eA) \longrightarrow eAe$ is a K-algebra isomorphism.

(e) The map $M \otimes_A Ae \longrightarrow Me$, $m \otimes x \mapsto mx$, is an isomorphism of right eAe-modules, and it is functorial in M.

10. Assume that A is a finite dimensional K-algebra. Prove that A is local if and only if every element of A is invertible or nilpotent.

11. Let KI be the incidence K-algebra of a poset $(I, \preceq)$ (see (1.5)(d)) and let B be the K-subalgebra of KI consisting of the matrices $\lambda = [\lambda_{ij}] \in KI$ such that $\lambda_{ii} = \lambda_{jj}$ for all $i, j \in I$. Prove the following statements:

(a) The algebra KI is basic, and KI is semisimple if and only if $a_i \npreceq a_j$ for every pair of elements $a_i \neq a_j$ of I.

(b) The algebra KI is local if and only if $|I| = 1$.

(c) The subalgebra B of KI is local.

(d) The algebra B is noncommutative if and only if there is a triple a_i, a_j, a_s of pairwise different elements of I such that $a_i \prec a_j \prec a_s$.

12. Let M be a module in $\mathrm{mod}\, A$. Prove that there is a functorial isomorphism $\mathrm{soc}\, DM \xrightarrow{\;\simeq\;} D(M/\mathrm{rad}\, M)$, where D is the standard duality.

13. Let $A = \mathbb{M}_n(K)$, where $n \geq 1$, and let M be an indecomposable A-module. Show that $\ell(M) = 1$ and $\dim_K M = n$.

14. Let A be a basic finite dimensional algebra over an algebraically closed field K, and let M be a finite dimensional right A-module. Show that $\ell(M) = \dim_K M$.

15. Let A be a finite dimensional K-algebra over an algebraically closed field K. Prove that the following three conditions are equivalent:

(a) The algebra A is basic.

(b) Every simple right A-module is one-dimensional.

(c) $\dim_K M = \ell(M)$, for any module M in $\operatorname{mod} A$.

Hint: Apply (6.2).

16. Let A be any of the two subalgebras

$$\begin{bmatrix} K & 0 & 0 & 0 \\ K & K & 0 & 0 \\ K & 0 & K & 0 \\ K & K & K & K \end{bmatrix} \subset \begin{bmatrix} K & 0 & 0 & 0 \\ K & K & 0 & 0 \\ K & K & K & 0 \\ K & K & K & K \end{bmatrix}$$

of the full matrix algebra $\mathbb{M}_4(K)$ defined in Examples 1.1(c) and 1.1(d). Let $e_1 = e_{11}, e_2 = e_{22}, e_3 = e_{33}, e_4 = e_{44}$ be the standard complete set of primitive orthogonal idempotents in A. Show that

(a) the algebra A is basic,

(b) there is an isomorphism $Ae_1 \cong D(e_4 A)$ of left A-modules, where D is the standard duality,

(c) the right ideal $S(1) = e_1 A$ of A is simple and $\operatorname{soc} A_A = \begin{bmatrix} K & 0 & 0 & 0 \\ K & 0 & 0 & 0 \\ K & 0 & 0 & 0 \\ K & 0 & 0 & 0 \end{bmatrix}$,

and

(d) the indecomposable projective right ideal $P(4) = e_4 A$ is an injective envelope of $S(1)$, and the indecomposable projective right ideals $P(1) = e_1 A$, $P(2) = e_2 A$ and $P(3) = e_3 A$ are not injective.

17. Assume that A is a finite dimensional K-algebra, $f : M \to N$ is a homomorphism in $\operatorname{mod} A$, and $M \neq 0$. Prove the following statements:

(a) The socle $\operatorname{soc} M$ of M is a nonzero semisimple submodule of M and $f(\operatorname{soc} M) \subseteq \operatorname{soc} N$.

(b) If $f(\operatorname{soc} M) \neq 0$, then $f \neq 0$.

(c) The inclusion homomorphism $\operatorname{soc} M \subseteq M$ induces an A-module isomorphism $E(\operatorname{soc} M) \xrightarrow{\cong} E(M)$ of the injective envelopes $E(\operatorname{soc} M)$ and $E(M)$ of $\operatorname{soc} M$ and M, respectively.

(d) The module M is indecomposable if and only if the injective envelope $E(M)$ of M is indecomposable.

Chapter II

Quivers and algebras

In this chapter, we show that to each finite dimensional algebra over an algebraically closed field K corresponds a graphical structure, called a quiver, and that, conversely, to each quiver corresponds an associative K-algebra, which has an identity and is finite dimensional under some conditions. Similarly, as will be seen in the next chapter, using the quiver associated to an algebra A, it will be possible to visualise a (finitely generated) A-module as a family of (finite dimensional) K-vector spaces connected by linear maps (see Examples (I.2.4)–(I.2.6)). The idea of such a graphical representation seems to go back to the late forties (see Gabriel [70], Grothendieck [82], and Thrall [167]) but it became widespread in the early seventies, mainly due to Gabriel [72], [73]. In an explicit form, the notions of quiver and linear representation of quiver were introduced by Gabriel in [72]. It was the starting point of the modern representation theory of associative algebras.

II.1. Quivers and path algebras

This first section is devoted to defining the graphical structures we are interested in and introducing the related terminology. We shall then be able to show how one can associate an algebra to each such graphical structure and study its properties.

1.1. Definition. A **quiver** $Q = (Q_0, Q_1, s, t)$ is a quadruple consisting of two sets: Q_0 (whose elements are called **points**, or **vertices**) and Q_1 (whose elements are called **arrows**), and two maps $s, t : Q_1 \to Q_0$ which associate to each arrow $\alpha \in Q_1$ its **source** $s(\alpha) \in Q_0$ and its **target** $t(\alpha) \in Q_0$, respectively.

An arrow $\alpha \in Q_1$ of source $a = s(\alpha)$ and target $b = t(\alpha)$ is usually denoted by $\alpha : a \to b$. A quiver $Q = (Q_0, Q_1, s, t)$ is usually denoted briefly by $Q = (Q_0, Q_1)$ or even simply by Q.

Thus, a quiver is nothing but an oriented graph without any restriction as to the number of arrows between two points, to the existence of loops

41

or oriented cycles. There are two main reasons for using the term quiver rather than graph: the first one is that the former has become generally accepted by specialists; the second is that the latter is used in so many different contexts and even senses (a graph can be oriented or not, with or without multiple arrows or loops) that it may lead, for our purposes at least, to certain ambiguities. When drawing a quiver, we agree to represent each point by an open dot, and each arrow will be pointing towards its target. With these conventions, the following are examples of quivers:

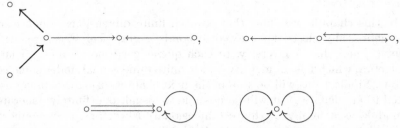

A **subquiver** of a quiver $Q = (Q_0, Q_1, s, t)$ is a quiver $Q' = (Q'_0, Q'_1, s', t')$ such that $Q'_0 \subseteq Q_0, Q'_1 \subseteq Q_1$ and the restrictions $s \mid_{Q'_1}, t \mid_{Q'_1}$ of s, t to Q'_1 are respectively equal to s', t' (that is, if $\alpha : a \to b$ is an arrow in Q_1 such that $\alpha \in Q'_1$ and $a, b \in Q'_0$, then $s'(\alpha) = a$ and $t'(\alpha) = b$). Such a subquiver is called **full** if Q'_1 equals the set of all those arrows in Q_1 whose source and target both belong to Q'_0, that is,

$$Q'_1 = \{\alpha \in Q_1 \mid s(\alpha) \in Q'_0 \quad \text{and} \quad t(\alpha) \in Q'_0\}.$$

In particular, a full subquiver is uniquely determined by its set of points.

A quiver Q is said to be **finite** if Q_0 and Q_1 are finite sets. The **underlying graph** $\overline{Q}$ of a quiver Q is obtained from Q by forgetting the orientation of the arrows. The quiver Q is said to be **connected** if $\overline{Q}$ is a connected graph.

Let $Q = (Q_0, Q_1, s, t)$ be a quiver and $a, b \in Q_0$. A **path of length** $\ell \geq 1$ with source a and target b (or, more briefly, from a to b) is a sequence

$$(a \mid \alpha_1, \alpha_2, \ldots, \alpha_\ell \mid b),$$

where $\alpha_k \in Q_1$ for all $1 \leq k \leq \ell$, and we have $s(\alpha_1) = a$, $t(\alpha_k) = s(\alpha_{k+1})$ for each $1 \leq k < \ell$, and finally $t(\alpha_\ell) = b$. Such a path is denoted briefly by $\alpha_1 \alpha_2 \ldots \alpha_\ell$ and may be visualised as follows

$$a = a_0 \xrightarrow{\alpha_1} a_1 \xrightarrow{\alpha_2} a_2 \longrightarrow \cdots \xrightarrow{\alpha_\ell} a_\ell = b.$$

We denote by Q_ℓ the set of all paths in Q of length ℓ. We also agree to associate with each point $a \in Q_0$ a path of length $\ell = 0$, called the **trivial** or **stationary path** at a, and denoted by

$$\varepsilon_a = (a \parallel a).$$

Thus the paths of lengths 0 and 1 are in bijective correspondence with the elements of Q_0 and Q_1, respectively. A path of length $\ell \geq 1$ is called a **cycle** whenever its source and target coincide. A cycle of length 1 is called a **loop**. A quiver is called **acyclic** if it contains no cycles.

We also need a notion of unoriented path, or a walk. To each arrow $\alpha : a \to b$ in a quiver Q, we associate a formal reverse $\alpha^{-1} : b \to a$, with the source $s(\alpha^{-1}) = b$ and the target $t(\alpha^{-1}) = a$. A **walk** of length $\ell \geq 1$ from a to b in Q is, by definition, a sequence $w = \alpha_1^{\varepsilon_1} \alpha_2^{\varepsilon_2} \ldots \alpha_\ell^{\varepsilon_\ell}$ with $\varepsilon_j \in \{-1, 1\}$, $s(\alpha_1^{\varepsilon_1}) = a$, $t(\alpha_\ell^{\varepsilon_\ell}) = b$ and $t(\alpha_j^{\varepsilon_j}) = s(\alpha_{j+1}^{\varepsilon_{j+1}})$, for all j such that $1 \leq j \leq \ell$.

If there exists in Q a path from a to b, then a is said to be a **predecessor** of b, and b is said to be a **successor** of a. In particular, if there exists an arrow $a \to b$, then a is said to be a **direct** (or **immediate**) **predecessor** of b, and b is said to be a **direct** (or **immediate**) **successor** of a. For $a \in Q_0$, we denote by a^- (or by a^+) the set of all direct predecessors (or successors, respectively) of a. The elements of $a^+ \cup a^-$ are called the **neighbours** of a.

Clearly, the composition of paths is a partially defined operation on the set of all paths in a quiver. We use it to define an algebra.

1.2. Definition. Let Q be a quiver. The **path algebra** KQ of Q is the K-algebra whose underlying K-vector space has as its basis the set of all paths $(a \mid \alpha_1, \ldots, \alpha_\ell \mid b)$ of length $\ell \geq 0$ in Q and such that the product of two basis vectors $(a \mid \alpha_1, \ldots, \alpha_\ell \mid b)$ and $(c \mid \beta_1, \ldots, \beta_k \mid d)$ of KQ is defined by

$$(a \mid \alpha_1, \ldots, \alpha_\ell \mid b)(c \mid \beta_1, \ldots, \beta_k \mid d) = \delta_{bc}(a \mid \alpha_1, \ldots, \alpha_\ell, \beta_1, \ldots, \beta_k \mid d),$$

where δ_{bc} denotes the Kronecker delta. In other words, the product of two paths $\alpha_1 \ldots \alpha_\ell$ and $\beta_1 \ldots \beta_k$ is equal to zero if $t(\alpha_\ell) \neq s(\beta_1)$ and is equal to the composed path $\alpha_1 \ldots \alpha_\ell \beta_1 \ldots \beta_k$ if $t(\alpha_\ell) = s(\beta_1)$. The product of basis elements is then extended to arbitrary elements of KQ by distributivity.

In other words, there is a direct sum decomposition

$$KQ = KQ_0 \oplus KQ_1 \oplus KQ_2 \oplus \ldots \oplus KQ_\ell \oplus \ldots$$

of the K-vector space KQ, where, for each $\ell \geq 0$, KQ_ℓ is the subspace of KQ generated by the set Q_ℓ of all paths of length ℓ. It is easy to see that $(KQ_n) \cdot (KQ_m) \subseteq KQ_{n+m}$ for all $n, m \geq 0$, because the product in KQ of a path of length n by a path of length m is either zero or a path of length $n+m$. This is expressed sometimes by saying that the decomposition defines a **grading** on KQ or that KQ is a **graded K-algebra**.

1.3. Examples. (a) Let Q be the quiver

consisting of a single point and a single loop. The defining basis of the path algebra KQ is $\{\varepsilon_1, \alpha, \alpha^2, \ldots, \alpha^\ell, \ldots\}$ and the multiplication of basis vectors is given by

$$\varepsilon_1 \alpha^\ell = \alpha^\ell \varepsilon_1 = \alpha^\ell \qquad \text{for all} \quad \ell \geq 0, \text{ and}$$
$$\alpha^\ell \alpha^k = \alpha^{\ell+k} \qquad \text{for all} \quad \ell, k \geq 0,$$

where $\alpha^0 = \varepsilon_1$. Thus KQ is isomorphic to the polynomial algebra $K[t]$ in one indeterminate t, the isomorphism being induced by the K-linear map such that

$$\varepsilon_1 \mapsto 1 \qquad \text{and} \quad \alpha \mapsto t.$$

(b) Let Q be the quiver

consisting of a single point and two loops α and β. The defining basis of KQ is the set of all words on $\{\alpha, \beta\}$, with the empty word equal to ε_1: this is the identity of the path algebra KQ. Also, the multiplication of basis vectors reduces to the multiplication in the free monoid over $\{\alpha, \beta\}$. Thus KQ is isomorphic to the free associative algebra in two noncommuting indeterminates $K\langle t_1, t_2 \rangle$, the isomorphism being the K-linear map such that

$$\varepsilon_1 \mapsto 1, \qquad \alpha \mapsto t_1, \quad \text{and} \quad \beta \mapsto t_2.$$

More generally, let $Q = (Q_0, Q_1)$ be a quiver such that Q_0 has only one element, then each $\beta \in Q_1$ is a loop and we have similarly that KQ is isomorphic to the free associative algebra in the indeterminates $(X_\beta)_{\beta \in Q_1}$.

(c) Let Q be the quiver

The path algebra KQ has as its defining basis the set $\{\varepsilon_1, \varepsilon_2, \alpha\}$ with the multiplication table

	ε_1	ε_2	α
ε_1	ε_1	0	0
ε_2	0	ε_2	α
α	α	0	0

Clearly, KQ is isomorphic to the 2×2 lower triangular matrix algebra

$$\mathbb{T}_2(K) = \left[\begin{matrix} K & 0 \\ K & K \end{matrix} \right] = \left\{ \left[\begin{smallmatrix} a & 0 \\ b & c \end{smallmatrix} \right] \mid a, b, c \in K \right\}$$

where the isomorphism is induced by the K-linear map such that

$$\varepsilon_1 \mapsto \left[\begin{smallmatrix} 1 & 0 \\ 0 & 0 \end{smallmatrix} \right], \quad \varepsilon_2 \mapsto \left[\begin{smallmatrix} 0 & 0 \\ 0 & 1 \end{smallmatrix} \right], \quad \alpha \mapsto \left[\begin{smallmatrix} 0 & 0 \\ 1 & 0 \end{smallmatrix} \right].$$

(d) Let Q be the quiver

One can easily show, as above, that there is a K-algebra isomorphism

$$KQ \cong \left[\begin{matrix} K & 0 & 0 & 0 \\ K & K & 0 & 0 \\ K & 0 & K & 0 \\ K & 0 & 0 & K \end{matrix} \right].$$

1.4. Lemma. *Let Q be a quiver and KQ be its path algebra. Then*
(a) *KQ is an associative algebra,*
(b) *KQ has an identity element if and only if Q_0 is finite, and*
(c) *KQ is finite dimensional if and only if Q is finite and acyclic.*

Proof. (a) This follows directly from the definition of multiplication because the product of basis vectors is the composition of paths, which is associative.

(b) Clearly, each stationary path $\varepsilon_a = (a \parallel a)$ is an idempotent of KQ. Thus, if Q_0 is finite, $\sum_{a \in Q_0} \varepsilon_a$ is an identity for KQ. Conversely, suppose that Q_0 is infinite, and suppose to the contrary that $1 = \sum_{i=1}^{m} \lambda_i w_i$ is an identity element of KQ (where the λ_i are nonzero scalars and the w_i are paths in Q). The set Q_0' of the sources of the w_i has at most m elements and in particular is finite. Let thus $a \in Q_0 \backslash Q_0'$, then $\varepsilon_a \cdot 1 = 0$, a contradiction.

(c) If Q is infinite, then so is the basis of KQ, which is therefore infinite dimensional. If $w = \alpha_1 \alpha_2 \dots \alpha_\ell$ is a cycle in Q then, for each $t \geq 0$, we have a basis vector $w^t = (\alpha_1 \alpha_2 \ldots \alpha_\ell)^t$, so that KQ is again infinite dimensional. Conversely, if Q is finite and acyclic, it contains only finitely many paths and so KQ is finite dimensional. $\qquad\square$

1.5. Corollary. *Let Q be a finite quiver. The element $1 = \sum_{a \in Q_0} \varepsilon_a$ is the identity of KQ and the set $\{\varepsilon_a \mid a \in Q_0\}$ of all the stationary paths $\varepsilon_a = (a \parallel a)$ is a complete set of primitive orthogonal idempotents for KQ.*

Proof. It follows from the definition of multiplication that the ε_a are orthogonal idempotents for KQ. Because the set Q_0 is finite, the element $1 = \sum_{a \in Q_0} \varepsilon_a$ is the identity of KQ. There remains to show that the ε_a are primitive or, what amounts to the same, that the only idempotents of the algebra $\varepsilon_a (KQ) \varepsilon_a$ are 0 and ε_a; see (I.4.7). Indeed, any idempotent ε of $\varepsilon_a (KQ) \varepsilon_a$ can be written in the form $\varepsilon = \lambda \varepsilon_a + w$, where $\lambda \in K$ and w is a linear combination of cycles through a of length ≥ 1. The equality

$$0 = \varepsilon^2 - \varepsilon = (\lambda^2 - \lambda)\varepsilon_a + (2\lambda - 1)w + w^2$$

gives $w = 0$ and $\lambda^2 = \lambda$, thus $\lambda = 0$ or $\lambda = 1$. In the former case, $\varepsilon = 0$ and in the latter $\varepsilon = \varepsilon_a$. $\qquad\square$

Clearly, the set $\{\varepsilon_a \mid a \in Q_0\}$ is usually not the unique complete set of primitive orthogonal idempotents for KQ. For instance, in Example 1.3 (c), besides the set $\{\varepsilon_1, \varepsilon_2\}$, the set $\{\varepsilon_1 + \alpha, \varepsilon_2 - \alpha\}$ is also a complete set of primitive orthogonal idempotents for KQ.

The following lemma reduces the connectedness of an algebra to a partition of a complete set of primitive orthogonal idempotents for this algebra. It will allow us to characterise connected path algebras, then, in Section 2, connected quotients of path algebras.

1.6. Lemma. *Let A be an associative algebra with an identity and assume that $\{e_1, \ldots, e_n\}$ is a (finite) complete set of primitive orthogonal idempotents. Then A is a connected algebra if and only if there does not exist a nontrivial partition $I \dot\cup J$ of the set $\{1, 2, \ldots, n\}$ such that $i \in I$ and $j \in J$ imply $e_i A e_j = 0 = e_j A e_i$.*

Proof. Assume that there exists such a partition and let $c = \sum_{j \in J} e_j$. Because the partition is nontrivial, $c \neq 0, 1$. Because the e_j are orthogonal idempotents, c is an idempotent. Moreover, $ce_i = e_i c = 0$ for each $i \in I$, and $ce_j = e_j c = e_j$ for each $j \in J$. Let now $a \in A$ be arbitrary. By hypothesis, $e_i a e_j = 0 = e_j a e_i$ whenever $i \in I$ and $j \in J$. Consequently

$$
\begin{aligned}
ca &= (\sum_{j \in J} e_j)a &&= (\sum_{j \in J} e_j a) \cdot 1 = (\sum_{j \in J} e_j a)(\sum_{i \in I} e_i + \sum_{k \in J} e_k) \\
&= \sum_{j,k \in J} e_j a e_k &&= (\sum_{j \in J} e_j + \sum_{i \in I} e_i)a(\sum_{k \in J} e_k) = ac.
\end{aligned}
$$

Thus c is a central idempotent, and $A = cA \times (1-c)A$ is a nontrivial product decomposition of A. Conversely, if A is not connected, it contains a central idempotent $c \neq 0, 1$. We have

$$
c = 1 \cdot c \cdot 1 = (\sum_{i=1}^{n} e_i)c(\sum_{j=1}^{n} e_j) = \sum_{i,j=1}^{n} e_i c e_j = \sum_{i=1}^{n} e_i c e_i,
$$

because c is central. Let $c_i = e_i c e_i \in e_i A e_i$. Then $c_i^2 = (e_i c e_i)(e_i c e_i) = e_i c^2 e_i = c_i$, so that c_i is an idempotent of $e_i A e_i$. Because e_i is primitive, $c_i = 0$ or $c_i = e_i$. Let $I = \{i \mid c_i = 0\}$ and $J = \{j \mid c_j = e_j\}$. Because $c \neq 0, 1$, this is indeed a nontrivial partition of $\{1, 2, \ldots, n\}$. Moreover, if $i \in I$, we have $e_i c = c e_i = 0$ and, if $j \in J$, we have $e_j c = c e_j = e_j$. Therefore, if $i \in I$ and $j \in J$, we have $e_i A e_j = e_i A c e_j = e_i c A e_j = 0$ and similarly $e_j A e_i = 0$. $\qquad\square$

1.7. Lemma. *Let Q be a finite quiver. The path algebra KQ is connected if and only if Q is a connected quiver.*

Proof. Assume that Q is not connected and let Q' be a connected component of Q. Denote by Q'' the full subquiver of Q having as set of points $Q_0'' = Q_0 \backslash Q_0'$. By hypothesis, neither Q' nor Q'' is empty. Let $a \in Q_0'$ and $b \in Q_0''$. Because Q is not connected, an arbitrary path w in Q is entirely contained in either Q' or (a connected component of) Q''. In the former case, we have $w \varepsilon_b = 0$ and hence $\varepsilon_a w \varepsilon_b = 0$. In the latter case, we have $\varepsilon_a w = 0$ and hence again $\varepsilon_a w \varepsilon_b = 0$. This shows that $\varepsilon_a(KQ)\varepsilon_b = 0$. Similarly, $\varepsilon_b(KQ)\varepsilon_a = 0$. By (1.6), KQ is not connected.

Suppose now that Q is connected but KQ is not. By (1.6), there exists a disjoint union partition $Q_0 = Q_0' \dot\cup Q_0''$ such that, if $x \in Q_0'$ and $y \in Q_0''$, then $\varepsilon_x(KQ)\varepsilon_y = 0 = \varepsilon_y(KQ)\varepsilon_x$. Because Q is connected, there exist $a \in Q_0'$ and $b \in Q_0''$ that are neighbours. Without loss of generality, we may suppose that there exists an arrow $\alpha : a \to b$. But then we have

$$
\alpha = \varepsilon_a \alpha \varepsilon_b \in \varepsilon_a(KQ)\varepsilon_b = 0,
$$

a contradiction that completes the proof of the lemma. $\qquad\square$

To summarise, we have shown that if Q is a finite connected quiver, the path algebra KQ of Q is a connected associative K-algebra with an identity, which admits $\{\varepsilon_a = (a \parallel a) \mid a \in Q_0\}$ as a complete set of primitive orthogonal idempotents. We shall now characterise it by a universal property.

1.8. Theorem. *Let Q be a finite connected quiver and A be an associative K-algebra with an identity. For any pair of maps $\varphi_0 : Q_0 \to A$ and $\varphi_1 : Q_1 \to A$ satisfying the folowing conditions:*

(i) $1 = \sum\limits_{a \in Q_0} \varphi_0(a)$, $\varphi_0(a)^2 = \varphi_0(a)$, and $\varphi_0(a) \cdot \varphi_0(b) = 0$, for all $a \neq b$,

(ii) *if $\alpha : a \to b$ then $\varphi_1(\alpha) = \varphi_0(a)\varphi_1(\alpha)\varphi_0(b)$,*

there exists a unique K-algebra homomorphism $\varphi : KQ \to A$ such that $\varphi(\varepsilon_a) = \varphi_0(a)$ for any $a \in Q_0$ and $\varphi(\alpha) = \varphi_1(\alpha)$ for any $\alpha \in Q_1$.

Proof. Indeed, assume there exists a homomorphism $\varphi : KQ \to A$ of K-algebras extending φ_0 and φ_1, and let $\alpha_1\alpha_2 \ldots \alpha_\ell$ be a path in Q. Because φ is a K-algebra homomorphism, we have

$$
\begin{aligned}
\varphi(\alpha_1\alpha_2 \ldots \alpha_\ell) &= \varphi(\alpha_1)\varphi(\alpha_2) \ldots \varphi(\alpha_\ell) \\
&= \varphi_1(\alpha_1)\varphi_1(\alpha_2) \ldots \varphi_1(\alpha_\ell).
\end{aligned}
$$

This shows uniqueness. On the other hand, this formula clearly defines a K-linear mapping from KQ to A that is compatible with the composition of paths (thus preserves the product) and is such that

$$
\varphi(1) = \varphi\left(\sum_{a \in Q_0} \varepsilon_a \right) = \sum_{a \in Q_0} \varphi(\varepsilon_a) = \sum_{a \in Q_0} \varphi_0(a) = 1,
$$

that is, it preserves the identity. It is therefore a K-algebra homomorphism. $\square$

We now calculate the radical of the path algebra of a finite, connected, and acyclic quiver. We need the following definition.

1.9. Definition. Let Q be a finite and connected quiver. The two-sided ideal of the path algebra KQ generated (as an ideal) by the arrows of Q is called the **arrow ideal** of KQ and is denoted by R_Q. Whenever this can be done without ambiguity we shall use the notation R instead of R_Q.

Note that there is a direct sum decomposition

$$
R_Q = KQ_1 \oplus KQ_2 \oplus \ldots \oplus KQ_\ell \oplus \ldots
$$

of the K-vector space R_Q, where KQ_ℓ is the subspace of KQ generated by the set Q_ℓ of all paths of length ℓ. In particular, the underlying K-vector space of R_Q is generated by all paths in Q of length $\ell \geq 1$. This implies that, for each $\ell \geq 1$,

$$
R_Q^\ell = \bigoplus_{m \geq \ell} KQ_m
$$

and therefore R_Q^ℓ is the ideal of KQ generated, as a K-vector space, by the set of all paths of length $\geq \ell$. Consequently, the K-vector space $R_Q^\ell/R_Q^{\ell+1}$ is generated by the residual classes of all paths in Q of length (exactly) equal to ℓ and there is an isomorphism of K-vector spaces $R_Q^\ell/R_Q^{\ell+1} \cong KQ_\ell$.

1.10. Proposition. *Let Q be a finite connected quiver, R be the arrow ideal of KQ and $\varepsilon_a = (a \parallel a)$ for $a \in Q_0$. The set $\{\bar{\varepsilon}_a = \varepsilon_a + R \mid a \in Q_0\}$ is a complete set of primitive orthogonal idempotents for KQ/R, and the latter is isomorphic to a product of copies of K. If, in addition, Q is acyclic, then $\operatorname{rad} KQ = R$ and KQ is a finite dimensional basic algebra.*

Proof. Clearly, there is a direct sum decomposition

$$KQ/R = \bigoplus_{a,b \in Q_0} \bar{\varepsilon}_a(KQ/R)\bar{\varepsilon}_b$$

as a K-vector space. Because R contains all paths of length ≥ 1, this becomes

$$KQ/R = \bigoplus_{a \in Q_0} \bar{\varepsilon}_a(KQ/R)\bar{\varepsilon}_a.$$

Then KQ/R is generated, as a K-vector space, by the residual classes of the paths of length zero, that is, by the set $\{\bar{\varepsilon}_a = \varepsilon_a + R \mid a \in Q_0\}$. Clearly, this set is a complete set of primitive orthogonal idempotents of the quotient algebra KQ/R. Moreover, for each $a \in Q_0$, the algebra $\bar{\varepsilon}_a(KQ/R)\bar{\varepsilon}_a$ is generated, as a K-vector space, by $\bar{\varepsilon}_a$ and consequently is isomorphic, as a K-algebra, to K. This shows that the quotient algebra KQ/R is isomorphic to a product of $|Q_0|$ copies of K.

Assume now that Q is acyclic (so that, by (1.4), KQ is a finite dimensional algebra). There exists a largest $\ell \geq 1$ such that Q contains a path of length ℓ. But this implies that any product of $\ell + 1$ arrows is zero, that is, $R^{\ell+1} = 0$. Consequently, the ideal R is nilpotent and hence, by (I.1.4), $R \subseteq \operatorname{rad} KQ$. Because KQ/R is isomorphic to a product of copies of K, it follows from (I.1.4) and (I.6.2) that $\operatorname{rad} KQ = R$ and the algebra KQ is basic. $\qquad\square$

We remark that if Q is not acyclic, it is generally not true that $\operatorname{rad} KQ = R_Q$. For instance, let Q be the quiver

As we have seen before, $KQ \cong K[t]$. Thus $\operatorname{rad} KQ = 0$, because the field K is algebraically closed (and hence infinite); then the set $\{t - \lambda \mid \lambda \in K\}$ is an infinite set of irreducible polynomials, which generates an infinite set of

maximal ideals with zero intersection. On the other hand, $R_Q = \bigoplus_{\ell>0} K\alpha^\ell$
as a K-vector space and thus is certainly nonzero.

We summarise our findings in the following corollary.

1.11. Corollary. *Let Q be a finite, connected, and acyclic quiver. The path algebra KQ is a basic and connected associative finite dimensional K-algebra with an identity, having the arrow ideal as radical, and the set $\{\varepsilon_a = (a \parallel a) \mid a \in Q_0\}$ as a complete set of primitive orthogonal idempotents.*

Proof. The statement collects results from (1.4), (1.5), (1.7), and (1.10). □

We now give a construction showing that an algebra as in (1.11) can always be realised as an algebra of lower triangular matrices. We start by recalling a classical construction for generalised matrix algebras. Let $(A_i)_{1\le i\le n}$ be a family of K-algebras and $(M_{ij})_{1\le i,j\le n}$ be a family of A_i-A_j-bimodules such that $M_{ii} = A_i$, for each i. Moreover, assume that we have for each triple (i,j,k) an A_i-A_k-bimodule homomorphism

$$\varphi^j_{ik} : M_{ij} \otimes M_{jk} \to M_{ik}$$

satisfying, for each quadruple (i,j,k,ℓ), the "associativity" condition

$$\varphi^k_{i\ell}\left(\varphi^j_{ik} \otimes 1\right) = \varphi^j_{i\ell}(1 \otimes \varphi^k_{j\ell}),$$

that is, the following square is commutative:

$$
\begin{array}{ccc}
M_{ij} \otimes M_{jk} \otimes M_{kl} & \xrightarrow{\;1\otimes\varphi^k_{jl}\;} & M_{ij} \otimes M_{jl} \\
\downarrow{\scriptstyle\varphi^j_{ik}\otimes 1} & & \downarrow{\scriptstyle\varphi^j_{il}} \\
M_{ik} \otimes M_{kl} & \xrightarrow{\;\varphi^k_{il}\;} & M_{il}
\end{array}
$$

Then it is easily verified that the K-vector space of $n \times n$ matrices

$$A = \begin{bmatrix} M_{11} & M_{12} & \cdots & M_{1n} \\ M_{21} & M_{22} & \cdots & M_{2n} \\ \vdots & \vdots & & \vdots \\ M_{n1} & M_{n2} & \cdots & M_{nn} \end{bmatrix} = \{[x_{ij}] \mid x_{ij} \in M_{ij} \quad \text{for all} \quad 1 \le i,j \le n\}$$

becomes a K-algebra if we define its multiplication by the formula

$$[x_{ij}] \cdot [y_{ij}] = \left[\sum_{k=1}^n \varphi^k_{ij}(x_{ik} \otimes y_{kj})\right].$$

Assume that Q is a finite and acyclic quiver. Let $n = |Q_0|$ be the number of points in Q. It is easy to see that we may number the points of Q from 1 to n such that, if there exists a path from i to j, then $j \leq i$.

1.12. Lemma. *Let Q be a connected, finite, and acyclic quiver with $Q_0 = \{1, 2, \ldots, n\}$ such that, for each $i, j \in Q_0$, $j \leq i$ whenever there exists a path from i to j in Q. Then the path algebra KQ is isomorphic to the triangular matrix algebra*

$$
A = \begin{bmatrix}
\varepsilon_1(KQ)\varepsilon_1 & 0 & \cdots & 0 \\
\varepsilon_2(KQ)\varepsilon_1 & \varepsilon_2(KQ)\varepsilon_2 & \cdots & 0 \\
\vdots & \vdots & & \vdots \\
\varepsilon_n(KQ)\varepsilon_1 & \varepsilon_n(KQ)\varepsilon_2 & \cdots & \varepsilon_n(KQ)\varepsilon_n
\end{bmatrix},
$$

where $\varepsilon_a = (a \parallel a)$ for any $a \in Q_0$, the addition is the obvious one, and the multiplication is induced from the multiplication of KQ.

Proof. Because $\{\varepsilon_a = (a \parallel a) \mid a \in Q_0\}$ is a complete set of primitive orthogonal idempotents for KQ (by (1.11)), we have a K-vector space decomposition of KQ

$$
KQ = \bigoplus_{a,b \in Q_0} \varepsilon_a(KQ)\varepsilon_b.
$$

It follows from the hypothesis that if $\varepsilon_i(KQ)\varepsilon_j \neq 0$, then $j \leq i$. For each point $i \in Q_0$, the absence of cycles through i implies that the algebra $\varepsilon_i(KQ)\varepsilon_i$ is isomorphic to K. The definition of the multiplication in KQ implies that, for each pair (j, i) such that $j \leq i$, $\varepsilon_i(KQ)\varepsilon_j$ is an $\varepsilon_i(KQ)\varepsilon_i$-$\varepsilon_j(KQ)\varepsilon_j$-bimodule and, for each triple (k, j, i) such that $k \leq j \leq i$, there exists a K-linear map

$$
\varphi_{ik}^j : \varepsilon_i(KQ)\varepsilon_j \otimes \varepsilon_j(KQ)\varepsilon_k \to \varepsilon_i(KQ)\varepsilon_k,
$$

where the tensor product is taken over $\varepsilon_j(KQ)\varepsilon_j$. It is easily seen that the φ_{ik}^j are actually $\varepsilon_i(KQ)\varepsilon_i$-$\varepsilon_k(KQ)\varepsilon_k$-bimodule homomorphisms satisfying the "associativity" conditions $\varphi_{i\ell}^k(\varphi_{ik}^j \otimes 1) = \varphi_{i\ell}^j(1 \otimes \varphi_{j\ell}^k)$ whenever $i \leq j \leq k \leq \ell$. We may thus construct a generalised matrix algebra as done earlier. Now, by associating to each path from i to j in KQ the corresponding element of A (that is, basis element of the bimodule $\varepsilon_i(KQ)\varepsilon_j$), we get a K-algebra isomorphism $KQ \cong A$. Indeed, the algebras A and KQ are clearly isomorphic as K-vector spaces and the bijection between the basis vectors is compatible with the algebra multiplications (by definition of the φ_{ik}^j), thus this vector space isomorphism is a K-algebra isomorphism. $\square$

In particular, if Q has no multiple arrows and its underlying graph is a tree, then there is at most one path between two given points of Q so that, for all $j \leq i$, we have $\dim_K(\varepsilon_i(KQ)\varepsilon_j) \leq 1$. Consequently, KQ is isomorphic to a subalgebra of the full lower triangular matrix algebra

$$\mathbb{T}_n(K) = \begin{bmatrix} K & 0 & \cdots & 0 \\ K & K & \cdots & 0 \\ \vdots & \vdots & \ddots & \vdots \\ K & K & \cdots & K \end{bmatrix}.$$

1.13. Examples. (a) Let Q be the quiver

$$\underset{1}{\circ}\longleftarrow\underset{2}{\circ}\longleftarrow\underset{3}{\circ}\longleftarrow \cdots \longleftarrow\underset{n-1}{\circ}\longleftarrow\underset{n}{\circ}$$

This construction gives the algebra isomorphism $KQ \cong \mathbb{T}_n(K)$.

(b) Let Q be the Kronecker quiver

$$1\ \circ \rightrightarrows \circ\ 2$$

Then there is an algebra isomorphism

$$KQ \cong \begin{bmatrix} K & 0 \\ K^2 & K \end{bmatrix},$$

where K^2 is considered as a K-K-bimodule in the obvious way

$$a \cdot (x,y) = (ax, ay), \qquad (x,y) \cdot b = (xb, yb)$$

for all $a, b, x, y \in K$. The path algebra of the Kronecker quiver is called the **Kronecker algebra**. Its module category is studied in detail later (see also (I.2.5)).

We remark that the expression of KQ as an algebra of lower triangular matrices (1.12) is not unique. For instance, the Kronecker algebra is isomorphic to the subalgebra

$$A = \left\{ \begin{bmatrix} a & 0 & 0 \\ b & d & 0 \\ c & 0 & d \end{bmatrix} \mid a,b,c,d \in K \right\}$$

of $\mathbb{T}_3(K)$. An algebra isomorphism between A and the Kronecker algebra is given by

$$\begin{bmatrix} a & 0 & 0 \\ b & d & 0 \\ c & 0 & d \end{bmatrix} \mapsto \begin{bmatrix} a & 0 \\ (b,c) & d \end{bmatrix}.$$

(c) Let Q be the quiver $\underset{1}{\circ}\longleftarrow\underset{2}{\circ}\rightrightarrows\underset{4}{\circ}\longrightarrow\underset{3}{\circ}$. Then

$$KQ \cong \begin{bmatrix} K & 0 & 0 & 0 \\ K & K & 0 & 0 \\ 0 & 0 & K & 0 \\ K^3 & K^3 & K & K \end{bmatrix},$$

where the multiplication is defined in a manner analogous to the one used in example (b).

II.2. Admissible ideals and quotients of the path algebra

Let Q be a finite quiver. By (1.4), the path algebra KQ of Q is an associative algebra with an identity and is finite dimensional if and only if Q is acyclic. Our objective in this section is to study the finite dimensional quotients of a not necessarily finite dimensional path algebra. We see in particular that they correspond to certain ideals we call admissible.

2.1. Definition. Let Q be a finite quiver and R_Q be the arrow ideal of the path algebra KQ. A two-sided ideal $\mathcal{I}$ of KQ is said to be **admissible** if there exists $m \geq 2$ such that

$$R_Q^m \subseteq \mathcal{I} \subseteq R_Q^2.$$

If $\mathcal{I}$ is an admissible ideal of KQ, the pair $(Q, \mathcal{I})$ is said to be a **bound quiver**. The quotient algebra $KQ/\mathcal{I}$ is said to be the algebra of the bound quiver $(Q, \mathcal{I})$ or, simply, a **bound quiver algebra**.

It follows directly from the definition that an ideal $\mathcal{I}$ of KQ, contained in R_Q^2, is admissible if and only if it contains all paths whose length is large enough. It can be shown that this is the case if and only if, for each cycle σ in Q, there exists $s \geq 1$ such that $\sigma^s \in \mathcal{I}$.

If, in particular, Q is acyclic, any ideal contained in R_Q^2 is admissible.

2.2. Examples. (a) For any finite quiver Q and any $m \geq 2$, the ideal R_Q^m is admissible.

(b) The zero ideal is admissible in KQ if and only if Q is acyclic. Indeed, the zero ideal is admissible if and only if there exists $m \geq 2$ such that $R_Q^m = 0$, that is, any product of m arrows in KQ is zero. This is the case if and only if Q is acyclic.

(c) Let Q be the quiver

$$
\begin{array}{ccc}
 & \overset{2}{\circ} & \\
\beta \nearrow & & \nwarrow \alpha \\
1\circ \;\overset{\lambda}{\underset{}{\longleftarrow}}\; & & \circ 4 \\
\delta \searrow & & \swarrow \gamma \\
 & \underset{3}{\circ} &
\end{array}
$$

The ideal $\mathcal{I}_1 = \langle \alpha\beta - \gamma\delta \rangle$ of the K-algebra KQ is admissible, but $\mathcal{I}_2 = \langle \alpha\beta - \lambda \rangle$ is not; indeed, $\alpha\beta - \lambda \notin R_Q^2$.

(d) Let Q be the quiver

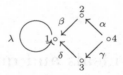

The ideal $\mathcal{I} = \langle \alpha\beta - \gamma\delta, \beta\lambda, \lambda^3 \rangle$ is admissible. Indeed, it is clear that $\mathcal{I} \subseteq R_Q^2$. We show that $R_Q^4 \subseteq \mathcal{I}$. Every path of length ≥ 4 and source 1, 2, or 3 contains the product λ^3 and hence lies in $\mathcal{I}$. The paths of length ≥ 4 and source 4 contain a path of the form $\alpha\beta\lambda^2$ or $\gamma\delta\lambda^2$ and hence lie in $\mathcal{I}$, in the first case, because $\beta\lambda \in \mathcal{I}$, and in the second, because $\gamma\delta\lambda^2 = (\gamma\delta - \alpha\beta)\lambda^2 + \alpha\beta\lambda^2 \in \mathcal{I}$. This completes the proof that $\mathcal{I} = \langle \alpha\beta - \gamma\delta, \beta\lambda, \lambda^3 \rangle$ is admissible. Another example of an admissible ideal is $\langle \lambda^5 \rangle$. On the other hand, $\langle \beta\lambda, \alpha\beta - \gamma\delta \rangle$ is not admissible.

(e) Let Q be the quiver $\underset{1}{\circ} \overset{\beta}{\underset{\gamma}{\rightleftarrows}} \underset{2}{\circ} \overset{\alpha}{\longleftarrow} \underset{3}{\circ}$. Each of the ideals $\mathcal{I}_1 = \langle \alpha\beta \rangle$ and $\mathcal{I}_2 = \langle \alpha\beta - \alpha\gamma \rangle$ is clearly admissible. The bound quiver algebras $KQ/\mathcal{I}_1$ and $KQ/\mathcal{I}_2$ are isomorphic under the isomorphism $KQ/\mathcal{I}_1 \to KQ/\mathcal{I}_2$ induced by the correspondence $\varepsilon_i \mapsto \varepsilon_i$ for $i = 1, 2, 3$; $\alpha \mapsto \alpha$, $\beta \mapsto \beta - \gamma$, and $\gamma \mapsto \gamma$.

The preceding examples show that it is convenient to define an admissible ideal in terms of its generators. These are called relations.

2.3. Definition. Let Q be a quiver. A **relation** in Q with coefficients in K is a K-linear combination of paths of length at least two having the same source and target. Thus, a relation ρ is an element of KQ such that

$$\rho = \sum_{i=1}^{m} \lambda_i w_i,$$

where the λ_i are scalars (not all zero) and the w_i are paths in Q of length at least 2 such that, if $i \neq j$, then the source (or the target, respectively) of w_i coincides with that of w_j.

If $m = 1$, the preceding relation is called a **zero relation** or a **monomial relation**. If it is of the form $w_1 - w_2$ (where w_1, w_2 are two paths), it is called a **commutativity relation**.

If $(\rho_j)_{j \in J}$ is a set of relations for a quiver Q such that the ideal they generate $\langle \rho_j \mid j \in J \rangle$ is admissible, we say that the quiver Q is **bound by the relations** $(\rho_j)_{j \in J}$ or by the relations $\rho_j = 0$ for all $j \in J$.

For instance, in Example 2.2 (d), the ideal $\mathcal{I}$ is generated by one commutativity relation $\rho_1 = \alpha\beta - \gamma\delta$ and two zero relations $\rho_2 = \beta\lambda$ and $\rho_3 = \lambda^3$;

we thus say that Q is bound by the relations $\alpha\beta = \gamma\delta$, $\beta\lambda = 0$, and $\lambda^3 = 0$.

2.4. Lemma. *Let Q be a finite quiver and $\mathcal{I}$ be an admissible ideal of KQ. The set $\{e_a = \varepsilon_a + \mathcal{I} \mid a \in Q_0\}$ is a complete set of primitive orthogonal idempotents of the bound quiver algebra $KQ/\mathcal{I}$.*

Proof. Because e_a is the image of ε_a under the canonical homomorphism $KQ \to KQ/\mathcal{I}$, it follows from (1.5) that the given set is indeed a complete set of orthogonal idempotents. There remains to check that each e_a is primitive, that is, the only idempotents of $e_a(KQ/\mathcal{I})e_a$ are 0 and e_a. Indeed, any idempotent e of $e_a(KQ/\mathcal{I})e_a$ can be written in the form $e = \lambda\varepsilon_a + w + \mathcal{I}$, where $\lambda \in K$ and w is a linear combination of cycles through a of length ≥ 1. The equality $e^2 = e$ gives

$$(\lambda^2 - \lambda)\varepsilon_a + (2\lambda - 1)w + w^2 \in \mathcal{I}.$$

Let R_Q be the arrow ideal of KQ. Because $\mathcal{I} \subseteq R_Q^2$, we must have $\lambda^2 - \lambda = 0$, so that $\lambda = 0$ or $\lambda = 1$. Assume that $\lambda = 0$, then $e = w + \mathcal{I}$, where w is idempotent modulo $\mathcal{I}$. On the other hand, because $R_Q^m \subseteq \mathcal{I}$ for some $m \geq 2$, we must have $w^m \in \mathcal{I}$, that is, w is also nilpotent modulo $\mathcal{I}$. Consequently, $w \in \mathcal{I}$ and e is zero. On the other hand, if $\lambda = 1$, then $e_a - e = -w + \mathcal{I}$ is also an idempotent in $e_a(KQ/\mathcal{I})e_a$ so that w is again idempotent modulo $\mathcal{I}$. Because, as before, it is also nilpotent modulo $\mathcal{I}$, it must belong to $\mathcal{I}$. Consequently, $e_a = e$. $\qquad\square$

2.5. Lemma. *Let Q be a finite quiver and $\mathcal{I}$ be an admissible ideal of KQ. The bound quiver algebra $KQ/\mathcal{I}$ is connected if and only if Q is a connected quiver.*

Proof. If Q is not a connected quiver, KQ is not a connected algebra (by (1.7)). Hence KQ contains a central idempotent γ not equal to 0 or 1 that may, by the proof of (1.6), be chosen to be a sum of paths of length zero, that is, of points. But then $c = \gamma + \mathcal{I}$ is not equal to $\mathcal{I}$. On the other hand, $c = 1 + \mathcal{I}$ implies $1 - \gamma \in \mathcal{I}$, which is also impossible (because $\mathcal{I} \subseteq R_Q^2$). Because it is clear that c is a central idempotent of $KQ/\mathcal{I}$, we infer that the latter is not a connected algebra.

The reverse implication is shown exactly as in (1.7). Assume that Q is a connected quiver but that $KQ/\mathcal{I}$ is not a connected algebra. By (1.6) (and (2.4)), there exists a nontrivial partition $Q_0 = Q_0' \cup Q_0''$ such that $x \in Q_0'$ and $y \in Q_0''$ imply $e_x(KQ/\mathcal{I})e_y = 0 = e_y(KQ/\mathcal{I})e_x$. Because Q is connected, there exist $a \in Q_0'$ and $b \in Q_0''$ that are neighbours. Without loss of generality, we may suppose that there exists an arrow $\alpha : a \to b$. But then $\alpha = \varepsilon_a \alpha \varepsilon_b$ implies that $\overline{\alpha} = \alpha + \mathcal{I}$ satisfies $\overline{\alpha} = e_a \overline{\alpha} e_b \in e_a(KQ/\mathcal{I})e_b = 0$.

As $\overline{\alpha} \neq \mathcal{I}$ (because $\mathcal{I} \subseteq R_Q^2$), we have reached a contradiction.   □

2.6. Proposition. *Let Q be a finite quiver and $\mathcal{I}$ be an admissible ideal of KQ. The bound quiver algebra $KQ/\mathcal{I}$ is finite dimensional.*

Proof. Because $\mathcal{I}$ is admissible, there exists $m \geq 2$ such that $R^m \subseteq I$, where R is the arrow ideal R_Q of KQ. But then there exists a surjective algebra homomorphism $KQ/R^m \to KQ/\mathcal{I}$. Thus it suffices to prove that KQ/R^m is finite dimensional. Now the residual classes of the paths of length less than m form a basis of KQ/R^m as a K-vector space. Because there are only finitely many such paths, our statement follows.   □

If $\mathcal{I}$ is not admissible, the algebra $KQ/\mathcal{I}$ is generally not finite dimensional or even not right noetherian, that is, it may contain a right ideal that is not finitely generated. The following classical example, due to J. Dieudonné (see [48], p. 16) shows a finitely generated (even cyclic) module that has a submodule that is not finitely generated.

2.7. Example. Let Q be the quiver

and $\mathcal{I} = \langle \beta\alpha, \beta^2 \rangle$. It is clear that $\mathcal{I}$ is not admissible, because $\alpha^m \notin \mathcal{I}$ for any $m \geq 1$. Let $A = KQ/\mathcal{I}$ and J be the subspace of A (considered as a K-vector space) generated by the elements of the form $\overline{\alpha}^n \overline{\beta}$, for all $n \geq 1$ (where, as usual, $\overline{\alpha} = \alpha + \mathcal{I}, \overline{\beta} = \beta + \mathcal{I}$). Then J is a right ideal of A. Indeed, it suffices to show that $J\overline{\alpha} \subseteq J$ and $J\overline{\beta} \subseteq J$, and this follows from the equalities $\overline{\alpha}^n \overline{\beta}\overline{\alpha} = 0$ and $\overline{\alpha}^n \overline{\beta}^2 = 0$ for all $n \geq 1$. In particular, J_A is a submodule of the cyclic module A_A but is not finitely generated (indeed, let m be the largest exponent of $\overline{\alpha}$ among the elements of a finite set $\mathcal{J}$ of generators of J, then $\overline{\alpha}^{m+1}\overline{\beta} \in J$ cannot be a K-linear combination of elements from $\mathcal{J}$).

2.8. Lemma. *Let Q be a finite quiver. Every admissible ideal $\mathcal{I}$ of KQ is finitely generated.*

Proof. Let R be the arrow ideal of KQ and $m \geq 2$ be an integer such that $R^m \subseteq \mathcal{I}$. We have a short exact sequence $0 \to R^m \to \mathcal{I} \to \mathcal{I}/R^m \to 0$ of KQ-modules.

It thus suffices to show that R^m and $\mathcal{I}/R^m$ are finitely generated as KQ-modules. Obviously, R^m is the KQ-module generated by the paths of length m. Because there are only finitely many such paths, R^m is finitely generated. On the other hand, $\mathcal{I}/R^m$ is an ideal of the finite dimensional algebra KQ/R^m (see (2.6)). Therefore $\mathcal{I}/R^m$ is a finite dimensional K-vector space, hence a finitely generated KQ-module.   □

2.9. Corollary. *Let Q be a finite quiver and $\mathcal{I}$ be an admissible ideal of KQ. There exists a finite set of relations $\{\rho_1, \ldots, \rho_m\}$ such that $\mathcal{I} = \langle \rho_1, \ldots, \rho_m \rangle$.*

Proof. By (2.8), an admissible ideal $\mathcal{I}$ of KQ always has a finite generating set $\{\sigma_1, \ldots, \sigma_t\}$. The elements σ_i of such a set are generally not relations, because the paths composing σ_i do not necessarily have the same sources and targets. On the other hand, for any i such that $1 \leq i \leq t$ and $a, b \in Q_0$, the term $\varepsilon_a \sigma_i \varepsilon_b$ is either zero or a relation. Because $\sigma_i = \sum\limits_{a,b \in Q_0} \varepsilon_a \sigma_i \varepsilon_b$, for $i \leq t$, the nonzero elements among the set $\{\varepsilon_a \sigma_i \varepsilon_b \mid 1 \leq i \leq t; a, b \in Q_0\}$ form a finite set of relations generating $\mathcal{I}$.   □

2.10. Lemma. *Let Q be a finite quiver, R_Q be the arrow ideal of KQ, and $\mathcal{I}$ be an admissible ideal of KQ. Then $\mathrm{rad}(KQ/\mathcal{I}) = R_Q/\mathcal{I}$. Moreover, the bound quiver algebra $KQ/\mathcal{I}$ is basic.*

Proof. Because $\mathcal{I}$ is an admissible ideal of KQ, there exists $m \geq 2$ such that $R^m \subseteq \mathcal{I}$, where $R = R_Q$. Consequently, $(R/\mathcal{I})^m = 0$ and $R/\mathcal{I}$ is a nilpotent ideal of $KQ/\mathcal{I}$. On the other hand, the algebra $(KQ/\mathcal{I})/(R/\mathcal{I}) \cong KQ/R$ is isomorphic to a direct product of copies of K, by (1.10). This implies both assertions, by (I.1.4).   □

2.11. Corollary. *For each $\ell \geq 1$, we have $\mathrm{rad}^{\ell}(KQ/\mathcal{I}) = (R_Q/\mathcal{I})^{\ell}$.* □

It follows from Lemma 2.10 and Corollary 2.11 that the K-vector space

$$\mathrm{rad}(KQ/\mathcal{I})/\mathrm{rad}^2(KQ/\mathcal{I}) = (R_Q/\mathcal{I})/(R_Q/\mathcal{I})^2 \cong R_Q/R_Q^2$$

admits as basis the set $\overline{\alpha} + \mathrm{rad}^2(KQ/\mathcal{I})$, where $\overline{\alpha} = \alpha + KQ/\mathcal{I}$ and $\alpha \in Q_1$. This remark is crucial for the understanding of Section 3.

We summarise our findings in the following corollary.

2.12. Corollary. *Let Q be a finite connected quiver, R_Q be the arrow ideal of KQ, and $\mathcal{I}$ be an admissible ideal of KQ. The bound quiver algebra $KQ/\mathcal{I}$ is a basic and connected finite dimensional algebra with an identity, having $R_Q/\mathcal{I}$ as radical and $\{e_a \mid a \in Q_0\}$ as complete set of primitive orthogonal idempotents.*

Proof. The statement collects results from (2.4), (2.5), (2.6), and (2.10).   □

2.13. Examples. (a) Let Q be the quiver $\underset{1}{\circ} \xleftarrow{\ \ \beta\ \ } \underset{2}{\circ} \xleftarrow{\ \ \alpha\ \ } \underset{3}{\circ}$.

We have seen in (1.13)(a) that

$$KQ \cong \mathbb{T}_3(K) = \begin{bmatrix} K & 0 & 0 \\ K & K & 0 \\ K & K & K \end{bmatrix}.$$

The ideal $\mathcal{I} = \langle \alpha\beta \rangle$ is admissible and actually equal to R_Q^2, that is,

$$\mathcal{I} \cong \mathrm{rad}^2 \mathbb{T}_3(K) = \begin{bmatrix} 0 & 0 & 0 \\ 0 & 0 & 0 \\ K & 0 & 0 \end{bmatrix}.$$

Thus $KQ/\mathcal{I}$ is isomorphic to the quotient of $\mathbb{T}_3(K)$ by the square of its radical.

(b) Let Q be the quiver

The ideal $\mathcal{I}$ of KQ generated by the commutativity relation $\alpha\beta - \gamma\delta$ is admissible. Thus $KQ/\mathcal{I}$ is a finite dimensional K-algebra, and $\{e_1, e_2, e_3, e_4, \overline{\alpha}, \overline{\beta}, \overline{\gamma}, \overline{\delta}, \overline{\alpha\beta}\}$ is its K-vector space basis. Using the construction in (1.12), we see that

$$KQ/\mathcal{I} \cong \begin{bmatrix} K & 0 & 0 & 0 \\ K & K & 0 & 0 \\ K & 0 & K & 0 \\ K & K & K & K \end{bmatrix}$$

under the isomorphism defined by

$$e_1 \mapsto \begin{bmatrix} 1 & 0 & 0 & 0 \\ 0 & 0 & 0 & 0 \\ 0 & 0 & 0 & 0 \\ 0 & 0 & 0 & 0 \end{bmatrix}, \quad e_2 \mapsto \begin{bmatrix} 0 & 0 & 0 & 0 \\ 0 & 1 & 0 & 0 \\ 0 & 0 & 0 & 0 \\ 0 & 0 & 0 & 0 \end{bmatrix}, \quad e_3 \mapsto \begin{bmatrix} 0 & 0 & 0 & 0 \\ 0 & 0 & 0 & 0 \\ 0 & 0 & 1 & 0 \\ 0 & 0 & 0 & 0 \end{bmatrix},$$

$$e_4 \mapsto \begin{bmatrix} 0 & 0 & 0 & 0 \\ 0 & 0 & 0 & 0 \\ 0 & 0 & 0 & 0 \\ 0 & 0 & 0 & 1 \end{bmatrix}, \quad \overline{\alpha} \mapsto \begin{bmatrix} 0 & 0 & 0 & 0 \\ 0 & 0 & 0 & 0 \\ 0 & 0 & 0 & 0 \\ 0 & 1 & 0 & 0 \end{bmatrix}, \quad \overline{\beta} \mapsto \begin{bmatrix} 0 & 0 & 0 & 0 \\ 1 & 0 & 0 & 0 \\ 0 & 0 & 0 & 0 \\ 0 & 0 & 0 & 0 \end{bmatrix},$$

$$\overline{\gamma} \mapsto \begin{bmatrix} 0 & 0 & 0 & 0 \\ 0 & 0 & 0 & 0 \\ 0 & 0 & 0 & 0 \\ 0 & 0 & 1 & 0 \end{bmatrix}, \quad \overline{\delta} \mapsto \begin{bmatrix} 0 & 0 & 0 & 0 \\ 0 & 0 & 0 & 0 \\ 1 & 0 & 0 & 0 \\ 0 & 0 & 0 & 0 \end{bmatrix}, \quad \overline{\alpha\beta} \mapsto \begin{bmatrix} 0 & 0 & 0 & 0 \\ 0 & 0 & 0 & 0 \\ 0 & 0 & 0 & 0 \\ 1 & 0 & 0 & 0 \end{bmatrix}.$$

(c) Let Q be the quiver

We have seen in (1.3)(a) that $KQ \cong K[t]$ (which is infinite dimensional). For each $m \geq 2$, the ideal $\langle \alpha^m \rangle$ is admissible (and actually any admissible ideal of KQ is of this form). Thus $KQ/\mathcal{I} \cong K[t]/\langle t^m \rangle$ is m-dimensional.

(d) Let Q be the quiver

We have seen in (1.3)(b) that $KQ \cong K\langle t_1, t_2 \rangle$. The ideal $\mathcal{I}$ generated by $\alpha\beta - \beta\alpha, \beta^2, \alpha^2$ is admissible. Indeed, it is clear that $\mathcal{I} \subseteq R_Q^2$. On the other hand, any path of length 3 belongs to $\mathcal{I}$ (and consequently $R_Q^3 \subseteq \mathcal{I}$). Indeed, such a path either contains a term of the form α^2 or β^2 or is of one of the forms $\alpha\beta\alpha$ or $\beta\alpha\beta$; because $\alpha\beta\alpha = (\alpha\beta - \beta\alpha)\alpha + \beta\alpha^2 \in \mathcal{I}$ and $\beta\alpha\beta = (\beta\alpha - \alpha\beta)\beta + \alpha\beta^2 \in \mathcal{I}$, we are done. The bound quiver algebra $KQ/\mathcal{I}$ is four-dimensional, with basis given by $\{e_1, \overline{\alpha}, \overline{\beta}, \overline{\alpha\beta}\}$. In fact, $KQ/\mathcal{I} \cong K[t_1, t_2]/\langle t_1^2, t_2^2 \rangle$, under the isomorphism defined by the formulas

$$e_1 \mapsto 1 + \langle t_1^2, t_2^2 \rangle, \ \overline{\alpha} \mapsto t_1 + \langle t_1^2, t_2^2 \rangle, \ \overline{\beta} \mapsto t_2 + \langle t_1^2, t_2^2 \rangle, \ \overline{\alpha\beta} \mapsto t_1 t_2 + \langle t_1^2, t_2^2 \rangle.$$

II.3. The quiver of a finite dimensional algebra

Let A be a finite dimensional (associative) algebra (with an identity) over an algebraically closed field K. As seen in (I.6.10), it may be assumed, from the point of view of studying the representation theory of A, that A is basic and connected. We now show that, under these hypotheses, A is isomorphic to a bound quiver algebra $KQ/\mathcal{I}$, where Q is a finite connected quiver and $\mathcal{I}$ is an admissible ideal of KQ. We start by associating, in a natural manner, a finite quiver to each basic and connected finite dimensional algebra A.

3.1. Definition. Let A be a basic and connected finite dimensional K-algebra and $\{e_1, e_2, \ldots, e_n\}$ be a complete set of primitive orthogonal idempotents of A. The **(ordinary) quiver** of A, denoted by Q_A, is defined as follows:

(a) The points of Q_A are the numbers $1, 2, \ldots, n$, which are in bijective correspondence with the idempotents $e_1, e_2, \ldots, e_n$.

(b) Given two points $a, b \in (Q_A)_0$, the arrows $\alpha : a \to b$ are in bijective correspondence with the vectors in a basis of the K-vector space $e_a(\operatorname{rad} A/\operatorname{rad}^2 A)e_b$.

Because A is finite dimensional, so is every vector space of the form $e_a(\operatorname{rad} A/\operatorname{rad}^2 A)e_b$ (with $a, b \in (Q_A)_0$). Consequently, Q_A is finite. The term "ordinary quiver," sometimes used for Q_A, comes from the fact that other quivers are also used to study A, as will be seen later. Now, Q_A

is constructed starting from a given complete set of primitive orthogonal idempotents. We must thus show that it does not depend on the particular set we have chosen.

3.2. Lemma. *Let A be a finite dimensional, basic, and connected algebra.*

(a) *The quiver Q_A of A does not depend on the choice of a complete set of primitive orthogonal idempotents in A.*

(b) *For any pair e_a, e_b of primitive orthogonal idempotents of A the K-linear map $\psi : e_a(\operatorname{rad} A)e_b/e_a(\operatorname{rad}^2 A)e_b \longrightarrow e_a(\operatorname{rad} A/\operatorname{rad}^2 A)e_b$, defined by the formula $e_a x e_b + e_a(\operatorname{rad}^2 A)e_b \mapsto e_a(x + \operatorname{rad}^2 A)e_b$, is an isomorphism.*

Proof. (a) The number of points in Q_A is uniquely determined, because it equals the number of indecomposable direct summands of A_A, and the latter is unique by the unique decomposition theorem (I.4.10). On the other hand, the same theorem says that the factors of this decomposition are uniquely determined up to isomorphism, that is, if

$$A_A = \bigoplus_{a=1}^{n} e_a A = \bigoplus_{b=1}^{n} e_b' A$$

then we can renumber the factors so that $e_a A \cong e_a' A$, for each a with $1 \leq a \leq n$. We must show that this implies $\dim_K e_a(\operatorname{rad} A/\operatorname{rad}^2 A)e_b = \dim_K e_a'(\operatorname{rad} A/\operatorname{rad}^2 A)e_b'$, for every pair (a, b). A routine calculation shows that the A-module homomorphism $\varphi : e_a(\operatorname{rad} A) \to e_a(\operatorname{rad} A/\operatorname{rad}^2 A)$ given by $e_a x \mapsto e_a(x + \operatorname{rad}^2 A)$ admits $e_a(\operatorname{rad}^2 A)$ as a kernel. Consequently

$$e_a(\operatorname{rad} A/\operatorname{rad}^2 A) \cong e_a(\operatorname{rad} A)/e_a(\operatorname{rad}^2 A) \cong \operatorname{rad}(e_a A)/\operatorname{rad}^2(e_a A).$$

We thus have a sequence of K-vector space isomorphisms

$$
\begin{aligned}
e_a(\operatorname{rad} A/\operatorname{rad}^2 A)e_b &\cong [\operatorname{rad}(e_a A)/\operatorname{rad}^2(e_a A)]e_b \\
&\cong \operatorname{Hom}_A(e_b A, \operatorname{rad}(e_a A)/\operatorname{rad}^2(e_a A)) \\
&\cong \operatorname{Hom}_A(e_b' A, \operatorname{rad}(e_a' A)/\operatorname{rad}^2(e_a' A)] \\
&\cong [\operatorname{rad}(e_a' A)/\operatorname{rad}^2(e_a' A)]e_b' \\
&\cong e_a'(\operatorname{rad} A/\operatorname{rad}^2 A)e_b'.
\end{aligned}
$$

(b) It is obvious that the K-linear map $e_a(\operatorname{rad} A)e_b \to e_a(\operatorname{rad} A/\operatorname{rad}^2 A)e_b$ defined by the formula $e_a x e_b \mapsto e_a(x + \operatorname{rad}^2 A)e_b$ admits $e_a(\operatorname{rad}^2 A)e_b$ as a kernel. Hence we conclude that the map ψ defined in the statement is an isomorphism. This finishes the proof. $\square$

We now show that the connectedness of the algebra A implies that of its quiver Q_A. By definition, there exists a basis $\{\overline{x}_\alpha\}_\alpha$ of $\operatorname{rad} A/\operatorname{rad}^2 A$, where α ranges over the set $(Q_A)_1$ of arrows of Q_A. For each $\alpha \in (Q_A)_1$, let $x_\alpha \in \operatorname{rad} A$ be such that $\overline{x}_\alpha = x_\alpha + \operatorname{rad}^2 A$. We show that we can express all the elements of $\operatorname{rad} A$ in terms of the x_α and the paths in Q_A.

3.3. Lemma. *For each arrow* $\alpha : i \to j$ *in* $(Q_A)_1$, *let* $x_\alpha \in e_i(\operatorname{rad} A)e_j$ *be such that the set* $\{x_\alpha + \operatorname{rad}^2 A \mid \alpha : i \to j\}$ *is a basis of* $e_i(\operatorname{rad} A/\operatorname{rad}^2 A)e_j$ *(see (3.2)(a)). Then*

(a) for any two points $a, b \in (Q_A)_0$, *every element* $x \in e_a(\operatorname{rad} A)e_b$ *can be written in the form:* $x = \sum x_{\alpha_1} x_{\alpha_2} \dots x_{\alpha_\ell} \lambda_{\alpha_1\alpha_2\dots\alpha_\ell}$, *where* $\lambda_{\alpha_1\alpha_2\dots\alpha_\ell} \in K$ *and the sum is taken over all paths* $\alpha_1\alpha_2\dots\alpha_\ell$ *in* Q_A *from* a *to* b; *and*

(b) for each arrow $\alpha : i \to j$, *the element* x_α *uniquely determines a nonzero nonisomorphism* $\widetilde{x}_\alpha \in \operatorname{Hom}_A(e_j A, e_i A)$ *such that* $\widetilde{x}_\alpha(e_j) = x_\alpha$, $\operatorname{Im} \widetilde{x}_\alpha \subseteq e_i(\operatorname{rad} A)$ *and* $\operatorname{Im} \widetilde{x}_\alpha \not\subseteq e_i(\operatorname{rad}^2 A)$.

Proof. (a) Because, as a K-vector space, $\operatorname{rad} A \cong (\operatorname{rad} A/\operatorname{rad}^2 A) \oplus \operatorname{rad}^2 A$, we have $e_a(\operatorname{rad} A)e_b \cong e_a(\operatorname{rad} A/\operatorname{rad}^2 A)e_b \oplus e_a(\operatorname{rad}^2 A)e_b$. Thus x can be written in the form

$$x = \sum_{\alpha : a \to b} x_\alpha \lambda_\alpha \quad \text{modulo} \quad e_a(\operatorname{rad}^2 A)e_b$$

(where $\lambda_\alpha \in K$ for every arrow α from a to b) or, more formally,

$$x' = x - \sum_{\alpha : a \to b} x_\alpha \lambda_\alpha \in e_a(\operatorname{rad}^2 A)e_b.$$

The decomposition $\operatorname{rad} A = \bigoplus_{i,j} e_i(\operatorname{rad} A)e_j$ implies that

$$e_a(\operatorname{rad}^2 A)e_b = \sum_{c \in (Q_A)_0} [e_a(\operatorname{rad} A)e_c][e_c(\operatorname{rad} A)e_b]$$

so that $x' = \sum_{c \in (Q_A)_0} x'_c y'_c$ where $x'_c \in e_a(\operatorname{rad} A)e_c$ and $y'_c \in e_c(\operatorname{rad} A)e_b$. By the preceding discussion, we have expressions of the form $x'_c = \sum_{\beta : a \to c} x_\beta \lambda_\beta$ and $y'_c = \sum_{\gamma : c \to b} x_\gamma \lambda_\gamma$ modulo $\operatorname{rad}^2 A$, where $\lambda_\beta, \lambda_\gamma \in K$. Hence

$$x = \sum_{\alpha : a \to b} x_\alpha \lambda_\alpha + \sum_{\beta : a \to c} \sum_{\gamma : c \to b} x_\beta x_\gamma \lambda_\beta \lambda_\gamma \quad \text{modulo} \quad e_a(\operatorname{rad}^3 A)e_b.$$

We complete the proof by an obvious induction using the fact that $\operatorname{rad} A$ is nilpotent.

(b) By our assumption, the element $x_\alpha \in e_i(\operatorname{rad} A)e_j$ is nonzero and maps to a nonzero element $\widetilde{x}_\alpha$ by the K-linear isomorphism $e_i(\operatorname{rad} A)e_j \cong$

$\mathrm{Hom}_A(e_j A, e_i(\mathrm{rad}A))$ (I.4.3). It follows that $\widetilde{x}_\alpha(e_j) = x_\alpha$, $\mathrm{Im}\,\widetilde{x}_\alpha \subseteq e_i(\mathrm{rad}A)$, and $\mathrm{Im}\,\widetilde{x}_\alpha \not\subseteq e_i(\mathrm{rad}^2 A)$. This finishes the proof. $\qquad\square$

3.4. Corollary. *If A is a basic and connected finite dimensional algebra, then the quiver Q_A of A is connected.*

Proof. If this is not the case, then the set $(Q_A)_0$ of points of Q_A can be written as the disjoint union of two nonempty sets Q_0' and Q_0'' such that the points of Q_0' are not connected to those of Q_0''. We show that, if $i \in Q_0'$ and $j \in Q_0''$, we have $e_i A e_j = 0$ and $e_j A e_i = 0$. Then (1.6) will imply that A is not connected, a contradiction. Because $i \neq j$, (I.4.2) yields

$$\begin{aligned} e_i A e_j &\cong \mathrm{Hom}_A(e_j A, e_i A) &&\cong \mathrm{Hom}_A(e_j A, \mathrm{rad}\,e_i A) \\ &\cong (\mathrm{rad}\,e_i A)e_j &&\cong e_i(\mathrm{rad}\,A)e_j. \end{aligned}$$

The conclusion follows at once from (3.3). $\qquad\square$

3.5. Examples. (a) If $A = K[t]/\langle t^m \rangle$, where $m \geq 1$, then Q_A has only one point, because the only nonzero idempotent of A is its identity. We have $\mathrm{rad}\,A = \langle \bar{t} \rangle$, where $\bar{t} = t + \langle t^m \rangle$; indeed, $\langle \bar{t} \rangle^m = 0$ and $A/\langle \bar{t} \rangle \cong K$. Consequently, $\mathrm{rad}^2 A = \langle \bar{t}^2 \rangle$ and $\dim_K(\mathrm{rad}\,A/\mathrm{rad}^2 A) = 1$. A basis of $\mathrm{rad}\,A/\mathrm{rad}^2 A$ is given by the class of $\bar{t}$ in the quotient $\langle \bar{t} \rangle/\langle \bar{t}^2 \rangle$. Thus Q_A is the quiver

$$1 \circlearrowright \alpha$$

(b) Let $A = \begin{bmatrix} K & 0 & 0 \\ K & K & 0 \\ K & 0 & K \end{bmatrix}$ be the algebra of the lower triangular matrices $[\lambda_{ij}] \in \mathbb{M}_3(K)$, with $\lambda_{32} = 0$ and $\lambda_{pq} = 0$, for $p > q$. An obvious complete set of primitive orthogonal idempotents of A is given by the three matrix idempotents:

$$e_1 = \begin{bmatrix} 1 & 0 & 0 \\ 0 & 0 & 0 \\ 0 & 0 & 0 \end{bmatrix}, \; e_2 = \begin{bmatrix} 0 & 0 & 0 \\ 0 & 1 & 0 \\ 0 & 0 & 0 \end{bmatrix}, \; e_3 = \begin{bmatrix} 0 & 0 & 0 \\ 0 & 0 & 0 \\ 0 & 0 & 1 \end{bmatrix}.$$

As in Example 3.5 (a), we show that $\mathrm{rad}\,A = \begin{bmatrix} 0 & 0 & 0 \\ K & 0 & 0 \\ K & 0 & 0 \end{bmatrix}$ and $\mathrm{rad}^2 A = 0$.

A straightforward calculation shows that $e_2(\mathrm{rad}\,A)e_1$ and $e_3(\mathrm{rad}\,A)e_1$ are one-dimensional and all remaining spaces of the form $e_i(\mathrm{rad}\,A)e_j$ are zero (because $\dim_K(\mathrm{rad}\,A) = 2$). Therefore Q_A is the quiver

$$2 \circ \xrightarrow{\;\alpha\;} \overset{1}{\circ} \xleftarrow{\;\beta\;} \circ 3$$

(c) An obvious generalisation of (b) is as follows. Let A be the algebra of $n \times n$ lower triangular matrices

$$A = \begin{bmatrix} K & 0 & 0 & \cdots & 0 \\ K & K & 0 & \cdots & 0 \\ K & 0 & K & \cdots & 0 \\ \vdots & \vdots & & \ddots & \vdots \\ K & 0 & 0 & \cdots & K \end{bmatrix},$$

that is, an element of A might have a nonzero coefficient only in the first column or the main diagonal and has zero everywhere else. Then Q_A is the quiver

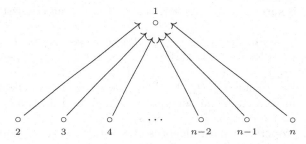

(d) Let A be the algebra of 3×3 lower triangular matrices

$$A = \left\{ \begin{bmatrix} a & 0 & 0 \\ c & b & 0 \\ e & d & a \end{bmatrix} \mid a, b, c, d, e \in K \right\}$$

and $\mathcal{I}$ be the ideal

$$\mathcal{I} = \left\{ \begin{bmatrix} 0 & 0 & 0 \\ 0 & 0 & 0 \\ e & 0 & 0 \end{bmatrix} \mid e \in K \right\}.$$

A complete set of primitive orthogonal idempotents for the algebra $B = A/\mathcal{I}$ consists of two elements

$$e_1 = \begin{bmatrix} 1 & 0 & 0 \\ 0 & 0 & 0 \\ 0 & 0 & 1 \end{bmatrix} + \mathcal{I} \quad \text{and} \quad e_2 = \begin{bmatrix} 0 & 0 & 0 \\ 0 & 1 & 0 \\ 0 & 0 & 0 \end{bmatrix} + \mathcal{I}.$$

Also, $\operatorname{rad} B = \left\{ \begin{bmatrix} 0 & 0 & 0 \\ c & 0 & 0 \\ 0 & d & 0 \end{bmatrix} + \mathcal{I} \mid c, d \in K \right\}$ and $\operatorname{rad}^2 B = 0$. Thus the K-vector spaces $e_2(\operatorname{rad} B)e_1$ and $e_1(\operatorname{rad} B)e_2$ are both one-dimensional and Q_B is the quiver $1 \circ \underset{\beta}{\overset{\alpha}{\rightleftarrows}} \circ 2$.

3.6. Lemma. *Let Q be a finite connected quiver, $\mathcal{I}$ be an admissible ideal of KQ, and $A = KQ/\mathcal{I}$. Then $Q_A = Q$.*

Proof. By (2.4), the set $\{e_a = \varepsilon_a + \mathcal{I} \mid a \in Q_0\}$ is a complete set of primitive orthogonal idempotents of $A = KQ/\mathcal{I}$. Thus the points of

Q_A are in bijective correspondence with those of Q. On the other hand, by (2.11) and the remark following it, the arrows from a to b in Q are in bijective correspondence with the vectors in a basis of the K-vector space $e_a(\operatorname{rad} A/\operatorname{rad}^2 A)e_b$, thus with the arrows from a to b in Q_A.   □

3.7. Theorem. *Let A be a basic and connected finite dimensional K-algebra. There exists an admissible ideal $\mathcal{I}$ of KQ_A such that $A \cong KQ_A/\mathcal{I}$.*

Proof. We first construct an algebra homomorphism $\varphi : KQ_A \to A$, then we show that φ is surjective and its kernel $\mathcal{I} = \operatorname{Ker} \varphi$ is an admissible ideal of KQ_A.

For each arrow $\alpha : i \to j$ in $(Q_A)_1$, let $x_\alpha \in \operatorname{rad} A$ be chosen so that $\{x_\alpha + \operatorname{rad}^2 A \mid \alpha : i \to j\}$ forms a basis of $e_i(\operatorname{rad} A/\operatorname{rad}^2 A)e_j$. Let $\varphi_0 : (Q_A)_0 \to A$ be the map defined by $\varphi_0(a) = e_a$ for $a \in (Q_A)_0$, and $\varphi_1 : (Q_A)_1 \to A$ be the map defined by $\varphi_1(\alpha) = x_\alpha$ for $\alpha \in (Q_A)_1$. Thus the elements $\varphi_0(a)$ form a complete set of primitive orthogonal idempotents in A, and if $\alpha : a \to b$, we have $\varphi_0(a)\varphi_1(\alpha)\varphi_0(b) = e_a x_\alpha e_b = x_\alpha = \varphi_1(\alpha)$.

By the universal property of path algebras (1.8), there exists a unique K-algebra homomorphism $\varphi : KQ_A \to A$ that extends φ_0 and φ_1.

We claim that φ is surjective. Because its image is clearly generated by the elements e_a (for $a \in (Q_A)_0$) and x_α (for $\alpha \in (Q_A)_1$), it suffices to show that these same elements generate A. Because K is algebraically closed, it follows from the Wedderburn–Malcev theorem (I.1.6) that the canonical homomorphism $A \to A/\operatorname{rad} A$ splits, that is, A is a split extension of the semisimple algebra $A/\operatorname{rad} A$ by $\operatorname{rad} A$. Because the former is clearly generated by the e_a, it suffices to show that each element of $\operatorname{rad} A$ can be written as a polynomial in the x_α and this follows from (3.3).

There remains to show that $\mathcal{I} = \operatorname{Ker} \varphi$ is admissible. Let R denote the arrow ideal of the algebra KQ_A. By definition of φ, we have $\varphi(R) \subseteq \operatorname{rad} A$ and hence $\varphi(R^\ell) \subseteq \operatorname{rad}^\ell A$ for each $\ell \geq 1$. Because $\operatorname{rad} A$ is nipotent, there exists $m \geq 1$ such that $\operatorname{rad}^m A = 0$ and consequently $R^m \subseteq \operatorname{Ker} \varphi = \mathcal{I}$. We now prove that $\mathcal{I} \subseteq R^2$. If $x \in \mathcal{I}$, then we can write

$$x = \sum_{a \in (Q_A)_0} \varepsilon_a \lambda_a + \sum_{\alpha \in (Q_A)_1} \alpha \mu_\alpha + y,$$

where $\lambda_a, \mu_\alpha \in K$ and $y \in R^2$. Now $\varphi(x) = 0$ gives

$$0 = \sum_{a \in (Q_A)_0} e_a \lambda_a + \sum_{\alpha \in (Q_A)_1} x_\alpha \mu_\alpha + \varphi(y).$$

Hence $\sum\limits_{a \in (Q_A)_0} e_a \lambda_a = - \sum\limits_{\alpha \in (Q_A)_1} x_\alpha \mu_\alpha - \varphi(y) \in \operatorname{rad} A$. Because $\operatorname{rad} A$ is nilpotent, and the e_a are orthogonal idempotents, we infer that $\lambda_a = 0$, for any $a \in (Q_A)_0$. Similarly $\sum\limits_{\alpha \in (Q_A)_1} x_\alpha \mu_\alpha = -\varphi(y) \in \mathrm{rad}^2 A$. Hence the equality $\sum\limits_{\alpha \in (Q_A)_1} (x_\alpha + \mathrm{rad}^2 A)\mu_\alpha = 0$ holds in $\mathrm{rad}\, A/\mathrm{rad}^2 A$. But the set $\{x_\alpha + \mathrm{rad}^2 A \mid \alpha \in (Q_A)_1\}$ is, by construction, a basis of $\mathrm{rad}\, A/\mathrm{rad}^2 A$. Therefore $\mu_\alpha = 0$ for each $\alpha \in (Q_A)_1$ and so $x = y \in R^2$. $\square$

3.8. Definition. Let A be a basic and connected finite dimensional K-algebra. An isomorphism $A \cong KQ_A/\mathcal{I}$, where $\mathcal{I}$ is an admissible ideal of KQ_A (such as the one constructed in Theorem 3.7) is called a **presentation** of the algebra A (as a bound quiver algebra).

3.9. Examples. (a) In Example 3.5 (a), the K-algebra homomorphism $\varphi : KQ_A \to A$ is defined by $\varphi(\varepsilon_1) = 1, \varphi(\alpha) = \bar{t}$. Clearly, φ is surjective, and $\mathrm{Ker}\, \varphi = \langle \alpha^m \rangle$.

(b) In Example 3.5 (b), the K-algebra homomorphism $\varphi : KQ_A \to A$ is defined by

$$\varphi(\varepsilon_1) = \begin{bmatrix} 1 & 0 & 0 \\ 0 & 0 & 0 \\ 0 & 0 & 0 \end{bmatrix}, \ \varphi(\varepsilon_2) = \begin{bmatrix} 0 & 0 & 0 \\ 0 & 1 & 0 \\ 0 & 0 & 0 \end{bmatrix}, \ \varphi(\varepsilon_3) = \begin{bmatrix} 0 & 0 & 0 \\ 0 & 0 & 0 \\ 0 & 0 & 1 \end{bmatrix},$$

$$\varphi(\alpha) = \begin{bmatrix} 0 & 0 & 0 \\ 1 & 0 & 0 \\ 0 & 0 & 0 \end{bmatrix}, \ \varphi(\beta) = \begin{bmatrix} 0 & 0 & 0 \\ 0 & 0 & 0 \\ 1 & 0 & 0 \end{bmatrix}.$$

Here, φ is an isomorphism so that $A \cong KQ_A$. Later we characterise the algebras (such as A) that are isomorphic to the path algebras of their ordinary quivers.

(c) In Example 3.5 (d), the K-algebra homomorphism $\varphi : KQ_B \to B$ is defined by

$$\varphi(\varepsilon_1) = \begin{bmatrix} 1 & 0 & 0 \\ 0 & 0 & 0 \\ 0 & 0 & 1 \end{bmatrix} + \mathcal{I}, \quad \varphi(\varepsilon_2) = \begin{bmatrix} 0 & 0 & 0 \\ 0 & 1 & 0 \\ 0 & 0 & 0 \end{bmatrix} + \mathcal{I},$$

$$\varphi(\alpha) = \begin{bmatrix} 0 & 0 & 0 \\ 1 & 0 & 0 \\ 0 & 0 & 0 \end{bmatrix} + \mathcal{I}, \quad \varphi(\beta) = \begin{bmatrix} 0 & 0 & 0 \\ 0 & 0 & 0 \\ 0 & 1 & 0 \end{bmatrix} + \mathcal{I}.$$

We see that $\mathrm{Ker}\, \varphi = \langle \alpha\beta, \beta\alpha \rangle = R_Q^2$ and hence $B \cong KQ_B/R_Q^2$, where $Q = Q_B$.

3.10. Remark. Usually, an algebra has more than one presentation as a bound quiver algebra; see, for instance, Example 2.2 (e).

II.4. Exercises

1. Let $Q = (Q_0, Q_1, s, t)$ be a quiver. The opposite quiver Q^{op} is the quiver $Q^{\mathrm{op}} = (Q_0, Q_1, s', t')$ where, for $\alpha \in Q_1$, $s'(\alpha) = t(\alpha)$ and $t'(\alpha) = s(\alpha)$. Show that $(KQ)^{\mathrm{op}} \cong KQ^{\mathrm{op}}$.

2. Let Q be a finite quiver. Show that:

(a) KQ is semisimple if and only if $|Q_1| = 0$,

(b) KQ is simple if and only if $|Q_0| = 1$ and $|Q_1| = 0$.

If, moreover, Q is connected, show that:

(c) KQ is local only if $|Q_0| = 1$ and $|Q_1| = 0$,

(d) KQ is commutative if and only if $|Q_0| = 1$ and $|Q_1| \le 1$.

3. For each of the following quivers, give a basis of the path algebra, then write the multiplication table of this basis, and finally write the path algebra as a triangular matrix algebra:

(a)

(b)   o ⎯⎯⎯⎯→ o ←⎯⎯⎯⎯ o

(c)   o ⇇⎯⎯⎯⎯ o ←⎯⎯⎯ o

(d)   o ⎯⎯⎯→ o ←⎯⎯⎯ o ←⎯⎯⎯ o

(e)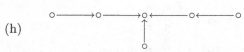

(f)

(g)

(h)

4. Let $E = \{1, 2, \ldots, n\}$ be partially ordered as follows: $1 \preceq i$ for all $1 \le i \le n$, and for each pair (i, j) with $2 \le i, j \le n$, we have $i \preceq j$ if and only if $i = j$. Show that the incidence K-algebra of $(E, \preceq)$ is isomorphic to the path algebra KQ of a quiver Q (to be determined).

5. Let $Q = (Q_0, Q_1)$ be a finite and acyclic quiver. Show that KQ is connected if and only if KQ/R^2 is connected, where R is the arrow ideal of KQ.

6. Let Q be the quiver

$$1 \, o \, \overset{\alpha}{\circlearrowright}$$

Show that the arrow ideal R_Q of the path K-algebra KQ is infinite dimensional, and $\operatorname{rad} KQ = 0$.

7. Let $Q = (Q_0, Q_1)$ be the quiver

Show that each of the following ideals of KQ is admissible:
(a) $\mathcal{I}_1 = \langle \alpha^2 - \beta\gamma, \gamma\beta - \gamma\alpha\beta, \alpha^4 \rangle$,
(b) $\mathcal{I}_2 = \langle \alpha^2 - \beta\gamma, \gamma\beta, \alpha^4 \rangle$.

8. Let Q be a quiver and $\mathcal{I}$ an admissible ideal in KQ. Construct an admissible ideal $\mathcal{I}^{\mathrm{op}}$ of KQ^{op} such that $KQ^{\mathrm{op}}/\mathcal{I}^{\mathrm{op}} \cong (KQ/\mathcal{I})^{\mathrm{op}}$.

9. Let $Q' = (Q_0', Q_1')$ be a full subquiver of $Q = (Q_0, Q_1)$ such that if $\alpha : a \to b$ is an arrow in Q with $a \in Q_0'$, then $b \in Q_0'$ and $\alpha \in Q_1'$. Let $\mathcal{I}$ be an admissible ideal of KQ and $\varepsilon = \sum_{a \in Q_0'} \varepsilon_a$.
(a) Show that $KQ' = \varepsilon(KQ)\varepsilon$ and that $\mathcal{I}' = \varepsilon I \varepsilon$ is an admissible ideal of KQ'.
(b) Show that $A' = KQ'/\mathcal{I}'$ is isomorphic to the quotient of $A = KQ/\mathcal{I}$ by $J = \langle \varepsilon_a + \mathcal{I} \mid a \notin Q_0' \rangle$.

10. Let A be an algebra such that $\mathrm{rad}^2 A = 0$. Show that if $\{e_1, \ldots, e_n\}$ is a complete set of primitive orthogonal idempotents, then $e_i A e_j \neq 0$ for $i \neq j$ if and only if there exists an arrow $i \to j$ in Q_A.

11. Describe, up to isomorphism, all basic three-dimensional algebras.

12. Let $A = \begin{bmatrix} K[t]/(t^2) & 0 \\ K[t]/(t^2) & K \end{bmatrix}$ and view A as a K-algebra with the usual matrix multiplication. Show that $A \cong KQ/\mathcal{I}$, where Q is the quiver

and $\mathcal{I}$ is the ideal of KQ generated by one zero relation β^2.

13. Let $A = \begin{bmatrix} K & 0 & 0 \\ 0 & K & 0 \\ K & K & K \end{bmatrix}$ be the K-subalgebra of $\mathbb{M}_3(K)$ defined in (I.1.1)(c) and let B be the subalgebra of A consisting of all matrices $\lambda = \begin{bmatrix} \lambda_{11} & 0 & 0 \\ 0 & \lambda_{22} & 0 \\ \lambda_{31} & \lambda_{32} & \lambda_{33} \end{bmatrix}$ in A such that $\lambda_{11} = \lambda_{22} = \lambda_{33}$. Show that the algebra B is commutative and local and that $\mathrm{rad}\, B$ consists of all matrices $\lambda = \begin{bmatrix} 0 & 0 & 0 \\ 0 & 0 & 0 \\ \lambda_{31} & \lambda_{32} & 0 \end{bmatrix}$ in B. Prove that there are K-algebra isomorphisms $B \cong K[t_1, t_2]/(t_1, t_2)^2 \cong KQ/\mathcal{I}$, where Q is the quiver

and $\mathcal{I} = \langle \alpha^2, \beta^2, \alpha\beta, \beta\alpha \rangle$.

14. Let $A = \mathbb{T}_3(K) = \begin{bmatrix} K & 0 & 0 \\ K & K & 0 \\ K & K & K \end{bmatrix}$ be as in (I.1.1) and let C be the subalgebra of A consisting of all matrices $\lambda = \begin{bmatrix} \lambda_{11} & 0 & 0 \\ \lambda_{21} & \lambda_{22} & 0 \\ \lambda_{31} & \lambda_{32} & \lambda_{33} \end{bmatrix}$ in A such that $\lambda_{11} = \lambda_{22} = \lambda_{33}$. Show that the algebra C is noncommutative and local and that there are K-algebra isomorphisms $C \cong K\langle t_1, t_2 \rangle / (t_1^2, t_2^2, t_2 t_1) = KQ/\mathcal{I}$, where Q is the quiver

and $\mathcal{I} = (\alpha^2, \beta^2, \beta\alpha)$.

15. Write a bound quiver presentation of each of the following algebras:

$$\begin{bmatrix} K & 0 & 0 & 0 & 0 \\ K & K & 0 & 0 & 0 \\ K & 0 & K & 0 & 0 \\ K & 0 & K & K & 0 \\ K & K & K & K & K \end{bmatrix}, \quad \begin{bmatrix} K & 0 & 0 & 0 & 0 \\ K & K & 0 & 0 & 0 \\ K & 0 & K & 0 & 0 \\ K & 0 & K & K & 0 \\ K & K & K & K & K \end{bmatrix}, \quad \begin{bmatrix} K & 0 & 0 & 0 & 0 \\ 0 & K & 0 & 0 & 0 \\ K & K & K & 0 & 0 \\ K & 0 & 0 & K & 0 \\ K & K & K & K & K \end{bmatrix}.$$

16. The hypothesis that the base field is algebraically closed is necessary for Theorem 3.7 to be valid. **Hint:** Show that the $\mathbb{R}$-algebra $\mathbb{C}$ is two-dimensional, basic, and connected but that there is no quiver Q such that $\mathbb{C} \cong \mathbb{R}Q/\mathcal{I}$ with $\mathcal{I}$ an admissible ideal of $\mathbb{R}Q$.

17. The following three examples show that generators of an admissible ideal are not uniquely determined in general:
(a) Let $Q = (Q_0, Q_1)$ be the quiver

and $\mathcal{I}_1 = \langle \alpha\beta + \gamma\delta \rangle$, $\mathcal{I}_2 = \langle \alpha\beta - \gamma\delta \rangle$ two-sided ideal of KQ. Show that $\mathcal{I}_1$ and $\mathcal{I}_2$ are admissible and distinct and that there is a K-algebra isomorphism $KQ/\mathcal{I}_1 \cong KQ/\mathcal{I}_2$, if char $K \neq 2$.
(b) Same exercise with $Q = (Q_0, Q_1)$, $\mathcal{I}_1$, $\mathcal{I}_2$ as in Exercise 7, char $K \neq 2$.
(c) Same exercise with $Q = (Q_0, Q_1)$ of the form $\circ \xleftarrow{\quad \gamma \quad} \circ \underset{\beta}{\overset{\alpha}{\underset{\longleftarrow}{\longleftarrow}}} \circ$, $\mathcal{I}_1 = \langle \alpha\gamma - \beta\gamma \rangle$, $\mathcal{I}_2 = \langle \alpha\gamma \rangle$, but the characteristic of K is arbitrary.

18. Let A be a finite dimensional commutative algebra. Show that A is a finite product of commutative local algebras.

19. Let A be a finite dimensional basic and connected algebra. Show that $Q_{A^{\mathrm{op}}} = (Q_A)^{\mathrm{op}}$ and that there exists an admissible ideal $\mathcal{I}^{\mathrm{op}}$ of $KQ_{A^{\mathrm{op}}}$ such that $A^{\mathrm{op}} \cong (KQ_{A^{\mathrm{op}}})/\mathcal{I}^{\mathrm{op}}$.

Chapter III

Representations and modules

As we saw in Chapter II, quivers provide a convenient way to visualise finite dimensional algebras. In this chapter we explain how quivers may be used to visualise modules. This idea has been illustrated by Examples (I.2.4)–(I.2.6).

Using a bound quiver $(Q, \mathcal{I})$ associated to an algebra A, we visualise any (finite dimensional) A-module M as a K-linear representation of $(Q, \mathcal{I})$, that is, a family of (finite dimensional) K-vector spaces M_a, with $a \in Q_0$, connected by K-linear maps $\varphi_\alpha : M_a \longrightarrow M_b$ corresponding to arrows $\alpha : a \longrightarrow b$ in Q, and satisfying some relations induced by $\mathcal{I}$. This description of A-modules is a powerful tool in the study of A-modules and is playing a fundamental rôle in the modern representation theory of finite dimensional algebras.

III.1. Representations of bound quivers

1.1. Definition. Let Q be a finite quiver. A K-**linear representation** or, more briefly, a **representation** M of Q is defined by the following data:

(a) To each point a in Q_0 is associated a K-vector space M_a.

(b) To each arrow $\alpha : a \longrightarrow b$ in Q_1 is associated a K-linear map $\varphi_\alpha : M_a \longrightarrow M_b$.

Such a representation is denoted as $M = (M_a, \varphi_\alpha)_{a \in Q_0, \alpha \in Q_1}$, or simply $M = (M_a, \varphi_\alpha)$. It is called **finite dimensional** if each vector space M_a is finite dimensional.

Let $M = (M_a, \varphi_\alpha)$ and $M' = (M'_a, \varphi'_\alpha)$ be two representations of Q. A **morphism** (of representations) $f : M \to M'$ is a family $f = (f_a)_{a \in Q_0}$ of K-linear maps $(f_a : M_a \longrightarrow M'_a)_{a \in Q_0}$ that are compatible with the structure maps φ_α, that is, for each arrow $\alpha : a \longrightarrow b$, we have $\varphi'_\alpha f_a = f_b \varphi_\alpha$ or, equivalently, the following square is commutative:

$$
\begin{array}{ccc}
M_a & \xrightarrow{\;\varphi_\alpha\;} & M_b \\
\downarrow{\scriptstyle f_a} & & \downarrow{\scriptstyle f_b} \\
M'_a & \xrightarrow{\;\varphi'_\alpha\;} & M'_b
\end{array}
$$

Let $f : M \to M'$ and $g : M' \to M''$ be two morphisms of representations of Q, where $f = (f_a)_{a \in Q_0}$ and $g = (g_a)_{a \in Q_0}$. Their composition is defined to be the family $gf = (g_a f_a)_{a \in Q_0}$. Then gf is easily seen to be a morphism from M to M''.

We have thus defined a category $\mathrm{Rep}(Q)$ of K-linear representations of Q. We denote by $\mathrm{rep}(Q)$ the full subcategory of $\mathrm{Rep}(Q)$ consisting of the finite dimensional representations.

1.2. Example. Let Q be the Kronecker quiver $1 \circ \underset{\beta}{\overset{\alpha}{\rightleftarrows}} \circ 2$.
A representation M of Q is given by

$$K^2 \xleftarrow[\begin{smallmatrix}0\\1\end{smallmatrix}]{\begin{smallmatrix}1\\0\end{smallmatrix}} K$$

Another representation M' is given by

$$K^2 \xleftarrow[\begin{smallmatrix}0&0\\1&0\end{smallmatrix}]{\begin{smallmatrix}1&0\\0&1\end{smallmatrix}} K^2$$

Both are finite dimensional. We have a morphism $M \to M'$ defined by

$$\begin{array}{ccc}
K^2 & \xleftarrow[\begin{smallmatrix}0\\1\end{smallmatrix}]{\begin{smallmatrix}1\\0\end{smallmatrix}} & K \\
\Big\downarrow{\scriptstyle[\begin{smallmatrix}1&0\\0&1\end{smallmatrix}]} & & \Big\downarrow{\scriptstyle[\begin{smallmatrix}1\\0\end{smallmatrix}]} \\
K^2 & \xleftarrow[\begin{smallmatrix}0&0\\1&0\end{smallmatrix}]{\begin{smallmatrix}1&0\\0&1\end{smallmatrix}} & K^2
\end{array}$$

Indeed, it is readily verified that

$$\begin{bmatrix}1&0\\0&1\end{bmatrix}\begin{bmatrix}1\\0\end{bmatrix} = \begin{bmatrix}1&0\\0&1\end{bmatrix}\begin{bmatrix}1\\0\end{bmatrix} \quad \text{and} \quad \begin{bmatrix}1&0\\0&1\end{bmatrix}\begin{bmatrix}0\\1\end{bmatrix} = \begin{bmatrix}0&0\\1&0\end{bmatrix}\begin{bmatrix}1\\0\end{bmatrix}.$$

We now prove that the categories $\mathrm{Rep}_K(Q)$ and $\mathrm{rep}_K(Q)$ are abelian. As we will show later, this is not surprising because they are equivalent to module categories. The straightforward verification will, however, allow us to describe the main features of these categories.

1.3. Lemma. *Let Q be a finite quiver. Then $\mathrm{Rep}_K(Q)$ and $\mathrm{rep}_K(Q)$ are abelian K-categories.*

Proof. (a) Let $f : M \to M'$ and $g : M \to M'$ be two morphisms in $\mathrm{Rep}_K(Q)$, with $f = (f_a)_{a \in Q_0}$ and $g = (g_a)_{a \in Q_0}$. The formula $f + g = (f_a + g_a)_{a \in Q_0}$ clearly defines a morphism from M to M'. With this definition, the set of all morphisms from M to M' becomes an abelian group. Further, this addition is compatible with the composition of morphisms, that is, $h'(f + g) = h'f + h'g$ for each morphism h' of source M', and $(f + g)h = fh + gh$ for each morphism h of target M.

(b) Given two representations $M = (M_a, \varphi_a)$ and $M' = (M'_a, \varphi'_a)$ of Q, the representation

$$M \oplus M' = \left(M_a \oplus M'_a, \begin{bmatrix} \varphi_\alpha & 0 \\ 0 & \varphi'_\alpha \end{bmatrix} \right)$$

is easily verified to be the direct sum of M and M' in $\mathrm{Rep}_K(Q)$.

(c) Let $f : M \to M'$ be a morphism in $\mathrm{Rep}_K(Q)$, where $M = (M_a, \varphi_a)$ and $M' = (M'_a, \varphi'_a)$. For each $a \in Q_0$, let L_a denote the kernel of $f_a : M_a \to M'_a$ and, for each arrow $\alpha : a \to b$, let $\psi_\alpha : L_a \to L_b$ denote the restriction of φ_α to L_a. Then the representation $L = (L_a, \psi_\alpha)$ is the kernel of f in $\mathrm{Rep}_K(Q)$ and similarly for the cokernel of f.

(d) The construction in (c) implies that a morphism $f : M \to M'$ is a monomorphism (or an epimorphism) if and only if each $f_a : M_a \to M'_a$ is injective (or surjective, respectively). Thus every morphism in $\mathrm{Rep}_K(Q)$ admits a canonical factorisation. We have shown that $\mathrm{Rep}_K(Q)$ is an abelian K-category.

If M and M' belong to $\mathrm{rep}_K(Q)$ (that is, $\dim_K M_a < \infty$ and $\dim_K M'_a < \infty$, for each $a \in Q_0$), the representation $M \oplus M'$ also belongs to $\mathrm{rep}_K(Q)$. Moreover, if $f : M \to M'$ is a morphism between objects in $\mathrm{rep}_K(Q)$, the construction in (c) shows that the kernel and the cokernel of f also belong to $\mathrm{rep}_K(Q)$. Therefore $\mathrm{rep}_K(Q)$ is also an abelian K-category. □

1.4. Definition. Let Q be a finite quiver and $M = (M_a, \varphi_a)$ be a representation of Q. For any nontrivial path $w = \alpha_1 \alpha_2 \ldots \alpha_\ell$ from a to b in Q, we define the **evaluation** of M on the path w to be the K-linear map from M_a to M_b defined by

$$\varphi_w = \varphi_{\alpha_\ell} \varphi_{\alpha_{\ell-1}} \cdots \varphi_{\alpha_2} \varphi_{\alpha_1}.$$

The definition of evaluation extends to K-linear combinations of paths with a common source and a common target; thus let

$$\rho = \sum_{i=1}^{m} \lambda_i w_i$$

be such a combination, where $\lambda_i \in K$ and w_i is a path in Q, for each i, then

$$\varphi_\rho = \sum_{i=1}^{m} \lambda_i \varphi_{w_i}.$$

We are now able to define a notion of representation of a bound quiver. Let thus Q be a finite quiver and $\mathcal{I}$ be an admissible ideal of KQ. A representation $M = (M_a, \varphi_a)$ of Q is said to be **bound by** $\mathcal{I}$, or to **satisfy the relations** in $\mathcal{I}$, if we have

$$\varphi_\rho = 0, \quad \text{for all relations} \quad \rho \in \mathcal{I}.$$

If $\mathcal{I}$ is generated by the finite set of relations $\{\rho_1, \ldots, \rho_m\}$, the representation M is bound by $\mathcal{I}$ if and only if $\varphi_{\rho_j} = 0$, for all j such that $1 \leq j \leq m$.

We denote by $\mathrm{Rep}_K(Q, \mathcal{I})$ (or by $\mathrm{rep}_K(Q, \mathcal{I})$ the full subcategory of $\mathrm{Rep}_K(Q)$ (or of $\mathrm{rep}_K(Q)$, respectively) consisting of the representations of Q bound by $\mathcal{I}$.

1.5. Example. Let Q be the quiver

$$
\begin{array}{c}
\overset{3}{\circ} \\
\beta \nearrow \quad \nwarrow \alpha \\
1\circ \xleftarrow{\lambda} \overset{2}{\circ} \qquad \qquad \circ 5 \\
\delta \searrow \quad \nearrow \gamma \\
\underset{4}{\circ}
\end{array}
$$

bound by the commutativity relation $\alpha\beta = \gamma\delta$. We consider the representations M and N of Q given by

$$
\begin{array}{c}
K \\
{\scriptstyle\begin{bmatrix}1\\0\end{bmatrix}} \nearrow \quad \nwarrow \\
K \xleftarrow{[1\,1]} K^2 \qquad \qquad 0 \quad \text{and} \\
{\scriptstyle\begin{bmatrix}0\\1\end{bmatrix}} \searrow \quad \nearrow \\
K
\end{array}
\qquad
\begin{array}{c}
K \\
1 \nearrow \quad \nwarrow 1 \\
K \xleftarrow{1} K \qquad \qquad K \\
1 \searrow \quad \nearrow 1 \\
K
\end{array}
$$

respectively. It is clear that M and N are bound by $\alpha\beta = \gamma\delta$. On the other hand, the following representation of Q is not bound by $\alpha\beta = \gamma\delta$

$$
\begin{array}{c}
0 \\
\nearrow \quad \nwarrow \\
K \xleftarrow{1} K \qquad \qquad K \\
1 \searrow \quad \nearrow 1 \\
K
\end{array}
$$

We are now in a position to justify the introduction of the preceding concepts. Our objective is to study the category $\mathrm{mod}\,A$, where A is a finite dimensional K-algebra, which we can assume, without loss of generality, to be basic and connected. We have seen that there exists a finite connected quiver Q_A and an admissible ideal $\mathcal{I}$ of KQ_A such that $A \cong KQ_A/\mathcal{I}$. We now show that the category $\mathrm{mod}\,A$ of finitely generated right A-modules is equivalent to the category $\mathrm{rep}_K(Q_A, \mathcal{I})$ of finite dimensional K-linear representations of Q_A bound by $\mathcal{I}$.

1.6. Theorem. *Let $A = KQ/\mathcal{I}$, where Q is a finite connected quiver and $\mathcal{I}$ is an admissible ideal of KQ. There exists a K-linear equivalence of categories*

$$F : \mathrm{Mod}\,A \xrightarrow{\;\simeq\;} \mathrm{Rep}_K(Q, \mathcal{I})$$

that restricts to an equivalence of categories $F : \mathrm{mod}\ A \xrightarrow{\;\simeq\;} \mathrm{rep}_K(Q, \mathcal{I})$.

Proof. (a) Construction of a functor $F : \mathrm{Mod}\,A \to \mathrm{Rep}_K(Q, \mathcal{I})$. Let M_A be an A-module. We define the K-linear representation $F(M) = (M_a, \varphi_\alpha)_{a \in Q_0, \alpha \in Q_1}$ of $(Q, \mathcal{I})$ as follows: if a belongs to Q_0, let $e_a = \varepsilon_a + \mathcal{I}$ be the corresponding primitive idempotent in $A = KQ/\mathcal{I}$, then set $M_a = Me_a$; if $\alpha : a \to b$ belongs to Q_1 and $\bar{\alpha} = \alpha + \mathcal{I}$ is its class modulo $\mathcal{I}$, define $\varphi_\alpha : M_a \to M_b$ by $\varphi_\alpha(x) = x\bar{\alpha}(= xe_a\bar{\alpha}e_b)$ for $x \in M_a$. Because M is an A-module, φ_α is a K-linear map. Then $F(M)$ is bound by $\mathcal{I}$: let $\rho = \sum_{i=1}^{m} \lambda_i w_i$ be a relation from a to b in $\mathcal{I}$, where $w_i = \alpha_{i,1}\alpha_{i,2}\ldots\alpha_{i,\ell_i}$; we have

$$
\begin{aligned}
\varphi_\rho(x) &= \sum_{i=1}^{m} \lambda_i \varphi_{w_i}(x) \\
&= \sum_{i=1}^{m} \lambda_i \varphi_{\alpha_{i,\ell_i}} \ldots \varphi_{\alpha_{i,1}}(x) \\
&= \sum_{i=1}^{m} \lambda_i(x\bar{\alpha}_{i,1} \ldots \bar{\alpha}_{i,\ell_i}) \\
&= x \cdot \sum_{i=1}^{m} \lambda_i(\bar{\alpha}_{i,1} \ldots \bar{\alpha}_{i,\ell_i}) \\
&= x \cdot \bar{\rho} = x0 = 0.
\end{aligned}
$$

This defines our functor on the objects.

Let $f : M_A \to M'_A$ be an A-module homomorphism. We want to define a morphism $F(f) : F(M) \to F(M')$ of $\mathrm{Rep}_K(Q,\mathcal{I})$. For $a \in Q_0$ and $x = xe_a \in Me_a = M_a$, we have $f(xe_a) = f(xe_a^2) = f(xe_a)e_a \in M'e_a = M'_a$. Thus the restriction f_a of f to M_a is a K-linear map $f_a : M_a \to M'_a$. We then put $F(f) = (f_a)_{a \in Q_0}$. We now verify that for any arrow $\alpha : a \to b$, we have $\varphi'_\alpha f_a = f_b\varphi_\alpha$; this will show that $F(f)$ is indeed a morphism of representations. Let $x \in M_a$, then

$$f_b\varphi_\alpha(x) = f_b(x\bar{\alpha}) = f(x\bar{\alpha}) = f(x)\bar{\alpha} = f_a(x)\bar{\alpha} = \varphi'_\alpha f_a(x).$$

Finally, it is trivially checked that $F : \mathrm{Mod}\, A \to \mathrm{Rep}_K(Q,\mathcal{I})$ is a K-linear functor and that F restricts to a K-linear functor $\mathrm{mod}\, A \longrightarrow \mathrm{rep}_K(Q,\mathcal{I})$.

(b) We construct a K-linear functor

$$G : \mathrm{Rep}_K(Q,\mathcal{I}) \to \mathrm{Mod}\, A, \qquad (1.6')$$

which is a quasi-inverse of F as follows. Let $M = (M_a, \varphi_\alpha)$ be an object of $\mathrm{Rep}_K(Q,\mathcal{I})$. We set $G(M) = \bigoplus_{a \in Q_0} M_a$, and we define an A-module structure on the K-vector space $G(M)$ as follows. Because $A = KQ/\mathcal{I}$, we start by defining a KQ-module structure of $G(M)$, then show it is annihilated by $\mathcal{I}$. Let thus $x = (x_a)_{a \in Q_0}$ belong to $G(M)$. To define a KQ-module structure on $G(M)$, it suffices to define the products of the form xw, where w is a path in Q. If $w = \varepsilon_a$ is the stationary path in a, we put

$$xw = x\varepsilon_a = x_a.$$

If $w = \alpha_1\alpha_2\ldots\alpha_\ell$ is a nontrivial path from a to b, we consider the K-linear map $\varphi_w = \varphi_{\alpha_\ell}\ldots\varphi_{\alpha_1} : M_a \to M_b$. We put

$$(xw)_c = \delta_{bc}\varphi_w(x_a),$$

where δ_{bc} denotes the Kronecker delta. In other words, xw is the element of $G(M) = \bigoplus\limits_{a \in Q_0} M_a$ whose only nonzero coordinate is $(xw)_b = \varphi_w(x_a) \in M_b$. This shows that $G(M)$ is a KQ-module. Moreover, it follows from the definition of $G(M)$ that, for each $\rho \in \mathcal{I}$ and $x \in G(M)$, we have $x\rho = 0$. Thus $G(M)$ becomes a $KQ/\mathcal{I}$-module under the assignment $x(v + \mathcal{I}) = xv$ for $x \in G(M)$ and $v \in KQ$. This defines our functor G on the objects.

Let now $(f_a)_{a \in Q_0}$ be a morphism from $M = (M_a, \varphi_\alpha)$ to $M' = (M'_a, \varphi'_\alpha)$ in $\mathrm{Rep}_K(Q, \mathcal{I})$. We want to construct a homomorphism $f : G(M) \to G(M')$ of A-modules. Because $G(M) = \bigoplus\limits_{a \in Q_0} M_a$ and $G(M') = \bigoplus\limits_{a \in Q_0} M'_a$ as K-vector spaces, there exists a K-linear map $f = \bigoplus\limits_{a \in Q_0} f_a : G(M) \to G(M')$. We claim that f is an A-module homomorphism, that is, for any $x \in G(M)$ and any $\overline{w} \in KQ/\mathcal{I}$, we have $f(x\overline{w}) = f(x)\overline{w}$. It suffices to show the statement for $x = x_a \in M_a$ and $\overline{w} = w + \mathcal{I}$, where w is a path from a to b in Q. Then

$$f(x\overline{w}) = f(x_a\overline{w}) = f_b\varphi_w(x_a) = \varphi'_w f_a(x_a) = f_a(x_a)\overline{w} = f(x)\overline{w}$$

and our claim follows.

Finally, it is evident that G is a K-linear functor and that G restricts to a K-linear functor $\mathrm{mod}\, A \longrightarrow \mathrm{rep}_K(Q, \mathcal{I})$. It is easy to check that $FG \cong 1_{\mathrm{Rep}_K(Q, \mathcal{I})}$ and $GF \cong 1_{\mathrm{Mod}\, A}$. The second statement of the theorem follows from the fact that, because Q is finite, for a K-linear representation $M = (M_a, \varphi_\alpha)$ of $(Q, \mathcal{I})$, we have $\dim_K(\bigoplus\limits_{a \in Q_0} M_a) < \infty$ if and only if $\dim_K M_a < \infty$ for all $a \in Q_0$. $\qquad\square$

1.7. Corollary. *Let Q be a finite, connected, and acyclic quiver. There exists an equivalence of categories* $\mathrm{Mod}\, KQ \cong \mathrm{Rep}_K(Q)$ *that restricts to an equivalence* $\mathrm{mod}\, KQ \cong \mathrm{rep}_K(Q)$.

Proof. Because Q is acyclic, by (II.1.4), the algebra KQ is finite dimensional. The statement follows by letting $\mathcal{I} = 0$ in Theorem 1.6. $\qquad\square$

Another consequence of the theorem is the (trivial) remark that the categories $\mathrm{Rep}_K(Q, \mathcal{I})$ and $\mathrm{rep}_K(Q, \mathcal{I})$ are abelian.

We conclude this section with an example showing how one can verify whether a given representation of a quiver is indecomposable. By (I.4.8), it suffices to verify whether its endomorphism algebra is local.

In the following example and throughout this book we denote by $J_{m,\lambda}$ the $m \times m$ Jordan block corresponding to the eigenvalue $\lambda \in K$, that is,

$$J_{m,\lambda} = \begin{bmatrix} \lambda & \cdots & \cdots & 0 \\ 1 & \ddots & & \vdots \\ \vdots & \ddots & \ddots & \vdots \\ 0 & \cdots & 1 & \lambda \end{bmatrix}.$$

1.8. Example. Let Q be the Kronecker quiver $1 \circ \underset{\beta}{\overset{\alpha}{\rightleftarrows}} \circ 2$ and M be the representation of Q defined by $K^3 \underset{J_{3,0}}{\overset{1}{\rightleftarrows}} K^3$, where 1 denotes, as usual, the identity and $J_{3,0} = \begin{bmatrix} 0 & 0 & 0 \\ 1 & 0 & 0 \\ 0 & 1 & 0 \end{bmatrix}$ the 3×3 nilpotent Jordan block (identified with a linear map $K^3 \to K^3$ defined by $J_{3,0}$ in the standard basis of K^3). We claim that M is indecomposable. An endomorphism of M is given by a pair of 3×3 matrices (f_1, f_2) compatible with the structure maps. Writing down the two compatibility conditions, we obtain $f_1 \cdot 1 = 1 \cdot f_2$ and $f_1 \cdot J_{3,0} = J_{3,0} \cdot f_2$. The first one says that

$$f_1 = f_2 = \begin{bmatrix} a_1 & a_2 & a_3 \\ b_1 & b_2 & b_3 \\ c_1 & c_2 & c_3 \end{bmatrix} \quad \text{(say),}$$

whereas the second gives the matrix equation

$$\begin{bmatrix} a_1 & a_2 & a_3 \\ b_1 & b_2 & b_3 \\ c_1 & c_2 & c_3 \end{bmatrix} \begin{bmatrix} 0 & 0 & 0 \\ 1 & 0 & 0 \\ 0 & 1 & 0 \end{bmatrix} = \begin{bmatrix} 0 & 0 & 0 \\ 1 & 0 & 0 \\ 0 & 1 & 0 \end{bmatrix} \begin{bmatrix} a_1 & a_2 & a_3 \\ b_1 & b_2 & b_3 \\ c_1 & c_2 & c_3 \end{bmatrix},$$

that is,

$$\begin{bmatrix} a_2 & a_3 & 0 \\ b_2 & b_3 & 0 \\ c_2 & c_3 & 0 \end{bmatrix} = \begin{bmatrix} 0 & 0 & 0 \\ a_1 & a_2 & a_3 \\ b_1 & b_2 & b_3 \end{bmatrix}.$$

Thus $a_2 = a_3 = b_3 = 0, a_1 = b_2 = c_3 = a$ (say) and $b_1 = c_2 = b$ (say). Setting $c_1 = c$, we get

$$f_1 = f_2 = \begin{bmatrix} a & 0 & 0 \\ b & a & 0 \\ c & b & a \end{bmatrix}.$$

We have thus shown that

$$\operatorname{End} M \cong \left\{ \begin{bmatrix} a & 0 & 0 \\ b & a & 0 \\ c & b & a \end{bmatrix} \mid a, b, c \in K \right\}.$$

The ideal

$$\mathcal{I} = \left\{ \begin{bmatrix} 0 & 0 & 0 \\ b & 0 & 0 \\ c & b & 0 \end{bmatrix} \mid b, c \in K \right\}$$

of $\operatorname{End} M$ satisfies $\mathcal{I}^3 = 0$. Because $(\operatorname{End} M)/\mathcal{I} \cong K$, then $\mathcal{I}$ is a maximal ideal of $\operatorname{End} M$. By (I.1.4), $\mathcal{I} = \operatorname{rad}(\operatorname{End} M)$ and $\operatorname{End} M$ is local, and from (I.4.8), it follows that M is indecomposable.

We observe that we have a K-algebra isomorphism $\operatorname{End} M \cong K[t]/\langle t^3 \rangle$ given by $\begin{bmatrix} a & 0 & 0 \\ b & a & 0 \\ c & b & a \end{bmatrix} \mapsto a + b\bar{t} + c\bar{t}^2$, where $\bar{t} = t + \langle t^3 \rangle$.

One shows exactly as earlier that, for any $m \geq 1$, the representation of Q defined by $K^m \underset{J_{m,0}}{\overset{1}{\rightleftarrows}} K^m$ is indecomposable, where 1 denotes the identity map on K^m and $J_{m,0}$ is the nilpotent $m \times m$ Jordan block corresponding to the eigenvalue $\lambda = 0$.

III.2. The simple, projective, and injective modules

Throughout this section, $(Q,\mathcal{I})$ will always denote a finite connected quiver Q having $\mid Q_0 \mid = n$ points and bound by an admissible ideal $\mathcal{I}$ of KQ. We denote by A the bound quiver algebra $A = KQ/\mathcal{I}$. As seen in (II.2.12), A is a basic and connected finite dimensional K-algebra with an identity, having $R/\mathcal{I}$ as radical (where R denotes, as usual, the arrow ideal of KQ) and $\{e_a \mid a \in Q_0\}$ as complete set of primitive orthogonal idempotents. Throughout, we identify A-modules and K-linear representations of $(Q,\mathcal{I})$ along the functor F defined in (1.6). The aim of this section is to present an explicit computation of the simple, the indecomposable projective, and the indecomposable injective A-modules as bound representations of $(Q,\mathcal{I})$. We also deduce several interesting consequences of this description.

Let $a \in Q_0$; we denote by $S(a)$ the representation $(S(a)_b, \varphi_\alpha)$ of Q defined as follows

$$S(a)_b = \begin{cases} 0 & \text{if } b \neq a \\ K & \text{if } b = a, \end{cases}$$

$$\varphi_\alpha = 0 \qquad \text{for all } \alpha \in Q_1.$$

Clearly, $S(a)$ is a bound representation of $(Q,\mathcal{I})$ (for any $\mathcal{I}$), and we have the following lemma.

2.1. Lemma. Let $A = KQ/\mathcal{I}$ be the bound quiver algebra of $(Q,\mathcal{I})$.

(a) For any $a \in Q_0$, $S(a)$ viewed as an A-module is isomorphic to the top of the indecomposable projective A-module $e_a A$.

(b) The set $\{S(a) \mid a \in Q_0\}$ is a complete set of representatives of the isomorphism classes of the simple A-modules.

Proof. For any $a \in Q_0$, the K-vector space $S(a)$ is one-dimensional and hence defines a simple representation of $(Q,\mathcal{I})$ and a simple A-module. Because by the proof of (1.6), we have $\operatorname{Hom}_A(e_a A, S(a)) \cong S(a)e_a \cong S(a)_a \neq 0$, then there exists a nonzero A-module homomorphism from the indecomposable projective A-module $e_a A$ onto the simple A-module $S(a)$. This proves (a), because $e_a A$ has a simple top (by (I.4.5)). On the other hand, if $a \neq b$, it is clear that $\mathrm{Hom}_A(S(a), S(b)) = 0$ and in particular $S(a) \not\cong S(b)$. Thus the simple modules $S(a), a \in Q_0$, are pairwise nonisomorphic. Because, by (I.5.17), there exists a bijection between a complete set of primitive orthogonal idempotents and a complete set of pairwise nonisomorphic simple A-modules given by $e_a \mapsto \mathrm{top}(e_a A)$, (b) follows. $\square$

We say in the sequel that $S(a)$ is the simple A-module corresponding to the point $a \in Q_0$.

We warn the reader that, in contrast to the description of the simple modules of (finite dimensional) bound quiver algebras $KQ/\mathcal{I}$ given in (2.1)(b), any path algebra $A = KQ$ of a finite quiver Q with an oriented cycle has infinitely many pairwise nonisomorphic simple modules of finite dimension, distinct from the modules $S(a)$, with $a \in Q_0$ (see Exercise 14).

An example of such an algebra is the path algebra $A = KQ$ of the quiver $Q : 1 \circ \underset{\beta}{\overset{\alpha}{\rightleftarrows}} \circ 2$. Indeed, the A-modules $S(1) = (K \underset{0}{\overset{0}{\rightleftarrows}} 0)$, $S(2) = (0 \underset{0}{\overset{0}{\rightleftarrows}} K)$, and $S_\lambda = (K \underset{\lambda}{\overset{1}{\rightleftarrows}} K)$, with $\lambda \in K$, are all simple, and one easily checks that $S_\lambda \not\cong S_\mu$ whenever $\lambda \neq \mu$.

2.2. Lemma. *Let $M = (M_a, \varphi_\alpha)$ be a bound representation of $(Q, \mathcal{I})$.*

(a) *M is semisimple if and only if $\varphi_\alpha = 0$ for every $\alpha \in Q_1$.*

(b) *$\mathrm{soc}\, M = N$, where $N = (N_a, \psi_\alpha)$ with $N_a = M_a$ if a is a sink, whereas*

$$N_a = \bigcap_{\alpha: a \to b} \mathrm{Ker}(\varphi_\alpha : M_a \to M_b)$$

if a is not a sink, and $\psi_\alpha = \varphi_\alpha |_{N_a} = 0$ for every arrow α of source a.

(c) *$\mathrm{rad}\, M = J$, where $J = (J_a, \gamma_\alpha)$ with $J_a = \sum_{\alpha: b \to a} \mathrm{Im}(\varphi_\alpha : M_b \to M_a)$ and $\gamma_\alpha = \varphi_\alpha |_{J_a}$ for every arrow α of source a.*

(d) *$\mathrm{top}\, M = L$, where $L = (L_a, \psi_\alpha)$ with $L_a = M_a$ if a is a source, whereas $L_a = \sum_{\alpha: b \to a} \mathrm{Coker}(\psi_\alpha : M_b \to M_a)$ if a is not a source and $\psi_\alpha = 0$ for every arrow α of source a.*

Proof. (a) The first part follows easily from the fact that $\varphi_\alpha = 0$ for every $\alpha \in Q_1$ if and only if $M \cong \bigoplus_{a \in Q_0} S(a)^{\dim_K M_a}$.

(b) Because $\psi_\alpha = \varphi_\alpha |_{N_a}$, N is a submodule of M. Because $\psi_\alpha = 0$ for each α, N is semisimple. Let S_A be a simple submodule of M. There exists $a \in Q_0$ such that $S \cong S(a)$. We thus have, for each $\alpha : a \to b$, a commutative square:

$$K = S(a)_a \longrightarrow S(a)_b = 0$$

$$\downarrow \qquad\qquad \downarrow$$

$$M_a \xrightarrow{\;\varphi_\alpha\;} M_b$$

Hence $S(a)_a \subseteq \operatorname{Ker} \varphi_\alpha$ for each $\alpha : a \to b$, and so $S(a)_a \subseteq N_a$. This shows that $S(a) \subseteq N$ and therefore $N = \operatorname{soc} M$.

(c) Let R be the arrow ideal of KQ. Because $\operatorname{rad} A = R/\mathcal{I}$ is generated as a two-sided ideal by the residual classes modulo $\mathcal{I}$ of the arrows $\alpha \in Q_1$, it follows from (I.3.7) that

$$J = \operatorname{rad} M = M \cdot \operatorname{rad} A = M \cdot (R/\mathcal{I}) = \sum_{\alpha \in Q_1} M\overline{\alpha},$$

where $\overline{\alpha} = \alpha + \mathcal{I}$. Hence, for any $a \in Q_0$, we have $J_a = \sum_{\alpha:b\to a} M\overline{\alpha}$, where the sum is taken over all arrows of target a. For such an arrow $\alpha : b \to a$, the definition of the functor F in (1.6) yields $M\overline{\alpha} = Me_b\overline{\alpha} = M_b\overline{\alpha} = \varphi_\alpha(M_b) = \operatorname{Im} \varphi_\alpha$, because the action of φ_α corresponds to the right multiplication by $\overline{\alpha}$. Hence $J_a = \sum_{\alpha:b\to a} \operatorname{Im}(\varphi_\alpha : M_b \to M_a)$. Because J is a submodule of M, we have $\gamma_\alpha = \varphi_\alpha \mid_{J_a}$.

(d) Follows from (c), because $L = M/(M \operatorname{rad} A) = M/\operatorname{rad} M$.   □

2.3. Examples. (a) Let Q be the Kronecker quiver $1 \circ \underset{\beta}{\overset{\alpha}{\rightleftarrows}} \circ 2$. The simple KQ-modules are given by the representations

$$S(1) = (K \rightleftarrows 0) \quad \text{and} \quad S(2) = (0 \rightleftarrows K)$$

Let M be given by the representation $K^{m-1} \underset{\pi_\beta}{\overset{\pi_\alpha}{\rightleftarrows}} K^m$, where $m \geq 2$ and π_α, π_β are the projections given by the $(m-1) \times m$ matrices

$$\pi_\alpha = \begin{bmatrix} 1 & 0 & 0 & 0 & \cdots & 0 \\ 0 & 0 & 1 & 0 & \cdots & 0 \\ 0 & 0 & 0 & 1 & \cdots & 0 \\ \vdots & \vdots & \vdots & \vdots & & \vdots \\ 0 & 0 & 0 & 0 & \cdots & 1 \end{bmatrix} \quad \text{and} \quad \pi_\beta = \begin{bmatrix} 0 & 1 & 0 & 0 & \cdots & 0 \\ 0 & 0 & 1 & 0 & \cdots & 0 \\ 0 & 0 & 0 & 1 & \cdots & 0 \\ \vdots & \vdots & \vdots & \vdots & & \vdots \\ 0 & 0 & 0 & 0 & \cdots & 1 \end{bmatrix}.$$

Then $\operatorname{soc} M = \operatorname{rad} M = \left(K^{m-1} \rightleftarrows 0 \right) = S(1)^{m-1}$, while $\operatorname{top} M = (0 \rightleftarrows K^m) = S(2)^m$.

(b) Let Q be the quiver $1 \circ \underset{\delta}{\overset{\beta}{\rightleftarrows}} \circ_2 \underset{\gamma}{\overset{\alpha}{\rightleftarrows}} \circ 3$, bound by $\alpha\beta = 0, \gamma\delta = 0$, and let M be the bound quiver representation

$$K \underset{[0\,0\,1]}{\overset{[0\,1\,0]}{\longleftarrow}} K^3 \underset{\left[\begin{smallmatrix}0\\1\\0\end{smallmatrix}\right]}{\overset{\left[\begin{smallmatrix}1\\0\\0\end{smallmatrix}\right]}{\longleftarrow}} K$$

Then

$$\operatorname{soc} M \;=\; \left(K \underset{0}{\overset{0}{\rightleftharpoons}} K \rightleftharpoons 0 \right) \;\cong\; S(1) \oplus S(2),$$

$$\operatorname{rad} M \;=\; \left(K \underset{[0\,0]}{\overset{[0\,1]}{\leftrightharpoons}} K^2 \rightleftharpoons 0 \right) \;\cong\; S(2) \oplus \left(K \underset{0}{\overset{1}{\rightleftharpoons}} K \rightleftharpoons 0 \right),$$

$$\operatorname{top} M \;=\; \left(0 \rightleftharpoons K \underset{0}{\overset{0}{\rightleftharpoons}} K \right) \;\cong\; S(2) \oplus S(3).$$

Moreover, an easy computation (as in example (1.8)) shows that $\operatorname{End} M$ is local so that M is indecomposable. However, $\operatorname{End} M$ is not a field, because $S(2)$ occurs as a summand of both the top and the socle of M, so there exist nonzero morphisms $p : M \to S(2)$ and $j : S(2) \to M$, and hence the composition $jp : M \to M$ is a nonzero endomorphism that is not invertible.

We now show how to compute the indecomposable projective A-modules. Because A is basic and $\{e_a \mid a \in Q_0\}$ is a complete set of primitive orthogonal idempotents of A, the decomposition $A_A = \bigoplus_{a \in Q_0} e_a A$ is a decomposition of A_A as a direct sum of pairwise nonisomorphic indecomposable projective A-modules. We wish to describe the modules $P(a) = e_a A$, with $a \in Q_0$.

2.4. Lemma. *Let $(Q, \mathcal{I})$ be a bound quiver, $A = KQ/\mathcal{I}$, and $P(a) = e_a A$, where $a \in Q_0$.*

(a) If $P(a) = (P(a)_b, \varphi_\beta)$, then $P(a)_b$ is the K-vector space with basis the set of all the $\overline{w} = w + \mathcal{I}$, with w a path from a to b and, for an arrow $\beta : b \to c$, the K-linear map $\varphi_\beta : P(a)_b \to P(a)_c$ is given by the right multiplication by $\overline{\beta} = \beta + \mathcal{I}$.

(b) Let $\operatorname{rad} P(a) = (P'(a)_b, \varphi'_\beta)$. Then $P'(a)_b = P(a)_b$ for $b \neq a$, $P'(a)_a$ is the K-vector space with basis the set of all $\overline{w} = w + \mathcal{I}$, with w a non-stationary path from a to a, $\varphi'_\beta = \varphi_\beta$ for any arrow β of source $b \neq a$ and $\varphi'_\alpha = \varphi_\alpha|_{P'(a)_a}$ for any arrow α of source a.

Proof. (a) It follows from the definition of the functor F in (1.6) that the representation corresponding via F to the A-module $P(a)_A = e_a A$ is such that, for each $b \in Q_0$, we have

$$P(a)_b = P(a)e_b = e_a A e_b = e_a (KQ/\mathcal{I})e_b = (\varepsilon_a (KQ)\varepsilon_b)/(\varepsilon_a \mathcal{I}\varepsilon_b).$$

Moreover, if $\beta : b \to c$ is an arrow of Q, then $\varphi_\beta : e_a A e_b \to e_a A e_c$ is given by the right multiplication by the residual class $\overline{\beta} = \beta + \mathcal{I}$, that is, if $\overline{w}$ is the residual class of a path w from a to b, then $\varphi_\beta(\overline{w}) = \overline{w}\overline{\beta}$.

The statement (b) is a consequence of (a) and (2.2). $\qquad\square$

We say in the sequel that $P(a)$ is the indecomposable projective A-module corresponding to the point $a \in Q_0$. It follows from (2.1) that $S(a)$ is isomorphic to the simple top of $P(a)$, and from (2.4)(b) that the radical of

$P(a)$ is given by $(P'(a)_b, \varphi'_\beta)$, where $P'(a)_b$ is the subspace of $P(a)_b$ spanned by the residual classes of paths of length at least one, and $\varphi'_\beta = \varphi_\beta \mid_{P'(a)_b}$. An important particular case is when Q is acyclic and $\mathcal{I} = 0$. In this case, $P(a)_b$ is equal to the vector space having as basis the set of all paths from a to b.

2.5. Examples. (a) Let Q be the quiver

The indecomposable projective KQ-modules are given by

$$P(1) = S(1) = \overset{K}{\underset{0}{\diagup}} \overset{}{\underset{0}{\diagdown}}, \quad P(2) = \overset{K}{\underset{K}{\diagup}} \overset{}{\underset{0}{\diagdown}} \quad \text{and} \quad P(3) = \overset{K}{\underset{0}{\diagup}} \overset{}{\underset{K}{\diagdown}} 1$$

Here $\operatorname{rad} P(1) = 0$, whereas $\operatorname{rad} P(2) \cong \operatorname{rad} P(3) \cong P(1)$.

(b) Let Q be the quiver $1 \circ \underset{\delta}{\overset{\beta}{\rightleftarrows}} \circ 2 \underset{\gamma}{\overset{\alpha}{\leftleftarrows}} \circ 3$ bound by $\alpha\beta = 0, \gamma\delta = 0$. The indecomposable projective A-modules are given by $P(1) = S(1)$,

$$P(2) = (K^2 \underset{\left[\begin{smallmatrix}0\\1\end{smallmatrix}\right]}{\overset{\left[\begin{smallmatrix}1\\0\end{smallmatrix}\right]}{\rightleftarrows}} K \leftleftarrows 0) \quad \text{and} \quad P(3) = (K^2 \underset{\left[\begin{smallmatrix}0&0\\1&0\end{smallmatrix}\right]}{\overset{\left[\begin{smallmatrix}0&1\\0&0\end{smallmatrix}\right]}{\rightleftarrows}} K^2 \underset{\left[\begin{smallmatrix}0\\1\end{smallmatrix}\right]}{\overset{\left[\begin{smallmatrix}1\\0\end{smallmatrix}\right]}{\leftleftarrows}} K).$$

Here, $\operatorname{rad} P(1) = 0$, $\operatorname{rad} P(2) = S(1)^2$, whereas

$$\operatorname{rad} P(3) \cong (K \underset{0}{\overset{1}{\rightleftarrows}} K \leftleftarrows 0) \oplus (K \underset{1}{\overset{0}{\rightleftarrows}} K \leftleftarrows 0).$$

We note that the two indecomposable summands of $P(3)$ are not isomorphic.

(c) Let Q be the quiver $1 \circ \underset{\beta}{\overset{\alpha}{\rightleftarrows}} \circ 2$, bound by $\alpha\beta = 0$, $\beta\alpha = 0$. Then

$$P(1) = (K \underset{0}{\overset{1}{\rightleftarrows}} K) \quad \text{and} \quad P(2) = (K \underset{1}{\overset{0}{\rightleftarrows}} K).$$

Here $\operatorname{rad} P(1) \cong S(2)$, while $\operatorname{rad} P(2) \cong S(1)$.

(d) Let Q be the quiver

bound by $\alpha\beta = \gamma\delta$, $\beta\lambda = 0$, and $\lambda^3 = 0$. Then

$$P(1) = \begin{bmatrix} 0\,1\,0 \\ 0\,0\,1 \\ 0\,0\,0 \end{bmatrix} \; K^3 \begin{smallmatrix} & & 0 & \\ & 0 \nearrow & \nwarrow & 0 \\ & & & 0, \\ & 0 \searrow & & \swarrow 0 \\ & & 0 & \end{smallmatrix} \qquad \operatorname{rad} P(1) = \begin{bmatrix} 0\,1 \\ 0\,0 \end{bmatrix} \; K^2 \begin{smallmatrix} & & 0 & \\ & 0 \nearrow & \nwarrow & 0 \\ & & & 0; \\ & 0 \searrow & & \swarrow 0 \\ & & 0 & \end{smallmatrix}$$

$$P(2) = \quad 0 \; K \begin{smallmatrix} & & K & \\ & 1 \nearrow & \nwarrow & 0 \\ & & & 0, \\ & 0 \searrow & & \swarrow 0 \\ & & 0 & \end{smallmatrix} \qquad \operatorname{rad} P(2) \cong S(1);$$

$$P(3) = \begin{bmatrix} 0\,1\,0 \\ 0\,0\,1 \\ 0\,0\,0 \end{bmatrix} \; K^3 \begin{smallmatrix} & & 0 & \\ & 0 \nearrow & \nwarrow & 0 \\ & & & 0, \\ & \left[\begin{smallmatrix}0\\0\\1\end{smallmatrix}\right] \nwarrow & & \swarrow 0 \\ & & K & \end{smallmatrix} \qquad \operatorname{rad} P(3) \cong P(1);$$

$$P(4) = \quad 0 \; K \begin{smallmatrix} & & K & \\ & 1 \nearrow & \nwarrow & 1 \\ & & & K, \\ & 1 \searrow & & \swarrow 1 \\ & & K & \end{smallmatrix} \qquad \operatorname{rad} P(4) \cong \quad 0 \; K \begin{smallmatrix} & & K & \\ & 1 \nearrow & \nwarrow & 0 \\ & & & 0. \\ & 1 \searrow & & \swarrow 0 \\ & & K & \end{smallmatrix}$$

In this example, we note that for each indecomposable projective module P, the module $\operatorname{rad} P$ is also indecomposable.

We now describe explicitly the indecomposable injective A-modules. By (I.5.17), a complete list of pairwise nonisomorphic indecomposable injective A-modules is given by the modules $I(a) = D(Ae_a)$ (with $a \in Q_0$), where $D = \operatorname{Hom}_K(-, K)$ denotes, as usual, the standard duality between the right and left A-modules.

2.6. Lemma. (a) *Given $a \in Q_0$, the simple module $S(a)$ is isomorphic to the simple socle of $I(a)$.*

(b) *If $I(a) = (I(a)_b, \varphi_\beta)$, then $I(a)_b$ is the dual of the K-vector space with basis the set of all $\overline{w} = w + \mathcal{I}$, with w a path from b to a and, for an arrow $\beta : b \to c$, the K-linear map $\varphi_\beta : I(a)_b \to I(a)_c$ is given by the dual of the left multiplication by $\overline{\beta} = \beta + \mathcal{I}$.*

(c) *Let $I(a)/S(a) = (L_b, \psi_\beta)$. Then L_b is the quotient space of $I(a)_b$ spanned by the residual classes of paths from b to a of length at most one, and ψ_β is the induced map.*

Proof. (a) We can apply (2.2)(b), or by dualising (2.1)(a) we get the isomorphisms $\operatorname{soc} I(a) \cong P(a)/\operatorname{rad} P(a) \cong S(a)$ of right A-modules.

(b) Because there are isomorphisms

$$I(a)_b = I(a)e_b = D(Ae_a)e_b \cong D(e_b Ae_a) \cong D(\varepsilon_b(KQ)\varepsilon_a/\varepsilon_b I \varepsilon_a),$$

the first statement follows from (2.4). Similarly, if $\beta : b \to c$ is an arrow, the

K-linear map $\varphi_\beta : D(\varepsilon_b(KQ)\varepsilon_a/\varepsilon_b\mathcal{I}\varepsilon_a) \to D(\varepsilon_c(KQ)\varepsilon_a/\varepsilon_c\mathcal{I}\varepsilon_a)$ is defined as follows: let $\mu_\beta : (\varepsilon_c(KQ)\varepsilon_a/\varepsilon_c\mathcal{I}\varepsilon_a) \to (\varepsilon_b(KQ)\varepsilon_a/\varepsilon_b\mathcal{I}\varepsilon_a)$ be the left multiplication $\overline{w} \mapsto \overline{\beta w}$, then $\varphi_\beta = D(\mu_\beta)$ is given by $\varphi_\beta(f) = f\mu_\beta$ for $f \in D(\varepsilon_b(KQ)\varepsilon_a/\varepsilon_b\mathcal{I}\varepsilon_a)$. In other words, $\varphi_\beta(f)(\overline{w}) = f(\overline{\beta w})$. The statement (c) is a consequence of (b). $\qquad\square$

We say in the sequel that $I(a)$ is the indecomposable injective A-module corresponding to the point $a \in Q_0$. An important particular case is when Q is acyclic and $\mathcal{I} = 0$. In this case, $I(a)_b$ is nothing but the dual of the vector space with basis the set of all paths from b to a.

2.7. Examples. (a) Let Q be the quiver

The indecomposable injective KQ-modules are $I(2) = S(2)$, $I(3) = S(3)$, and

Thus $I(2)/S(2) = 0, I(3)/S(3) = 0$, whereas $I(1)/S(1) \cong S(2) \bigoplus S(3)$.

(b) Let Q be the quiver $1 \circ \xrightleftharpoons[\delta]{\beta} \underset{2}{\circ} \xrightleftharpoons[\gamma]{\alpha} \circ 3$, bound by $\alpha\beta = 0, \gamma\delta = 0$. The indecomposable injective KQ-modules are given by $I(3) = S(3)$,

$$I(2) = (0 \Longleftarrow K \underset{[0\,1]}{\overset{[1\,0]}{\Longleftarrow}} K^2) \quad \text{and} \quad I(1) = (K \underset{[0\,1]}{\overset{[1\,0]}{\Longleftarrow}} K^2 \underset{[\substack{0\,1 \\ 0\,0}]}{\overset{[\substack{0\,0 \\ 1\,0}]}{\Longleftarrow}} K^2).$$

Here, $I(3)/S(3) = 0, I(2)/S(2) \cong S(3)^2$, and

$$I(1)/S(1) \cong (0 \Longleftarrow K \underset{0}{\overset{1}{\Longleftarrow}} K) \oplus (0 \Longleftarrow K \underset{1}{\overset{0}{\Longleftarrow}} K).$$

Again, the two indecomposable summands of $I(1)/S(1)$ are not isomorphic.

(c) Let Q be the quiver $1 \circ \underset{\beta}{\overset{\alpha}{\rightleftarrows}} \circ 2$, bound by $\alpha\beta = 0, \beta\alpha = 0$, then $I(1) \cong P(2), I(2) \cong P(1), I(1)/S(1) \cong S(2)$, and $I(2)/S(2) \cong S(1)$. This shows that $A = KQ/\mathcal{I}$ is a self-injective algebra, that is, the module A_A is injective.

(d) Let Q be the quiver

bound by $\alpha\beta = \gamma\delta, \beta\lambda = 0, \lambda^3 = 0$. Then $I(4) = S(4), I(4)/S(4) = 0$, and

$$I(3) = {}_0\circlearrowleft 0 \quad \begin{array}{c} 0 \\ {}_0\swarrow \quad \nwarrow {}_0 \\ K \\ {}_0\nwarrow \quad \swarrow {}_1 \\ K \end{array} K, \qquad I(3)/S(3) \cong S(4);$$

$$I(2) = {}_0\circlearrowleft 0 \quad \begin{array}{c} K \\ {}_0\swarrow \quad \nwarrow {}_1 \\ K \\ {}_0\nwarrow \quad \swarrow {}_0 \\ 0 \end{array} K, \qquad I(2)/S(2) \cong S(4);$$

$$I(1) = \begin{bmatrix}0&1&0\\0&0&1\\0&0&0\end{bmatrix} \circlearrowleft K^3 \quad \begin{array}{c} K \\ {}_{\left[\begin{smallmatrix}1\\0\\0\end{smallmatrix}\right]}\swarrow \quad \nwarrow {}_1 \\ K \\ {}_1\nwarrow \quad \swarrow {}_{\left[\begin{smallmatrix}1\\0\\0\end{smallmatrix}\right]} \\ K^3 \end{array} K, \qquad I(1)/S(1) \cong \begin{bmatrix}0&1\\0&0\end{bmatrix} \circlearrowleft K^2 \quad \begin{array}{c} K \\ {}_0\swarrow \quad \nwarrow {}_1 \\ K^2 \\ {}_{\left[\begin{smallmatrix}0&1&0\\0&0&1\end{smallmatrix}\right]}\nwarrow \quad \swarrow {}_{\left[\begin{smallmatrix}1\\0\end{smallmatrix}\right]} \\ K^3 \end{array} K.$$

In particular, $I(1)/S(1)$ is easily seen to be the direct sum of two indecomposable representations given respectively by

$$\begin{bmatrix}0&0\\1&0\end{bmatrix} \circlearrowleft K^2 \quad \begin{array}{c} 0 \\ {}_0\swarrow \quad \nwarrow {}_0 \\ 0 \\ {}_1\nwarrow \quad \swarrow {}_0 \\ K^2 \end{array} 0 \qquad \text{and} \qquad {}_0\circlearrowleft 0 \quad \begin{array}{c} K \\ {}_0\swarrow \quad \nwarrow {}_1 \\ 0 \\ {}_0\nwarrow \quad \swarrow {}_1 \\ K \end{array} K$$

The previous results show that to each point $a \in Q_0$ correspond an indecomposable projective A-module $P(a)$ and an indecomposable injective A-module $I(a)$. The connection between them can be expressed by means of an endofunctor of the module category.

2.8. Definition. The **Nakayama functor** of $\operatorname{mod} A$ is defined to be the endofunctor $\nu = D\operatorname{Hom}_A(-, A) : \operatorname{mod} A \to \operatorname{mod} A$.

There is another possible definition for the Nakayama functor.

2.9. Lemma. *The Nakayama functor ν is right exact and is functorially isomorphic to $- \otimes_A DA$.*

Proof. The right exactness of ν follows from the fact that ν is equal to the composition of two contravariant left exact functors. Consider the functorial morphism $\phi : - \otimes_A DA \to \nu = D\operatorname{Hom}_A(-, A)$, defined on an A-module M by

$$\phi_M : M \otimes_A DA \to D\operatorname{Hom}_A(M, A), \quad x \otimes f \mapsto (\varphi \mapsto f(\varphi(x))),$$

for $x \in M, f \in DA$, and $\varphi \in \operatorname{Hom}_A(M, A)$. Clearly, ϕ_M is an isomorphism if $M_A = A_A$. Because both functors are K-linear, ϕ_M is an isomorphism if

M_A is a projective A-module. Let now M be arbitrary, and

$$P_1 \xrightarrow{p_1} P_0 \xrightarrow{p_0} M \longrightarrow 0$$

be a projective presentation for M. Because $-\otimes_A DA$ and ν are both right exact, we have a commutative diagram with exact rows:

$$
\begin{array}{ccccccc}
P_1 \otimes_A DA & \xrightarrow{p_1 \otimes DA} & P_0 \otimes_A DA & \xrightarrow{p_0 \otimes DA} & M \otimes_A DA & \longrightarrow & 0 \\
\downarrow{\scriptstyle \phi_{P_1}} & & \downarrow{\scriptstyle \phi_{P_0}} & & \downarrow{\scriptstyle \phi_M} & & \\
\nu P_1 & \xrightarrow{\nu p_1} & \nu P_0 & \xrightarrow{\nu p_0} & \nu M & \longrightarrow & 0
\end{array}
$$

Because ϕ_{P_1} and ϕ_{P_0} are isomorphisms, so is ϕ_M.   □

2.10. Proposition. *The restriction of the Nakayama functor ν :* mod $A \to$ mod A *to the full subcategory* proj A *of* mod A *whose objects are the projective modules induces an equivalence between* proj A *and the full subcategory* inj A *of* mod A *whose objects are the injective modules. The quasi-inverse of this restriction is given by $\nu^{-1} = \operatorname{Hom}_A(D(_AA), -)$:* inj $A \to$ proj A.

Proof. For any $a \in Q_0$, we have $\nu P(a) = D\operatorname{Hom}_A(e_a A, A) \cong D(Ae_a) = I(a)$. Hence the image of proj A under ν lies in inj A. On the other hand,

$$
\begin{aligned}
\operatorname{Hom}_A(D(_AA), I(a)) &= \operatorname{Hom}_A(D(_AA), D(Ae_a)) \\
&\cong \operatorname{Hom}_{A^{\mathrm{op}}}(Ae_a, A) \cong e_a A = P(a). \quad \square
\end{aligned}
$$

2.11. Lemma. *Let $A = KQ/\mathcal{I}$ be a bound quiver algebra. For every A-module M and $a \in Q_0$, the K-linear map (I.4.3) induces functorial isomorphisms of K-vector spaces*

$$\operatorname{Hom}_A(P(a), M) \xrightarrow{\sim} Me_a \xrightarrow{\sim} D\operatorname{Hom}_A(M, I(a)).$$

Proof. By (I.4.2), the K-linear map $\operatorname{Hom}_A(P(a), M) \xrightarrow{\sim} Me_a$ given by the formula $f \mapsto f(e_a)$ is a functorial isomorphism. The second isomorphism is the composition

$$
\begin{aligned}
D\operatorname{Hom}_A(M, I(a)) = D\operatorname{Hom}_A(M, D(Ae_a)) &\cong D\operatorname{Hom}_{A^{\mathrm{op}}}(Ae_a, DM) \\
&\cong D(e_a DM) \cong D(DM)e_a \cong Me_a. \quad \square
\end{aligned}
$$

As a consequence, we obtain an expression of the quiver of A in terms of the extensions between simple modules.

2.12. Lemma. *Let $A = KQ/\mathcal{I}$ be a bound quiver algebra and let $a, b \in Q_0$.*

(a) *There exists an isomorphism of K-vector spaces*

$$\operatorname{Ext}^1_A(S(a), S(b)) \;\cong\; e_a(\operatorname{rad} A/\operatorname{rad}^2 A)e_b.$$

(b) *The number of arrows in Q from a to b is equal to the dimension* $\dim_K \operatorname{Ext}^1_A(S(a), S(b))$ *of* $\operatorname{Ext}^1_A(S(a), S(b))$.

Proof. (a) Let $\ldots \longrightarrow P_2 \xrightarrow{p_2} P_1 \xrightarrow{p_1} P_0 \xrightarrow{p_0} S \longrightarrow 0$ be a minimal projective resolution of the simple module S. We wish to compute $\operatorname{Ext}^1_A(S, S')$, where S' is another simple module. Using the definition of $\operatorname{Ext}^1_A(-, S')$ as a right-derived functor, we consider the deleted complex $\ldots \longrightarrow P_2 \xrightarrow{p_2} P_1 \xrightarrow{p_1} P_0 \longrightarrow 0$ to which we apply the functor $\operatorname{Hom}_A(-, S')$, thus obtaining the complex

$$0 \longrightarrow \operatorname{Hom}_A(P_0, S') \xrightarrow{\operatorname{Hom}_A(p_1,S')} \operatorname{Hom}_A(P_1, S') \xrightarrow{\operatorname{Hom}_A(p_2,S')}$$

$$\operatorname{Hom}_A(P_2, S') \xrightarrow{\operatorname{Hom}_A(p_3,S')} \operatorname{Hom}_A(P_3, S') \xrightarrow{\operatorname{Hom}_A(p_4,S')} \ldots$$

We claim that $\operatorname{Hom}_A(p_{i+1}, S') = 0$ for every $i \geq 0$. Let $f \in \operatorname{Hom}_A(P_i, S')$ be a nonzero homomorphism. Because S' is simple, f is surjective so there exists an indecomposable summand P' of P_i such that f equals the composition of the canonical projection $P_i \longrightarrow P'$, the canonical homomorphism $P' \longrightarrow P'/\operatorname{rad} P'$, and an isomorphism $P'/\operatorname{rad} P' \cong S'$. Now $\operatorname{Im} p_{i+1} = \operatorname{Ker} p_i \subseteq \operatorname{rad} P_i$, by definition of the minimal projective resolution. Hence

$$\operatorname{Hom}_A(p_{i+1}, S')(f)(x) = (fp_{i+1})(x) \in f(\operatorname{Im} p_{i+1}) \subseteq f(\operatorname{rad} P_i) = 0,$$

for any $x \in P_i$. Therefore $\operatorname{Hom}_A(p_{i+1}, S')(f) = 0$ and our claim follows. In particular, we get $\operatorname{Ext}^1_A(S, S') \cong \operatorname{Ker} \operatorname{Hom}_A(p_2, S')/\operatorname{Im} \operatorname{Hom}_A(p_1, S') \cong \operatorname{Hom}_A(P_1, S')$.

If $S = S(a)$ and we write $\operatorname{rad} P(a)/\operatorname{rad}^2 P(a) = \bigoplus_{c \in Q_0} S(c)^{n_c}$, a minimal projective resolution of $S(a)$ is of the form

$$\ldots \to \bigoplus_{c \in Q_0} P(c)^{n_c} \to P(a) \to S(a) \to 0,$$

so that

$$\begin{aligned}
\operatorname{Ext}^1_A(S(a), S(b)) &\cong \operatorname{Hom}_A\Big(\bigoplus_{c \in Q_0} P(c)^{n_c}, S(b)\Big)\\
&\cong \operatorname{Hom}_A(\operatorname{rad} P(a)/\operatorname{rad}^2 P(a), S(b))\\
&\cong \operatorname{Hom}_A(\operatorname{rad} P(a)/\operatorname{rad}^2 P(a), I(b))\\
&\cong D\operatorname{Hom}_A(P(b), \operatorname{rad} P(a)/\operatorname{rad}^2 P(a))\\
&\cong D\operatorname{Hom}_A(e_b A, e_a(\operatorname{rad} A/\operatorname{rad}^2 A))\\
&\cong D(e_a(\operatorname{rad} A/\operatorname{rad}^2 A)e_b)\\
&\cong e_a(\operatorname{rad} A/\operatorname{rad}^2 A)e_b.
\end{aligned}$$

(b) By definition, the number of arrows from a to b in the quiver Q is equal to $\dim_K(e_a(\operatorname{rad} A/\operatorname{rad}^2 A)e_b)$. Then (b) follows from (a). $\qquad\square$

III.3. The dimension vector of a module and the Euler characteristic

In this section, we attach to each A-module a vector with integral coordinates, called its dimension vector. This will allow us to use methods of linear algebra when studying modules over finite dimensional algebras.

Let A be a basic and connected finite dimensional K-algebra and $A \cong KQ/\mathcal{I}$ be a bound quiver presentation of A, where Q is a finite, connected quiver and $\mathcal{I}$ is an admissible ideal of KQ. Throughout this section, we assume that the points of the quiver Q of A are numbered as $\{1, \ldots, n\}$. As usual, we denote by e_j the primitive idempotent of A corresponding to $j \in Q_0$ and by $P(j) = e_j A$ (or $I(j) = D(Ae_j)$, or $S(j) = \operatorname{top}(e_j A)$) the corresponding indecomposable projective A-module (or indecomposable injective, or simple, respectively), where D is the standard duality. In particular, there is an indecomposable decomposition $A_A = e_1 A \oplus \cdots \oplus e_n A$.

We recall from (1.6) and (2.11) that if M is viewed as a K-linear representation (M_j, φ_β) of the bound quiver $(Q, \mathcal{I})$, then we have K-vector space isomorphisms $M_j = Me_j \cong \operatorname{Hom}_A(P(j), M) \cong D\operatorname{Hom}_A(M, I(j))$. This leads us to the following definition.

3.1. Definition. Let $A \cong KQ/\mathcal{I}$ be a K-algebra and let M be a module in $\operatorname{mod} A$. The **dimension vector** of M is defined to be the vector

$$\mathbf{dim}\, M = \begin{bmatrix} \dim_K Me_1 \\ \vdots \\ \dim_K Me_n \end{bmatrix} = [\dim_K Me_1 \ \ldots \ \dim_K Me_n]^t$$

in $\mathbb{Z}^n$, where $e_1, \ldots, e_n$ are primitive orthogonal idempotents of A corresponding to the points $1, \ldots, n$ of Q_0.

Thus, the dimension vector of the simple module $S(j)$ is the jth canonical basis vector of the group $\mathbb{Z}^n$. Note also that (2.11) yields

$$\mathbf{dim}\, M = \begin{bmatrix} \dim_K \operatorname{Hom}_A(P(1), M) \\ \vdots \\ \dim_K \operatorname{Hom}_A(P(n), M) \end{bmatrix} = \begin{bmatrix} \dim_K \operatorname{Hom}_A(M, I(1)) \\ \vdots \\ \dim_K \operatorname{Hom}_A(M, I(n)) \end{bmatrix}.$$

It follows from the unique decomposition theorem (I.4.10) that the vector $\mathbf{dim}\, M$ does not depend on the choice of a complete set $\{e_1, \ldots, e_n\}$ of primitive orthogonal idempotents of A, up to permutation of its coordinates.

Throughout, by **dim** M, we mean the dimension vector of M defined with respect to a given complete set $\{e_1, \dots, e_n\}$ of primitive orthogonal idempotents of A.

3.2. Example. In Examples 2.5 (d) and 2.7 (d), the dimension vectors of the indecomposable projective and injective modules are the vectors

$$\begin{aligned}
\mathbf{dim}\, P(1) &= [3\ 0\ 0\ 0]^t, & \mathbf{dim}\, I(1) &= [3\ 1\ 3\ 1]^t, \\
\mathbf{dim}\, P(2) &= [1\ 1\ 0\ 0]^t, & \mathbf{dim}\, I(2) &= [0\ 1\ 0\ 1]^t, \\
\mathbf{dim}\, P(3) &= [3\ 0\ 1\ 0]^t, & \mathbf{dim}\, I(3) &= [0\ 0\ 1\ 1]^t, \\
\mathbf{dim}\, P(4) &= [1\ 1\ 1\ 1]^t, & \mathbf{dim}\, I(4) &= [0\ 0\ 0\ 1]^t.
\end{aligned}$$

It is sometimes convenient to represent dimension vectors in a more suggestive way, following the shape of the quiver, as follows

$$\mathbf{dim}\, I(1) \;=\; 3\,{}^{1}_{3}\,1 \qquad\qquad \mathbf{dim}\, I(4) \;=\; 0\,{}^{0}_{0}\,1$$

3.3. Lemma. *If* $A \cong KQ/\mathcal{I}$ *and* $0 \to L \to M \to N \to 0$ *is a short exact sequence of* A*-modules, then* $\mathbf{dim}\, M = \mathbf{dim}\, L + \mathbf{dim}\, N$.

Proof. By applying the exact functor $\mathrm{Hom}_A(P(j), -)$ to the given short exact sequence $0 \to L \to M \to N \to 0$ we get the exact sequence of K-vector spaces $0 \to Le_j \to Me_j \to Ne_j \to 0$. Hence $\dim_K Me_j = \dim_K Le_j + \dim_K Ne_j$ for each $j \in Q_0$ and the statement follows. $\square$

The property of the previous lemma is sometimes expressed by saying that **dim** is an **additive function**. This brings us to another interpretation of the dimension vector of a module in terms of the Grothendieck group of mod A in the following sense.

3.4. Definition. Let A be a K-algebra. The **Grothendieck group** of A (or more precisely, of mod A), is the abelian group $K_0(A) = \mathcal{F}/\mathcal{F}'$, where $\mathcal{F}$ is the free abelian group having as basis the set of the isomorphism classes $\widetilde{M}$ of modules M in mod A and $\mathcal{F}'$ is the subgroup of $\mathcal{F}$ generated by the elements $\widetilde{M} - \widetilde{L} - \widetilde{N}$ corresponding to all exact sequences

$$0 \to L \to M \to N \to 0$$

in mod A. We denote by $[M]$ the image of the isomorphism class $\widetilde{M}$ of the module M under the canonical group epimorphism $\mathcal{F} \to \mathcal{F}/\mathcal{F}'$.

We remark that $\mathcal{F}$ is a set, because each A-module M of a given dimension m admits an A-module epimorphism $A^m \to M$.

Now we show that the group $K_0(A)$ is itself free and in fact isomorphic to the free group $\mathbb{Z}^n$.

3.5. Theorem. *Let* A *be a basic finite dimensional* K*-algebra and let* $S(1), \dots, S(n)$ *be a complete set of the isomorphism classes of simple right* A*-modules. Then the Grothendieck group* $K_0(A)$ *of* A *is a free abelian group* *having as a basis the set* $\{[S(1)], \ldots, [S(n)]\}$ *and there exists a unique group isomorphism* $\mathbf{dim} : K_0(A) \to \mathbb{Z}^n$ *such that* $\mathbf{dim}\,[M] = \mathbf{dim}\,M$ *for each* A-*module* M.

Proof. We first show that the set $\{[S(1)], \ldots, [S(n)]\}$ generates the group $K_0(A)$. Let M be a module in $\mathrm{mod}\,A$ and let $0 = M_0 \subset M_1 \subset M_2 \subset \cdots \subset M_t = M$ be a composition series for M. By the definition of $K_0(A)$, we have

$$[M] = [M_t/M_{t-1}] + [M_{t-1}] = \cdots = \sum_{j=1}^{n}[M_j/M_{j-1}] = \sum_{i=1}^{n} \mathbf{c}_i(M)[S(i)],$$

where $\mathbf{c}_i(M)$ is the number of composition factors M_j/M_{j-1} of M that are isomorphic to $S(i)$. This shows that $\{[S(1)], \ldots, [S(n)]\}$ generates the group $K_0(A)$.

It is clear that $M \cong N$ implies $\mathbf{dim}\,M = \mathbf{dim}\,N$. Moreover, the additivity of $\mathbf{dim}$ (see (3.3)) implies the existence of a unique group homomorphism $\mathbf{dim} : K_0(A) \to \mathbb{Z}^n$ such that $\mathbf{dim}\,[M] = \mathbf{dim}\,M$ for all M in $\mathrm{mod}\,A$. Because the image of the generating set $\{[S(1)], \ldots, [S(n)]\}$ under the homomorphism $\mathbf{dim}$ is the canonical basis of the free group $\mathbb{Z}^n$, this set is $\mathbb{Z}$-linearly independent in $K_0(A)$. It follows that $K_0(A)$ is free and that the homomorphism $\mathbf{dim} : K_0(A) \to \mathbb{Z}^n$ is an isomorphism. $\square$

As a consequence, we show that the dimension vector of a module M can also be regarded as a record of the number of simple composition factors of M that are isomorphic to each simple module.

3.6. Corollary. *Let* $A \cong KQ/\mathcal{I}$ *be a* K-*algebra and let* $S(j)$, *with* $j \in Q_0$, *be a fixed simple* A-*module. For any module* M *in* $\mathrm{mod}\,A$ *the number* $\mathbf{c}_j(M)$ *of simple composition factors of* M *that are isomorphic to* $S(j)$ *is* $\dim_K Me_j$, *and the composition length* $\ell(M)$ *of* M *is given by* $\ell(M) = \sum_{j \in Q_0} \dim_K Me_j = \dim_K M$.

Proof. As we have seen, the equality $[M] = \sum_{i=1}^{n} \mathbf{c}_i(M)[S(i)]$ holds. Hence we get $\mathbf{dim}\,M = \mathbf{dim}\,[M] = \sum_{i=1}^{n} \mathbf{c}_i(M)\mathbf{dim}\,[S(i)] = \sum_{i=1}^{n} \mathbf{c}_i(M)\mathbf{dim}\,S(i)$. Because $\{\mathbf{dim}\,S(1), \ldots, \mathbf{dim}\,S(n)\}$ is the canonical basis of the abelian group $\mathbb{Z}^n$, we get, by equating coordinates, the required equality $\mathbf{c}_j(M) = \dim_K Me_j$. This also yields $\ell(M) = \sum_{j \in Q_0} \mathbf{c}_i(M) = \sum_{j \in Q_0} \dim_K Me_j = \dim_K M$. $\square$

In particular, putting together the dimension vectors of the indecomposable projective (or injective) A-modules yields a square matrix with integral coefficients, called the Cartan matrix of A.

3.7. Definition. Let A be a basic finite dimensional K-algebra with a complete set $\{e_1, \ldots, e_n\}$ of primitive orthogonal idempotents. The **Cartan matrix** of A is the $n \times n$ matrix

$$
\mathbf{C}_A = \begin{bmatrix} c_{11} & \cdots & c_{1n} \\ \vdots & \ddots & \vdots \\ c_{n1} & \cdots & c_{nn} \end{bmatrix} \in \mathbb{M}_n(\mathbb{Z}),
$$

where $c_{ji} = \dim_K e_i A e_j$, for $i, j = 1, \ldots, n$.

It follows from the unique decomposition theorem (I.4.10) that if $\mathbf{C}'_A$ is the Cartan matrix of A with respect to another complete set $\{e'_1, \ldots, e'_n\}$ of primitive orthogonal idempotents of A, then $\mathbf{C}'_A$ is obtained from $\mathbf{C}_A$ by a permutation of its rows and columns and therefore the matrices $\mathbf{C}_A$ and $\mathbf{C}'_A$ are $\mathbb{Z}$-conjugate. Throughout, by the Cartan matrix of A we mean the Cartan matrix defined with respect to a given complete set $\{e_1, \ldots, e_n\}$ of primitive orthogonal idempotents of A.

Because we have, by (2.10) and (2.11), K-vector space isomorphisms $e_b A e_a \cong \mathrm{Hom}_A(P(a), P(b)) \cong \mathrm{Hom}_A(I(a), I(b))$, the Cartan matrix of A records the number of linearly independent homomorphisms between the indecomposable projective A-modules and the number of linearly independent homomorphisms between the indecomposable injective A-modules.

We record some elementary facts on the Cartan matrix in the following result.

3.8. Proposition. *Let $\mathbf{C}_A$ be the Cartan matrix of a basic K-algebra $A \cong KQ/\mathcal{I}$.*
  (a) *The ith column of $\mathbf{C}_A$ is $\mathbf{dim}\, P(i)$.*
  (b) *The ith row of $\mathbf{C}_A$ is $[\mathbf{dim}\, I(i)]^t$.*
  (c) $\mathbf{dim}\, P(i) = \mathbf{C}_A \cdot \mathbf{dim}\, S(i)$.
  (d) $\mathbf{dim}\, I(i) = \mathbf{C}_A^t \cdot \mathbf{dim}\, S(i)$.

Proof. The statement (a) follows from the definition and the obvious equality $e_i A e_j = P(i)e_j$ for all i, j. The statement (b) follows from the definition and from the equalities $\dim_K I(i)e_j = \dim_K e_j A e_i = c_{ij}$ for all i, j (apply (2.11)). The equalities (c) and (d) follow from (a), (b), and the fact that the vectors $\mathbf{dim}\, S(1), \ldots, \mathbf{dim}\, S(n)$ form the standard basis of the free abelian group $\mathbb{Z}^n$, where $n = |Q_0|$. $\square$

3.9. Examples. (a) The Cartan matrix of the Kronecker algebra $A = \begin{pmatrix} K & 0 \\ K^2 & K \end{pmatrix}$ has the form $\mathbf{C}_A = \begin{pmatrix} 1 & 2 \\ 0 & 1 \end{pmatrix}$.
  (b) If A is given by the quiver of (2.5)(a), (2.5)(b), (2.5)(c), or (2.5)(d) respectively, then the Cartan matrix $\mathbf{C}_A$ of A is, respectively, the matrix

$$\begin{bmatrix} 1 & 1 & 1 \\ 0 & 1 & 0 \\ 0 & 0 & 1 \end{bmatrix}, \quad \begin{bmatrix} 1 & 2 & 2 \\ 0 & 1 & 2 \\ 0 & 0 & 1 \end{bmatrix}, \quad \begin{bmatrix} 1 & 1 \\ 1 & 1 \end{bmatrix}, \quad \text{or} \quad \begin{bmatrix} 3 & 1 & 3 & 1 \\ 0 & 1 & 0 & 1 \\ 0 & 0 & 1 & 1 \\ 0 & 0 & 0 & 1 \end{bmatrix}.$$

3.10. Proposition. *Let $A \cong KQ/\mathcal{I}$ be an algebra of finite global dimension. Then $\det \mathbf{C}_A \in \{-1, 1\}$. In particular the Cartan matrix $\mathbf{C}_A$ of A is invertible in the matrix ring $\mathbb{M}_n(\mathbb{Z})$, that is, $\mathbf{C}_A \in \mathrm{Gl}(n, \mathbb{Z}) = \{A \in \mathbb{M}_n(\mathbb{Z}); \ \det A \in \{-1, 1\}\}$.*

Proof. Let $n = |Q_0|$ and $a \in Q_0$. By our hypothesis, the simple module $S(a)$ has a projective resolution $0 \to P_{m_a} \to \cdots \to P_1 \to P_0 \to S(a) \to 0$ in $\mathrm{mod}\, A$, where m_a is finite. It follows that $\mathbf{dim}\, S(a) = \sum_{i=0}^{m_a} (-1)^j \mathbf{dim}\, P_j$. By the unique decomposition theorem (I.4.10), each of the modules P_j is the direct sum of finitely many copies of the modules $P(1), \ldots, P(n)$. Therefore the ath standard basis vector $\mathbf{dim}\, S(a)$ of $\mathbb{Z}^n$ is a linear combination of the vectors $\mathbf{dim}\, P(1), \ldots, \mathbf{dim}\, P(n) \in \mathbb{Z}^n$ with integral coefficients. It follows from (3.8)(a) that there exists $B \in \mathbb{M}_n(\mathbb{Z})$ such that

$$E = [\mathbf{dim}\, S(1) \mid \cdots \mid \mathbf{dim}\, S(n)] = [\mathbf{dim}\, P(1) \mid \cdots \mid \mathbf{dim}\, P(n)]B = \mathbf{C}_A B,$$

where E is the identity matrix, and we denote by $[v_1 \mid \ldots \mid v_n]$ the matrix having as respective columns the vectors $v_1, \ldots, v_n \in \mathbb{Z}^n$. Consequently, $\mathbf{C}_A \cdot B = E$ and the result follows. $\qquad \square$

We now use the Cartan matrix $\mathbf{C}_A$ to define a nonsymmetric $\mathbb{Z}$-bilinear form on the group $\mathbb{Z}^n$.

3.11. Definition. Let A be a basic K-algebra of finite global dimension, and let $\mathbf{C}_A$ be the Cartan matrix of A with respect to a complete set $\{e_1, \ldots, e_n\}$ of primitive orthogonal idempotents of A.

The **Euler characteristic** of A is the $\mathbb{Z}$-bilinear (nonsymmetric) form $\langle -, - \rangle_A : \mathbb{Z}^n \times \mathbb{Z}^n \longrightarrow \mathbb{Z}$ defined by $\langle \mathbf{x}, \mathbf{y} \rangle_A = \mathbf{x}^t (\mathbf{C}_A^{-1})^t \mathbf{y}$, for $\mathbf{x}, \mathbf{y} \in \mathbb{Z}^n$.

The **Euler quadratic form of an algebra** A is the quadratic form $q_A : \mathbb{Z}^n \longrightarrow \mathbb{Z}$ defined by $q_A(\mathbf{x}) = \langle \mathbf{x}, \mathbf{x} \rangle_A$, for $\mathbf{x} \in \mathbb{Z}^n$.

The definition makes sense, because the matrix $\mathbf{C}_A$ is invertible in the matrix ring $\mathbb{M}_n(\mathbb{Z})$, by (3.10).

3.12. Examples. (a) If $A = \begin{pmatrix} K & 0 \\ K^2 & K \end{pmatrix}$ is the Kronecker algebra, then $n = 2$, $\mathbf{C}_A = \begin{pmatrix} 1 & 2 \\ 0 & 1 \end{pmatrix}$, $(\mathbf{C}_A^{-1})^t = \begin{pmatrix} 1 & 0 \\ -2 & 1 \end{pmatrix}$, and the Euler characteristic of A is given by $\langle \mathbf{x}, \mathbf{y} \rangle_A = x_1 y_1 + x_2 y_2 - 2 x_1 y_2$.

(b) Let A and B be as in Examples 2.5 (a) and 2.5 (b), respectively. Then $n = 3$,

$$(\mathbf{C}_A^{-1})^t = \begin{bmatrix} 1 & 0 & 0 \\ -1 & 1 & 0 \\ -1 & 0 & 1 \end{bmatrix}, \quad \text{and} \quad (\mathbf{C}_B^{-1})^t = \begin{bmatrix} 1 & 0 & 0 \\ -2 & 1 & 0 \\ 2 & -2 & 1 \end{bmatrix}.$$

Hence the Euler characteristics of A and B are given by

$$
\begin{aligned}
\langle \mathbf{x}, \mathbf{y} \rangle_A &= x_1 y_1 + x_2 y_2 + x_3 y_3 - x_2 y_1 - x_3 y_1, \\
\langle \mathbf{x}, \mathbf{y} \rangle_B &= x_1 y_1 + x_2 y_2 + x_3 y_3 - 2x_2 y_1 + 2x_3 y_1 - 2x_3 y_2.
\end{aligned}
$$

(c) The algebras of Examples 2.5 (c) and 2.5 (d) have infinite global dimension and hence their Euler characteristics are not defined. This follows from (3.10) or directly from the fact that in (2.5)(c), the minimal projective resolution of the simple module $S(1)$ is infinite and has the form

$$
\ldots \to P(1) \to P(2) \to P(1) \to P(2) \to P(1) \to S(1) \to 0.
$$

Similarly, in (2.5)(d), the minimal projective resolution of the simple module $S(1)$ is infinite and has the form

$$
\ldots \to P(1) \to P(1) \to P(1) \to P(1) \to P(1) \to S(1) \to 0.
$$

We also note that the Cartan matrices of these algebras are not invertible over $\mathbb{Z}$.

The following proposition gives a homological interpretation of the Euler characteristic.

3.13. Proposition. *Let A be a basic K-algebra of finite global dimension and $\langle -, - \rangle_A$ be the Euler characteristic of A. Then, for any pair M, N of modules in $\mathrm{mod}\, A$, we have*

(a) $\langle \mathbf{dim}\, M, \mathbf{dim}\, N \rangle_A = \sum\limits_{j=0}^{\infty} (-1)^j \dim_K \mathrm{Ext}_A^j(M, N)$, *and*

(b) $q_A(\mathbf{dim}\, M) = \sum\limits_{j=0}^{\infty} (-1)^j \dim_K \mathrm{Ext}_A^j(M, M)$.

Proof. Because $q_A(\mathbf{dim}\, M) = \langle \mathbf{dim}\, M, \mathbf{dim}\, M \rangle_A$, it is sufficient to prove the statement (a). We prove it by induction on $d = \mathrm{pd}\, M < \infty$. Because both sides of the required equality are additive, we may, without loss of generality, assume that M is indecomposable.

Assume that $d = 0$. Then M is projective, say $M \cong P(i) = e_i A$ for some $i \in \{1, \ldots, n\}$. By (3.8) and (2.11), we have

$$
\begin{aligned}
\langle \mathbf{dim}\, M, \mathbf{dim}\, N \rangle_A &= \langle \mathbf{dim}\, P(i), \mathbf{dim}\, N \rangle_A \\
&= [\mathbf{dim}\, P(i)]^t (\mathbf{C}_A^{-1})^t \mathbf{dim}\, N \\
&= [(\mathbf{C}_A^{-1}) \mathbf{dim}\, P(i)]^t \mathbf{dim}\, N \\
&= [\mathbf{dim}\, S(i)]^t \mathbf{dim}\, N \\
&= \dim_K N e_i \\
&= \dim_K \mathrm{Hom}_A(P(i), N).
\end{aligned}
$$

This shows the statement (a) if $d = 0$. Assume now that $d \geq 1$ and that the result holds for all modules M' with $\mathrm{pd}\, M' = d - 1$. Consider a short exact sequence $0 \to L \to P \to M \to 0$ with P projective. It follows that $\mathrm{pd}\, L = d-1$ and, according to (A.4.5) of the Appendix, the sequence induces a long exact Ext-sequence

$$
\begin{array}{llll}
0 & \xrightarrow{} & \mathrm{Hom}_A(M,N) & \longrightarrow & \mathrm{Hom}_A(P,N) & \longrightarrow & \mathrm{Hom}_A(L,N) \\
& \xrightarrow{\delta_0} & \mathrm{Ext}^1_A(M,N) & \longrightarrow & \mathrm{Ext}^1_A(P,N) & \longrightarrow & \mathrm{Ext}^1_A(L,N) \\
& & \vdots & & \vdots & & \vdots \\
\cdots & \xrightarrow{\delta_{m-1}} & \mathrm{Ext}^m_A(M,N) & \longrightarrow & \mathrm{Ext}^m_A(P,N) & \longrightarrow & \mathrm{Ext}^m_A(L,N) \\
& \xrightarrow{\delta_m} & \mathrm{Ext}^{m+1}_A(M,N) & \longrightarrow & \cdots \, .
\end{array}
$$

Counting dimensions and using the induction hypothesis yields

$$
\begin{aligned}
\langle \dim M, \dim N \rangle_A &= \langle \dim P - \dim L, \dim N \rangle_A \\
&= \langle \dim P, \dim N \rangle_A - \langle \dim L, \dim N \rangle_A \\
&= \sum_{j=0}^{\infty} (-1)^j \dim_K \mathrm{Ext}^j_A(P,N) \\
&\quad - \sum_{j=0}^{\infty} (-1)^j \dim_K \mathrm{Ext}^j_A(L,N) \\
&= \sum_{j=0}^{\infty} (-1)^j \dim_K \mathrm{Ext}^j_A(M,N),
\end{aligned}
$$

because $\dim M = \dim P - \dim L$, by (3.3). This finishes the proof. $\qquad\square$

Another matrix with integral coefficients is useful for us. This is the Coxeter matrix, defined as follows.

3.14. Definition. Let A be a basic K-algebra of finite global dimension, and let $\mathbf{C}_A$ be the Cartan matrix of A with respect to a complete set $\{e_1, \ldots, e_n\}$ of primitive orthogonal idempotents of A. The **Coxeter matrix** of A is the matrix

$$
\boldsymbol{\Phi}_A = -\mathbf{C}^t_A \mathbf{C}^{-1}_A.
$$

The group homomorphism $\boldsymbol{\Phi}_A : \mathbb{Z}^n \longrightarrow \mathbb{Z}^n$ defined by the formula $\boldsymbol{\Phi}_A(\mathbf{x}) = \boldsymbol{\Phi}_A \cdot \mathbf{x}$, for all $\mathbf{x} = [x_1 \ldots x_n]^t \in \mathbb{Z}^n$, is called the **Coxeter transformation** of A.

3.15. Examples. (a) If $A = \left(\begin{smallmatrix} K & 0 \\ K^2 & K \end{smallmatrix} \right)$ is the Kronecker K-algebra, then $\boldsymbol{\Phi}_A = \left(\begin{smallmatrix} -1 & 2 \\ -2 & 3 \end{smallmatrix} \right)$.

(b) Let A be as in Examples 2.5 (a) or 2.5 (b). Then $\boldsymbol{\Phi}_A$ is the matrix

$$
\begin{bmatrix} -1 & 1 & 1 \\ -1 & 0 & 1 \\ -1 & 1 & 0 \end{bmatrix} \quad \text{or} \quad \begin{bmatrix} -1 & 2 & -2 \\ -2 & 3 & -2 \\ -2 & 2 & -1 \end{bmatrix},
$$

respectively.

(c) The algebras of Examples 2.5 (c) and 2.5 (d) have infinite global dimension, hence their Coxeter matrices are not defined.   □

We record some elementary properties of the Coxeter matrix in the following lemma.

3.16. Lemma. (a) $\Phi_A \cdot \mathbf{dim}\, P(i) = -\mathbf{dim}\, I(i)$, for each $i \in \{1, \ldots, n\}$.
(b) $\langle \mathbf{x}, \mathbf{y} \rangle_A = -\langle \mathbf{y}, \Phi_A \mathbf{x} \rangle_A = \langle \Phi_A \mathbf{x}, \Phi_A \mathbf{y} \rangle_A$, for all $\mathbf{x}, \mathbf{y} \in \mathbb{Z}^n$.

Proof. (a) By applying (3.8), we get $\mathbf{dim}\, S(i) = \mathbf{C}_A^{-1} \mathbf{dim}\, P(i)$ and hence $\mathbf{dim}\, I(i) = \mathbf{C}_A^t \mathbf{dim}\, S(i) = \mathbf{C}_A^t \mathbf{C}_A^{-1} \mathbf{dim}\, P(i) = -\Phi_A \cdot \mathbf{dim}\, P(i)$.
(b) $\langle \mathbf{x}, \mathbf{y} \rangle_A = \mathbf{x}^t (\mathbf{C}_A^{-1})^t \mathbf{y} = ((\mathbf{y}^t \mathbf{C}_A^{-1}) \mathbf{x})^t = \mathbf{y}^t \mathbf{C}_A^{-1} \mathbf{x} = \mathbf{y}^t (\mathbf{C}_A^{-1})^t \mathbf{C}_A^t \mathbf{C}_A^{-1} \mathbf{x}$
$= \mathbf{y}^t (\mathbf{C}_A^{-1})^t (-\Phi_A) \mathbf{x} = -\langle \mathbf{y}, \Phi_A \mathbf{x} \rangle_A$. This gives the first equality. The second follows on applying the first twice.   □

Part (a) of (3.16) can be expressed by means of the Nakayama functor ν; see (2.8). Because, according to (2.10), for each $i \in Q_0$, we have $\nu P(i) \cong I(i)$, we deduce that $\Phi_A \cdot \mathbf{dim}\, P = -\mathbf{dim}\, \nu P$, for every projective A-module P.

An application of the Coxeter transformation Φ_A in Auslander–Reiten theory is presented in (IV.2.8) and (IV.2.9) of Chapter IV.

III.4 Exercises

1. Let $M = (M_a, \varphi_\alpha)$ be a K-linear representation of the bound quiver $(Q, \mathcal{I})$. The **support** supp M of M is the full subquiver of Q such that $(\text{supp}\, M)_0 = \{b \in Q_0 \mid M_b \neq 0\}$. Show that if M is indecomposable, then supp M is connected (but the converse is not true).

2. Let Q be a not necessarily acyclic quiver. Show that
(a) There exists an equivalence of categories $\text{Mod}\, KQ \cong \text{Rep}_K(Q)$.
(b) This equivalence restricts to an equivalence $\text{mod}\, KQ \cong \text{rep}_K(Q)$ if and only if Q is acyclic.

3. Let $(Q, \mathcal{I})$ be a bound quiver, $A = KQ/\mathcal{I}$ and Q^{op}, $\mathcal{I}^{\text{op}}$ be as in Exercises 1 and 8 of Chapter II. We have two equivalences of categories $G : \text{rep}_K(Q, \mathcal{I}) \to \text{mod}\, A$, $F : \text{mod}\, A^{\text{op}} \to \text{rep}_K(Q^{\text{op}}, \mathcal{I}^{\text{op}})$ so that we have a duality $FDG : \text{rep}_K(Q, \mathcal{I}) \to \text{rep}_K(Q^{\text{op}}, \mathcal{I}^{\text{op}})$ (with $D = \text{Hom}_K(-, K)$, which we also denote by D).
(a) Let $M = (M_a, \varphi_\alpha)$ be an object in $\text{rep}_K(Q, \mathcal{I})$. For each $a \in Q$ let $M_a^* = \text{Hom}_K(M_a, K)$ be the dual space and, for each $\alpha \in Q_1$, let $\varphi_\alpha^* = \text{Hom}_K(\varphi_\alpha, K)$. Show that $DM \cong (M_a^*, \varphi_\alpha^*)$.
(b) Let $f : M \to M'$ be a morphism in $\text{rep}_K(Q, \mathcal{I})$. Describe the morphism $Df : DM' \to DM$ in $\text{rep}_K(Q^{\text{op}}, \mathcal{I}^{\text{op}})$.

4. In each of the following examples, describe the simple modules, the indecomposable projectives and their radicals, and the indecomposable injectives and their quotients by their socle.

(a) Q : $\mathcal{I} = 0$

(b) Q : $\alpha\beta = 0$

(c) Q : $\mathcal{I} = \mathrm{rad}^2 KQ$

(d) Q : $\gamma\varepsilon = 0 = \delta\varepsilon$

(e) Q : $\mu\alpha = 0, \quad \mu\gamma = 0,$
$\lambda\alpha = 0, \quad \alpha\beta = \gamma\delta$

(f) Q : $\alpha\mu = \nu\gamma, \quad \beta\lambda = \mu\delta,$
$\alpha\beta = 0, \qquad \gamma\delta = 0$

(g) Q : $\alpha\beta = \alpha\gamma$

(h) Q : $\gamma\delta = 0, \quad \beta\gamma = 0,$
$\alpha\beta = 0$

5. Let Q be the quiver

and M be the representation

of Q. Compute top M, soc M, and rad M. Show that the algebra End M is not a field, but that M is indecomposable.

6. Let Q be the quiver $\circ \Longleftarrow \circ$, $n \geq 1$, and $M^{(n)}$ be the representation

$$K[T]/\langle T^n \rangle \xleftarrow[\;\;\chi\;\;]{\;\;1\;\;} K[T]/\langle T^n \rangle$$

of Q, where χ is the K-linear map defined by $\chi(f + \langle T^n \rangle) = T \cdot f + \langle T^n \rangle$ for $f \in K[T]$. Show that End $M^{(n)} \cong K[T]/\langle T^n \rangle$ (hence $M^{(n)}$ is indecomposable).

7. Let Q be the quiver

bound by $\alpha\beta = 0$. Show that the representation

is indecomposable.

8. Let Q be the Kronecker quiver $\circ \Longleftarrow \circ$. We define the representation H_λ of Q by $K \xleftarrow[\;\;\lambda\;\;]{\;\;1\;\;} K$, for every $\lambda \in K$. Show that, for every $\lambda \in K$, H_λ is indecomposable and that $H_\lambda \cong H_\mu$ if and only if $\lambda = \mu$.

9. Let $a \in Q_0$ be a point in a finite quiver $Q = (Q_0, Q_1)$.

(a) Show that the projective KQ-module $P(a)$ is simple if and only if a is a sink.

(b) Show that the injective KQ-module $I(a)$ is simple if and only if a is a source.

(c) Characterise the points $a \in Q_0$ such that rad $P(a)$ is simple.

(d) Characterise the points $a \in Q_0$ such that $I(a)/S(a)$ is simple.

10. Let Q be the quiver $\circ \underset{\beta}{\overset{\alpha}{\rightleftarrows}} \circ$ bound by $\mathcal{I} = \langle \alpha\beta, \beta\alpha \rangle$. Show that the global dimension of the bound quiver algebra $A = KQ/\mathcal{I}$ is infinite, by completing the arguments given in (3.12)(c).

11. Let $Q = (Q_0, Q_1)$ be a finite quiver, $\mathcal{I}$ be an admissible ideal of KQ, and $A = KQ/\mathcal{I}$. For each $a \in Q_0$, let $P(a) = e_a A$. Show that

(a) the top of $P(b)$ is a composition factor of $P(a)$ if and only if there exists a path $w : a \to \cdots \to b$ with $w \notin \mathcal{I}$, and

(b) $a, b \in Q_0$ are in the same connected component of Q if and only if there exists a sequence $a = a_1, a_2, \ldots, a_t = b$ $(t > 1)$ of vertices in Q such that, for each $1 \le i < t$, $P(a_i)$ and $P(a_{i+1})$ have some composition factor in common.

12. Compute the global dimension and the Cartan matrix of each of the algebras of Exercise 4.

13. Let $A = KQ$, where Q is the quiver $1 \circ \underset{\beta}{\overset{\alpha}{\rightleftarrows}} \circ\, 2$. Show that

(a) the A-modules $S(1) = K \underset{0}{\overset{0}{\rightleftarrows}} 0$, $S(2) = 0 \underset{0}{\overset{0}{\rightleftarrows}} K$, and $S(1,2)_\lambda = K \underset{\lambda}{\overset{1}{\rightleftarrows}} K$, with $\lambda \in K$, are simple and that $S(1,2)_\lambda \not\cong S(1,2)_\mu$ whenever $\lambda \ne \mu$, and

(b) every finite dimensional and simple right A-module is isomorphic to $S(1)$, $S(2)$, or to $S(1,2)_\lambda$, where $\lambda \in K$.

Hint: The field K is algebraically closed.

14. Let Q be a finite quiver with at least one cycle. Show that the path algebra $A = KQ$ has infinitely many pairwise nonisomorphic simple modules of finite dimension.

15. Let A be the path K-algebra of the Kronecker quiver $\circ \underset{\beta}{\overset{\alpha}{\rightleftarrows}} \circ$ and M_A be the representation $K[t] \underset{\varphi_\beta}{\overset{\varphi_\alpha}{\longleftarrow}} K[t]$ viewed as a right A-module, where φ_α is the identity map and φ_β is the multiplication by the indeterminate t. Show that the infinite dimensional A-module M_A is indecomposable and the algebra $\operatorname{End} M$ is not local.

Hint: Find K-algebra isomorphisms $\operatorname{End} M \cong \operatorname{End} K[t] \cong K[t]$ and note that the algebra $K[t]$ is not local and has only two idempotents 0 and 1.

16. Assume that Q is a finite and acyclic quiver.

(a) Let $P(a) = (P(a)_b, \varphi_\beta)$ be the indecomposable projective corresponding to $a \in Q_0$. Show that, for each arrow β, the map φ_β is injective.

(b) Dually, let $I(a) = (I(a)_b, \psi_\beta)$ be the indecomposable injective corresponding to $a \in Q_0$. Show that for each arrow β the map ψ_β is surjective.

17. Determine the Coxeter matrix of the K-algebra $A = \left(\begin{smallmatrix} K & K^2 \\ 0 & K \end{smallmatrix} \right)$. Compare it with Example 3.15.

18. Determine the Coxeter matrix of the K-algebras defined in Exercise 15 of Chapter II.

Chapter IV

Auslander–Reiten theory

As we saw in the previous chapter, quiver-theoretical techniques provide a convenient way to visualise finite dimensional algebras and their modules. However, to actually compute the indecomposable modules and the homomorphisms between them, we need other tools. Particularly useful in this context are the notions of irreducible morphisms and almost split sequences. These were introduced by Auslander [13] and Auslander and Reiten [19], [20] while presenting a categorical proof of the first Brauer–Thrall conjecture (see Section 5 and [136] for a historical account). Their main theorem may be stated as follows.

Let A be a finite dimensional K-algebra and N_A be a finite dimensional indecomposable nonprojective A-module. Then there exists a nonsplit short exact sequence

$$0 \longrightarrow L \stackrel{f}{\longrightarrow} M \stackrel{g}{\longrightarrow} N \longrightarrow 0$$

in mod A *such that*

(a) *L is indecomposable noninjective;*

(b) *if $u : L \to U$ is not a section, then there exists $u' : M \to U$ such that $u = u'f$; and*

(c) *if $v : V \to N$ is not a retraction, then there exists $v' : V \to M$ such that $v = gv'$.*

Further, the sequence is uniquely determined up to isomorphism. Dually, if L_A is indecomposable noninjective, a nonsplit short exact sequence as preceding exists, with N indecomposable nonprojective and satisfying the properties (b) and (c). It is again unique up to isomorphism.

Such a sequence is called an almost split sequence ending with N (or starting with L). In this chapter, we introduce the notions of irreducible morphisms and almost split morphisms, then prove the preceding existence theorem for almost split sequences in module categories. This allows us to define a new quiver, called the Auslander–Reiten quiver, which can be considered as a first approximation for the module category. We then apply these results to prove the first Brauer–Thrall conjecture.

Throughout this chapter, we let A denote a finite dimensional K-algebra, K denote an algebraically closed field, and all A-modules are, unless otherwise specified, right finite dimensional A-modules.

IV.1. Irreducible morphisms and almost split sequences

This first section is devoted to introducing the notions of irreducible, minimal, and almost split morphisms in the category $\operatorname{mod} A$ of finite dimensional right A-modules. We recall that the ultimate aim of the representation theory of algebras is, given an algebra A, to describe the finite dimensional A-modules and the homomorphisms between them.

By the unique decomposition theorem (I.4.10), any module in $\operatorname{mod} A$ is a direct sum of indecomposable modules and such a decomposition is unique up to isomorphism and a permutation of its indecomposable summands. It thus suffices to describe the latter and the A-module homomorphisms between them.

Before stating the following definitions, we recall that an A-homomorphism is a section (or a retraction) whenever it admits a left inverse (or a right inverse, respectively).

1.1. Definition. Let L, M, N be modules in $\operatorname{mod} A$.

(a) An A-module homomorphism $f : L \to M$ is called **left minimal** if every $h \in \operatorname{End} M$ such that $hf = f$ is an automorphism.

(b) An A-module homomorphism $g : M \to N$ is called **right minimal** if every $k \in \operatorname{End} M$ such that $gk = g$ is an automorphism.

(c) An A-module homomorphism $f : L \to M$ is called **left almost split** if

(i) f is not a section and

(ii) for every A-homomorphism $u : L \to U$ that is not a section there exists $u' : M \to U$ such that $u'f = u$, that is, u' makes the following triangle commutative

$$
\begin{array}{ccc}
L & \overset{f}{\longrightarrow} & M \\
{\scriptstyle u}\big\downarrow & \swarrow {\scriptstyle u'} & \\
U & &
\end{array}
$$

(d) An A-homomorphism $g : M \to N$ is called **right almost split** if

(i) g is not a retraction and

(ii) for every A-homomorphism $v : V \to N$ that is not a retraction, there exists $v' : V \to M$ such that $gv' = v$, that is, v' makes the following triangle commutative

$$
\begin{array}{ccc}
& & V \\
& {\scriptstyle v'}\swarrow & \big\downarrow {\scriptstyle v} \\
M & \overset{g}{\longrightarrow} & N
\end{array}
$$

(e) An A-module homomorphism $f : L \to M$ is called **left minimal almost split** if it is both left minimal and left almost split.

(f) An A-module homomorphism $g : M \to N$ is called **right minimal almost split** if it is both right minimal and right almost split.

Clearly, each "right-hand" notion is the dual of the corresponding "left-hand" notion. As a first observation, we prove that left (or right) minimal almost split morphisms uniquely determine their targets (or sources, respectively).

1.2. Proposition. (a) *If the A-module homomorphisms $f : L \to M$ and $f' : L \to M'$ are left minimal almost split, then there exists an isomorphism $h : M \to M'$ such that $f' = hf$.*

(b) *If the A-module homomorphisms $g : M \to N$ and $g' : M' \to N$ are right minimal almost split, then there exists an isomorphism $k : M \to M'$ such that $g = g'k$.*

Proof. We only prove (a); the proof of (b) is similar. Because f and f' are almost split, there exist $h : M \to M'$ and $h' : M' \to M$ such that $f' = hf$ and $f = h'f'$. Hence $f = h'hf$ and $f' = hh'f'$. Because f and f' are minimal, hh' and $h'h$ are automorphisms. Consequently, h is an isomorphism. $\qquad\square$

We now see that almost split morphisms are closely related to indecomposable modules.

1.3. Lemma. (a) *If $f : L \to M$ is a left almost split morphism in $\mathrm{mod}\, A$, then the module L is indecomposable.*

(b) *If $g : M \to N$ is a right almost split morphism in $\mathrm{mod}\, A$, then the module N is indecomposable.*

Proof. We only prove (a); the proof of (b) is similar. Assume that $L = L_1 \oplus L_2$, with both L_1 and L_2 nonzero and let $p_i : L \to L_i$ (with $i = 1, 2$) denote the corresponding projections. For any i (with $i = 1, 2$), the homomorphism p_i is not a section. Hence there exists a homomorphism $u_i : M \to L_i$ such that $u_i f = p_i$. But then $u = \begin{bmatrix} u_1 \\ u_2 \end{bmatrix} : M \to L$ satisfies $uf = 1_L$, and this contradicts the fact that f is not a section. $\qquad\square$

1.4. Definition. A homomorphism $f : X \to Y$ in $\mathrm{mod}\, A$ is said to be **irreducible** provided:

(a) f is neither a section nor a retraction and

(b) if $f = f_1 f_2$, either f_1 is a retraction or f_2 is a section

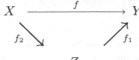

Clearly, this notion is self-dual. An irreducible morphism in $\operatorname{mod} A$ is either a proper monomorphism or a proper epimorphism: indeed, if f : $X \to Y$ is irreducible but is not a proper epimorphism, and $f = jp$ is its canonical factorisation through $\operatorname{Im} f$, then j is not a retraction, and consequently p is a section, so that f is a proper monomorphism. The same argument shows that the irreducible morphisms are precisely those that admit no nontrivial factorisation.

1.5. Example. (a) Let $e \in A$ be a primitive idempotent. Then the right A-module eA is indecomposable and the inclusion $\operatorname{rad} eA \hookrightarrow eA$ is right almost split and is an irreducible morphism. Indeed, if $v \in \operatorname{Hom}_A(V, eA)$ and v is not a retraction, then $\operatorname{Im} v$ is a proper submodule of eA. It follows from (I.4.5)(c) that $\operatorname{Im} v \subseteq \operatorname{rad} eA$, that is, $v : V \to eA$ factors through $\operatorname{rad} eA$, and consequently, $\operatorname{rad} eA \hookrightarrow eA$ is right almost split. It follows from the maximality of $\operatorname{rad} eA$ in eA that $\operatorname{rad} eA \hookrightarrow eA$ is an irreducible morphism.

(b) Let S be a simple A-module, and let $E = E_A(S)$ be the injective envelope of S in $\operatorname{mod} A$. Then the canonical epimorphism $p : E \to E/S$ is left almost split and is an irreducible morphism. This follows from (a) by applying the duality functor $D : \operatorname{mod} A \longrightarrow \operatorname{mod} A^{\operatorname{op}}$ and (I.5.13).

We now reformulate the definition of irreducible morphisms using the notion of **radical** rad_A **of the category** $\operatorname{mod} A$ introduced in Section A.3 of the Appendix.

We recall that $\operatorname{rad}_A = \operatorname{rad}_{\operatorname{mod} A}$ denotes the radical $\operatorname{rad}_{\mathcal{C}}$ of the category $\mathcal{C} = \operatorname{mod} A$. If X and Y are indecomposable modules in $\operatorname{mod} A$, then $\operatorname{rad}_A(X, Y)$ is the K-vector space of all noninvertible homomorphisms from X to Y. Thus, if X is indecomposable, $\operatorname{rad}_A(X, X)$ is just the radical of the local algebra $\operatorname{End} X$. Further, if X and Y are arbitrary modules in $\operatorname{mod} A$, then $\operatorname{rad}_A(X, Y)$ is an $\operatorname{End} Y$–$\operatorname{End} X$-submodule of $\operatorname{Hom}_A(X, Y)$. This implies that $\operatorname{rad}_A(-, -)$ is a subfunctor of the bifunctor $\operatorname{Hom}_A(-, -)$.

Similarly, if X and Y are modules in $\operatorname{mod} A$, we define $\operatorname{rad}_A^2(X, Y)$ to consist of all A-module homomorphisms of the form gf, where $f \in \operatorname{rad}_A(X, Z)$ and $g \in \operatorname{rad}_A(Z, Y)$ for some (not necessarily indecomposable) object Z in $\operatorname{mod} A$. It is clear that $\operatorname{rad}_A^2(X, Y) \subseteq \operatorname{rad}_A(X, Y)$ and even that $\operatorname{rad}_A^2(X, Y)$ is an $\operatorname{End} Y$–$\operatorname{End} X$-submodule of $\operatorname{rad}_A(X, Y)$.

The next lemma shows that the quotient space $\operatorname{rad}_A(X, Y)/\operatorname{rad}_A^2(X, Y)$ measures the number of irreducible morphisms between indecomposable modules X and Y.

1.6. Lemma. *Let X, Y be indecomposable modules in $\operatorname{mod} A$. A morphism $f : X \to Y$ is irreducible if and only if $f \in \operatorname{rad}_A(X, Y) \setminus \operatorname{rad}_A^2(X, Y)$.*

Proof. Assume that f is irreducible. Then, clearly, $f \in \text{rad}_A(X, Y)$. If $f \in \text{rad}_A^2(X, Y)$, then f can be written as $f = gh$, where $h \in \text{rad}_A(X, Z)$ and $g \in \text{rad}_A(Z, Y)$ for some Z in mod A. Decomposing Z into indecomposable summands as $Z = \bigoplus_{i=1}^{t} Z_i$, we can write $h = \begin{bmatrix} h_1 \\ \vdots \\ h_t \end{bmatrix} : X \longrightarrow \bigoplus_{i=1}^{t} Z_i$ and $g = [g_1 \ldots g_t] : \bigoplus_{i=1}^{t} Z_i \longrightarrow Y$. Because f is irreducible, h is a section or g is a retraction. Assume the former, and let $h' = [h'_1 \ldots h'_t] : \bigoplus_{i=1}^{t} Z_i \longrightarrow X$ be such that $1_X = h'h = \sum_{i=1}^{t} h'_i h_i$. Because h_i is not invertible (for any i), $h'_i h_i$ is not invertible either, and so $h'_i h_i \in \text{rad}_A(X, X) = \text{rad End } X$. Because $\text{End } X$ is local, we infer that $1_X \in \text{rad End } X$, a contradiction. Consequently, h is not a section. Similarly, g is not a retraction. This contradiction shows that $f \notin \text{rad}_A^2(X, Y)$.

Conversely, assume that $f \in \text{rad}_A(X, Y) \backslash \text{rad}_A^2(X, Y)$. Because X, Y are indecomposable and f is not an isomorphism, it is clearly neither a section nor a retraction. Suppose that $f = gh$, where $h : X \to Z$, $g : Z \to Y$. Decompose Z into indecomposable summands as $Z = \bigoplus_{i=1}^{t} Z_i$ and write

$$h = \begin{bmatrix} h_1 \\ \vdots \\ h_t \end{bmatrix} : X \longrightarrow \bigoplus_{i=1}^{t} Z_i \text{ and } g = [g_1 \ldots g_t] : \bigoplus_{i=1}^{t} Z_i \longrightarrow Y$$

so that $f = \sum_{i=1}^{t} g_i h_i$. Because $f \notin \text{rad}_A^2(X, Y)$, there is either an index i such that h_i is invertible or an index j such that g_j is invertible. In the first case, h is a section; in the second, g is a retraction. $\square$

In the following lemma, we characterise irreducible monomorphisms (or epimorphisms) in mod A by means of their cokernels (or kernels, respectively).

1.7 Lemma. *Let $0 \to L \xrightarrow{f} M \xrightarrow{g} N \to 0$ be a nonsplit short exact sequence in* mod A.

(a) *The homomorphism $f : L \to M$ is irreducible if and only if, for every homomorphism $v : V \to N$, there exists $v_1 : V \to M$ such that $v = gv_1$ or $v_2 : M \to V$ such that $g = vv_2$.*

(b) *The homomorphism $g : M \to N$ is irreducible if and only if, for every homomorphism $u : L \to U$, there exists $u_1 : M \to U$ such that*

$u = u_1 f$ *or* $u_2 : U \to M$ *such that* $f = u_2 u$.

Proof. We only prove (a); the proof of (b) is similar. Assume first that $f : L \to M$ is irreducible, and let $v : V \to N$ be arbitrary. We have a commutative diagram

$$
\begin{array}{ccccccccc}
0 & \longrightarrow & L & \xrightarrow{f'} & E & \xrightarrow{g'} & V & \longrightarrow & 0 \\
& & \downarrow{\scriptstyle 1_L} & & \downarrow{\scriptstyle u} & & \downarrow{\scriptstyle v} & & \\
0 & \longrightarrow & L & \xrightarrow{f} & M & \xrightarrow{g} & N & \longrightarrow & 0
\end{array}
$$

with exact rows, where E denotes the fibered product of V and M over N. Because $f = uf'$ is irreducible, f' is a section or u is a retraction. In the first case, g' is a retraction and there exists $v_1 : V \to M$ such that $gv_1 = v$. If $u' : V \to E$ is such that $g'u' = 1_V$, then $v_1 = uu'$ satisfies $gv_1 = v$. In the second case, there exists $v_2 : M \to V$ such that $g = vv_2$.

Conversely, assume that the stated condition is satisfied. Because the given sequence is not split, f is neither a section nor a retraction. Suppose that $f = f_1 f_2$, where $f_2 : L \to U$, $f_1 : U \to M$. Because f is a monomorphism, so is f_2 and we have a commutative diagram

$$
\begin{array}{ccccccccc}
0 & \longrightarrow & L & \xrightarrow{f_2} & U & \xrightarrow{u} & V & \longrightarrow & 0 \\
& & \downarrow{\scriptstyle 1_L} & & \downarrow{\scriptstyle f_1} & & \downarrow{\scriptstyle v} & & \\
0 & \longrightarrow & L & \xrightarrow{f} & M & \xrightarrow{g} & N & \longrightarrow & 0
\end{array}
$$

with exact rows, where $V = \operatorname{Coker} f_2$. In particular, by (A.5.3) of the Appendix, the module U is isomorphic to the fibered product of V and M over N. If there exists $v_1 : V \to M$ such that $v = gv_1$, then the universal property of the fibered product implies that u is a retraction and so f_2 is a section. If there exists $v_2 : M \to V$ such that $g = vv_2$, then, similarly, f_1 is a retraction. This shows that f is irreducible. $\qquad\square$

As a first application of Lemma 1.7, we show that irreducible morphisms provide a useful method to construct indecomposable modules.

1.8. Corollary. (a) *If $f : L \to M$ is an irreducible monomorphism, then $N = \operatorname{Coker} f$ is indecomposable.*

(b) *If $g : M \to N$ is an irreducible epimorphism, then $L = \operatorname{Ker} g$ is indecomposable.*

Proof. We only prove (a); the proof of (b) is similar. Let $g : M \to N$ be the cokernel of f and assume that $N = N_1 \oplus N_2$ with N_1 and N_2 nonzero. Let $q_i : N_i \to N$ (with $i = 1, 2$) denote the corresponding inclusions. If there exists a morphism $u_i : M \to N_i$ such that $g = q_i u_i$, then, because g is an epimorphism, q_i is also an epimorphism and hence an isomorphism, contrary to the fact that $N_1 \neq 0$ and $N_2 \neq 0$. Then, by (1.7), there exists, for each $i = 1, 2$, a homomorphism $v_i : N_i \to M$ such that $gv_i = q_i$. Thus $v = [v_1 \; v_2] : N_1 \oplus N_2 \to M$ satisfies $gv = 1_N$, so that g is a retraction. But then f is a section, and this contradicts the fact that f is irreducible. $\square$

The following easy lemma is needed in the proof of the next theorem.

1.9. Lemma. (a) *Let $f : L \to M$ be a nonzero A-module homomorphism, with L indecomposable. Then f is not a section if and only if* $\operatorname{Im} \operatorname{Hom}_A(f, L) \subseteq \operatorname{rad} \operatorname{End} L$.

(b) *Let $g : M \to N$ be a nonzero A-module homomorphism, with N indecomposable. Then g is not a retraction if and only if* $\operatorname{Im} \operatorname{Hom}_A(N, g) \subseteq \operatorname{rad} \operatorname{End} N$.

Proof. We prove (a); the proof of (b) is similar. Because L is indecomposable, $\operatorname{End} L$ is local. If $\operatorname{Im} \operatorname{Hom}_A(f, L) \not\subseteq \operatorname{rad} \operatorname{End} L$, there exists $h : M \to L$ such that $k = \operatorname{Hom}_A(f, L)(h) = hf$ is invertible. But then $k^{-1}hf = 1_L$ shows that f is a section. Conversely, if there exists h such that $hf = 1_L$, then $\operatorname{Hom}_A(f, L)(h) = 1_L$ shows that $\operatorname{Hom}_A(f, L)$ is an epimorphism. $\square$

We now relate the previous notions, showing that one may think of irreducible morphisms as components of minimal almost split morphisms.

1.10. Theorem. (a) *Let $f : L \to M$ be left minimal almost split in* mod A. *Then f is irreducible. Further, a homomorphism $f' : L \to M'$ of A-modules is irreducible if and only if $M' \neq 0$ and there exists a direct sum decomposition $M \cong M' \oplus M''$ and a homomorphism $f'' : L \to M''$ such that*
$$\begin{bmatrix} f' \\ f'' \end{bmatrix} : L \longrightarrow M' \oplus M'' \text{ is left minimal almost split.}$$

(b) *Let $g : M \to N$ be right minimal almost split in* mod A. *Then g is irreducible. Further, a homomorphism $g' : M' \to N$ of A-modules is irreducible if and only if $M' \neq 0$ and there exists a direct sum decomposition $M \cong M' \oplus M''$ and a homomorphism $g'' : M'' \to N$ such that $[g' \; g''] : M' \oplus M'' \longrightarrow N$ is right minimal almost split.*

Proof. We prove (a); the proof of (b) is similar. Let $f : L \to M$ be a left minimal almost split homomorphism in mod A. By definition, f is not a section. Because, by (1.3), L is indecomposable and f is not an isomorphism, f is not a retraction either. Assume that $f = f_1 f_2$, where $f_2 : L \to X$ and $f_1 : X \to M$. We suppose that f_2 is not a section and prove that f_1 is a retraction. Because f is left almost split, there exists $f_2' : M \to X$ such that $f_2 = f_2' f$. Hence $f = f_1 f_2 = f_1 f_2' f$. Because f is left minimal, $f_1 f_2'$ is an automorphism and so f_1 is a retraction. This proves the first statement.

Let now $f' : L \to M'$ be an irreducible morphism in $\operatorname{mod} A$. Then clearly, $M' \neq 0$. Also, f' is not a section, hence there exists $h : M \to M'$ such that $f' = hf$. Because f' is irreducible and f is not a section, h is a retraction. Let $M'' = \operatorname{Ker} h$. Then there exists a homomorphism $q : M \to M''$ such that $\begin{bmatrix} h \\ q \end{bmatrix} : M \to M' \oplus M''$ is an isomorphism. It follows that $\begin{bmatrix} h \\ q \end{bmatrix} f = \begin{bmatrix} f' \\ qf \end{bmatrix} : L \to M' \oplus M''$ is left minimal almost split.

Conversely, assume that f' satisfies the stated condition; we must show that it is irreducible. Because L is indecomposable and f' is not an isomorphism, f' is not a retraction. On the other hand, if there exists h such that $hf' = 1_L$, then $[h \ 0] \begin{bmatrix} f' \\ f'' \end{bmatrix} = 1_L$ implies that $\begin{bmatrix} f' \\ f'' \end{bmatrix}$ is a section, a contradiction. Thus, f' is not a section. Assume that $f' = f_1 f_2$, where $f_2 : L \to X$ and $f_1 : X \to M'$. We suppose that f_2 is not a section and show that f_1 is a retraction. We have $\begin{bmatrix} f' \\ f'' \end{bmatrix} = \begin{bmatrix} f_1 & 0 \\ 0 & 1 \end{bmatrix} \begin{bmatrix} f_2 \\ f'' \end{bmatrix}$, where $\begin{bmatrix} f_2 \\ f'' \end{bmatrix} : L \to X \oplus M''$ and $\begin{bmatrix} f_1 & 0 \\ 0 & 1 \end{bmatrix} : X \oplus M'' \to M' \oplus M''$. Because f_2 is not a section, it follows from (1.9) that $\operatorname{Im} \operatorname{Hom}_A(f_2, L) \subseteq \operatorname{rad} \operatorname{End} L$. Similarly $\operatorname{Im} \operatorname{Hom}_A(f'', L) \subseteq \operatorname{rad} \operatorname{End} L$. Consequently, $\operatorname{Im} \operatorname{Hom}_A(\begin{bmatrix} f_2 \\ f'' \end{bmatrix}, L) \subseteq \operatorname{rad} \operatorname{End} L$, hence, again by (1.9), $\begin{bmatrix} f_2 \\ f'' \end{bmatrix}$ is not a section. Because $\begin{bmatrix} f' \\ f'' \end{bmatrix}$ is left minimal almost split and hence irreducible, $\begin{bmatrix} f_1 & 0 \\ 0 & 1 \end{bmatrix}$ is a retraction, and this implies that f_1 is a retraction. The proof is now complete. $\square$

We now define a particular type of short exact sequence, which is particularly useful in the representation theory of algebras.

1.11. Definition. A short exact sequence in $\operatorname{mod} A$

$$0 \longrightarrow L \xrightarrow{f} M \xrightarrow{g} N \longrightarrow 0$$

is called an **almost split sequence** provided:
   (a) f is left minimal almost split and
   (b) g is right minimal almost split.

While the existence of almost split sequences is far from obvious, it follows from (1.3) that if such a sequence exists, then L and N are indecomposable modules. Also, an almost split sequence is never split (because f is not a section and g is not a retraction) so that L is not injective, and N is not projective. Finally, an almost split sequence is uniquely determined (up to isomorphism) by each of its end terms; indeed, if $0 \to L \to M \to N \to 0$ and $0 \to L' \to M' \to N' \to 0$ are two almost split sequences in mod A, then (1.2) implies that the following assertions are equivalent:

(a) The two sequences are isomorphic.

(b) There is an isomorphism $L \cong L'$ of A-modules.

(c) There is an isomorphism $N \cong N'$ of A-modules.

1.12. Lemma. *Let*

$$
\begin{array}{ccccccccc}
0 & \longrightarrow & L & \xrightarrow{f} & M & \xrightarrow{g} & N & \longrightarrow & 0 \\
 & & \downarrow{\scriptstyle u} & & \downarrow{\scriptstyle v} & & \downarrow{\scriptstyle w} & & \\
0 & \longrightarrow & L & \xrightarrow{f} & M & \xrightarrow{g} & N & \longrightarrow & 0
\end{array}
$$

be a commutative diagram in mod A, *where the rows are exact and not split.*

(a) *If L is indecomposable and w is an automorphism, then u and hence v are automorphisms.*

(b) *If N is indecomposable and u is an automorphism, then w and hence v are automorphisms.*

Proof. We only prove (a); the proof of (b) is similar. We may suppose that $w = 1_N$. If u is not an isomorphism, it must be nilpotent (because End L is local) and so there exists m such that $u^m = 0$. Then $v^m f = f u^m = 0$ and so v^m factors through the cokernel N of f, that is, there exists $h : N \to M$ such that $v^m = hg$. Because $gv^m = g$, we deduce that $ghg = g$ and consequently $gh = 1_N$ (because g is an epimorphism). This contradicts the fact that the given sequence is not split. $\qquad\square$

We end this section by giving several equivalent characterisations of almost split sequences.

1.13. Theorem. *Let* $0 \longrightarrow L \xrightarrow{f} M \xrightarrow{g} N \longrightarrow 0$ *be a short exact sequence in* mod A. *The following assertions are equivalent:*

(a) *The given sequence is almost split.*

(b) *L is indecomposable, and g is right almost split.*

(c) *N is indecomposable, and f is left almost split.*

(d) *The homomorphism f is left minimal almost split.*

(e) *The homomorphism g is right minimal almost split.*

(f) *L and N are indecomposable, and f and g are irreducible.*

Proof. By definition of almost split sequence, (a) implies (d) and (e). By (1.3), (a) implies (b) and (c). By (1.10) and (1.3), (a) implies (f) as well. To prove the equivalence of the first five conditions, we start by proving that (e) implies (b). Dually, (d) implies (c). Thus, the equivalence of the first three conditions implies that of the first five conditions. We show that (b) implies (c); the proof that (c) implies (b) is similar, and we prove that both conditions together imply (a). Finally, we show that (f) implies (b), which will complete the proof of the theorem.

Assume (e), that is, g is right minimal almost split. By (1.10), g is irreducible. Hence, by (1.8), $L = \operatorname{Ker} g$ is indecomposable. Thus, (e) implies (b).

To show that (b) implies (c), it suffices, by (1.3), to show that f is left almost split. Because g is not a retraction, f is not a section. Let $u : L \to U$ be such that $u'f \neq u$ for all $u' : M \to U$. We must prove that u is a section. It follows from (A.5.3) of the Appendix that there exists a commutative diagram

$$\begin{array}{ccccccccc} 0 & \longrightarrow & L & \xrightarrow{\ f\ } & M & \xrightarrow{\ g\ } & N & \longrightarrow & 0 \\ & & \downarrow u & & \downarrow v & & \downarrow 1_N & & \\ 0 & \longrightarrow & U & \xrightarrow{\ h\ } & V & \xrightarrow{\ k\ } & N & \longrightarrow & 0 \end{array}$$

with exact rows, where V is the amalgamed sum. The lower sequence is not split and hence k is not a retraction. Because g is right almost split, there exists $\overline{v} : V \to M$ such that $k = g\overline{v}$, and hence we get a commutative diagram

$$\begin{array}{ccccccccc} 0 & \longrightarrow & L & \xrightarrow{\ f\ } & M & \xrightarrow{\ g\ } & N & \longrightarrow & 0 \\ & & \downarrow \overline{u}u & & \downarrow \overline{v}v & & \downarrow 1_N & & \\ 0 & \longrightarrow & L & \xrightarrow{\ f\ } & M & \xrightarrow{\ g\ } & N & \longrightarrow & 0 \end{array}$$

with exact rows, where $\overline{u}$ is derived from $\overline{v}$ and 1_N by passing to the kernels. By (1.12), $\overline{u}u$ is an automorphism. Hence u is a section.

Now, assume that both (b) and (c) hold; we must prove that f and g are minimal. To prove that f is left minimal, let $h \in \operatorname{End} M$ be such that $hf = f$. We have a commutative diagram

$$\begin{array}{ccccccccc} 0 & \longrightarrow & L & \xrightarrow{\ f\ } & M & \xrightarrow{\ g\ } & N & \longrightarrow & 0 \\ & & \downarrow 1_L & & \downarrow h & & \downarrow 1_N & & \\ 0 & \longrightarrow & L & \xrightarrow{\ f\ } & M & \xrightarrow{\ g\ } & N & \longrightarrow & 0 \end{array}$$

with exact rows. By (1.12), h is an automorphism. Hence f is left minimal. Similarly, g is right minimal.

We now prove that (f) implies (b). By hypothesis, L is indecomposable and g is not a retraction. Assume that $v : V \to N$ is not a retraction. We may suppose that V is indecomposable (replacing it, if necessary, by one of its indecomposable summands). Because f is irreducible, (1.7) gives $v' : V \to M$ such that $v = gv'$ (and then we are done), or else $h : M \to V$ such that $g = vh$. But in this case, because g is irreducible and v is not a retraction, h must be a section. Because V is indecomposable, h is an isomorphism. But then $v' = h^{-1}$ satisfies $v = gv'$ and we have completed the proof of our theorem. $\square$

IV.2. The Auslander–Reiten translations

In this section and the next, we prove the existence of almost split sequences in the category mod A of finite dimensional A-modules, for A a finite dimensional K-algebra. We first consider the A-dual functor

$$(-)^t = \mathrm{Hom}_A(-, A) : \quad \mathrm{mod}\, A \longrightarrow \mathrm{mod}\, A^{\mathrm{op}}.$$

We note that if P_A is a projective right A-module, then $P^t = \mathrm{Hom}_A(P, A)$ is a projective left A-module; indeed, if $P_A \cong eA$, with $e \in A$ a primitive idempotent, then $P^t = \mathrm{Hom}_A(eA, A) \cong Ae$, and our statement thus follows from the additivity of $(-)^t$. Moreover, one shows easily that the evaluation homomorphism $\epsilon_M : M \to M^{tt}$ defined by $\epsilon_M(z)(f) = f(z)$ (for $z \in M$ and $f \in M^t$) is functorial in M and is an isomorphism whenever M is projective. Thus, the functor $(-)^t$ induces a duality, also denoted by $(-)^t$, between the category proj A of projective right A-modules, and the category proj A^{op} of projective left A-modules. We use this new duality to define a duality on an appropriate quotient of mod A, and this duality is called the transposition.

We start by approximating each module M_A by projective modules. Let thus
$$P_1 \xrightarrow{p_1} P_0 \xrightarrow{p_0} M \longrightarrow 0$$
be a minimal projective presentation of M, that is, an exact sequence such that $p_0 : P_0 \to M$ and $p_1 : P_1 \to \mathrm{Ker}\, p_0$ are projective covers. Applying the (left exact, contravariant) functor $(-)^t$, we obtain an exact sequence of left A-modules
$$0 \longrightarrow M^t \xrightarrow{p_0^t} P_0^t \xrightarrow{p_1^t} P_1^t \longrightarrow \mathrm{Coker}\, p_1^t \longrightarrow 0.$$
We denote $\mathrm{Coker}\, p_1^t$ by $\mathrm{Tr}\, M$ and call it the **transpose** of M.

We observe that the left A-module $\mathrm{Tr}\, M$ is uniquely determined up to isomorphism; this indeed follows from the fact that projective covers (and hence minimal projective presentations) are uniquely determined up to isomorphism.

We now summarise the main properties of the transpose Tr.

2.1. Proposition. *Let M be an indecomposable module in* mod A.
(a) *The left A-module* $\mathrm{Tr}\, M$ *has no nonzero projective direct summands.*
(b) *If M is not projective, then the sequence*
$$P_0^t \xrightarrow{p_1^t} P_1^t \longrightarrow \mathrm{Tr}\, M \longrightarrow 0$$
induced from the minimal projective presentation $P_1 \xrightarrow{p_1} P_0 \xrightarrow{p_0} M \to 0$ *of M is a minimal projective presentation of the left A-module* $\mathrm{Tr}\, M$.

(c) M *is projective if and only if* $\operatorname{Tr} M = 0$. *If* M *is not projective, then* $\operatorname{Tr} M$ *is indecomposable and* $\operatorname{Tr}(\operatorname{Tr} M) \cong M$.

(d) *If* M *and* N *are indecomposable nonprojective, then* $M \cong N$ *if and only if* $\operatorname{Tr} M \cong \operatorname{Tr} N$.

Proof. If M is projective, then the term P_1 in the minimal projective presentation of M is zero, and therefore $\operatorname{Tr} M = 0$. Conversely, if $\operatorname{Tr} M = 0$, then p_1^t is an epimorphism, hence a retraction (because $_A(P_1^t)$ is projective). Thus, p_1 is a section, and M is projective. This shows the first part of (c).

Assume that M is not projective. Then $\operatorname{Tr} M \neq 0$. The sequence given in (b) is certainly a projective presentation of the left module $\operatorname{Tr} M$. We claim it is minimal. Indeed, if this is not the case, there exist nontrivial direct sum decompositions $P_0^t = E_0' \oplus E_0''$, $P_1^t = E_1' \oplus E_1''$ and an isomorphism $v : E_0'' \xrightarrow{\cong} E_1''$ such that this sequence is isomorphic to the sequence

$$E_0' \oplus E_0'' \xrightarrow{\begin{bmatrix} u & 0 \\ 0 & v \end{bmatrix}} E_1' \oplus E_1'' \longrightarrow \operatorname{Tr} M \longrightarrow 0,$$

where $u : E_0' \to E_1'$ is a homomorphism of left A-modules. But then applying $(-)^t$ yields a projective presentation of M of the form

$$E_1'^t \xrightarrow{u^t} E_0'^t \longrightarrow M \longrightarrow 0,$$

and this contradicts the minimality of the projective presentation of M. This shows our claim. Moreover, if $\operatorname{Tr} M$ has a nonzero projective direct summand, the homomorphism p_1^t has a direct summand of the form $(0 \to E)$, with $_AE$ projective. But, as earlier, this implies that p_1 has a direct summand of the form $(E^t \to 0)$, and we obtain another contradiction. We have thus shown (a) and (b).

Applying now $(-)^t$ to the exact sequence in (b), we get a commutative diagram

$$
\begin{array}{ccccccc}
P_1 & \xrightarrow{p_1} & P_0 & \xrightarrow{p_0} & M & \longrightarrow & 0 \\
{\scriptstyle \varepsilon_{P_1}} \downarrow {\scriptstyle \cong} & & {\scriptstyle \varepsilon_{P_0}} \downarrow {\scriptstyle \cong} & & & & \\
P_1^{tt} & \xrightarrow{p_1^{tt}} & P_0^{tt} & \xrightarrow{p_0^{tt}} & \operatorname{Tr}\operatorname{Tr} M & \longrightarrow & 0
\end{array}
$$

with exact rows. Hence there is an isomorphism $M \cong \operatorname{Tr}\operatorname{Tr} M$ making the right square commutative. This proves (c), and (d) follows immediately. $\square$

We have shown that the transpose Tr maps modules of $\operatorname{mod} A$ to modules of $\operatorname{mod} A^{\mathrm{op}}$ but does not define a duality $\operatorname{mod} A \to \operatorname{mod} A^{\mathrm{op}}$, because it annihilates the projectives. In order to make this correspondence a duality, we thus need to annihilate the projectives from $\operatorname{mod} A$ and $\operatorname{mod} A^{\mathrm{op}}$. This motivates the following construction.

For two A-modules M, N, let $\mathcal{P}(M, N)$ denote the subset of $\operatorname{Hom}_A(M, N)$ consisting of all homomorphisms that factor through a projective A-module.

We claim that this defines an ideal $\mathcal{P}$ in the category $\operatorname{mod} A$. First, for two modules M, N, the set $\mathcal{P}(M, N)$ is a subspace of the K-vector space $\operatorname{Hom}_A(M, N)$; indeed, if $f, f' \in \mathcal{P}(M, N)$, then f and f' can be respectively written as $f = hg$ and $f' = h'g'$, where the targets P of g and P' of g' are projective; consequently

$$f + f' = hg + h'g' = [h \ h'] \begin{bmatrix} g \\ g' \end{bmatrix}$$

factors through the projective module $P \oplus P'$. On the other hand, if $\lambda \in K$ and $f \in \mathcal{P}(M, N)$, then $\lambda f \in \mathcal{P}(M, N)$. Next, if $f \in \mathcal{P}(L, M)$ and $g \in \operatorname{Hom}_A(M, N)$, then $gf \in \mathcal{P}(L, N)$ and similarly, if $f \in \operatorname{Hom}_A(L, M)$ and $g \in \mathcal{P}(M, N)$, then $gf \in \mathcal{P}(L, N)$. This completes the proof that $\mathcal{P}$ is an ideal of $\operatorname{mod} A$.

We may thus consider the quotient category

$$\underline{\operatorname{mod}} \ A = \operatorname{mod} A / \mathcal{P}$$

called the **projectively stable category**. Its objects are the same as those of $\operatorname{mod} A$, but the K-vector space $\underline{\operatorname{Hom}}_A(M, N)$ of morphisms from M to N in $\underline{\operatorname{mod}} \ A$ is defined to be the quotient vector space

$$\underline{\operatorname{Hom}}_A(M, N) = \operatorname{Hom}_A(M, N) / \mathcal{P}(M, N)$$

of $\operatorname{Hom}_A(M, N)$ with the composition of morphisms induced from the composition in $\operatorname{mod} A$. There clearly exists a functor $\operatorname{mod} A \to \underline{\operatorname{mod}} \ A$ that is the identity on objects and associates to a homomorphism $f : M \to N$ in $\operatorname{mod} A$ its residual class modulo $\mathcal{P}(M, N)$ in $\underline{\operatorname{mod}} \ A$.

Dually, one may construct an ideal $\mathcal{I}$ in $\operatorname{mod} A$ by considering, for each pair (M, N) of A-modules, the K-subspace $\mathcal{I}(M, N)$ of $\operatorname{Hom}_A(M, N)$ consisting of all homomorphisms that factor through an injective A-module. The quotient category

$$\overline{\operatorname{mod}} \ A = \operatorname{mod} A / \mathcal{I}$$

is called the **injectively stable category**. Its objects are the same as those of $\operatorname{mod} A$, but the K-vector space $\overline{\operatorname{Hom}}_A(M, N)$ of morphisms from M to N in $\overline{\operatorname{mod}} \ A$ is given by the quotient vector space

$$\overline{\operatorname{Hom}}_A(M, N) = \operatorname{Hom}_A(M, N) / \mathcal{I}(M, N)$$

of $\operatorname{Hom}_A(M, N)$ with the composition of morphisms induced from the composition in $\operatorname{mod} A$. One again defines in the obvious way the residual class functor $\operatorname{mod} A \to \overline{\operatorname{mod}} \ A$.

We now see that, although the correspondence $M \mapsto \operatorname{Tr} M$ does not define a duality between $\operatorname{mod} A$ and $\operatorname{mod} A^{\mathrm{op}}$, it does define one between the quotient categories $\underline{\mathrm{mod}}\, A$ and $\underline{\mathrm{mod}}\, A^{\mathrm{op}}$.

2.2. Proposition. *The correspondence* $M \mapsto \mathrm{Tr}\, M$ *induces a K-linear duality functor* $\mathrm{Tr} : \underline{\mathrm{mod}}\, A \longrightarrow \underline{\mathrm{mod}}\, A^{\mathrm{op}}$.

Proof. To construct this duality, we start by giving an alternative construction of $\underline{\mathrm{mod}}\, A$ as a quotient category. Let $\overrightarrow{\mathrm{proj}}\, A$ denote the category whose objects are the triples (P_1, P_0, f), where P_1, P_0 are projective A-modules, and $f : P_1 \to P_0$ is a homomorphism in $\mathrm{mod}\, A$. (The notation $\overrightarrow{\mathrm{proj}}\, A$ is meant to suggest that we are dealing with homomorphisms between projective modules.) We define a morphism $(P_1, P_0, f) \longrightarrow (P_1', P_0', f')$ to be a pair (u_1, u_0) of homomorphisms in $\mathrm{mod}\, A$ such that $u_1 : P_1 \to P_1'$ and $u_0 : P_0 \to P_0'$ satisfy $f'u_1 = u_0 f$, that is, the following square is commutative

$$
\begin{array}{ccc}
P_1 & \xrightarrow{\;f\;} & P_0 \\
\downarrow{\scriptstyle u_1} & & \downarrow{\scriptstyle u_0} \\
P_1' & \xrightarrow{\;f'\;} & P_0'
\end{array}
$$

The composition of the morphisms $(u_1, u_0) : (P_1, P_0, f) \longrightarrow (P_1', P_0', f')$ and $(u_1', u_0') : (P_1', P_0', f') \longrightarrow (P_1'', P_0'', f'')$ in the category $\overrightarrow{\mathrm{proj}}\, A$ is defined by the formula $(u_1', u_0')(u_1, u_0) = (u_1'u_1, u_0'u_0)$.

Let now $F : \overrightarrow{\mathrm{proj}}\, A \longrightarrow \underline{\mathrm{mod}}\, A$ denote the composition of the cokernel functor $\overrightarrow{\mathrm{proj}}\, A \longrightarrow \mathrm{mod}\, A$, given by $(P_1, P_0, f) \mapsto \mathrm{Coker}\, f$, with the residual class functor $\mathrm{mod}\, A \longrightarrow \underline{\mathrm{mod}}\, A$. Let $(u_1, u_0) : (P_1, P_0, f) \longrightarrow (P_1', P_0', f')$ be a morphism in $\overrightarrow{\mathrm{proj}}\, A$. We claim that $F(u_1, u_0) = 0$ if and only if there exists $w : P_0 \to P_1'$ such that $f'wf = u_0 f$. The situation can be visualised in the following diagram

$$
\begin{array}{ccc}
P_1 & \xrightarrow{\;f\;} & P_0 \\
\downarrow{\scriptstyle u_1} & {\scriptstyle w} & \downarrow{\scriptstyle u_0} \\
P_1' & \xrightarrow{\;f'\;} & P_0'
\end{array}
$$

Indeed, assume that such a homomorphism w exists and consider the commutative diagram

$$
\begin{array}{ccccccccc}
P_1 & \xrightarrow{\;f\;} & P_0 & \xrightarrow{\;g\;} & M & \longrightarrow & 0 \\
\downarrow{\scriptstyle u_1} & {\scriptstyle w} & \downarrow{\scriptstyle u_0} & {\scriptstyle v} & \downarrow{\scriptstyle u} \\
P_1' & \xrightarrow{\;f'\;} & P_0' & \xrightarrow{\;g'\;} & M' & \longrightarrow & 0
\end{array}
$$

with exact rows, where M and M' denote the cokernels of f and f', respectively, and u is induced from u_1 and u_0 by passing to the cokernels. Because $(u_0 - f'w)f = 0$, there exists $v : M \to P_0'$ such that $u_0 - f'w = vg$. But then $g'vg = g'u_0 = ug$ gives $g'v = u$ (because g is an epimorphism). Hence $u \in \mathcal{P}(M, M')$ and $F(u_1, u_0) = 0$. Conversely, assume that $F(u_1, u_0) = 0$. This means that the homomorphism u induced from u_1 and u_0 by passing to the respective cokernels of f and f' factors through a projective module. Because g' is an epimorphism, this implies the existence of $v : M \to P_0'$ such that $u = g'v$. But then $g'(u_0 - vg) = g'u_0 - g'vg = g'u_0 - ug = 0$ and there exists $w : P_0 \to P_1'$ such that $f'w = u_0 - vg$. Hence $f'wf = u_0 f$ and we have proved our claim.

This implies at once that the class $\overrightarrow{\mathrm{proj}}_1 A$ of those morphisms (u_1, u_0) in $\overrightarrow{\mathrm{proj}}\, A$ such that $F(u_1, u_0) = 0$ forms an ideal in $\overrightarrow{\mathrm{proj}}\, A$. To see this, assume that $(u_1, u_0) : (P_1, P_0, f) \to (P_1', P_0', f')$ is a morphism in $\overrightarrow{\mathrm{proj}}_1 A$ and let $(v_1, v_0) : (P_1', P_0', f') \to (P_1'', P_0'', f'')$ be any morphism in $\overrightarrow{\mathrm{proj}}\, A$. It follows from the preceding claim that there exists $w : P_0 \to P_1'$ such that $f'wf = u_0 f$. But then $v_1 w : P_0 \to P_1''$ satisfies $f''(v_1 w)f = (f''v_1)wf = (v_0 f')wf = (v_0 u_0)f$ so that $(v_1 u_1, v_0 u_0)$ belongs to $\overrightarrow{\mathrm{proj}}_1 A$. Similarly, if (u_1, u_0) is as earlier and $(w_1, w_0) : (Q_1, Q_0, g) \to (P_1, P_0, f)$ is any morphism in $\overrightarrow{\mathrm{proj}}\, A$, then $(u_1 w_1, u_0 w_0)$ belongs to $\overrightarrow{\mathrm{proj}}_1 A$.

The foregoing considerations imply that the category $\underline{\mathrm{mod}}\, A$ is equivalent to the quotient of $\overrightarrow{\mathrm{proj}}\, A$ modulo $\overrightarrow{\mathrm{proj}}_1 A$. Indeed, if M is an object in $\underline{\mathrm{mod}}\, A$, then we can write $M = F(P_1, P_0, f)$, where $P_1 \xrightarrow{f} P_0 \longrightarrow M \longrightarrow 0$ is a minimal projective presentation of M and, given a morphism $u : M \to M'$ in $\underline{\mathrm{mod}}\, A$, where $M = F(P_1, P_0, f)$ and $M' = F(P_1', P_0', f')$, there exists a morphism $(u_1, u_0) : (P_1, P_0, f) \to (P_1', P_0', f')$ in $\overrightarrow{\mathrm{proj}}\, A$ making the following diagram commutative

$$
\begin{array}{ccccccc}
P_1 & \xrightarrow{\ f\ } & P_0 & \longrightarrow & M & \longrightarrow & 0 \\
\downarrow{\scriptstyle u_1} & & \downarrow{\scriptstyle u_0} & & \downarrow{\scriptstyle u} & & \\
P_1' & \xrightarrow{\ f'\ } & P_0' & \longrightarrow & M' & \longrightarrow & 0
\end{array}
$$

(where the rows are minimal projective presentations), that is, $u = F(u_1, u_0)$. The morphism u equals zero in $\underline{\mathrm{mod}}\, A$ if and only if $F(u_1, u_0) = 0$, that is, if and only if (u_1, u_0) belongs to $\overrightarrow{\mathrm{proj}}_1 A$. This shows that we have an "exact" sequence

$$
0 \longrightarrow \overrightarrow{\mathrm{proj}}_1 A \longrightarrow \overrightarrow{\mathrm{proj}}\, A \xrightarrow{\ F\ } \underline{\mathrm{mod}}\, A \longrightarrow 0.
$$

We are now in a position to construct a duality $\underline{\mathrm{mod}}\, A \to \underline{\mathrm{mod}}\, A^{\mathrm{op}}$ induced by the correspondence $M \mapsto \mathrm{Tr}\, M$.

The duality $(-)^t : \mathrm{proj}\, A \xrightarrow{\ F\ } \mathrm{proj}\, A^{\mathrm{op}}$ induces obviously a duality $\overrightarrow{\mathrm{proj}}\, A \xrightarrow{\ F\ } \overrightarrow{\mathrm{proj}}\, A^{\mathrm{op}}$ given by the formula $(P_1, P_0, f) \mapsto (P_0^t, P_1^t, f^t)$. We also denote this duality by $(-)^t$. Now we claim that the restriction of $(-)^t$ to $\overrightarrow{\mathrm{proj}}_1 A$ induces a duality $\overrightarrow{\mathrm{proj}}_1 A \longrightarrow \overrightarrow{\mathrm{proj}}_1 A^{\mathrm{op}}$. Indeed, let $(u_1, u_0) : (P_1, P_0, f) \to (P_1', P_0', f')$ belong to $\overrightarrow{\mathrm{proj}}_1 A$; we must show that $(u_0^t, u_1^t) : (P_0'^t, P_1'^t, f'^t) \to (P_0^t, P_1^t, f^t)$ belongs to $\overrightarrow{\mathrm{proj}}_1 A^{\mathrm{op}}$. But the hypothesis implies the existence of a homomorphism $w : P_0 \to P_1'$ such that $f'wf = u_0 f$. Hence $f^t w^t f'^t = f^t u_0^t = u_1^t f'^t$, and the conclusion follows.

We thus have a diagram with "exact rows" and commutative left square

$$
\begin{array}{ccccccccc}
0 & \longrightarrow & \overrightarrow{\mathrm{proj}_1}\,A & \longrightarrow & \overrightarrow{\mathrm{proj}}\,A & \longrightarrow & \underline{\mathrm{mod}}\,A & \longrightarrow & 0 \\
& & \downarrow{\scriptstyle(-)^t} & & \downarrow{\scriptstyle(-)^t} & & \big\downarrow & & \\
0 & \longrightarrow & \overrightarrow{\mathrm{proj}_1}\,A^{\mathrm{op}} & \longrightarrow & \overrightarrow{\mathrm{proj}}\,A^{\mathrm{op}} & \longrightarrow & \underline{\mathrm{mod}}\,A^{\mathrm{op}} & \longrightarrow & 0
\end{array}
$$

We define $\mathrm{Tr} : \underline{\mathrm{mod}}\,A \longrightarrow \underline{\mathrm{mod}}\,A^{\mathrm{op}}$ to be the unique functor that makes the right square commutative, namely, if $M = F(P_1, P_0, f)$, we set $\mathrm{Tr}\,M = F(P_0^t, P_1^t, f^t)$ and if $u : M \to M'$ is a morphism in $\underline{\mathrm{mod}}\,A$, where $M = F(P_1, P_0, f)$ and $M' = F(P_1', P_0', f')$, there exists a commutative diagram

$$
\begin{array}{ccccccc}
P_1 & \xrightarrow{\ f\ } & P_0 & \longrightarrow & M & \longrightarrow & 0 \\
\downarrow{\scriptstyle u_1} & & \downarrow{\scriptstyle u_0} & & \downarrow{\scriptstyle u} & & \\
P_1' & \xrightarrow{\ f'\ } & P_0' & \longrightarrow & M' & \longrightarrow & 0
\end{array}
$$

with exact rows. Applying the functor $(-)^t$ yields a commutative diagram

$$
\begin{array}{ccccccc}
P_0^t & \xrightarrow{\ f^t\ } & P_1^t & \longrightarrow & \mathrm{Tr}\,M & \longrightarrow & 0 \\
\uparrow{\scriptstyle u_0^t} & & \uparrow{\scriptstyle u_1^t} & & \big\uparrow & & \\
P_0'^t & \xrightarrow{\ f'^t\ } & P_1'^t & \longrightarrow & \mathrm{Tr}\,M' & \longrightarrow & 0
\end{array}
$$

with exact rows and a commutative left square. Let $\mathrm{Tr}\,u : \mathrm{Tr}\,M' \to \mathrm{Tr}\,M$ be the unique homomorphism that makes the right square commutative. It follows easily from these considerations that

$$
\mathrm{Tr} : \underline{\mathrm{mod}}\,A \longrightarrow \underline{\mathrm{mod}}\,A^{\mathrm{op}}
$$

is a well-defined functor and, in fact, a duality. $\qquad\square$

The duality Tr defined in (2.2) is called the **transposition**. It transforms right A-modules into left A-modules and conversely. Thus, if we wish to define an endofunctor of $\mathrm{mod}\,A$, we need to compose it with another duality between right and left A-modules, namely the standard duality $D = \mathrm{Hom}_K(-, K)$.

2.3. Definition. The **Auslander–Reiten translations** are defined to be the compositions of D with Tr, namely, we set

$$
\tau = D\mathrm{Tr} \qquad \text{and} \qquad \tau^{-1} = \mathrm{Tr}\,D.
$$

In view of the importance of the translations in the sequel, we present in the following proposition a construction method for the Auslander–Reiten translate of a module.

We first recall that the **Nakayama functor** (see (III.2.8)),
$$\nu = D(-)^t = D\mathrm{Hom}_A(-, A) : \mathrm{mod}\, A \longrightarrow \mathrm{mod}\, A,$$
induces two equivalences of categories $\mathrm{proj}\, A \underset{\nu^{-1}}{\overset{\nu}{\rightleftarrows}} \mathrm{inj}\, A$, where $\nu^{-1} = \mathrm{Hom}_A(DA, -)$ is quasi-inverse to ν.

2.4. Proposition. (a) *Let* $P_1 \xrightarrow{p_1} P_0 \xrightarrow{p_0} M \longrightarrow 0$ *be a minimal projective presentation of an A-module M. Then there exists an exact sequence*
$$0 \longrightarrow \tau M \longrightarrow \nu P_1 \xrightarrow{\nu p_1} \nu P_0 \xrightarrow{\nu p_0} \nu M \longrightarrow 0.$$

(b) *Let* $0 \longrightarrow N \xrightarrow{i_0} E_0 \xrightarrow{i_1} E_1$ *be a minimal injective presentation of an A-module N. Then there exists an exact sequence*
$$0 \longrightarrow \nu^{-1}N \xrightarrow{\nu^{-1}i_0} \nu^{-1}E_0 \xrightarrow{\nu^{-1}i_1} \nu^{-1}E_1 \longrightarrow \tau^{-1}N \longrightarrow 0.$$

Proof. (a) Applying successively the functors $(-)^t$ and D to the given minimal projective presentation of M, we obtain an exact sequence
$$0 \longrightarrow D\mathrm{Tr}\, M \longrightarrow \nu P_1 \xrightarrow{\nu p_1} \nu P_0 \xrightarrow{\nu p_0} \nu M \longrightarrow 0$$
and (a) follows.

(b) Applying successively the functors D and $(-)^t$ to the given minimal injective presentation of N, we obtain an exact sequence
$$0 \longrightarrow (DN)^t \xrightarrow{(Di_0)^t} (DE_0)^t \xrightarrow{(Di_1)^t} (DE_1)^t \longrightarrow \mathrm{Tr}\, DN \longrightarrow 0.$$
For any A-module X we have a composed functorial isomorphism
$$(DX)^t \cong \mathrm{Hom}_{A^{\mathrm{op}}}(DX, A) \cong \mathrm{Hom}_A(DA, DDX) \cong \mathrm{Hom}_A(DA, X) \cong \nu^{-1}X.$$
This isomorphism induces a commutative diagram
$$\begin{array}{ccccccc}
0 \longrightarrow & (DN)^t & \xrightarrow{(Di_0)^t} & (DE_0)^t & \xrightarrow{(Di_1)^t} & (DE_1)^t & \longrightarrow \mathrm{Tr}\, DN \longrightarrow 0 \\
& \downarrow{\cong} & & \downarrow{\cong} & & \downarrow{\cong} & \\
0 \longrightarrow & \nu^{-1}N & \xrightarrow{\nu^{-1}i_0} & \nu^{-1}E_0 & \xrightarrow{\nu^{-1}i_1} & \nu^{-1}E_1 &
\end{array}$$
with exact rows. Hence (b) follows. $\qquad\square$

2.5. Example. Let A be given by the Kronecker quiver $1 \circ \underset{\beta}{\overset{\alpha}{\rightleftarrows}} \circ 2$ and M_A be the representation $K \underset{0}{\overset{1}{\rightleftarrows}} K$, where 1 denotes, as usual, the identity homomorphism and 0 the zero homomorphism. Then M is indecomposable; indeed, an endomorphism f of M is given by a pair (a_1, a_2) of scalars such that $a_1 \cdot 1 = 1 \cdot a_2$ and $a_1 \cdot 0 = 0 \cdot a_2$. These two conditions yield $f = a \cdot 1_M$, where $a = a_1 = a_2 \in K$. Thus $\mathrm{End}\, M_A \cong K$ and so M is indecomposable. A minimal projective presentation of M_A is given by
$$0 \longrightarrow P(1) \xrightarrow{p_1} P(2) \xrightarrow{p_2} M_A \longrightarrow 0,$$

where $P(1) = S(1) = (K \xleftarrow{\hspace{2cm}} 0)$ and $P(2) = (K^2 \xleftarrow[\hspace{0.3cm}\begin{smallmatrix}[0]\\[1]\end{smallmatrix}\hspace{0.3cm}]{\begin{smallmatrix}[1]\\[0]\end{smallmatrix}} K)$
are the indecomposable projective A-modules, p_1 is an isomorphism of $P(1)$ onto the direct summand of $\operatorname{rad} P(2)$ equal to $\begin{bmatrix}0\\1\end{bmatrix} K \xleftarrow{\hspace{1.5cm}} 0$, and p_2 is its cokernel homomorphism. Thus, in particular, M_A is not projective. By (2.4)(a), applying the Nakayama functor ν to this exact sequence, we get a short exact sequence

$$0 \longrightarrow \tau M \longrightarrow I(1) \xrightarrow{\nu p_1} I(2) \longrightarrow 0,$$

where $I(1) = (K \xleftarrow[{[0\,1]}]{[1\,0]} K^2)$ and $I(2) = S(2) = (0 \xleftarrow{\hspace{1.2cm}} K)$ are the indecomposable injective A-modules. An obvious computation shows that the homomorphism νp_1 induces an isomorphism of the quotient module of $I(1)$ defined by $0 \xleftarrow{\hspace{1cm}} \begin{bmatrix}0\\1\end{bmatrix} K)$ onto $I(2)$. Then $\tau M = \operatorname{Ker} \nu p_1$ is given by $K \xleftarrow[0]{1} K$, that is, $\tau M \cong M$.

2.6. Example. Let A be given by the quiver

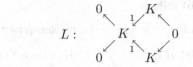

bound by $\alpha\beta = \gamma\delta$, $\delta\mu = 0$, and $\beta\lambda = 0$. Take the simple injective module

$$S(6): \quad \begin{array}{ccc} 0 & & 0 \\ & \searrow & \swarrow & \searrow \\ & 0 & & K \\ & \swarrow & \searrow & \swarrow \\ 0 & & 0 \end{array}$$

The projective cover of $S(6)$ is $P(6)$ and the kernel L of the canonical epimorphism $P(6) \to S(6)$ is the indecomposable module

$$L: \quad \begin{array}{ccc} 0 & & K \\ & \searrow & \xrightarrow{1} & \searrow \\ & K & & 0 \\ & \swarrow & \searrow & \swarrow \\ 0 & \xrightarrow{1} & K \end{array}$$

Because the top of L is isomorphic to $S(4) \oplus S(5)$, then the projective cover of L is isomorphic to $P(4) \oplus P(5)$ and therefore the module $S(6)$ has a minimal projective presentation of the form $P(4) \oplus P(5) \xrightarrow{p_1} P(6) \xrightarrow{p_2} S(6) \longrightarrow 0$ (see (I.5.8)). By (2.4)(a), applying the functor ν to the exact sequence, we get an exact sequence $0 \longrightarrow \tau S(6) \longrightarrow I(4) \oplus I(5) \xrightarrow{\nu p_1} I(6) \longrightarrow 0$, because $\nu p_1 \neq 0$ and $I(6) = S(6)$ is simple. Hence we get

$$\tau S(6): \quad \begin{array}{ccc} & 0 \quad K & \\ & \nwarrow \nearrow \nwarrow 1 & \\ 0 & & K \\ & \swarrow \nearrow \swarrow 1 & \\ & 0 \quad K & \end{array}$$

and obviously $\tau S(6) \not\cong S(6)$.

This proposition yields at once an easy and useful criterion for a module to have projective, or injective, dimension at most one.

2.7. Lemma. *Let M be a module in $\mod A$.*
(a) $\mathrm{pd}_A M \le 1$ *if and only if* $\mathrm{Hom}_A(DA, \tau M) = 0$.
(b) $\mathrm{id}_A M \le 1$ *if and only if* $\mathrm{Hom}_A(\tau^{-1} M, A) = 0$.

Proof. We only prove (a); the proof of (b) is similar. Applying the left exact functor $\nu^{-1} = \mathrm{Hom}_A(DA, -)$ to the exact sequence

$$0 \longrightarrow \tau M \longrightarrow \nu P_1 \xrightarrow{\nu p_1} \nu P_0 \xrightarrow{\nu p_0} \nu M \longrightarrow 0$$

given in (2.4) we obtain a commutative diagram

$$
\begin{array}{ccccccccc}
0 & \longrightarrow & \nu^{-1}\tau M & \longrightarrow & \nu^{-1}\nu P_1 & \longrightarrow & \nu^{-1}\nu P_0 & & \\
 & & & & \downarrow \cong & & \downarrow \cong & & \\
0 & \longrightarrow & \mathrm{Ker}\, p_1 & \longrightarrow & P_1 & \xrightarrow{p_1} & P_0 & \xrightarrow{p_0} M \longrightarrow 0
\end{array}
$$

with exact rows. Thus $\mathrm{Hom}_A(DA, \tau M) = \nu^{-1}\tau M \cong \mathrm{Ker}\, p_1$ vanishes if and only if $\mathrm{pd}\, M \le 1$. $\qquad\square$

The previous results yield formulas for the dimension vector of the Auslander–Reiten translate in terms of the Coxeter transformation $\mathbf{\Phi}_A :$ $\mathbb{Z}^n \longrightarrow \mathbb{Z}^n$ of any algebra A of finite global dimension (see (III.3.14)).

2.8. Lemma. (a) *Let M be an indecomposable nonprojective module in $\mod A$ and $P_1 \xrightarrow{p_1} P_0 \xrightarrow{p_0} M \longrightarrow 0$ be a minimal projective presentation of M. Then*

$$\mathbf{\dim}\, \tau M = \mathbf{\Phi}_A(\mathbf{\dim}\, M) - \mathbf{\Phi}_A(\mathbf{\dim}\, \mathrm{Ker}\, p_1) + \mathbf{\dim}\, \nu M.$$

(b) *Let N be an indecomposable noninjective module in $\mod A$ and let $0 \longrightarrow N \xrightarrow{i_0} E_0 \xrightarrow{i_1} E_1$ be a minimal injective presentation of N. Then*

$$\mathbf{\dim}\, \tau^{-1} N = \mathbf{\Phi}_A^{-1}(\mathbf{\dim}\, N) - \mathbf{\Phi}_A^{-1}(\mathbf{\dim}\, \mathrm{Coker}\, i_1) + \mathbf{\dim}\, \nu^{-1} N.$$

Proof. We only prove (a); the proof of (b) is similar. The exact sequence $0 \longrightarrow \mathrm{Ker}\, p_1 \longrightarrow P_1 \xrightarrow{p_1} P_0 \xrightarrow{p_0} M \longrightarrow 0$ yields

$$\mathbf{\dim}\, M - \mathbf{\dim}\, \mathrm{Ker}\, p_1 = -\mathbf{\dim}\, P_1 + \mathbf{\dim}\, P_0.$$

Applying the Coxeter transformation Φ_A and using (III.3.16)(a), we get

$$\Phi_A(\dim M) - \Phi_A(\dim \operatorname{Ker} p_1) = \dim \nu\, P_1 - \dim \nu\, P_0.$$

Then the injective presentation $0 \longrightarrow \tau M \longrightarrow \nu P_1 \longrightarrow \nu P_0 \longrightarrow \nu M \longrightarrow 0$
of τM yields $\dim \tau M = \dim \nu P_1 - \dim \nu P_0 + \dim \nu M = \Phi_A(\dim M) -$
$\Phi_A(\dim \operatorname{Ker} p_1) + \dim \nu M$. $\square$

2.9. Corollary. (a) *If M is an indecomposable module in* $\operatorname{mod} A$ *such
that* $\operatorname{pd}_A M \le 1$ *and* $\operatorname{Hom}_A(M, A) = 0$, *then* $\dim \tau M = \Phi_A(\dim M)$.

(b) *If N is an indecomposable module in* $\operatorname{mod} A$ *such that* $\operatorname{id}_A N \le 1$
and $\operatorname{Hom}_A(DA, N) = 0$, *then* $\dim \tau^{-1} N = \Phi_A^{-1}(\dim N)$.

Proof. We only prove (a); the proof of (b) is similar. By our assumption, M is not projective and $\nu M = D\operatorname{Hom}_A(M, A) = 0$. Then (a) is a
consequence of (2.8), because $\operatorname{pd}_A M \le 1$ implies $\operatorname{Ker} p_1 = 0$, in the notation of (2.8). $\square$

The following proposition records some of the most elementary properties of Auslander–Reiten translations.

2.10. Proposition. *Let M and N be indecomposable modules in*
$\operatorname{mod} A$.

(a) *The module τM is zero if and only if M is projective.*

(a′) *The module $\tau^{-1} N$ is zero if and only if N is injective.*

(b) *If M is a nonprojective module, then τM is indecomposable noninjective and $\tau^{-1}\tau M \cong M$.*

(b′) *If N is a noninjective module, then $\tau^{-1} N$ is indecomposable nonprojective and $\tau\tau^{-1} N \cong N$.*

(c) *If M and N are nonprojective, then $M \cong N$ if and only if there is
an isomorphism $\tau M \cong \tau N$.*

(c′) *If M and N are noninjective, then $M \cong N$ if and only if there is
an isomorphism $\tau^{-1} M \cong \tau^{-1} N$.*

Proof. Because the translations τ and τ^{-1} are compositions of the
transposition Tr and the duality D, the proposition follows directly from
(2.1), (I.5.13), and the definitions. A detailed proof is left as an exercise
(see (IV.7.25)). $\square$

2.11. Corollary. *The Auslander–Reiten translations τ and τ^{-1} induce
mutually inverse equivalences* $\underline{\operatorname{mod}} A \xrightleftharpoons[\tau^{-1}]{\tau} \overline{\operatorname{mod}} A$.

Proof. This follows directly from (2.2) and (2.10). $\square$

For an A-module X, we consider the functorial homomorphism

$$\varphi^X : (-) \otimes_A X^t \longrightarrow \operatorname{Hom}_A(X, -)$$

defined on a module Y_A by

$$\varphi_Y^X \;:\; \begin{array}{ccc} Y \otimes_A X^t & \longrightarrow & \mathrm{Hom}_A(X,Y) \\ y \otimes f & \mapsto & (x \mapsto yf(x)), \end{array}$$

where $x \in X$, $y \in Y$ and $f \in X^t$. It is easily seen that if X is projective, then φ^X is a functorial isomorphism and that if Y is projective, then φ_Y^X is an isomorphism. We prove that the cokernel of φ_Y^X coincides with $\underline{\mathrm{Hom}}_A(X,Y)$.

2.12. Lemma. *For any A-modules X and Y, there is an exact sequence*

$$Y \otimes_A X^t \xrightarrow{\;\varphi_Y^X\;} \mathrm{Hom}_A(X,Y) \longrightarrow \underline{\mathrm{Hom}}_A(X,Y) \longrightarrow 0$$

with all homomorphisms functorial in both variables.

Proof. For an A-module Y, let $f : P \to Y$ be an epimorphism with P projective. We claim that for any A-module X, there is an exact sequence

$$\mathrm{Hom}_A(X,P) \xrightarrow{\;\mathrm{Hom}_A(X,f)\;} \mathrm{Hom}_A(X,Y) \longrightarrow \underline{\mathrm{Hom}}_A(X,Y) \longrightarrow 0.$$

Indeed, it is sufficient to show that $\mathrm{Im}\,\mathrm{Hom}_A(X,f) = \mathcal{P}(X,Y)$. Because, clearly, $\mathrm{Im}\,\mathrm{Hom}_A(X,f) \subseteq \mathcal{P}(X,Y)$, we take $g \in \mathcal{P}(X,Y)$. By definition, there exist a projective module P_A' and homomorphisms $g_2 : X \to P'$, $g_1 : P' \to Y$ such that $g = g_1 g_2$. Because $f : P \to Y$ is an epimorphism and P' is projective, there exists $h : P' \to P$ such that $g_1 = fh$. Then $g = g_1 g_2 = fh g_2 = \mathrm{Hom}_A(X,f)(hg_2) \in \mathrm{Im}\,\mathrm{Hom}_A(X,f)$ and we have proved our claim.

Because $\varphi_P^X : P \otimes_A X^t \to \mathrm{Hom}_A(X,P)$ is an isomorphism and φ^X is functorial, we have a commutative diagram

$$\begin{array}{ccccc}
P \otimes_A X^t & \xrightarrow{\;f \otimes X^t\;} & Y \otimes_A X^t & \longrightarrow & 0 \\
{\scriptstyle \varphi_P^X}\downarrow {\scriptstyle \cong} & & {\scriptstyle \varphi_Y^X}\downarrow & & \\
\mathrm{Hom}_A(X,P) & \xrightarrow{\;\mathrm{Hom}_A(X,f)\;} & \mathrm{Hom}_A(X,Y) & \longrightarrow & \underline{\mathrm{Hom}}_A(X,Y) \longrightarrow 0
\end{array}$$

with exact rows. Consequently

$$\begin{aligned}
\mathrm{Im}\,\varphi_Y^X &= \varphi_Y^X(f \otimes X^t)(P \otimes X^t) \\
&= \mathrm{Hom}_A(X,f)\varphi_P^X(P \otimes X^t) \\
&\cong \mathrm{Im}\,\mathrm{Hom}_A(X,f) = \mathcal{P}(X,Y)
\end{aligned}$$

and therefore $\mathrm{Coker}\,\varphi_Y^X \cong \underline{\mathrm{Hom}}_A(X,Y)$. $\square$

2.13. Theorem (the Auslander–Reiten formulas). *Let A be a K-algebra and M, N be two A-modules in $\mathrm{mod}\,A$. Then there exist isomorphisms*

$$\operatorname{Ext}_A^1(M, N) \;\cong\; D\underline{\operatorname{Hom}}_A(\tau^{-1}N, M) \;\cong\; D\overline{\operatorname{Hom}}_A(N, \tau M)$$

that are functorial in both variables.

Proof. We only prove the first isomorphism; the proof of the second is similar. Clearly, it suffices to prove our claim for modules N having no injective direct summand. In view of (2.10), we can suppose that $N = \tau L$, where $L = \tau^{-1}N$. Let $P_1 \xrightarrow{p_1} P_0 \xrightarrow{p_0} L \longrightarrow 0$ be a minimal projective presentation of L. Applying the functor $\nu = D(-)^t$, we obtain the exact sequence (see (2.4)(a))

$$0 \longrightarrow \tau L \longrightarrow DP_1^t \xrightarrow{Dp_1^t} DP_0^t \xrightarrow{Dp_0^t} DL^t \longrightarrow 0,$$

where both DP_1^t and DP_0^t are injective. The functor $\operatorname{Hom}_A(M, -)$ yields the complex

$$0 \to \operatorname{Hom}_A(M, \tau L) \to \operatorname{Hom}_A(M, DP_1^t) \xrightarrow{\overline{p}_1} \operatorname{Hom}_A(M, DP_0^t) \xrightarrow{\overline{p}_0} \operatorname{Hom}_A(M, DL^t),$$

where, for brevity, we write $\overline{p}_1$ for $\operatorname{Hom}_A(M, Dp_1^t)$ and $\overline{p}_0$ for $\operatorname{Hom}_A(M, Dp_0^t)$. Thus we have

$$\operatorname{Ext}_A^1(M, N) = \operatorname{Ext}_A^1(M, \tau L) = \operatorname{Ker} \overline{p}_0 / \operatorname{Im} \overline{p}_1.$$

On the other hand, applying the right exact functor $D\operatorname{Hom}_A(-, M)$ to the minimal projective presentation of L yields an exact sequence

$$D\operatorname{Hom}_A(P_1, M) \xrightarrow{\widetilde{p}_1} D\operatorname{Hom}_A(P_0, M) \xrightarrow{\widetilde{p}_0} D\operatorname{Hom}_A(L, M) \longrightarrow 0,$$

where, for brevity, we write $\widetilde{p}_1$ for $D\operatorname{Hom}_A(p_1, M)$ and $\widetilde{p}_0$ for $D\operatorname{Hom}_A(p_0, M)$. Now associated to an A-module X there exists a functorial morphism $\varphi^X : (-) \otimes_A X^t \longrightarrow \operatorname{Hom}_A(X, -)$ introduced earlier. The composition of the dual homomorphism $D\varphi^X : D\operatorname{Hom}_A(X, -) \longrightarrow D((-) \otimes_A X^t)$ with the adjunction isomorphism $\eta^X : D((-) \otimes_A X^t) \xrightarrow{\simeq} \operatorname{Hom}_A(-, DX^t)$ yields a functorial morphism

$$\omega^X = \eta^X D\varphi^X : D\operatorname{Hom}_A(X, -) \longrightarrow \operatorname{Hom}_A(-, DX^t),$$

which is an isomorphism whenever X is projective. We thus have a commutative diagram with exact lower row

$$
\begin{array}{ccccc}
\operatorname{Hom}_A(M, DP_1^t) & \xrightarrow{\overline{p}_1} & \operatorname{Hom}_A(M, DP_0^t) & \xrightarrow{\overline{p}_0} & \operatorname{Hom}_A(M, DL^t) \\
\omega_M^{P_1} \big\uparrow \cong & & \omega_M^{P_0} \big\uparrow \cong & & \omega_M^{L} \big\uparrow \\
D\operatorname{Hom}_A(P_1, M) & \xrightarrow{\widetilde{p}_1} & D\operatorname{Hom}_A(P_0, M) & \xrightarrow{\widetilde{p}_0} & D\operatorname{Hom}_A(L, M) \longrightarrow 0
\end{array}
$$

The homomorphism $\widetilde{p}_0 (\omega_M^{P_0})^{-1}$ of A-modules induces a homomorphism $\psi : \operatorname{Ker} \overline{p}_0 \to \operatorname{Ker} \omega_M^{L}$. Because $\widetilde{p}_0$ is an epimorphism and $\omega_M^{P_0}$ an isomorphism,

ψ must be an epimorphism. Because $\operatorname{Ker}\widetilde{p}_0 = \operatorname{Im}\widetilde{p}_1$ and the maps $\omega_M^{P_0}$, $\omega_M^{P_1}$ are isomorphisms, we deduce that $\operatorname{Ker}\psi \cong \operatorname{Im}\overline{p}_1$. Consequently, we have

$$\begin{aligned}
\operatorname{Ker}\overline{p}_0/\operatorname{Im}\overline{p}_1 &\cong &\operatorname{Ker}\overline{p}_0/\operatorname{Ker}\psi &\cong &\operatorname{Ker}\omega_M^L\\
&= &\operatorname{Ker} D\varphi_M^L &\cong &D\operatorname{Coker}\varphi_M^L.
\end{aligned}$$

Thus there exist an isomorphism $\operatorname{Ext}_A^1(M,N) \cong D\operatorname{Coker}\varphi_M^L$ and, by (2.12), $\operatorname{Coker}\varphi_M^L \cong \underline{\operatorname{Hom}}_A(L,M) = \underline{\operatorname{Hom}}_A(\tau^{-1}N, M)$. The proof is complete. $\quad\square$

2.14. Corollary. *Let A be a K-algebra and M, N be two modules in* mod A.

(a) *If* pd $M \le 1$ *and N is arbitrary, then there exists a K-linear isomorphism*
$$\operatorname{Ext}_A^1(M,N) \cong D\operatorname{Hom}_A(N,\tau M).$$

(b) *If* id $N \le 1$ *and M is arbitrary, then there exists a K-linear isomorphism*
$$\operatorname{Ext}_A^1(M,N) \cong D\operatorname{Hom}_A(\tau^{-1}N, M).$$

Proof. The Auslander–Reiten formulas (2.13) give an isomorphism $\operatorname{Ext}_A^1(M,N) \cong D\overline{\operatorname{Hom}}_A(N,\tau M)$. Now pd $M \le 1$ gives $\operatorname{Hom}_A(DA,\tau M) = 0$ (by (2.7)). Hence $\mathcal{I}(N,\tau M) = 0$, because every injective module in mod A is a direct summand of $(DA)^s$, for some $s \ge 1$. Consequently, $\overline{\operatorname{Hom}}_A(N,\tau M) = \operatorname{Hom}_A(N,\tau M)$ and (a) follows. The proof of (b) is similar to that of (a). $\quad\square$

2.15. Corollary. *Let A be a K-algebra and M, N be two modules in* mod A.

(a) *If* pd $M \le 1$ *and* id $N \le 1$*, then there exists a K-linear isomorphism*

$$\operatorname{Hom}_A(N,\tau M) \cong \operatorname{Hom}_A(\tau^{-1}N, M).$$

(b) *If* pd $M \le 1$*,* id $\tau N \le 1$ *and N is indecomposable nonprojective, then there is a K-linear isomorphism*

$$\operatorname{Hom}_A(\tau N, \tau M) \cong \operatorname{Hom}_A(N, M).$$

(c) *If* pd $\tau^{-1}M \le 1$*,* id $N \le 1$ *and M is indecomposable noninjective, then there is a K-linear isomorphism*

$$\operatorname{Hom}_A(\tau^{-1}N, \tau^{-1}M) \cong \operatorname{Hom}_A(N, M).$$

Proof. The statement (a) is an immediate consequence of (2.14). Finally, (b) and (c) follow from (a) and (2.10). $\quad\square$

IV.3. The existence of almost split sequences

We are now able, using the results of Section 2, to prove the main existence theorem for almost split sequences, due to Auslander and Reiten. In this section, as in the previous one, we let A denote a fixed finite dimensional K-algebra, and we denote by rad_A the radical of the category $\mathrm{mod}\,A$.

3.1. Theorem. (a) *For any indecomposable nonprojective A-module M_A, there exists an almost split sequence $0 \to \tau M \to E \to M \to 0$ in* $\mathrm{mod}\,A$.

(b) *For any indecomposable noninjective A-module N_A, there exists an almost split sequence $0 \to N \to F \to \tau^{-1}N \to 0$ in* $\mathrm{mod}\,A$.

Proof. We only prove (a); the proof of (b) is similar. Let M be an indecomposable nonprojective A-module. By the Auslander–Reiten formulas (2.13), there exists an isomorphism

$$D\underline{\mathrm{Hom}}_A(L, M) \cong \mathrm{Ext}^1_A(M, \tau L)$$

for any indecomposable module L, which is functorial in both variables. Let $S(L, M) = \mathrm{Hom}_A(L, M)/\mathrm{rad}_A(L, M)$. Because $\mathcal{P}(L, M) \subseteq \mathrm{rad}_A(L, M)$, we have a canonical K-linear epimorphism $p_{L,M} : \underline{\mathrm{Hom}}_A(L, M) \to S(L, M)$ and hence a canonical monomorphism $Dp_{L,M} : DS(L, M) \to D\underline{\mathrm{Hom}}_A(L, M)$.

Now, M being indecomposable, $\mathrm{End}\,M$ and hence $\underline{\mathrm{End}}\,M$ are local. Because we have an epimorphism

$$p_{M,M} : \underline{\mathrm{End}}\,M \to S(M, M) = \mathrm{End}\,M/\mathrm{rad}\,\mathrm{End}\,M,$$

$S(M, M)$ is isomorphic to the simple top of $\underline{\mathrm{End}}\,M$ considered as a left or right $\mathrm{End}\,M$-module, and its image under $Dp_{M,M}$ is the simple socle of the $\mathrm{End}\,M$-module $D\underline{\mathrm{Hom}}_A(M, M)$. Let ξ' be a nonzero element in $DS(M, M)$ and ξ be its image in $\mathrm{Ext}^1_A(M, \tau M) \cong D\underline{\mathrm{Hom}}_A(M, M)$. We claim that if ξ is represented by the short exact sequence

$$0 \longrightarrow \tau M \xrightarrow{f} E \xrightarrow{g} M \longrightarrow 0,$$

then this sequence is almost split.

First, this sequence is not split, and by (2.10), the module τM is indecomposable. It suffices thus, by (1.13), to show that g is right almost split. Because ξ is a nonzero element in $\mathrm{Ext}^1_A(M, \tau M)$, g is not a retraction. Let $v : V \to M$ be a homomorphism that is not a retraction. We may assume that V is indecomposable. Then v is not an isomorphism. It follows from the functoriality that we have a commutative diagram

$$DS(M, M) \xrightarrow{D p_{M,M}} D\underline{\mathrm{Hom}}_A(M, M) \xrightarrow{\ \cong\ } \mathrm{Ext}^1_A(M, \tau M)$$

$$\Big\downarrow DS(M,v) \qquad\qquad \Big\downarrow D\underline{\mathrm{Hom}}_A(M,v) \qquad\qquad \Big\downarrow \mathrm{Ext}^1_A(v, \tau M)$$

$$DS(M, V) \xrightarrow{D p_{M,V}} D\underline{\mathrm{Hom}}_A(M, V) \xrightarrow{\ \cong\ } \mathrm{Ext}^1_A(V, \tau M)$$

where the vertical maps are induced by v. By hypothesis, $v \in \mathrm{rad}_A(V, M)$ and therefore $DS(M, v)(\xi') = 0$. Consequently, the image $\mathrm{Ext}^1_A(v, \tau M)(\xi)$ of ξ in $\mathrm{Ext}^1_A(V, \tau M)$ is zero, that is, there exists a commutative diagram

$$\mathrm{Ext}^1_A(v, \tau M)(\xi): \qquad 0 \longrightarrow \tau M \xrightarrow{f'} E' \xrightarrow{g'} V \longrightarrow 0$$

$$\Big\downarrow 1_{\tau M} \qquad \Big\downarrow w \qquad \Big\downarrow v$$

$$\xi: \qquad 0 \longrightarrow \tau M \xrightarrow{f} E \xrightarrow{g} M \longrightarrow 0$$

with exact rows, where the upper sequence splits. Let thus $g'' : V \to E'$ be such that $g'g'' = 1_V$. Then $v' = wg''$ satisfies $gv' = gwg'' = vg'g'' = v$. This completes the proof that g is right almost split and hence the proof of the theorem.   □

The next corollary provides examples of almost split sequences.

3.2. Corollary. (a) *If* $0 \to \tau M \to E \to M \to 0$ *is an almost split sequence in* $\mathrm{mod}\, A$ *then it represents a nonzero element* ξ *of the simple socle of the* $\mathrm{End}\, M$-$\mathrm{End}\, M$-*bimodule* $\mathrm{Ext}^1_A(M, \tau M) \cong D\underline{\mathrm{Hom}}_A(M, M)$.

(b) *Let* M *be an indecomposable nonprojective module in* $\mathrm{mod}\, A$. *Then* $\underline{\mathrm{End}}\, M$ *is a skew field if and only if* $\overline{\mathrm{End}}\, \tau M$ *is a skew field, and in this case, any nonsplit exact sequence* $0 \to \tau M \to E \to M \to 0$ *is almost split and* $\underline{\mathrm{End}}\, M \cong K$.

(c) *Let* N *be an indecomposable noninjective module in* $\mathrm{mod}\, A$. *Then* $\overline{\mathrm{End}}\, N$ *is a skew field if and only if* $\underline{\mathrm{End}}\, \tau^{-1} N$ *is a skew field, and in this case, any nonsplit exact sequence* $0 \to N \to F \to \tau^{-1} N \to 0$ *is almost split and* $\overline{\mathrm{End}}\, N \cong K$.

Proof. The statement (a) follows from the proof of (3.1). We only prove (b); the proof of (c) is similar. The first statement of (a) follows from (2.11). Assume that $\underline{\mathrm{End}}\, M$ is a skew field. Because $\dim_K \underline{\mathrm{End}}\, M$ is finite and the field K is algebraically closed, $\underline{\mathrm{End}}\, M \cong K$ and $\mathrm{Ext}^1_A(M, \tau M)$ is a one-dimensional K-vector space (because it has simple socle, by (a)). Hence, by the proof of (3.1), any nonsplit extension represents an element in the socle of $\mathrm{Ext}^1_A(M, \tau M)$ and thus is almost split.   □

3.3. Example. Let A be the K-algebra given by the Kronecker quiver $1 \circ \underset{\beta}{\overset{\alpha}{\rightleftarrows}} \circ 2$ and M be the representation $K \underset{0}{\overset{1}{\rightleftarrows}} K$. As we have seen before, $\mathrm{End}\, M \cong K$ and $\tau M \cong M$. It follows from (3.2) that any nonsplit extension $0 \to M \to E \to M \to 0$ is an almost split sequence. Let E be the representation

$$K^2 \underset{\begin{bmatrix} 0 & 0 \\ 1 & 0 \end{bmatrix}}{\overset{\begin{bmatrix} 1 & 0 \\ 0 & 1 \end{bmatrix}}{\longleftarrow}} K^2$$

The subrepresentation E' of E given by $\begin{bmatrix} 0 \\ 1 \end{bmatrix} K \underset{0}{\overset{1}{\longleftarrow}} \begin{bmatrix} 0 \\ 1 \end{bmatrix} K$ is clearly isomorphic to M, and moreover $E/E' \cong M$. We thus have a short exact sequence as required. To prove that it is almost split, we show it is not split, and it suffices to show that E is indecomposable. To do this, we observe that any endomorphism f of E is given by a pair of matrices $\begin{pmatrix} a & b \\ c & d \end{pmatrix}$, $\begin{pmatrix} a' & b' \\ c' & d' \end{pmatrix}$ such that

$$\begin{pmatrix} a & b \\ c & d \end{pmatrix}\begin{pmatrix} 1 & 0 \\ 0 & 1 \end{pmatrix} = \begin{pmatrix} 1 & 0 \\ 0 & 1 \end{pmatrix}\begin{pmatrix} a' & b' \\ c' & d' \end{pmatrix}, \text{ and } \begin{pmatrix} a & b \\ c & d \end{pmatrix}\begin{pmatrix} 0 & 0 \\ 1 & 0 \end{pmatrix} = \begin{pmatrix} 0 & 0 \\ 1 & 0 \end{pmatrix}\begin{pmatrix} a' & b' \\ c' & d' \end{pmatrix}.$$

These two conditions yield $a = a' = d = d'$, $b = b' = 0$, and $c = c'$. Thus $f = a \cdot 1_E + g$, where $a \in K$ and $g \in \operatorname{End} E$ is nilpotent. Let now $I = \{f \in \operatorname{End} E \mid a = 0\}$. Then I is a nilpotent ideal of $\operatorname{End} E$. Because moreover $(\operatorname{End} E)/I \cong K$, I is a maximal ideal of $\operatorname{End} E$. Therefore $I = \operatorname{rad} \operatorname{End} E$ and $\operatorname{End} E$ is local. Thus, E is indecomposable.

3.4. Example. Let A be the K-algebra given by the quiver

bound by $\alpha\beta = \gamma\delta$, $\delta\mu = 0$, $\beta\lambda = 0$. It was shown in Example 2.6 that there is an exact sequence $0 \longrightarrow \tau S(6) \longrightarrow I(4) \oplus I(5) \overset{\nu\, p_1}{\longrightarrow} I(6) \longrightarrow 0$. It is clear that $\operatorname{End} \tau S(6) \cong K$, hence $\overline{\operatorname{End}} \tau S(6) \cong K$. In view of the unique decomposition theorem (I.4.10), this sequence does not split. It then follows from (3.2)(b) that the sequence is almost split.

It also follows from (3.1) that there exists a right (or left) minimal almost split morphism ending (or starting, respectively) at any indecomposable nonprojective (or noninjective, respectively) module. We now want to show the existence of such a homomorphism ending (or starting) at an indecomposable projective (or injective, respectively) module.

3.5. Proposition. (a) *Let P be an indecomposable projective module in $\operatorname{mod} A$. An A-module homomorphism $g : M \to P$ is right minimal almost split if and only if g is a monomorphism with image equal to $\operatorname{rad} P$.*

(b) *Let I be an indecomposable injective module. An A-module homomorphism $f : I \to M$ is left minimal almost split if and only if f is an* *epimorphism with kernel equal to* soc I.

Proof. We only prove (a); the proof of (b) is similar. It suffices, by (1.2), to show that the inclusion homomorphism g : rad $P \to P$ is right minimal almost split. Because g is a monomorphism, g is right minimal. Clearly, g is not a retraction. Let thus $v : V \to P$ be a homomorphism that is not a retraction. Because P is projective, by (I.4.5), the module rad P is the unique maximal submodule of P. Because v is not an epimorphism, $v(V) \subseteq$ rad P, that is, v factors through g. □

3.6. Corollary. *Let X be an indecomposable module in* mod A.

(a) *There exists a right minimal almost split morphism $g : M \to X$. Moreover $M = 0$ if and only if X is simple projective.*

(b) *There exists a left minimal almost split morphism $f : X \to M$. Moreover, $M = 0$ if and only if X is simple injective.*

Proof. The proof follows directly from (3.1) and (3.5). □

3.7. Example. Let A be the K-algebra given by the quiver 1∘⟵——— ∘2. Consider the short exact sequence $0 \longrightarrow S(1) \overset{f}{\longrightarrow} P(2) \overset{g}{\longrightarrow} S(2) \longrightarrow 0$ in mod A, where f is the embedding of $S(1)$ as the radical of $P(2)$ and g is the canonical homomorphism of $P(2)$ onto its top. Because $P(2) = I(1)$, it follows from (3.5) that f is right minimal almost split and g is left minimal almost split. On the other hand, it will be shown in (3.11) that, because the middle term is projective-injective, the sequence is almost split (thus, f is also left minimal almost split and g is right minimal almost split).

3.8. Proposition. (a) *Let M be an indecomposable nonprojective module in* mod A. *There exists an irreducible morphism $f : X \to M$ if and only if there exists an irreducible morphism $f' : \tau M \to X$.*

(b) *Let N be an indecomposable noninjective module in* mod A. *There exists an irreducible morphism $g : N \to Y$ if and only if there exists an irreducible morphism $g' : Y \to \tau^{-1}N$.*

Proof. We only prove (a); the proof of (b) is similar. Assume that $f : X \to M$ is irreducible. By (1.10), there exists $h : Y \to M$ such that $[f\ h] : X \oplus Y \to M$ is right minimal almost split. But then $[f\ h]$ is an epimorphism, because M is not projective. Therefore, by (1.8), $L =$ Ker$[f\ h]$ is indecomposable, and thus, by (1.13), the short exact sequence

$$0 \longrightarrow L \overset{\left[\begin{smallmatrix} f' \\ h' \end{smallmatrix}\right]}{\longrightarrow} X \oplus Y \overset{[f\ h]}{\longrightarrow} M \longrightarrow 0$$

is almost split. Consequently, there exists an isomorphism $g : \tau M \overset{\simeq}{\longrightarrow} L$ and the homomorphism $f'g : \tau M \to X$ is irreducible. The proof of the converse is similar.   □

3.9. Corollary. (a) *Let S be a simple projective noninjective module in* $\operatorname{mod} A$. *If $f : S \to M$ is irreducible, then M is projective.*

(b) *Let S be a simple injective nonprojective module in* $\operatorname{mod} A$. *If $g : M \to S$ is irreducible, then M is injective.*

Proof. We only prove (a); the proof of (b) is similar. We may clearly assume M to be indecomposable. If M is not projective, there exists, by (3.8), an irreducible morphism $\tau M \to S$, and this contradicts (3.6).   □

This corollary allows us to construct examples of almost split sequences. Indeed, let S be simple projective noninjective and $f : S \to P$ be left minimal almost split. By (3.9), P is projective and by (3.5), for each indecomposable summand P' of P, the corresponding component $f' : S \to P'$ of f is a monomorphism with image a summand of $\operatorname{rad} P'$. It follows that, if P is the direct sum of all such indecomposable projectives P', then the sequence $0 \longrightarrow S \stackrel{f}{\longrightarrow} P \longrightarrow \operatorname{Coker} f \longrightarrow 0$ is almost split.

3.10. Example. Assume that A is a K-algebra given by the quiver
$\overset{1}{\circ}\longleftarrow\overset{2}{\circ}\longrightarrow\overset{3}{\circ}\longleftarrow\overset{4}{\circ}$. Then $S(3)$ is a simple projective noninjective summand of $\operatorname{rad} P(2)$ and is equal to $\operatorname{rad} P(4)$. Thus we have an almost split sequence

$$0 \longrightarrow S(3) \longrightarrow P(2) \oplus P(4) \longrightarrow (P(2) \oplus P(4))/S(3) \longrightarrow 0.$$

The preceding remark is essentially used in the next section. We conclude this section with a further example of an almost split sequence.

3.11. Proposition. *Let P be a nonsimple indecomposable projective-injective module, $S = \operatorname{soc} P$, and $R = \operatorname{rad} R$. Then the sequence*

$$0 \longrightarrow R \stackrel{[\begin{smallmatrix}q\\i\end{smallmatrix}]}{\longrightarrow} R/S \oplus P \stackrel{[-j\,p]}{\longrightarrow} P/S \longrightarrow 0$$

is almost split, where i, j are the inclusions and p, q the projections.

Proof. Because R has simple socle S, it is indecomposable. Hence $i : R \to P$ is, up to isomorphism, the unique nontrivial irreducible morphism ending in P (by (3.5)). Dually, the module P/S is indecomposable and $p : P \to P/S$ is, up to isomorphism, the unique nontrivial irreducible morphism starting with P. It follows from (3.8) that $R \cong \tau(P/S)$. Because the given exact sequence is not split, it remains to show (by (1.13)) that the monomorphism $[\begin{smallmatrix}q\\i\end{smallmatrix}] : R \to R/S \oplus P$ is left almost split. Assume that $u : R \to U$ is not a section. If u is a monomorphism, then, because P is injective, u factors through P and we are done. If not, there exists a factorisation $u = u'u''$, with $u'' : R \to U'$ a proper epimorphism and $u' : U' \to U$ a monomorphism. Because $\operatorname{Ker} u \neq 0$, the simple socle S of R is contained in $\operatorname{Ker} u = \operatorname{Ker} u''$. Thus the epimorphism u'' factors through R/S, that is, there exists $u_1 : R/S \to U'$ such that $u'' = u_1 q$. Hence $\overline{u} = [u'u_1, 0]$ satisfies $\overline{u}\begin{bmatrix} q \\ i \end{bmatrix} = [u'u_1,\ 0]\begin{bmatrix} q \\ i \end{bmatrix} = u'u_1 q = u'u'' = u$. □

3.12. Example. Let A be the K-algebra given by the quiver

bound by the commutativity relations: $\alpha\beta = \gamma\delta$ and $\gamma\delta = \lambda\mu\nu$. The A-module $P(6) = I(1)$ is projective-injective and the almost split sequence described in (3.11) with $P = P(6)$ is of the form

$$0 \longrightarrow \operatorname{rad} P(6) \longrightarrow S(2) \oplus S(3) \oplus \frac{P(5)}{S(1)} \oplus P(6) \longrightarrow \frac{P(6)}{S(1)} \longrightarrow 0.$$

IV.4. The Auslander–Reiten quiver of an algebra

Let A be a finite dimensional K-algebra. We may wish to record the information we have on the category $\operatorname{mod} A$ in the form of a quiver. Then it seems clear that points should represent modules and arrows should represent homomorphisms. Because any module in $\operatorname{mod} A$ decomposes as the direct sum of indecomposable modules uniquely determined up to isomorphism, we should let the points represent isomorphism classes of indecomposable modules. Similarly, the homomorphisms that admit no nontrivial factorisation are the irreducible morphisms; thus our arrows should correspond to the irreducible morphisms. But to be more precise, we need some additional considerations on irreducible morphisms.

Let M and N be indecomposable modules in $\operatorname{mod} A$. We have seen in (1.6) that an A-homomorphism $f : M \to N$ is an irreducible morphism if and only if $f \in \operatorname{rad}_A(M, N) \setminus \operatorname{rad}_A^2(M, N)$. Thus the quotient

$$\operatorname{Irr}(M, N) = \operatorname{rad}_A(M, N)/\operatorname{rad}_A^2(M, N) \qquad (4.1)$$

of the K-vector spaces $\operatorname{rad}_A(M, N)$ and $\operatorname{rad}_A^2(M, N)$ measures the number of irreducible morphisms from M to N. It is called the **space of irreducible morphisms**. It is easily seen (see (1.6)) that $\operatorname{Irr}(M, N)$ is in fact an $\operatorname{End} N$–$\operatorname{End} M$-bimodule, annihilated on the left by $\operatorname{rad}_A(N, N) = \operatorname{rad} \operatorname{End} N$ and on the right by $\operatorname{rad}_A(M, M) = \operatorname{rad} \operatorname{End} M$.

We now give the relation between the space of irreducible morphisms and minimal almost split morphisms.

4.2. Proposition. *Let* $M = \bigoplus_{i=1}^{t} M_i^{n_i}$ *be a module in* $\operatorname{mod} A$, *with the* M_i *indecomposable and pairwise nonisomorphic.*

(a) *Let* $f : L \to M$ *be a homomorphism in* $\operatorname{mod} A$ *with* L *indecomposable,* $f = \begin{bmatrix} f_1 \\ \vdots \\ f_t \end{bmatrix}$, *where* $f_i = \begin{bmatrix} f_{i1} \\ \vdots \\ f_{in_i} \end{bmatrix} : L \longrightarrow M_i^{n_i}$. *Then* f *is left minimal almost split if and only if the* f_{ij} *belong to* $\operatorname{rad}_A(L, M_i)$ *and their residual classes* $\overline{f}_{i1}, \ldots, \overline{f}_{in_i}$ *modulo* $\operatorname{rad}_A^2(L, M_i)$ *form a* K-*basis of* $\operatorname{Irr}(L, M_i)$ *for all* i, *and if there is an indecomposable module* M' *in* $\operatorname{mod} A$ *such that* $\operatorname{Irr}(L, M') \neq 0$, *then* $M' \cong M_i$ *for some* i.

(b) *Let* $g : M \to N$ *be a homomorphism in* $\operatorname{mod} A$ *with* N *indecomposable,* $g = \begin{bmatrix} g_1 & \cdots & g_t \end{bmatrix}$, *where* $g_i = \begin{bmatrix} g_{i1} & \cdots & g_{in_i} \end{bmatrix} : M_i^{n_i} \longrightarrow N$. *Then* g *is right minimal almost split if and only if the* g_{ij} *belong to* $\operatorname{rad}_A(M_i, N)$ *and their residual classes* $\overline{g}_{i1}, \ldots, \overline{g}_{in_i}$ *modulo* $\operatorname{rad}_A^2(M_i, N)$ *form a* K-*basis of* $\operatorname{Irr}(M_i, N)$ *for all* i, *and, if there is an indecomposable module* M' *in* $\operatorname{mod} A$ *such that* $\operatorname{Irr}(M', N) \neq 0$, *then* $M' \cong M_i$ *for some* i.

Proof. We only prove (a); the proof of (b) is similar. Assume thus that f is left minimal almost split. Note that, by the statement (a) of (1.10), if $u : U \to V$ is irreducible and $v : V \to W$ is a retraction, then $vu : U \to W$ is irreducible. Because, again by (1.10), $f : L \to M$ is irreducible, this remark implies that each $f_{ij} : L \to M_i$ is irreducible and thus belongs to $\operatorname{rad}_A(L, M_i)$ (by (1.6)).

On the other hand, (1.10) also shows that if there is an indecomposable module M' such that $\operatorname{Irr}(L, M') \neq 0$, so that there is an irreducible morphism $L \to M'$, then $M' \cong M_i$ for some i. We now want to show that for each i, $\{\overline{f}_{i1}, \ldots \overline{f}_{in_i}\}$ is a K-basis of $\operatorname{Irr}(L, M_i)$.

Let $\overline{h} \in \operatorname{Irr}(L, M_i)$ be the residual class of $h \in \operatorname{rad}_A(L, M_i)$. Because h is not a section, it factors through f, that is, there exists a homomorphism $u = [u_1, \ldots, u_t] : \bigoplus_{k=1}^{t} M_k^{n_k} \to M_i$, with $u_k = [u_{k1}, \ldots, u_{kn_k}] : M_k^{n_k} \to M_i$ such that

$$h = uf = \sum_{k=1}^{t} \sum_{j=1}^{n_k} u_{kj} f_{kj}.$$

Any u_{ij} is an endomorphism of M_i. Because $\operatorname{End} M_i$ is local and the base field K is algebraically closed, we have that $\operatorname{End} M_i / \operatorname{rad} \operatorname{End} M_i \cong K$, so that $u_{ij} = \lambda_j \cdot 1_{M_i} + u'_{ij}$ with $\lambda_j \in K$ and $u'_{ij} \in \operatorname{rad}_A(M_i, M_i) = \operatorname{rad} \operatorname{End} M_i$. On the other hand, if $k \neq i$, then $u_{kj} \in \operatorname{rad}_A(M_k, M_i)$. Because $f_{kj} \in$

$\text{rad}_A(L, M_k)$, we have $u_{kj}f_{kj} \in \text{rad}_A^2(L, M_i)$ for $k \neq i$. Thus

$$\overline{h} = \sum_k \sum_j \overline{u}_{kj}\overline{f}_{kj} = \sum_j \lambda_j \cdot \overline{f}_{ij}.$$

This shows that $\{\overline{f}_{i1}, \ldots, \overline{f}_{in_i}\}$ generates $\text{Irr}(L, M_i)$ as a K-vector space. To prove the linear independence of this set, assume that $\sum_j \lambda_j \overline{f}_{ij} = 0$ in $\text{Irr}(L, M_i)$, where $\lambda_j \in K$. Thus the homomorphism $v = \sum_j \lambda_j f_{ij}$ belongs to $\text{rad}_A^2(L, M_i)$. Assume that $\lambda_j \neq 0$ for some j; then the homomorphism $l = [\lambda_1, \ldots, \lambda_{n_i}] : M_i^{n_i} \to M_i$ is a retraction, and, by the first remark, $v = lf_i$ is irreducible, a contradiction, because $v \in \text{rad}_A^2(L, M_i)$. Consequently, $\lambda_j = 0$. We have completed the proof that $\{\overline{f}_{i1}, \ldots, \overline{f}_{in_i}\}$ is a K-basis of $\text{Irr}(L, M_i)$ and thus of the necessity.

For the sufficiency, assume that for each j, $\{\overline{f}_{j1}, \ldots, \overline{f}_{jn_j}\}$ is a basis of the K-vector space $\text{Irr}(L, M_j)$ and consider a left minimal almost split morphism $f' : L \to U$ (see (3.6)). It follows that $f : L \to M$ is not a section and applying the necessity part to U yields that $U \cong M$. Indeed, let $U = \bigoplus_{k=1}^s U_k^{m_k}$ be a decomposition of U, where $U_1, \ldots, U_s$ are pairwise nonisomorphic indecomposable modules. For each k, $\text{Irr}(L, U_k) \neq 0$ yields $U_k \cong M_j$ for some j and $m_k = \dim_K \text{Irr}(L, U_k) = \dim_K \text{Irr}(L, M_j) = n_j$. Analogously, for each j, $\text{Irr}(L, M_j) \neq 0$ yields $M_j \cong U_k$ for some k. Hence we deduce that $U = \bigoplus_{k=1}^s U_k^{m_k} \cong \bigoplus_{j=1}^t M_j^{n_j} = M$.

Without loss of generality we may assume that $U = M$ and $f' : L \to M$ is left minimal almost split. Applying the necessity part to f' yields that $f' = [f'_{js}] : L \to \bigoplus_{j=1}^t M_j^{n_j}$ and, for each j, the set $\{\overline{f}'_{j1}, \ldots, \overline{f}'_{jn_j}\}$ is a basis of the K-vector space $\text{Irr}(L, M_j)$. Because f is not a section, there exists $h : M \to M$ such that $f = hf'$. Hence we conclude that h is an isomorphism. Consequently, f is a left minimal almost split morphism. $\square$

4.3. Remark. Let $P(a) = e_a A$ be an indecomposable projective A-module and $I(a) = D(Ae_a)$ be an indecomposable injective A-module.

(a) The embedding $\text{rad} P(a) \hookrightarrow P(a)$ is an irreducible morphism and is right minimal almost split. If $X_1, \ldots X_t$ are indecomposable and pairwise nonisomorphic A-modules such that $\text{rad} P(a) \cong X_1^{n_1} \oplus \cdots \oplus X_t^{n_t}$, then $n_j = \dim_K \text{Irr}(X_j, P(a))$ and every indecomposable A-module X with $\text{Irr}(X, P(a)) \neq 0$ is isomorphic to X_j for some j.

(b) The natural epimorphism $I(a) \to I(a)/\text{soc} I(a)$ is an irreducible morphism and is left minimal almost split. If $Y_1, \ldots Y_s$ are indecomposable and pairwise nonisomorphic such that $I(a)/\text{soc} I(a) \cong Y_1^{m_1} \oplus \cdots \oplus Y_t^{m_t}$, then $m_j = \dim_K \text{Irr}(I(a), Y_j)$ and every indecomposable A-module Y with $\text{Irr}(I(a), Y) \neq 0$ is isomorphic to Y_j for some j.

The first statement of (a) follows from (3.5)(a). The remaining part of (a) is a consequence of (4.2) and the unique decomposition theorem (I.4.10).

The first statement of (b) follows from (3.5)(b). The remaining part of (b) follows easily by applying the duality $D : \operatorname{mod} A^{\mathrm{op}} \to \operatorname{mod} A$.

We collect some of the previous results in the following useful corollary.

4.4. Corollary. *Let* $0 \longrightarrow L \stackrel{f}{\longrightarrow} \bigoplus_{i=1}^{t} M_i^{n_i} \stackrel{g}{\longrightarrow} N \longrightarrow 0$ *be a short exact sequence in* $\operatorname{mod} A$ *with* L, N *indecomposable and the* M_i *indecomposable and pairwise nonisomorphic. Write* $f = \begin{bmatrix} f_1 \\ \vdots \\ f_t \end{bmatrix}$ *and* $g = [g_1 \ldots g_t]$,

where $f_i = \begin{bmatrix} f_{i1} \\ \vdots \\ f_{in_i} \end{bmatrix} : L \longrightarrow M_i^{n_i}$ *and* $g = [g_{i1} \ldots g_{in_i}] : M_i^{n_i} \longrightarrow N$.

The following conditions are equivalent:

(a) *The given sequence is almost split.*

(b) *For each* i, *the homomorphisms* f_{ij} *belong to* $\operatorname{rad}_A(L, M_i)$, *their residual classes* $\overline{f}_{ij}$ *modulo* $\operatorname{rad}_A^2(L, M_i)$ *form a* K-*basis of* $\operatorname{Irr}(L, M_i)$, *and if there exists an indecomposable module* M' *with* $\operatorname{Irr}(L, M') \neq 0$, *then* $M' \cong M_i$ *for some* i.

(c) *For each* i, *the homomorphisms* g_{ij} *belong to* $\operatorname{rad}_A(M_i, N)$, *their residual classes* $\overline{g}_{ij}$ *modulo* $\operatorname{rad}_A^2(M_i, N)$ *form a* K-*basis of* $\operatorname{Irr}(M_i, N)$, *and if there exists an indecomposable module* M' *with* $\operatorname{Irr}(M', N) \neq 0$, *then* $M' \cong M_i$ *for some* i.

Further, if these equivalent conditions hold, then for each i,

$$\dim_K \operatorname{Irr}(L, M_i) = \dim_K \operatorname{Irr}(M_i, N).$$

Proof. The equivalence of these conditions follows from (4.2), and the last statement from (b) and (c). $\square$

4.5. Corollary. *Let* X *and* Y *be indecomposable modules in* $\operatorname{mod} A$.

(a) *If* $\tau X \neq 0$ *and* $\tau Y \neq 0$, *then there exists a* K-*linear isomorphism* $\operatorname{Irr}(\tau X, \tau Y) \cong \operatorname{Irr}(X, Y)$.

(b) *If* $\tau^- X \neq 0$ *and* $\tau^- Y \neq 0$, *then there exists a* K-*linear isomorphism* $\operatorname{Irr}(\tau^- X, \tau^- Y) \cong \operatorname{Irr}(X, Y)$.

Proof. We only prove (a); the proof of (b) is dual. Because $\tau X \neq 0$ and $\tau Y \neq 0$, X is not projective, Y is not projective, and there exist almost split sequences $0 \longrightarrow \tau X \longrightarrow U \stackrel{u}{\longrightarrow} X \longrightarrow 0$ and $0 \longrightarrow \tau Y \longrightarrow V \stackrel{v}{\longrightarrow} Y \longrightarrow 0$ in $\operatorname{mod} A$. First, we prove that $\operatorname{Irr}(X, Y) \neq 0$ implies $\operatorname{Irr}(\tau X, \tau Y) \cong \operatorname{Irr}(X, Y)$. Assume that $\operatorname{Irr}(X, Y) \neq 0$. Because v is a right minimal almost split morphism, according to (4.2)(b), the module X is isomorphic to a direct summand of V, and by (3.8) there is an irreducible morphism $\tau Y \to X$.

Then, by (4.4), there is a K-linear isomorphism $\operatorname{Irr}(\tau Y, X) \cong \operatorname{Irr}(X, Y)$. Because u is a right minimal almost split morphism and $\operatorname{Irr}(\tau Y, X) \neq 0$, then, according to (4.2)(b), the module τY is isomorphic to a direct summand of U and, according to (4.4), there is a K-linear isomorphism $\operatorname{Irr}(\tau Y, X) \cong \operatorname{Irr}(\tau X, \tau Y)$. Consequently, we get a K-linear isomorphism $\operatorname{Irr}(\tau X, \tau Y) \cong \operatorname{Irr}(X, Y)$.

Using these arguments, we also prove that $\operatorname{Irr}(\tau X, \tau Y) \cong \operatorname{Irr}(X, Y)$ if $\operatorname{Irr}(\tau X, \tau Y) \neq 0$. This finishes the proof. $\square$

We are now able to define the quiver of the category $\operatorname{mod} A$.

4.6. Definition. Let A be a basic and connected finite dimensional K-algebra The **quiver** $\Gamma(\operatorname{mod} A)$ of $\operatorname{mod} A$ is defined as follows:

(a) The points of $\Gamma(\operatorname{mod} A)$ are the isomorphism classes $[X]$ of indecomposable modules X in $\operatorname{mod} A$.

(b) Let $[M]$, $[N]$ be the points in $\Gamma(\operatorname{mod} A)$ corresponding to the indecomposable modules M, N in $\operatorname{mod} A$. The arrows $[M] \to [N]$ are in bijective correspondence with the vectors of a basis of the K-vector space $\operatorname{Irr}(M, N)$.

The quiver $\Gamma(\operatorname{mod} A)$ of the module category $\operatorname{mod} A$ is called the **Auslander–Reiten quiver** of A.

We may define in exactly the same way the quiver $\Gamma(\mathcal{C})$ of an arbitrary additive subcategory $\mathcal{C}$ of $\operatorname{mod} A$ that is closed under direct sums and summands. We leave to the reader the verification that if $\mathcal{C} = \operatorname{proj} A$, the quiver $\Gamma(\operatorname{proj} A)$ is the opposite of the ordinary quiver of A. In the rest of this section, we examine the combinatorial structure of the Auslander–Reiten quiver $\Gamma(\operatorname{mod} A)$ of A.

It follows from the definition that the points of $\Gamma(\operatorname{mod} A)$ are the isomorphism classes of indecomposable A-modules, and that there exists an arrow $[L] \to [M]$ if and only if $\operatorname{Irr}(L, M) \neq 0$, that is, if and only if there exists an irreducible morphism $L \to M$. By (4.2), (3.1), and (3.5), the set $[M]^-$ of the immediate predecessors of $[M]$ coincides with the set of those points $[L]$ such that L is either an indecomposable direct summand of $\operatorname{rad} M$, if M is projective, or an indecomposable direct summand of the middle term of the almost split sequence ending with M, if M is not projective. Similarly, the set $[M]^+$ of the immediate successors of M coincides with the set of those points $[N]$ such that N is either an indecomposable summand of $M/\operatorname{soc} M$, if M is injective, or an indecomposable direct summand of the middle term of the almost split sequence starting with M, if M is not injective. In particular, for every M, the sets $[M]^+$ and $[M]^-$ are finite. This shows that each point of $\Gamma(\operatorname{mod} A)$ has only finitely many neighbours.

A quiver having this property, that is, such that each point has only finitely may neighbours, is called **locally finite**.

An obvious consequence is that each connected component of an Auslander–Reiten quiver has at most countably many points. Indeed, let x be an arbitrary fixed point of a locally finite quiver Γ. Denote by N_1 the set of neighbours of x, and for each $i \geq 2$ define N_i to be the set of neighbours of points from N_{i-1}. Because Γ is locally finite, each N_i is finite. Because Γ is connected, the set $\Gamma_0 = \bigcup_{i \geq 1} N_i$ is a connected component consisting of at most countably many points.

It is clear that $\Gamma(\text{mod}\,A)$ is finite (or, equivalently, has finitely many points) if and only if A is representation–finite, that is, the number of the isomorphism classes of indecomposable finite dimensional right A-modules is finite (see (I.4.11)). In fact, we show in the next section that if $\Gamma(\text{mod}\,A)$ has a finite connected component Γ, then $\Gamma(\text{mod}\,A) = \Gamma$ and, consequently, A is representation–finite.

We recall that A is called representation–infinite if A is not representation–finite.

A second observation is that every irreducible morphism $f : M \to N$ is either a proper monomorphism or a proper epimorphism; see (1.4). Moreover, if $M = N$, then, because M is finite dimensional as a K-vector space, f should be an isomorphism. This shows that the source and the target of this homomorphism must be distinct and therefore an Auslander–Reiten quiver has no loops.

The Auslander–Reiten quiver is actually endowed with an additional structure. Let Γ_0' (or Γ_0'') denote the set of those points in $\Gamma(\text{mod}\,A)$ that correspond to a projective (or an injective, respectively) indecomposable module. For each $[N] \in \Gamma(\text{mod}\,A)_0 \setminus \Gamma_0'$, the Auslander–Reiten translate τN of N exists, and, by (2.10), we have $[\tau N] \in \Gamma(\text{mod}\,A)_0 \setminus \Gamma_0''$. This defines a bijection

$$\tau : \Gamma(\text{mod}\,A)_0 \setminus \Gamma_0' \longrightarrow \Gamma(\text{mod}\,A)_0 \setminus \Gamma_0'',$$

also denoted by τ. Thus, for each indecomposable nonprojective module N, we have $\tau[N] = [\tau N]$. The inverse bijection is denoted by

$$\tau^{-1} : \Gamma(\text{mod}\,A)_0 \setminus \Gamma_0'' \longrightarrow \Gamma(\text{mod}\,A)_0 \setminus \Gamma_0'$$

and, for each indecomposable noninjective module L, we have $\tau^{-1}[L] = [\tau^{-1}L]$. We say that τ is the **translation** of the quiver $\Gamma(\text{mod}\,A)$. Let thus N be an indecomposable nonprojective A-module, and let

$$0 \longrightarrow \tau N \longrightarrow \bigoplus_{i=1}^{t} M_i^{n_i} \longrightarrow N \longrightarrow 0$$

be an almost split sequence ending with N, with the M_i indecomposable and pairwise nonisomorphic. By (4.4), for each i, we have

$$n_i = \dim_K \text{Irr}(M_i, N) = \dim_K \text{Irr}(\tau N, M_i).$$

Hence, corresponding to this almost split sequence is the following "mesh" in $\Gamma(\operatorname{mod} A)$:

$$[M_1]$$

$$\alpha_{11} \nearrow \cdots \nearrow \alpha_{1n_1} \quad \vdots \quad \beta_{1n_1} \searrow \cdots \searrow \beta_{11}$$

$$[\tau N] \text{ - - - - - - - - - - - - - } [N]$$

$$\alpha_{t1} \searrow \cdots \searrow \alpha_{1n_t} \quad \vdots \quad \beta_{tn_t} \nearrow \cdots \nearrow \beta_{t1}$$

$$[M_t]$$

In particular, we see that $[\tau N]^+ = [N]^-$ and that for each $[M_i]$ in this set, there exists a bijection between the set $\{\alpha_{i1}, \ldots, \alpha_{in_i}\}$ of arrows from $[\tau N]$ to $[M_i]$ and the set $\{\beta_{i1}, \ldots, \beta_{in_i}\}$ of arrows from $[M_i]$ to $[N]$.

We may thus define a new combinatorial structure.

4.7. Definition. Let Γ be a locally finite quiver without loops and τ be a bijection whose domain and codomain are both subsets of Γ_0. The pair (Γ, τ) (or more briefly, Γ) is said to be a **translation quiver** if for every $x \in \Gamma_0$ such that τx exists, and every $y \in x^-$, the number of arrows from y to x is equal to the number of arrows from τx to y.

A **full translation subquiver** of a translation quiver (Γ, τ) is a translation quiver (Γ', τ') such that Γ' is a full subquiver of Γ and $\tau' x = \tau x$, whenever x is a vertex of Γ' such that τx belongs to Γ'.

It follows directly from the definition that, if $x \in \Gamma_0$ is such that τx exists, then $(\tau x)^+ = x^-$. The bijection τ is called the **translation** of Γ. The points of Γ, where τ (or τ^{-1}) is not defined are called **projective points** (or **injective points**, respectively). The full subquiver of Γ consisting of a nonprojective point $x \in \Gamma_0$, its translate τx, and the points of $(\tau x)^+ = x^-$ is called the **mesh** ending with x and starting with τx. Let Γ'_1 denote the subset of Γ_1 consisting of the arrows with nonprojective target. Because, for $x \in \Gamma_0$ nonprojective there exists a bijection between the arrows having x as target and those having τx as source, we can define an injective mapping $\sigma : \Gamma'_1 \to \Gamma_1$ such that if $\alpha \in \Gamma'_1$ has target x, then $\sigma \alpha$ has source τx. Such a mapping is called a **polarisation** of Γ. Clearly, if Γ has no multiple arrows, there exists a unique polarisation on Γ. Otherwise, there usually exist many polarisations. We have already proven the following lemma.

4.8. Lemma. *The Auslander–Reiten quiver $\Gamma(\operatorname{mod} A)$ of an algebra A is a translation quiver, the translation τ being defined for all points $[M]$ such that M is not a projective module, by $\tau[M] = [\tau M]$.* □

It is, of course, easy to construct examples of translation quivers that are not necessarily Auslander–Reiten quivers, for instance

$$
\begin{array}{c}
\circ z \\
\circ \tau^2 y \quad \circ \tau y \quad \circ y \\
\circ \tau^5 z \quad \circ \tau^4 z \quad \circ \tau^3 z \quad \circ \tau^2 z \quad \circ \tau z \quad \circ z \\
\circ \tau u = u \quad \quad \quad \quad \quad \quad \quad \circ \tau v \quad \circ v \\
\quad \quad \quad \quad \quad \circ \tau w \quad \circ w \\
\tau^2 w = w
\end{array}
$$

In most cases we consider the Auslander–Reiten quiver has no multiple arrows. This is the case for representation–finite algebras.

4.9. Proposition. *Let A be a representation–finite algebra. Then $\Gamma(\operatorname{mod} A)$ has no multiple arrows.*

Proof. We must show that, for each pair M, N of indecomposable A-modules, we have $\dim_K \operatorname{Irr}(M, N) \leq 1$. We assume that this is not the case, that is, that there exists a pair M, N such that $\dim_K \operatorname{Irr}(M, N) \geq 2$. In particular, $\operatorname{Irr}(M, N) \neq 0$. Because every irreducible morphism $M \to N$ is an epimorphism or a monomorphism, we must have $\dim_K M \neq \dim_K N$. Suppose $\dim_K M > \dim_K N$ (the other case is dual). In particular, N cannot be projective, and there exists an almost split sequence of the form $0 \longrightarrow \tau N \longrightarrow M^2 \oplus E \longrightarrow N \longrightarrow 0$. Hence we get

$$
\begin{aligned}
\dim_K \tau N \quad &= 2 \dim_K M + \dim_K E - \dim_K N \\
&> \dim_K M > \dim_K N.
\end{aligned}
$$

Furthermore, $\dim_K \operatorname{Irr}(\tau N, M) \geq 2$. An obvious induction shows that, for any two natural numbers i, j such that $i > j$, we have

$$
\dim_K \tau^i M > \dim_K \tau^i N > \dim_K \tau^j M > \dim_K \tau^j N.
$$

This implies that the mapping $\mathbb{N} \to \Gamma(\operatorname{mod} A)_0$ given by $i \mapsto \tau^i[N]$ is injective, and the connected component of $\Gamma(\operatorname{mod} A)$ containing $[N]$ is infinite, which contradicts the hypothesis that A is representation–finite. $\square$

We now turn to the construction of the Auslander–Reiten quiver of an algebra A. In many simple cases, it is possible to construct $\Gamma(\operatorname{mod} A)$ without constructing explicitly all the almost split sequences in $\operatorname{mod} A$. We illustrate the procedure with examples. In these examples, we agree to identify isomorphic modules and homomorphisms.

4.10. Example. Let A be the path K-algebra of the linear quiver $\underset{1}{\circ} \xleftarrow{\;\;\beta\;\;} \underset{2}{\circ} \xleftarrow{\;\;\alpha\;\;} \underset{3}{\circ}$. We have a complete list of the indecomposable projective or injective A-modules, given as representations (see (III.2)):

$$
\begin{aligned}
P(1) &= (K\longleftarrow 0\longleftarrow 0) = S(1)\\
P(2) &= (K\xleftarrow{\;1\;}K\longleftarrow 0)\\
P(3) &= (K\xleftarrow{\;1\;}K\xleftarrow{\;1\;}K) = I(1)\\
I(2) &= (0\longleftarrow K\xleftarrow{\;1\;}K)\\
I(3) &= (0\longleftarrow 0\longleftarrow K),
\end{aligned}
$$

and we also have a simple module $S(2)$, which is neither projective nor injective. Further, we have

$$
\begin{aligned}
P(1) &= \operatorname{rad} P(2) & P(2) &= \operatorname{rad} P(3)\\
I(3) &= I(2)/S(2) & I(2) &= I(1)/S(1) = P(3)/S(1).
\end{aligned}
$$

Because the A-module $P(1)$ is simple projective and noninjective, by (3.9), the target of each irreducible morphism starting with $P(1)$ is projective. Because $P(1) = \operatorname{rad} P(2)$, and $P(1)$ is not a summand of $\operatorname{rad} P(3)$, the inclusion $i : P(1) \to P(2)$ is the only such irreducible morphism and is actually the only right minimal almost split morphism ending with $P(2)$. Thus we have an almost split sequence $0 \longrightarrow P(1) \xrightarrow{\;i\;} P(2) \longrightarrow \operatorname{Coker} i \longrightarrow 0$. It is easily seen that $\operatorname{Coker} i = P(2)/P(1) = S(2)$.

Now consider $P(2)$. We have just seen that there exists an irreducible morphism $P(2) \to S(2)$. On the other hand $\operatorname{rad} P(3) = P(2)$, hence there exists an irreducible (inclusion) morphism $P(2) \to P(3)$. Now $P(3) = I(1)$ is projective-injective, hence, by (3.11), we have an almost split sequence of the form $0 \longrightarrow P(2) \longrightarrow P(3) \oplus S(2) \longrightarrow I(2) \longrightarrow 0$. On the other hand, the homomorphism $I(2) \to I(2)/S(2) = I(3) = S(3)$ is left minimal almost split, with kernel $S(2)$, so that we have an almost split sequence $0 \longrightarrow S(2) \longrightarrow I(2) \longrightarrow S(3) \longrightarrow 0$. Putting together the information we obtained, $\Gamma(\operatorname{mod} A)$ is the quiver

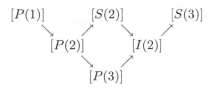

It is customary, when drawing $\Gamma(\operatorname{mod} A)$, to put the translate τx of a nonprojective point x on the same horizontal line as x. We always follow this convention.

4.11. Example. Let A be given by the quiver $\underset{1}{\circ}\xleftarrow{\;\gamma\;}\underset{2}{\circ}\xleftarrow{\;\beta\;}\underset{3}{\circ}\xleftarrow{\;\alpha\;}\underset{4}{\circ}$ bound by $\alpha\beta\gamma = 0$. We have the following list of indecomposable projective or injective A-modules (see (III.2)):

$$P(1) = S(1);$$
$$P(2) = (K \xleftarrow{1} K \longleftarrow 0 \longleftarrow 0);$$
$$P(3) = (K \xleftarrow{1} K \xleftarrow{1} K \longleftarrow 0) = I(1);$$
$$P(4) = (0 \longleftarrow K \xleftarrow{1} K \xleftarrow{1} K) = I(2);$$
$$I(3) = (0 \longleftarrow 0 \longleftarrow K \xleftarrow{1} K);$$
$$I(4) = S(4).$$

We thus have two right minimal almost split morphisms $P(1) \to P(2)$, $P(2) \to P(3)$ and two left minimal almost split morphisms $I(2) \to I(3)$, $I(3) \to I(4)$. Because $P(3)$ and $P(4)$ are projective-injective, we have almost split sequences (by (3.11))

$$0 \longrightarrow P(2) \longrightarrow P(3) \oplus \frac{P(2)}{S(1)} \longrightarrow \frac{P(3)}{S(1)} \longrightarrow 0 \ ;$$

$$0 \longrightarrow \operatorname{rad} P(4) \longrightarrow P(4) \oplus \frac{\operatorname{rad} P(4)}{S(2)} \longrightarrow \frac{P(4)}{S(2)} \longrightarrow 0.$$

Here we observe that $P(2)/S(1) = S(2)$, $P(4)/S(2) = I(3)$, and $\operatorname{rad} P(4) = P(3)/S(1)$ is the indecomposable module M in $\operatorname{mod} A$ given by the diagram $(0 \longleftarrow K \xleftarrow{1} K \longleftarrow 0)$, and $(\operatorname{rad} P(4))/S(2) = S(3)$. Computing successively kernels and cokernels, we obtain $\Gamma(\operatorname{mod} A)$ of the form

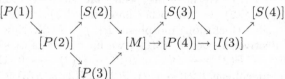

We remark that, if we replace each indecomposable module by its dimension vector, we obtain

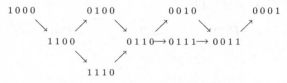

Thus, for each mesh of $\Gamma(\operatorname{mod} A)$ of the form

$$
\begin{array}{ccc}
 & [M_1] & \\
\nearrow & \vdots & \searrow \\
[\tau N] \ {-}{-}{-}{-}{-}{-}{-}{-} & & [N] \\
\searrow & \vdots & \nearrow \\
 & [M_t] &
\end{array}
$$

one has $\mathbf{dim}\,N + \mathbf{dim}\,\tau N = \sum_{i=1}^{t} \mathbf{dim}\,M_i$; this follows from the fact that the corresponding almost split sequence is exact. This seemingly innocent (and trivial) remark gives a method of construction we illustrate in the next example.

4.12. Example. Let A be the K-algebra given by the quiver

$$
\begin{array}{ccc}
 & \overset{2}{\circ} & \\
{}^{\beta}\swarrow & & \searrow{}^{\alpha} \\
1\,\circ & & \circ\,4 \\
{}_{\delta}\nwarrow & & \swarrow{}_{\gamma} \\
 & \underset{3}{\circ} & \searrow{}^{\varepsilon} \\
 & & \circ\,5
\end{array}
$$

bound by $\alpha\beta = \gamma\delta$, $\varepsilon\delta = 0$. Any algebra A whose ordinary quiver Q_A is acyclic admits at least one simple projective module. In our case, there exists only one, namely $P(1)$, whose dimension vector is $1\,{}^{0}_{0}\,{}^{0}_{0}$. We know that no arrow of $\Gamma(\mathrm{mod}\,A)$ ends in $P(1)$ and that the target of each arrow starting at $P(1)$ is projective. In our case, we find two such arrows, namely $[P(1)] \to [P(2)]$ and $[P(1)] \to [P(3)]$ (indeed, $P(1) = \mathrm{rad}\,P(2) = \mathrm{rad}\,P(3)$), which are our first two arrows. Moreover, these are the only arrows of targets $P(2)$ and $P(3)$, respectively. Because $P(1)$ is not injective, we have an almost split sequence

$$0 \longrightarrow P(1) \longrightarrow P(2) \oplus P(3) \longrightarrow \tau^{-1}P(1) \longrightarrow 0.$$

Moreover, $\mathbf{dim}\,\tau^{-1}P(1) = \mathbf{dim}\,P(2) + \mathbf{dim}\,P(3) - \mathbf{dim}\,P(1) = 1\,{}^{1}_{0}\,{}^{0}_{0} + 1\,{}^{0}_{1}\,{}^{0}_{0} - 1\,{}^{0}_{0}\,{}^{0}_{0} = 1\,{}^{1}_{1}\,{}^{0}_{0}$. We see at once that $\tau^{-1}P(1) = \mathrm{rad}\,P(4)$, and hence there is a unique arrow of target $P(4)$, namely $[\tau^{-1}P(1)] \to [P(4)]$. This gives us the beginning of $\Gamma(\mathrm{mod}\,A)$ (where the isomorphism classes of indecomposable A-modules are replaced by their dimension vectors):

$$
\begin{array}{ccccccc}
 & & 1\,{}^{1}_{0}\,{}^{0}_{0} & & & & \\
 & \nearrow & & \searrow & & & \\
1\,{}^{0}_{0}\,{}^{0}_{0} & & & & 1\,{}^{1}_{1}\,{}^{0}_{0} & \longrightarrow & 1\,{}^{1}_{1}\,{}^{1}_{0} \\
 & \searrow & & \nearrow & & & \\
 & & 1\,{}^{0}_{1}\,{}^{0}_{0} & & & &
\end{array}
$$

The calculation of the almost split sequences starting at $P(2)$ and $P(3)$, respectively, gives

$$
\begin{array}{ccccccc}
 & & \begin{smallmatrix}&1&\\1&&0\\&0&\\&0&\end{smallmatrix} & & & & \begin{smallmatrix}&0&\\0&&1\\&0&\\&0&\end{smallmatrix} \\[4pt]
 & \nearrow & & \searrow & & \nearrow & \\[2pt]
\begin{smallmatrix}&0&\\1&&0\\&0&\\&0&\end{smallmatrix} & & & & \begin{smallmatrix}&1&\\1&&0\\&1&\\&0&\end{smallmatrix} & \longrightarrow & \begin{smallmatrix}&1&\\1&&1\\&0&\\&&\end{smallmatrix} \\[2pt]
 & \searrow & & \nearrow & & \searrow & \\[4pt]
 & & \begin{smallmatrix}&1&\\1&&0\\&1&\\&0&\end{smallmatrix} & & & & \begin{smallmatrix}&0&\\0&&1\\&0&\\&0&\end{smallmatrix}
\end{array}
$$

Because $S(3) = \operatorname{rad} P(5)$, there exists a unique arrow of target $P(5)$, namely $[S(3)] \to [P(5)]$. In this way, all the projectives have been obtained. All other indecomposable modules are thus of the form $\tau^{-1}L$, with L indecomposable: to obtain the dimension vector of such a module, we consider the almost split sequence

$$0 \longrightarrow L \longrightarrow M_1 \oplus \ldots \oplus M_t \longrightarrow \tau^{-1}L \longrightarrow 0.$$

Because we can assume by induction that $\mathbf{dim}\, L$ and $\mathbf{dim}\, M_i$ (for all i with $1 \le i \le t$) are known, we deduce $\mathbf{dim}\, \tau^{-1}L = \sum_{i=1}^{t} \mathbf{dim}\, M_i - \mathbf{dim}\, L$. This allows us to construct the rest of $\Gamma(\operatorname{mod} A)$. The construction stops when we reach the injectives; indeed, the left minimal almost split morphism starting at an indecomposable injective $I(a)$ is the projection onto its socle factor $I(a)/S(a)$, and

$$\dim_K I(a) = 1 + \dim_K I(a)/S(a) > \dim_K I(a)/S(a).$$

Thus the previous method would give a dimension vector with negative coordinates, a contradiction. Continuing the construction yields the Auslander–Reiten quiver $\Gamma(\operatorname{mod} A)$

$$
\begin{array}{ccccccccc}
 & & & & \begin{smallmatrix}&0&\\0&&0\\&1&\\&1&\end{smallmatrix} & & & & \begin{smallmatrix}&0&\\0&&1\\&0&\\&&\end{smallmatrix} \\[4pt]
 & & & \nearrow & & \searrow & & \nearrow & \\[2pt]
 & & \begin{smallmatrix}&0&\\0&&0\\&1&\\&0&\end{smallmatrix} & & & & \begin{smallmatrix}&0&\\0&&1\\&1&\\&1&\end{smallmatrix} & & \begin{smallmatrix}&0&\\0&&1\\&0&\\&0&\end{smallmatrix} \\[2pt]
 \begin{smallmatrix}&1&\\1&&0\\&0&\\&0&\end{smallmatrix} & \nearrow & & \searrow & \nearrow & & \searrow & \nearrow & \\[2pt]
\begin{smallmatrix}&0&\\1&&0\\&0&\\&0&\end{smallmatrix} & & \begin{smallmatrix}&1&\\1&&0\\&1&\\&0&\end{smallmatrix} & \longrightarrow & \begin{smallmatrix}&1&\\1&&1\\&0&\\&&\end{smallmatrix} & \longrightarrow & \begin{smallmatrix}&0&\\0&&1\\&1&\\&0&\end{smallmatrix} & & \begin{smallmatrix}&0&\\0&&1\\&1&\\&1&\end{smallmatrix} \\[2pt]
 & \searrow & & \nearrow & & \searrow & & \nearrow & \searrow \\[2pt]
 & & \begin{smallmatrix}&1&\\1&&0\\&1&\\&0&\end{smallmatrix} & & \begin{smallmatrix}&0&\\0&&0\\&0&\\&0&\end{smallmatrix} & & \begin{smallmatrix}&0&\\0&&1\\&1&\\&0&\end{smallmatrix} & & \begin{smallmatrix}&0&\\0&&0\\&0&\\&1&\end{smallmatrix}
\end{array}
$$

4.13. Example. Let A be the K-algebra given by the quiver

$$
\begin{array}{ccccc}
\circ & & \circ & & \\
 & \overset{\lambda}{\nwarrow}\; \overset{\beta}{\nearrow} & & \overset{\alpha}{\nwarrow} & \\
 & \circ & & & \circ \\
 & \underset{\mu}{\swarrow}\; \underset{\delta}{\nwarrow} & & \underset{\gamma}{\swarrow} & \\
\circ & & \circ & &
\end{array}
$$

bound by $\alpha\beta = \gamma\delta$, $\delta\mu = 0$, and $\beta\lambda = 0$. Then $\Gamma(\mathrm{mod}\,A)$ can be constructed as earlier and is of the form

$$
\begin{array}{ccccccccc}
 & & & \begin{smallmatrix}0&1&0\\&1&\\1&&0\end{smallmatrix} & & & & & \\
 & & \nearrow & & \searrow & & & & \\
\begin{smallmatrix}1&0&0\\0&&0\end{smallmatrix} & & \begin{smallmatrix}0&0&0\\&1&0\\1&&0\end{smallmatrix} & & \begin{smallmatrix}0&1&1\\0&&0\end{smallmatrix} & & \begin{smallmatrix}0&0&0\\0&&0\end{smallmatrix} & & \begin{smallmatrix}0&1&1\\0&0&0\end{smallmatrix} \\
\searrow & \nearrow & \searrow & \nearrow & \searrow & \nearrow & \searrow & \nearrow & \\
 & \begin{smallmatrix}1&0&\\1&&0\\1&&0\end{smallmatrix} & & \begin{smallmatrix}0&0&0\\&1&0\\0&&0\end{smallmatrix} & & \begin{smallmatrix}0&1&0\\0&1&0\end{smallmatrix}\!\rightarrow\!\begin{smallmatrix}0&1&1\\0&1&1\end{smallmatrix}\!\rightarrow\!\begin{smallmatrix}0&0&1\\0&1&1\end{smallmatrix} & & & \begin{smallmatrix}0&0&0\\0&0&1\end{smallmatrix} \\
\nearrow & \searrow & \nearrow & \searrow & \nearrow & \searrow & \nearrow & \searrow & \\
\begin{smallmatrix}0&0&0\\1&0&0\end{smallmatrix} & & \begin{smallmatrix}1&1&0\\0&&0\end{smallmatrix} & & \begin{smallmatrix}0&1&0\\0&1&\end{smallmatrix} & & \begin{smallmatrix}0&0&1\\0&0&\end{smallmatrix} & & \begin{smallmatrix}0&0&0\\0&0&1\end{smallmatrix} \\
 & & \searrow & & \nearrow & & & & \\
 & & & \begin{smallmatrix}1&0&0\\0&1&\end{smallmatrix} & & & & & \\
\end{array}
$$

Let M, N, and L be the simple A-modules such that $\dim M = \begin{smallmatrix}&0&\\0&1&0\\&0&\end{smallmatrix}$, $\dim N = \begin{smallmatrix}&0&\\0&0&0\\&&1\end{smallmatrix}$, and $\dim L = \begin{smallmatrix}&0&\\1&0&0\\&0&\end{smallmatrix}$. Because $\dim \tau M = \begin{smallmatrix}&1&\\1&0&0\\&0&\end{smallmatrix}$, we get $\mathrm{Hom}_A(DA, \tau M) = 0$, and (2.7)(a) yields $\mathrm{pd}_A M = 1$.

On the other hand, $\mathrm{pd}_A N \geq 2$, because $\dim \tau N = \begin{smallmatrix}&0&\\0&0&1\\&1&\end{smallmatrix}$ and therefore there is a nonzero homomorphism from the indecomposable injective A-module E of dimension vector $\begin{smallmatrix}&0&\\1&1&1\\&&\end{smallmatrix}$ to the module τN. Then we get $\mathrm{Hom}_A(DA, \tau N) \neq 0$ and (2.7)(a) yields $\mathrm{pd}_A N \geq 2$. Actually, $\mathrm{pd}_A N = 2$, because the minimal projective resolution of N has the form

$$
0 \longrightarrow \begin{smallmatrix}&1&\\1&0&0\\&0&\end{smallmatrix} \longrightarrow \begin{smallmatrix}&1&\\0&0&1\\&0&\end{smallmatrix} \oplus \begin{smallmatrix}&0&\\1&1&0\\&0&\end{smallmatrix} \longrightarrow \begin{smallmatrix}&0&\\0&1&1\\&1&\end{smallmatrix} \longrightarrow \begin{smallmatrix}&0&\\0&0&1\\&&\end{smallmatrix} \longrightarrow 0
$$

Similarly, $\mathrm{id}_A L \geq 2$, because $\dim \tau^{-1}L = \begin{smallmatrix}&0&\\1&1&0\\&1&\end{smallmatrix}$ and there is a nonzero homomorphism from $\tau^{-1}L$ to the indecomposable projective module P of dimension vector $\begin{smallmatrix}&0&\\1&1&1\\&1&\end{smallmatrix}$. It follows that $\mathrm{Hom}_A(\tau^{-1}L, A) \neq 0$ and (2.7)(b) yields $\mathrm{id}_A L \geq 2$.

The method presented in these examples works perfectly well for all finite and acyclic Auslander–Reiten quivers. An interesting remark in this case is that, as suggested by the examples, every indecomposable module is (up to isomorphism) uniquely determined by its dimension vector. This is shown later.

4.14. Example. Let A be the K-algebra given by the quiver

$$
\begin{array}{ccc}
 & \overset{2}{\circ} & \\
{\scriptstyle \beta}\nearrow & & \nwarrow{\scriptstyle \alpha} \\
1\,\circ & \underset{\gamma}{\longleftarrow} & \circ\,3
\end{array}
$$

bound by $\alpha\beta = 0$. Then $\Gamma(\text{mod } A)$ is given by

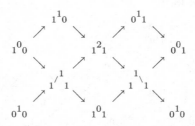

where modules are replaced by their dimension vectors and one must identify the two copies of $S(2) = {}_0 1_0$, thus forming a cycle. Here, ${}_1 {}^1 {}_1$ represents the indecomposable projective module $P(3) = {}_K \overset{0}{\underset{1}{\swarrow}} \overset{K}{\longleftarrow} \overset{1}{\searrow} {}_K$, while ${}_1 {}^1 \backslash_1$ represents the indecomposable injective module $I(1) = {}_K \overset{1}{\swarrow} \overset{K}{\underset{1}{\longleftarrow}} \overset{0}{\searrow} {}_K$. It follows that indecomposable modules are not uniquely determined by their dimension vectors, because $P(3) \not\cong I(1)$ and $\mathbf{dim}\, P(3) = \mathbf{dim}\, I(1)$.

IV.5. The first Brauer–Thrall conjecture

At the origin of many recent developments of representation theory are the following two conjectures attributed to Brauer and Thrall.

Conjecture 1. *A finite dimensional K-algebra is either representation–finite or there exist indecomposable modules with arbitrarily large dimension.*

Conjecture 2. *A finite dimensional algebra over an infinite field K is either representation–finite or there exists an infinite sequence of numbers $d_i \in \mathbb{N}$ such that, for each i, there exists an infinite number of nonisomorphic indecomposable modules with K-dimension d_i.*

The first statement has now been shown to hold true, whenever the field K is arbitrary (see [13], [14], [140], [147], [148], [151], [154], [170]), and the second one when K is algebraically closed (see [26], [27], [124], [140], [162], and for historical notes see [83]). Our objective in this section is to give a simple proof of the first conjecture.

Let A be a finite dimensional K-algebra. A sequence of irreducible morphisms in mod A of the form

$$M_0 \xrightarrow{f_1} M_1 \xrightarrow{f_2} \cdots \xrightarrow{f_t} M_t$$

with all the M_i indecomposables is called a **chain of irreducible morphisms** from M_0 to M_t of length t.

5.1. Lemma. *Let $t \in \mathbb{N}$ and let M and N be indecomposable right A-modules with $\mathrm{Hom}_A(M, N) \neq 0$. Assume that there exists no chain of irreducible morphisms from M to N of length $< t$.*

(a) *There exists a chain of irreducible morphisms*

$$M = M_0 \xrightarrow{f_1} M_1 \xrightarrow{f_2} M_2 \longrightarrow \cdots \xrightarrow{f_t} M_t$$

and a homomorphism $g : M_t \to N$ with $g f_t \ldots f_2 f_1 \neq 0$.

(b) *There exists a chain of irreducible morphisms*

$$N_t \xrightarrow{g_t} N_{t-1} \xrightarrow{g_{t-1}} \cdots \longrightarrow N_1 \xrightarrow{g_1} N_0 = N$$

and a homomorphism $f : M \to N_t$ with $g_1 \ldots g_t f \neq 0$.

Proof. We only prove (a); the proof of (b) is similar. We proceed by induction on t. For $t = 0$, there is nothing to show. Assume thus that M and N are given with $\mathrm{Hom}_A(M, N) \neq 0$ and that there is no chain of irreducible morphisms from M to N of length $< t + 1$. By the induction hypothesis, there exists a chain of irreducible morphisms

$$M = M_0 \xrightarrow{f_1} M_1 \xrightarrow{f_2} \cdots \xrightarrow{f_t} M_t$$

and a homomorphism $g : M_t \to N$ with $g f_t \ldots f_1 \neq 0$. The induction hypothesis implies that g cannot be an isomorphism. Because M_t and N are indecomposable, g is not a section. We consider the left minimal almost split morphism starting with M_t

$$h = \begin{bmatrix} h_1 \\ \vdots \\ h_1 \end{bmatrix} : M_t \longrightarrow \bigoplus_{j=1}^{s} L_j,$$

where the modules $L_1, \ldots, L_s$ are indecomposable. Then g factors through h, that is, there exists $u = [u_1, \ldots, u_s] : \bigoplus_{j=1}^{s} L_j \longrightarrow N$ such that $g = uh = \sum_{j=1}^{s} u_j h_j$. Thus, because $0 \neq g f_t \ldots f_1 = \sum_{j=1}^{s} u_j h_j f_t \ldots f_1$, there exists j such that $1 \leq j \leq s$ and $u_j h_j f_t \ldots f_1 \neq 0$. Setting $M_{t+1} = L_j$, $f_{t+1} = h_j$ and $g' = u_j$, our claim follows from the fact that h_j is irreducible. $\square$

5.2. Lemma (Harada and Sai). *For a natural number b, let*

$$M_1 \xrightarrow{f_1} M_2 \xrightarrow{f_2} M_3 \to \cdots \to M_{2^b-1} \xrightarrow{f_{2^b-1}} M_{2^b}$$

be a chain of nonzero nonisomorphisms in $\operatorname{mod} A$, *with all M_i indecomposables of length $\leq b$. Then $f_{2^b-1} \cdots f_2 f_1 = 0$.*

Proof. We show by induction on n that if

$$M_1 \xrightarrow{f_1} M_2 \xrightarrow{f_2} M_3 \to \cdots \to M_{2^n-1} \xrightarrow{f_{2^n-1}} M_{2^n}$$

is a sequence of nonzero nonisomorphisms between indecomposable modules of length $\leq b$, then the length of the image of the composite homomorphism $f_{2^n-1} \cdots f_2 f_1$ is $\leq b - n$. This will imply the statement upon setting $b = n$.

Let $n = 1$. If the length $\ell(\operatorname{Im} f_1)$ of $\operatorname{Im} f_1$ is equal to b, then f_1 is an isomorphism, a contradiction that shows that $\ell(\operatorname{Im} f_1) \leq b - 1$. Assume that the statement holds for n, and let

$$M_1 \xrightarrow{f_1} M_2 \xrightarrow{f_2} \cdots \to M_{2^n-1} \xrightarrow{f_{2^n-1}} M_{2^n} \xrightarrow{f_{2^n}} M_{2^n+1} \xrightarrow{f_{2^n+1}} \cdots \xrightarrow{f_{2^{n+1}-1}} M_{2^{n+1}}$$

be a sequence of nonzero nonisomorphisms between indecomposable modules of length $\leq b$. We consider the two homomorphisms $f = f_{2^n-1} \cdots f_2 f_1$ and $h = f_{2^{n+1}-1} \cdots f_{2^n+1}$. By the induction hypothesis, $\ell(\operatorname{Im} f) \leq b - n$ and $\ell(\operatorname{Im} h) \leq b - n$. If at least one of these two inequalities is strict, we are done. We may thus suppose that $\ell(\operatorname{Im} f) = \ell(\operatorname{Im} h) = b - n > 0$. Let $g = f_{2^n}$. We must show that $\ell(\operatorname{Im} hgf) \leq b - n - 1$.

We claim that if this is not the case, then g is an isomorphism, a contradiction that completes the proof. Assume thus that $\ell(\operatorname{Im} hgf) > b - n - 1$. Because $\ell(\operatorname{Im} hgf) \leq \ell(\operatorname{Im} f) = b - n$, this implies that $\ell(\operatorname{Im} hgf) = b - n$. Now

$$\ell(\operatorname{Im} hgf) = \ell\left(\frac{\operatorname{Im} f}{\operatorname{Im} f \cap \operatorname{Ker} hg}\right) = \ell(\operatorname{Im} f) - \ell(\operatorname{Im} f \cap \operatorname{Ker} hg).$$

This implies that $\ell(\operatorname{Im} f \cap \operatorname{Ker} hg) = 0$, hence $\operatorname{Im} f \cap \operatorname{Ker} hg = 0$. On the other hand, $\operatorname{Im} hgf \subseteq \operatorname{Im} hg \subseteq \operatorname{Im} h$ and $\ell(\operatorname{Im} hgf) = \ell(\operatorname{Im} h) = b - n$ give $\ell(\operatorname{Im} hg) = b - n$. Consequently,

$$\ell(\operatorname{Ker} hg) = \ell(M_{2^n}) - \ell(\operatorname{Im} hg) = \ell(M_{2^n}) - (b - n) = \ell(M_{2^n}) - \ell(\operatorname{Im} f).$$

This shows that $M_{2^n} = \operatorname{Im} f \oplus \operatorname{Ker} hg$. Because M_{2^n} is indecomposable and $f \neq 0$, we have $\operatorname{Ker} hg = 0$. Therefore hg is a monomorphism. Hence g itself is a monomorphism. Similarly, one shows that $\operatorname{Im} gf \cap \operatorname{Ker} h = 0$, hence that $M_{2^n+1} = \operatorname{Im} gf \oplus \operatorname{Ker} h$. Because $gf \neq 0$ and the module M_{2^n+1} is indecomposable then we get $M_{2^n+1} = \operatorname{Im} gf$, so that gf and therefore g are epimorphisms. This completes the proof that g is an isomorphism, and hence of the lemma. $\square$

The following example shows that the bounds given in the Harada–Sai lemma are the best bounds possible.

5.3. Example. Let A be given by the quiver

consisting of two loops α and β, bound by $\alpha^2 = 0$, $\beta^2 = 0$, $\alpha\beta = 0$, and $\beta\alpha = 0$.

We construct 7 indecomposable A-modules of length ≤ 3 and 6 nonisomorphisms between them with nonzero composition.

The algebra A admits a unique simple module S_A and any A-module can be written in a form of a triple $(V, \varphi_\alpha, \varphi_\beta)$, where V is a finite dimensional K-vector space and $\varphi_\alpha, \varphi_\beta : V \to V$ are K-linear endomorphisms satisfying the conditions $\varphi_\alpha^2 = 0$, $\varphi_\beta^2 = 0$, $\varphi_\alpha\varphi_\beta = \varphi_\beta\varphi_\alpha = 0$, and a morphism $(V, \varphi_\alpha, \varphi_\beta) \to (V', \varphi'_\alpha, \varphi'_\beta)$ is a K-linear map $f : V \to V'$ such that $\varphi'_\alpha f = f\varphi_\alpha$ and $\varphi'_\beta f = f\varphi_\beta$. Let thus

$$
\begin{aligned}
M_1 &= M_5 = A_A &&= (K^3, \begin{bmatrix} 0&0&0 \\ 1&0&0 \\ 0&0&0 \end{bmatrix}, \begin{bmatrix} 0&0&0 \\ 0&0&0 \\ 1&0&0 \end{bmatrix}), \\
M_2 &= M_6 = A_A/S &&= (K^2, \begin{bmatrix} 0&0 \\ 1&0 \end{bmatrix}, 0), \\
M_3 &= M_7 = (DA)_A &&= (K^3, \begin{bmatrix} 0&1&0 \\ 0&0&0 \\ 0&0&0 \end{bmatrix}, \begin{bmatrix} 0&0&1 \\ 0&0&0 \\ 0&0&0 \end{bmatrix}), \\
M_4 &= S_A = (K, 0, 0).
\end{aligned}
$$

Each of these modules has a simple top or a simple socle and hence is indecomposable. Let now

$$
f_1 = \begin{bmatrix} 1&0&0 \\ 0&0&1 \end{bmatrix} : M_1 \longrightarrow M_2, \qquad f_2 = \begin{bmatrix} 0&1 \\ 0&0 \\ 1&0 \end{bmatrix} : M_2 \longrightarrow M_3,
$$

$$
f_3 = \begin{bmatrix} 1&0&0 \end{bmatrix} : M_3 \longrightarrow M_4, \qquad f_4 = \begin{bmatrix} 0 \\ 0 \\ 1 \end{bmatrix} : M_4 \longrightarrow M_5,
$$

$$
f_5 = \begin{bmatrix} 1&0&0 \\ 0&0&1 \end{bmatrix} : M_5 \longrightarrow M_6, \qquad f_6 = \begin{bmatrix} 1&0 \\ 0&0 \\ 0&1 \end{bmatrix} : M_6 \longrightarrow M_7.
$$

It is easily checked that each of these matrices defines an A-module homomorphism, and $f_6 f_5 f_4 f_3 f_2 f_1 = \begin{bmatrix} 0&0&0 \\ 0&0&0 \\ 0&0&1 \end{bmatrix} \neq 0$.

We are now able to prove our criterion of representation–finiteness, which was announced in the previous section and implicitly used in the construction of Auslander–Reiten quivers.

5.4. Theorem. *Assume that A is a basic and connected finite dimensional K-algebra. If $\Gamma(\mathrm{mod}\, A)$ admits a connected component $\mathcal{C}$ whose* *modules are of bounded length, then* $\mathcal{C}$ *is finite and* $\mathcal{C} = \Gamma(\mathrm{mod}\, A)$. *In particular,* A *is representation–finite.*

Proof. Let b be a bound for the length of the indecomposable modules X with $[X]$ in $\mathcal{C}$. Let M, N be two indecomposable A-modules such that $\mathrm{Hom}_A(M, N) \neq 0$. If $[M] \in \mathcal{C}_0$, there exists a chain of irreducible morphisms from M to N of length smaller than $2^b - 1 = t$, and in particular $[N] \in \mathcal{C}_0$. Indeed, if this is not the case, there exists, by (5.1), a chain of irreducible morphisms

$$M = M_0 \xrightarrow{f_1} M_1 \xrightarrow{f_2} M_2 \to \cdots \to M_{t-1} \xrightarrow{f_t} M_t$$

and a homomorphism $g : M_t \to N$ with $g f_t \ldots f_1 \neq 0$. However, (5.2) yields $f_t \ldots f_1 = 0$, a contradiction that shows our claim. Similarly, if $[N] \in \mathcal{C}_0$, we have $[M] \in \mathcal{C}_0$.

Let now $[M] \in \mathcal{C}_0$ be arbitrary. There exists an indecomposable projective module P_A such that $\mathrm{Hom}_A(P, M) \neq 0$; hence we also have $[P] \in \mathcal{C}_0$. It follows from (II.3.4) and (I.5.17) that, for any other indecomposable projective P', there exists a sequence of indecomposable projective modules $P = P_0, P_1, \ldots, P_s = P'$ such that $\mathrm{Hom}_A(P_{i-1}, P_i) \neq 0$ or $\mathrm{Hom}_A(P_i, P_{i-1}) \neq 0$ for each $1 \leq i \leq s$, because the algebra A is connected, $P \cong e_a A$ and $P' \cong e_b A$ for some primitive orthogonal idempotents e_a, e_b of A, and (I.4.2) yields $\mathrm{Hom}_A(e_a A, e_b A) \cong e_b A e_a$. Hence $[P'] \in \mathcal{C}_0$. We deduce that any indecomposable A-module X corresponds to a point $[X]$ in $\mathcal{C}$, because there exists an indecomposable projective A-module P' such that $\mathrm{Hom}_A(P', X) \neq 0$. This shows that $\mathcal{C} = \Gamma(\mathrm{mod}\, A)$.

On the other hand, for each indecomposable projective A-module P and each indecomposable A-module M such that $\mathrm{Hom}_A(P, M) \neq 0$, we know that there exists a chain of irreducible morphisms from P to M of length smaller than $t = 2^b - 1$. Because there are only finitely many nonisomorphic indecomposable projectives, there are only finitely many nonisomorphic indecomposable modules corresponding to points in $\mathcal{C}$. Hence A is representation–finite. $\qquad\square$

As a consequence of (5.4) we get the validity of the first Brauer–Thrall conjecture.

5.5. Corollary. *Any algebra is either representation–finite or admits indecomposable modules of arbitrary length.* $\qquad\square$

We end this section with the following corollary, which underlines the importance of the irreducible morphisms and hence of the Auslander–Reiten quiver, for the description of the module category of a representation–finite algebra.

5.6. Corollary. *Let A be a representation–finite algebra. Any nonzero nonisomorphism between indecomposable modules in* mod A *is a sum of compositions of irreducible morphisms.*

Proof. Let M, N be indecomposable A-modules and $t \geq 1$. Denote by $\operatorname{rad}_A^t(M, N)$ the K-subspace of $\operatorname{rad}_A(M, N)$ consisting of the K-linear combinations of compositions $f_1 f_2 \ldots f_t$, where $f_1, f_2, \ldots, f_t$ are nonisomorphisms between indecomposable A-modules. Because A is representation–finite, the lengths of the indecomposable A-modules are bounded; hence, by the Harada–Sai lemma (5.2), there exists $m \geq 1$ such that $\operatorname{rad}_A^{m+1}(M, N) = 0$ for all M and N.

Let $g \in \operatorname{rad}_A(M, N)$ be nonzero. If $g \notin \operatorname{rad}_A^2(M, N)$, then g is irreducible and there is nothing to prove. If $g \in \operatorname{rad}_A^2(M, N)$, there exists s such that $2 \leq s \leq m$ and $g \in \operatorname{rad}_A^s(M, N) \setminus \operatorname{rad}_A^{s+1}(M, N)$.

We prove our statement by descending induction on s. If $s = m$, then g is a sum of nonzero compositions $g_1 \cdot g_2 \cdot \ldots \cdot g_m$ of nonisomorphisms $g_1, g_2, \ldots, g_m$ between indecomposable modules. Because $\operatorname{rad}_A^{m+1}(M, N) = 0$, the homomorphisms $g_1, \ldots, g_m$ do not belong to the square of the radical and therefore are irreducible. This proves the statement for $s = m$. Suppose that $s \leq m - 1$. Then g is a sum of nonzero compositions $g_1 g_2 \ldots g_s$ of nonisomorphisms between indecomposable modules. Let g' denote the sum of all the summands $g_1 g_2 \ldots g_s$ of g in which all the homomorphisms $g_1, g_2, \ldots, g_s$ are irreducible. Then $g'' = g - g' \in \operatorname{rad}_A^{s+1}(M, N)$. If $g'' = 0$, the statement is trivial. If $g'' \neq 0$, then, by the induction hypothesis, g'' is a sum of compositions of irreducible morphisms and therefore so is $g = g' + g''$. The proof is now complete. $\square$

IV.6. Functorial approach to almost split sequences

Let A be a finite dimensional K-algebra. We present in this section an interpretation of the almost split sequences in mod A in terms of the projective resolutions of the simple objects in the categories $\mathcal{F}un^{op} A$ and $\mathcal{F}un\, A$ of the contravariant, and covariant, respectively, K-linear functors from the category mod A of finitely generated right A-modules into the category mod K of finite dimensional K-vector spaces. These categories are defined in Section A.2 of the Appendix and are both seen to be abelian. We recall that, given a pair of functors F and G in the category $\mathcal{F}un^{op} A$ (or in $\mathcal{F}un\, A$), we denote by $\operatorname{Hom}(F, G)$ the set of functorial morphisms

$\varphi : F \to G$.

Of particular interest in our study is the following classical result.

6.1. Theorem (Yoneda's lemma). *Let $\mathcal{C}$ be an additive K-category and X be an object in $\mathcal{C}$.*

(a) *For any contravariant functor $F : \mathcal{C} \longrightarrow \operatorname{mod} K$, the correspondence $\pi : \varphi \mapsto \varphi_X(1_X)$ defines a bijection between the set $\operatorname{Hom}(\operatorname{Hom}_{\mathcal{C}}(-, X), F)$ of functorial morphisms $\varphi : \operatorname{Hom}_{\mathcal{C}}(-, X) \longrightarrow F$ and the set $F(X)$.*

(b) *For any covariant functor $F : \mathcal{C} \longrightarrow \operatorname{mod} K$, the correspondence $\pi : \varphi \mapsto \varphi_X(1_X)$ defines a bijection between the set $\operatorname{Hom}(\operatorname{Hom}_{\mathcal{C}}(X, -), F)$ of functorial morphisms $\varphi : \operatorname{Hom}_{\mathcal{C}}(X, -) \longrightarrow F$ and the set $F(X)$.*

Proof. We only prove (a); the proof of (b) is similar. For a functorial morphism $\varphi : \operatorname{Hom}_{\mathcal{C}}(-, X) \longrightarrow F$, we have $\varphi_X(1_X) \in F(X)$, so π defines a map $\operatorname{Hom}(\operatorname{Hom}_{\mathcal{C}}(-, X), F) \longrightarrow F(X)$. We now construct its inverse

$$\sigma : F(X) \longrightarrow \operatorname{Hom}(\operatorname{Hom}_{\mathcal{C}}(-, X), F).$$

Let $a \in F(X)$ and Y be an arbitrary object in $\mathcal{C}$. We define the map $\sigma(a)_Y : \operatorname{Hom}_{\mathcal{C}}(Y, X) \longrightarrow F(Y)$ to be given by $\sigma(a)_Y(f) = F(f)(a)$, for $f \in \operatorname{Hom}_{\mathcal{C}}(Y, X)$.

To show that $\sigma(a) : \operatorname{Hom}_{\mathcal{C}}(-, X) \longrightarrow F$ is a functorial morphism, we must show that, for any morphism $g : Y \to Z$, the following diagram is commutative

$$
\begin{array}{ccc}
\operatorname{Hom}_{\mathcal{C}}(Y, X) & \xrightarrow{\ \sigma(a)_Y\ } & F(Y) \\[4pt]
{\scriptstyle \operatorname{Hom}_{\mathcal{C}}(g, X)}\Big\uparrow & & \Big\uparrow{\scriptstyle F(g)} \\[4pt]
\operatorname{Hom}_{\mathcal{C}}(Z, X) & \xrightarrow{\ \sigma(a)_Z\ } & F(Z)
\end{array}
$$

Let thus $f \in \operatorname{Hom}_{\mathcal{C}}(Z, X)$; then $F(g)\sigma(a)_Z(f) = F(g)F(f)(a) = F(f \circ g)(a)$, while $\sigma(a)_Y \operatorname{Hom}_{\mathcal{C}}(g, X)(f) = \sigma(a)_Y(f \circ g) = F(f \circ g)(a)$.

It remains to show that π and σ are mutually inverse.

(i) Let $a \in F(X)$. To prove that $\pi\sigma(a) = a$, we note that

$$\pi\sigma(a) = \sigma(a)_X(1_X) = F(1_X)(a) = 1_{F(X)}(a) = a.$$

(ii) Let $\varphi \in \operatorname{Hom}(\operatorname{Hom}_{\mathcal{C}}(-, X), F)$. To prove that $\sigma\pi(\varphi) = \varphi$, we show that, for any object Y in $\mathcal{C}$, we have $\sigma\pi(\varphi)_Y = \varphi_Y$. By definition, for any $f \in \operatorname{Hom}_{\mathcal{C}}(Y, X)$, we have

$$\sigma\pi(\varphi)_Y(f) = F(f)(\pi(\varphi)) = F(f)\varphi_X(1_X).$$

Because φ is a functorial morphism, the following diagram is commutative:

$$\begin{array}{ccc}
\operatorname{Hom}_{\mathcal{C}}(X,X) & \xrightarrow{\varphi_X} & F(X) \\
\scriptstyle{\operatorname{Hom}_{\mathcal{C}}(f,X)}\downarrow & & \downarrow{\scriptstyle F(f)} \\
\operatorname{Hom}_{\mathcal{C}}(Y,X) & \xrightarrow{\varphi_Y} & F(Y)
\end{array}$$

That is, $F(f)\varphi_X = \varphi_Y \operatorname{Hom}_{\mathcal{C}}(f,X)$. Thus we have

$$\sigma\pi(\varphi)_Y(f) = \varphi_Y \operatorname{Hom}_{\mathcal{C}}(f,X)(1_X) = \varphi_Y(f)$$

and the proof is complete. □

6.2. Corollary. *Let $\mathcal{C}$ be an additive K-category and let X be an object in $\mathcal{C}$.*

(a) *Let F be a subfunctor of $\operatorname{Hom}_{\mathcal{C}}(-,X)$. The map $f \mapsto \operatorname{Hom}_{\mathcal{C}}(-,f)$ is a bijection $F(X) \cong \operatorname{Hom}(\operatorname{Hom}_{\mathcal{C}}(-,X),F)$. In particular, for any object Y in $\mathcal{C}$, the map $\operatorname{Hom}_{\mathcal{C}}(X,Y) \longrightarrow \operatorname{Hom}(\operatorname{Hom}_{\mathcal{C}}(-,X),\operatorname{Hom}_{\mathcal{C}}(-,Y))$ given by $f \mapsto \operatorname{Hom}_{\mathcal{C}}(-,f)$ is a bijection.*

(b) *Let F be a subfunctor of $\operatorname{Hom}_{\mathcal{C}}(X,-)$. The map $f \mapsto \operatorname{Hom}_{\mathcal{C}}(f,-)$ is a bijection $F(X) \cong \operatorname{Hom}(\operatorname{Hom}_{\mathcal{C}}(X,-),F)$. In particular, for any object Y in $\mathcal{C}$, the map $\operatorname{Hom}_{\mathcal{C}}(X,Y)\longrightarrow\operatorname{Hom}(\operatorname{Hom}_{\mathcal{C}}(Y,-),\operatorname{Hom}_{\mathcal{C}}(X,-))$ given by $f \mapsto \operatorname{Hom}_{\mathcal{C}}(f,-)$ is a bijection.*

Proof. We only prove (a); the proof of (b) is similar. Let $f \in F(X) \subseteq \operatorname{Hom}_{\mathcal{C}}(X,X)$. It was shown that the inverse of the bijection π in Yoneda's lemma 6.1 is given by $\sigma(f) : \operatorname{Hom}_{\mathcal{C}}(-,X) \longrightarrow F$. We show that $\sigma(f) = \operatorname{Hom}_{\mathcal{C}}(-,f)$. Indeed, let Y be an object in $\mathcal{C}$ and $g \in \operatorname{Hom}_{\mathcal{C}}(Y,X)$; then $\sigma(f)_Y(g) = F(g)(f) = f \circ g = \operatorname{Hom}_{\mathcal{C}}(Y,f)(g)$ because, by definition, $F(g) \in F(Y) \subseteq \operatorname{Hom}_{\mathcal{C}}(Y,X)$. This shows the first assertion. The second follows from the first applied to the functor $F = \operatorname{Hom}_{\mathcal{C}}(-,Y)$. □

In particular, it follows from (6.2) that the categories $\mathcal{F}un^{\mathrm{op}}A$ and $\mathcal{F}un\,A$ are not only abelian, they are also additive K-categories. As a second corollary, we now show that a Hom functor uniquely determines the representing object.

6.3. Corollary. *Let $\mathcal{C}$ be an additive K-category and let X, Y be two objects in $\mathcal{C}$.*

(a) *$X \cong Y$ if and only if $\operatorname{Hom}_{\mathcal{C}}(-,X) \cong \operatorname{Hom}_{\mathcal{C}}(-,Y)$.*

(b) *$X \cong Y$ if and only if $\operatorname{Hom}_{\mathcal{C}}(X,-) \cong \operatorname{Hom}_{\mathcal{C}}(Y,-)$.*

Proof. We only prove (a); the proof of (b) is similar. Clearly, $X \cong Y$ implies $\operatorname{Hom}_{\mathcal{C}}(-,X) \cong \operatorname{Hom}_{\mathcal{C}}(-,Y)$. Conversely, assume that there is an isomorphism $\operatorname{Hom}_{\mathcal{C}}(-,X) \cong \operatorname{Hom}_{\mathcal{C}}(-,Y)$ of functors. By (6.2), there exist morphisms $f : X \to Y$ and $g : Y \to X$ in $\mathcal{C}$ such that $\operatorname{Hom}_{\mathcal{C}}(-,f) : \operatorname{Hom}_{\mathcal{C}}(-,X)\to\operatorname{Hom}_{\mathcal{C}}(-,Y)$ and $\operatorname{Hom}_{\mathcal{C}}(-,g) : \operatorname{Hom}_{\mathcal{C}}(-,Y)\to\operatorname{Hom}_{\mathcal{C}}(-,X)$

are mutually inverse functorial isomorphisms. Thus the equalities
$$\mathrm{Hom}_{\mathcal{C}}(-, 1_X) = 1_{\mathrm{Hom}_{\mathcal{C}}(-,X)} = \mathrm{Hom}_{\mathcal{C}}(-, g) \circ \mathrm{Hom}_{\mathcal{C}}(-, f) = \mathrm{Hom}_{\mathcal{C}}(-, g \circ f)$$
give $g \circ f = 1_X$, by (6.2) again. Similarly, $f \circ g = 1_Y$. □

An object P in $\mathcal{F}un^{\mathrm{op}} A$ (or in $\mathcal{F}un\, A$) is said to be **projective** if for any functorial epimorphism $\varphi : F \to G$, the induced map of K-vector spaces $\mathrm{Hom}(P, \varphi) : \mathrm{Hom}(P, F) \longrightarrow \mathrm{Hom}(P, G)$, given by $\psi \mapsto \varphi\psi$, is surjective.

We now observe that Yoneda's lemma also gives projective objects in the categories $\mathcal{F}un^{\mathrm{op}} A$ and $\mathcal{F}un\, A$.

6.4. Corollary. *Let A be a K-algebra and M be a module in* $\mathrm{mod}\, A$.
(a) *The functor* $\mathrm{Hom}_A(-, M)$ *is a projective object in* $\mathcal{F}un^{\mathrm{op}} A$.
(b) *The functor* $\mathrm{Hom}_A(M, -)$ *is a projective object in* $\mathcal{F}un A$.

Proof. We only prove (a); the proof of (b) is similar. We must prove that, for any functorial epimorphism $\varphi : F \to G$, the induced map

$$\mathrm{Hom}(\mathrm{Hom}_A(-, M), \varphi) : \mathrm{Hom}(\mathrm{Hom}_A(-, M), F) \longrightarrow \mathrm{Hom}(\mathrm{Hom}_A(-, M), G)$$

given by $\psi \mapsto \varphi\psi$, is surjective. We claim that the following diagram

$$
\begin{array}{ccc}
\mathrm{Hom}(\mathrm{Hom}_A(-, M), F) & \xrightarrow{\ \mathrm{Hom}(\mathrm{Hom}_A(-,M),\varphi)\ } & \mathrm{Hom}(\mathrm{Hom}_A(-, M), G) \\
{\scriptstyle \pi^F}\downarrow {\scriptstyle \cong} & & {\scriptstyle \cong}\downarrow {\scriptstyle \pi^G} \\
F(M) & \xrightarrow{\qquad \varphi_M \qquad} & G(M)
\end{array}
$$

is commutative, where π^F and π^G denote the bijection π in Yoneda's lemma 6.1 applied to F and G, respectively. Indeed, let $\psi \in \mathrm{Hom}(\mathrm{Hom}_A(-, M), F)$, then

$$
\begin{aligned}
\varphi_M \pi^F(\psi) &= \varphi_M \psi_M(1_M) = (\varphi\psi)_M(1_M) = \pi^G(\varphi\psi) \\
&= \pi^G \mathrm{Hom}(\mathrm{Hom}_A(-, M), \varphi)(\psi).
\end{aligned}
$$

On the other hand, φ_M is surjective, because φ is a functorial epimorphism. Hence so is $\mathrm{Hom}(\mathrm{Hom}_A(-, M), \varphi)$. □

A functor F in $\mathcal{F}un^{\mathrm{op}} A$ (or in $\mathcal{F}un\, A$) is called **finitely generated** if F is isomorphic to a quotient of a functor of the form $\mathrm{Hom}_A(-, M)$ (or $\mathrm{Hom}_A(M, -)$, respectively) for some A-module M, that is, there exists a functorial epimorphism $\mathrm{Hom}_A(-, M) \longrightarrow F \longrightarrow 0$, (or a functorial epimorphism $\mathrm{Hom}_A(M, -) \longrightarrow F \longrightarrow 0$, respectively).

We now characterise the finitely generated projective objects in our functor categories $\mathcal{F}un^{\mathrm{op}} A$ and $\mathcal{F}un\, A$.

6.5. Lemma. (a) *An object in $\mathcal{F}un^{\mathrm{op}} A$ is finitely generated projective if and only if it is isomorphic to a functor of the form $\mathrm{Hom}_A(-, M)$, for*

M an A-module. Such a functor is indecomposable if and only if M is indecomposable.

(b) *An object in $\mathcal{F}un\, A$ is finitely generated projective if and only if it is isomorphic to a functor of the form $\mathrm{Hom}_A(M, -)$, for M an A-module. Such a functor is indecomposable if and only if M is indecomposable.*

Proof. We only prove (a); the proof of (b) is similar. The projectivity of the finitely generated functor $\mathrm{Hom}_A(-, M)$ follows from (6.4). Conversely, let F be a finitely generated projective object in $\mathcal{F}un^{\mathrm{op}}A$, then there exists a functorial epimorphism $\varphi : \mathrm{Hom}_A(-, X) \longrightarrow F$, for some A-module X. Because F is projective, φ is a retraction and so there exists a functorial monomorphism $\psi : F \longrightarrow \mathrm{Hom}_A(-, X)$ such that $\varphi\psi = 1_F$. Let $\pi = \psi\varphi : \mathrm{Hom}_A(-, X) \longrightarrow F \longrightarrow \mathrm{Hom}_A(-, X)$ (thus, $F = \mathrm{Im}\,\pi$). By (6.2), there exists an endomorphism f of X such that $\pi = \mathrm{Hom}_A(-, f)$. Because π is an idempotent, we have $\mathrm{Hom}_A(-, f^2) = \mathrm{Hom}_A(-, f)^2 = \pi^2 = \pi = \mathrm{Hom}_A(-, f)$ thus $f^2 = f$, again by (6.2), that is, f is an idempotent. Consequently, $M = \mathrm{Im}\,f$ is a direct summand of X. Because $\mathrm{Hom}_A(-, M)$ is the image of $\mathrm{Hom}_A(-, f)$, we deduce that $F \cong \mathrm{Hom}_A(-, M)$. The same argument shows the last assertion. $\square$

We now show that if M is an indecomposable module, the Hom functors $\mathrm{Hom}_A(-, M)$ and $\mathrm{Hom}_A(M, -)$ behave, in their respective categories, in a similar way to the finitely generated indecomposable projective modules over a finite dimensional algebra, in the sense that they have simple tops.

6.6. Lemma. *Let M be an indecomposable A-module.*

(a) *The functor $\mathrm{rad}_A(-, M)$ is the unique maximal subfunctor of the functor $\mathrm{Hom}_A(-, M)$.*

(b) *The functor $\mathrm{rad}_A(M, -)$ is the unique maximal subfunctor of the functor $\mathrm{Hom}_A(M, -)$.*

Proof. We only prove (a); the proof of (b) is similar. It suffices to show that any proper subfunctor F of $\mathrm{Hom}_A(-, M)$ is contained in $\mathrm{rad}_A(-, M)$, that is, for any indecomposable A-module N, we have $F(N) \subseteq \mathrm{rad}_A(N, M)$. If $N \not\cong M$, this follows from the fact that, by (A.3.5) of the Appendix, $\mathrm{rad}_A(N, M) = \mathrm{Hom}_A(N, M)$. Assume thus $N \cong M$ and let $f : M \to M$ belong to $F(M)$. By (6.2), $\mathrm{Hom}_A(-, f)$ maps $\mathrm{Hom}_A(-, M)$ to F, which is a proper subfunctor of $\mathrm{Hom}_A(-, M)$. Consequently, the functorial morphism $\mathrm{Hom}_A(-, f) : \mathrm{Hom}_A(-, M) \longrightarrow F \longrightarrow \mathrm{Hom}_A(-, M)$ is not an isomorphism. Hence neither is f and thus $f \in \mathrm{rad}_A(M, M)$. $\square$

A nonzero functor is called **simple** if it has no nontrivial subfunctor.

Lemma 6.6 thus implies the following corollary.

6.7. Corollary. *Let M be an indecomposable A-module.*
(a) *The functor $S^M = \operatorname{Hom}_A(-, M)/\operatorname{rad}_A(-, M)$ is simple in $\mathscr{F}un^{op}A$.*
(b) *The functor $S_M = \operatorname{Hom}_A(M, -)/\operatorname{rad}_A(M, -)$ is simple in $\mathscr{F}un\, A$.*

$\square$

In particular, $S^M(M) \cong S_M(M) \cong \operatorname{End} M/\operatorname{rad} \operatorname{End} M$ is a one-dimensional K-vector space (because the module M is indecomposable). By (6.2), this implies that $\operatorname{Hom}(\operatorname{Hom}_A(-, M), S^M)$ and $\operatorname{Hom}(\operatorname{Hom}_A(M, -), S_M)$ are also one-dimensional K-vector spaces and hence there exist nonzero functorial morphisms

$$\pi^M : \operatorname{Hom}_A(-, M) \longrightarrow S^M \quad \text{and} \quad \pi_M : \operatorname{Hom}_A(M, -) \longrightarrow S_M$$

that are uniquely determined up to a scalar multiple. Moreover, π^M and π_M are necessarily epimorphisms, because their targets are simple.

On the other hand, Corollary 6.7 also implies that if X is an indecomposable A-module not isomorphic to M, we have $S^M(X) = 0$ and $S_M(X) = 0$. Therefore the explicit expression of the functorial morphisms π^M and π_M follows from the proof of Yoneda's lemma, that is, if X is an indecomposable A-module, the morphisms $\pi^M(X) : \operatorname{Hom}_A(X, M) \longrightarrow S^M(X)$ and $\pi_M(X) : \operatorname{Hom}_A(M, X) \longrightarrow S_M(X)$ are both isomorphic to the canonical surjection $\operatorname{End} M \longrightarrow \operatorname{End} M/\operatorname{rad} \operatorname{End} M$ if $X \cong M$ and are zero otherwise.

Following (I.5.6), a functorial epimorphism $\varphi : F \to G$ in $\mathscr{F}un^{op}A$ (or in $\mathscr{F}un\, A$) is called **minimal** if, for each functorial morphism $\psi : H \to F$, the composite morphism $\varphi\psi$ is an epimorphism if and only if ψ is an epimorphism. A minimal functorial epimorphism $\varphi : F \to G$, with F projective, is called a **projective cover** of G.

An exact sequence $F_1 \xrightarrow{\varphi_1} F_0 \xrightarrow{\varphi_0} G \longrightarrow 0$ in $\mathscr{F}un^{op}A$ (or in $\mathscr{F}un A$) is called a **projective presentation** of G. If, in addition, $\varphi_0 : F_0 \longrightarrow G$ is a projective cover and $\varphi_1 : F_1 \xrightarrow{\varphi_1} \operatorname{Im} \varphi_1$ is a projective cover, the sequence is called a **minimal projective presentation** of G.

We now prove the converse of Corollary 6.7, namely, we show that any simple contravariant (or covariant) functor is of the form described in (a) (or in (b), respectively) of the corollary.

6.8. Lemma. (a) *Let S be a simple object in $\mathscr{F}un^{op}A$. There exists, up to isomorphism, a unique indecomposable A-module M such that $S(M) \neq 0$. Further, $S \cong S^M$, the functorial morphism $\pi^M : \operatorname{Hom}_A(-, M) \longrightarrow S^M$ is a projective cover and $S(X) \neq 0$ if and only if M is isomorphic to a direct summand of X.*

(b) *Let S be a simple object in $\mathscr{F}un\, A$. There exists, up to isomorphism, a unique indecomposable A-module M such that $S(M) \neq 0$. Further, $S \cong$*

S_M, the functorial morphism π_M : $\text{Hom}_A(M, -) \longrightarrow S_M$ is a projective cover, and $S(X) \neq 0$ if and only if M is isomorphic to a direct summand of X.

Proof. We only prove (a); the proof of (b) is similar. Let S be a simple functor. We first note that, by Yoneda's lemma (6.1), $S(X) \neq 0$ for some A-module X if and only if there exists a nonzero functorial morphism π^X : $\text{Hom}_A(-, X) \longrightarrow S$ that is necessarily an epimorphism, because S is simple. Because $S \neq 0$, there exists an indecomposable A-module M such that $S(M) \neq 0$. Let X be an arbitrary module such that $S(X) \neq 0$. We thus have functorial epimorphisms π^M : $\text{Hom}_A(-, M) \longrightarrow S$ and π^X : $\text{Hom}_A(-, X) \longrightarrow S$. By the projectivity of the functors $\text{Hom}_A(-, M)$ and $\text{Hom}_A(-, X)$ (see(6.4)), we obtain a commutative diagram with exact rows

$$
\begin{array}{ccccc}
\text{Hom}_A(-, M) & \xrightarrow{\;\pi^M\;} & S & \longrightarrow & 0 \\
{\scriptstyle \text{Hom}_A(-,f)}\downarrow & & \downarrow{\scriptstyle 1_S} & & \\
\text{Hom}_A(-, X) & \xrightarrow{\;\pi^X\;} & S & \longrightarrow & 0 \\
{\scriptstyle \text{Hom}_A(-,g)}\downarrow & & \downarrow{\scriptstyle 1_S} & & \\
\text{Hom}_A(-, M) & \xrightarrow{\;\pi^M\;} & S & \longrightarrow & 0
\end{array}
$$

where the existence of the morphisms $f : M \to X$ and $g : X \to M$ follows from (6.2). Because M is indecomposable, $\text{End}\, M$ is local, hence $gf \in \text{End}\, M$ must be nilpotent or invertible, by (I.4.6). However, if $(gf)^m = 0$ for some $m \geq 1$, we obtain $\pi^M = \pi^M \text{Hom}_A(-, (gf)^m) = 0$, a contradiction. Hence gf is invertible so that f is a section and g is a retraction. Consequently, the functorial morphism $\text{Hom}_A(-, g)$ is a retraction. This shows that π^M : $\text{Hom}_A(-, M) \longrightarrow S$ is a projective cover. The uniqueness up to isomorphism of the indecomposable module M follows from the uniqueness up to isomorphism of the projective cover and (6.4). Finally, because, by (6.6), $\text{Hom}_A(-, M)$ has $\text{rad}(-, M)$ as unique maximal subfunctor, we infer the existence of a functorial isomorphism $S \cong \text{Hom}_A(-, M)/\text{rad}_A(-, M) = S^M$. $\square$

We have thus exhibited a bijective correspondence $M \mapsto S^M$ (or $M \mapsto S_M$) between the isomorphism classes of indecomposable A-modules and of simple objects in $\mathcal{F}un^{\text{op}}A$ (or in $\mathcal{F}un\, A$, respectively). We now show that almost split morphisms in $\text{mod}\, A$ correspond to projective presentations of these simple objects.

6.9. Lemma. (a) *Let N be an indecomposable A-module. A homomorphism $g : M \to N$ of A-modules is a right almost split morphism if and only if the induced sequence of functors*

$$
\text{Hom}_A(-, M) \xrightarrow{\;\text{Hom}_A(-,g)\;} \text{Hom}_A(-, N) \xrightarrow{\;\pi^N\;} S^N \longrightarrow 0
$$

is a projective presentation of S^N in $\mathcal{F}un^{op}A$.

(b) *Let L be an indecomposable A-module. A homomorphism $f : L \to M$ of A-modules is a left almost split morphism if and only if the induced sequence of functors*

$$\operatorname{Hom}_A(M,-) \xrightarrow{\operatorname{Hom}_A(f,-)} \operatorname{Hom}_A(L,-) \xrightarrow{\;\pi_L\;} S_L \longrightarrow 0$$

is a projective presentation of S_L in $\mathcal{F}un\, A$.

Proof. We only prove (a); the proof of (b) is similar. Assume that g is right almost split. To prove that the induced sequence of functors is a projective presentation of S^N in $\mathcal{F}un^{op}A$, it suffices, by (6.4), to prove it is exact, or equivalently, by (6.7), to prove that $\operatorname{Im}\operatorname{Hom}_A(-,g) = \operatorname{rad}_A(-,N)$. Thus, we must show that, for every indecomposable A-module X, $\operatorname{Im}\operatorname{Hom}_A(X,g) = \operatorname{rad}_A(X,N)$.

Let $h \in \operatorname{rad}_A(X,N)$. Then $h : X \to N$ is not an isomorphism. Because g is a right almost split morphism, there exists $k : X \to M$ such that $h = gk = \operatorname{Hom}_A(X,g)(k)$. Thus $\operatorname{rad}_A(X,N) \subseteq \operatorname{Im}\operatorname{Hom}_A(X,g)$. For the reverse inclusion, assume first $X \not\cong N$, then $\operatorname{rad}_A(X,N) = \operatorname{Hom}_A(X,N)$ and clearly $\operatorname{Im}\operatorname{Hom}_A(X,g) \subseteq \operatorname{Hom}_A(X,N)$; on the other hand, if $X \cong N$, this follows from the fact that g is not a retraction and (1.9). We have thus shown the necessity.

For the sufficiency, assume that the given sequence of functors is exact. We must show that g is right almost split. Suppose first that g is a retraction and $g' : N \to M$ is such that $gg' = 1_N$. Then, for any $h \in \operatorname{End}N$, we have $h = gg'h = \operatorname{Hom}_A(N,g)(g'h) \in \operatorname{Im}\operatorname{Hom}_A(N,g) = \operatorname{Ker}\pi_N^N$. This implies that $S^N(N) = 0$, a contradiction. Hence g is not a retraction. Let X be indecomposable, and $h : X \to N$ be a nonisomorphism, that is, $h \in \operatorname{rad}_A(X,N)$. Because the given sequence of functors is exact, evaluating these functors at X yields $\operatorname{rad}_A(X,N) = \operatorname{Ker}\pi_X^N = \operatorname{Im}\operatorname{Hom}_A(X,g)$. Hence there exists $k : X \to M$ such that $h = \operatorname{Hom}_A(X,g)(k) = gk$. Thus g is right almost split. $\qquad\square$

Furthermore, minimal almost split morphisms in $\operatorname{mod}A$ correspond to minimal projective presentations of simple functors, as we show in the following lemma.

6.10. Lemma. (a) *Let N be an indecomposable A-module. A homomorphism $g : M \to N$ of A-modules is a right minimal almost split morphism if and only if the induced sequence of functors*

$$\operatorname{Hom}_A(-,M) \xrightarrow{\operatorname{Hom}_A(-,g)} \operatorname{Hom}_A(-,N) \xrightarrow{\;\pi^N\;} S^N \longrightarrow 0$$

is a minimal projective presentation of S^N in $\mathcal{F}un^{op}A$.

(b) *Let L be an indecomposable A-module. A homomorphism $f : L \to M$ of A-modules is a left minimal almost split morphism if and only if the induced sequence of functors*

$$\operatorname{Hom}_A(M, -) \xrightarrow{\operatorname{Hom}_A(f, -)} \operatorname{Hom}_A(L, -) \xrightarrow{\pi_L} S_L \longrightarrow 0$$

is a minimal projective presentation of S_L in $\mathcal{F}un\, A$.

Proof. We only prove (a); the proof of (b) is similar. Assume that g is right minimal almost split. It follows from (6.9) that the induced sequence of functors is a projective presentation. We claim it is minimal, that is, by (6.6), $\operatorname{Hom}_A(-, g) : \operatorname{Hom}_A(-, M) \longrightarrow \operatorname{rad}_A(-, N)$ is a projective cover. Let thus $\varphi : \operatorname{Hom}_A(-, X) \longrightarrow \operatorname{rad}_A(-, N)$ be a functorial epimorphism. It follows from (6.4) and (6.2) that there exist morphisms $u : M \to X$ and $v : X \to M$ such that we have a commutative diagram with exact rows

$$
\begin{array}{ccccc}
\operatorname{Hom}_A(-, M) & \xrightarrow{\operatorname{Hom}_A(-, g)} & \operatorname{rad}_A(-, N) & \longrightarrow & 0 \\
{\scriptstyle \operatorname{Hom}_A(-, u)} \downarrow & & \downarrow {\scriptstyle 1} & & \\
\operatorname{Hom}_A(-, X) & \xrightarrow{\varphi} & \operatorname{rad}_A(-, N) & \longrightarrow & 0 \\
{\scriptstyle \operatorname{Hom}_A(-, v)} \downarrow & & \downarrow {\scriptstyle 1} & & \\
\operatorname{Hom}_A(-, M) & \xrightarrow{\operatorname{Hom}_A(-, g)} & \operatorname{rad}(-, N) & \longrightarrow & 0
\end{array}
$$

that is, $\operatorname{Hom}_A(-, g) \circ \operatorname{Hom}_A(-, v) \circ \operatorname{Hom}_A(-, u) = \operatorname{Hom}_A(-, g)$. By (6.2) again, $g(vu) = g$. Because g is right minimal, vu is an automorphism. Consequently, v is a retraction and therefore $\operatorname{Hom}_A(-, v)$ is a retraction. This shows that $\operatorname{Hom}_A(-, g) : \operatorname{Hom}_A(-, M) \longrightarrow \operatorname{rad}_A(-, N)$ is a projective cover.

Conversely, if the shown sequence of functors is a minimal projective presentation, it follows from (6.9) that g is right almost split. We must show that it is right minimal. Assume $h : M \to M$ is such that $gh = g$. We have a commutative diagram with exact rows

$$
\begin{array}{ccccc}
\operatorname{Hom}_A(-, M) & \xrightarrow{\operatorname{Hom}_A(-, g)} & \operatorname{rad}_A(-, N) & \longrightarrow & 0 \\
{\scriptstyle \operatorname{Hom}_A(-, h)} \downarrow & & \downarrow {\scriptstyle 1} & & \\
\operatorname{Hom}_A(-, M) & \xrightarrow{\operatorname{Hom}_A(-, g)} & \operatorname{rad}(-, N) & \longrightarrow & 0
\end{array}
$$

Because $\operatorname{Hom}_A(-, g)$ is a projective cover, $\operatorname{Hom}_A(-, h)$ is an isomorphism and hence so is h. $\qquad\square$

We are now able to prove the main theorem of this section, which shows that almost split sequences in $\operatorname{mod} A$ correspond to minimal projective resolutions of simple functors in $\mathcal{F}un^{\mathrm{op}} A$ and in $\mathcal{F}un\, A$ defined in a usual way.

6.11. Theorem. (a) *Let N be an indecomposable A-module.*

(i) *N is projective, and $g : M \to N$ is right minimal almost split if and only if the induced sequence of functors*

$$0 \longrightarrow \operatorname{Hom}_A(-, M) \xrightarrow{\operatorname{Hom}_A(-,g)} \operatorname{Hom}_A(-, N) \xrightarrow{\pi^N} S^N \longrightarrow 0$$

is a minimal projective resolution of S^N in $\mathcal{F}un^{op}A$.

(ii) *N is not projective, and the sequence $0 \to L \xrightarrow{f} M \xrightarrow{g} N \to 0$ is exact and almost split if and only if the induced sequence of functors*

$$0 \longrightarrow \operatorname{Hom}_A(-, L) \xrightarrow{\operatorname{Hom}_A(-,f)} \operatorname{Hom}_A(-, M) \xrightarrow{\operatorname{Hom}_A(-,g)} \operatorname{Hom}_A(-, N)$$
$$\xrightarrow{\pi^N} S^N \longrightarrow 0$$

(where $L \neq 0$) is a minimal projective resolution of S^N in $\mathcal{F}un^{op}A$.

(b) *Let L be an indecomposable A-module.*

(i) *L is injective, and $f : L \to M$ is left minimal almost split if and only if the induced sequence of functors*

$$0 \longrightarrow \operatorname{Hom}_A(M, -) \xrightarrow{\operatorname{Hom}_A(f,-)} \operatorname{Hom}_A(L, -) \xrightarrow{\pi_L} S_L \longrightarrow 0$$

is a minimal projective resolution of S_L in $\mathcal{F}un\,A$.

(ii) *L is not injective, and the sequence $0 \longrightarrow L \xrightarrow{f} M \xrightarrow{g} N \longrightarrow 0$ is exact and almost split if and only if the induced sequence of functors*

$$0 \longrightarrow \operatorname{Hom}_A(N, -) \xrightarrow{\operatorname{Hom}_A(g,-)} \operatorname{Hom}_A(M, -) \xrightarrow{\operatorname{Hom}_A(f,-)} \operatorname{Hom}_A(L, -)$$
$$\xrightarrow{\pi_L} S_L \longrightarrow 0$$

(where $N \neq 0$) is a minimal projective resolution of S_L in $\mathcal{F}un\,A$.

Proof. We only prove (a); the proof of (b) is similar.

(i) Assume that N is projective, and $g : M \to N$ is right minimal almost split. By (3.5), g is a monomorphism with image equal to rad N. By the left exactness of the Hom functor, $\operatorname{Hom}_A(-, g) : \operatorname{Hom}_A(-, M) \longrightarrow \operatorname{Hom}_A(-, N)$ is a monomorphism. Thus, it follows from (6.10) that the induced sequence of functors

$$0 \longrightarrow \operatorname{Hom}_A(-, M) \xrightarrow{\operatorname{Hom}_A(-,g)} \operatorname{Hom}_A(-, N) \xrightarrow{\pi^N} S^N \longrightarrow 0$$

is a minimal projective resolution of S^N in $\mathcal{F}un^{op}A$. Conversely, if the sequence of functors is a minimal projective resolution of S^N in $\mathcal{F}un^{op}A$, it follows from (6.10) that g is right minimal almost split. Evaluating the sequence of functors at A_A yields that g is a monomorphism. But, by the description of right minimal almost split morphisms in (3.1) and (3.2), this implies that N is projective.

(ii) Assume that N is not projective, and let

$$0 \longrightarrow L \xrightarrow{\ f\ } M \xrightarrow{\ g\ } N \longrightarrow 0$$

be an almost split sequence. By the left exactness of the Hom functor, we derive an exact sequence of projective functors

$$0 \longrightarrow \mathrm{Hom}_A(-,L) \xrightarrow{\mathrm{Hom}_A(-,f)} \mathrm{Hom}_A(-,M) \xrightarrow{\mathrm{Hom}_A(-,g)} \mathrm{Hom}_A(-,N).$$

Because $g : M \to N$ is right minimal almost split, (6.10) yields that the induced sequence of functors

$$0 \longrightarrow \mathrm{Hom}_A(-,L) \xrightarrow{\mathrm{Hom}_A(-,f)} \mathrm{Hom}_A(-,M) \xrightarrow{\mathrm{Hom}_A(-,g)} \mathrm{Hom}_A(-,N)$$
$$\xrightarrow{\ \pi^N\ } S^N \longrightarrow 0$$

is a minimal projective resolution of S^N in $\mathcal{F}un^{\mathrm{op}} A$. Conversely, assume that the sequence of functors (where $L \neq 0$) is a minimal projective resolution of S^N in $\mathcal{F}un^{\mathrm{op}} A$. First, we claim that N is not projective. Indeed, if this were the case, then S^N has, by (a), a minimal projective resolution of the form

$$0 \longrightarrow \mathrm{Hom}_A(-,\mathrm{rad}\,N) \longrightarrow \mathrm{Hom}_A(-,N) \xrightarrow{\ \pi^N\ } S^N \longrightarrow 0,$$

where the first morphism is induced from the canonical inclusion of $\mathrm{rad}\,N$ into N. We thus have a short exact sequence of functors

$$0 \longrightarrow \mathrm{Hom}_A(-,L) \xrightarrow{\mathrm{Hom}_A(-,f)} \mathrm{Hom}_A(-,M) \longrightarrow \mathrm{Hom}_A(-,\mathrm{rad}\,N) \longrightarrow 0$$

that splits, because $\mathrm{Hom}_A(-,\mathrm{rad}\,N)$ is projective. In particular, the morphism $\mathrm{Hom}_A(-,f)$ is a section, a contradiction to the minimality of the given projective resolution. This shows our claim that N is not projective. In particular, N is not isomorphic to a direct summand of A_A hence, by (6.8), $S^N(A_A) = 0$. Evaluating the given projective resolution at A_A yields a short exact sequence of A-modules

$$0 \longrightarrow L \xrightarrow{\ f\ } M \xrightarrow{\ g\ } N \longrightarrow 0,$$

where, by (6.10), g is right minimal almost split. But this implies, by (1.13), that the sequence is almost split. $\qquad\square$

It is useful to observe that it follows from (6.11)(a) that, for any projective A-module P, there exists a functorial isomorphism $\mathrm{rad}_A(-,P) \cong \mathrm{Hom}_A(-,\mathrm{rad}\,P)$. Dually, for any injective A-module I, there exists a functorial isomorphism $\mathrm{rad}_A(I,-) \cong \mathrm{Hom}_A(I/\mathrm{soc}\,I,-)$.

IV.7. Exercises

1. Let $f : M \longrightarrow N$ be a homomorphism in $\operatorname{mod} A$. Show that the following conditions are equivalent:

(a) For every epimorphism $h : L \longrightarrow N$, there exists $g : M \longrightarrow L$ such that $f = hg$.

(b) For every epimorphism $h : L \longrightarrow N$ with L projective there exists $g : M \longrightarrow L$ such that $f = hg$.

(c) $f \in \mathcal{P}(M, N)$, that is, f factors through a projective A-module.

2. State and prove the dual of Exercise 1.

3. Let M be a left A-module without projective direct summand. Show that there is a functorial isomorphism $\underline{\operatorname{Hom}}_{A^{\mathrm{op}}}(M, -) \cong \operatorname{Tor}_1^{A^{\mathrm{op}}}(M, -)$.

4. Let p be a prime, $n > 0$, and $\mathbb{Z}_{p^j} = \mathbb{Z}/(p^j)$. Show that the exact sequence in $\operatorname{mod} \mathbb{Z}$

$$0 \longrightarrow \mathbb{Z}_{p^n} \xrightarrow{\left[\begin{smallmatrix} u_n \\ \pi_n \end{smallmatrix}\right]} \mathbb{Z}_{p^{n+1}} \oplus \mathbb{Z}_{p^{n-1}} \xrightarrow{[\pi_{n+1} \, u_{n-1}]} \mathbb{Z}_{p^n} \longrightarrow 0$$

is almost split, where $u_j : \mathbb{Z}_{p^j} \to \mathbb{Z}_{p^{j+1}}$ is the monomorphism given by $\overline{x} \mapsto \overline{px}$ and $\pi_j : \mathbb{Z}_{p^j} \to \mathbb{Z}_{p^{j-1}}$ is the canonical epimorphism.

5. Let M be an indecomposable nonprojective right A-module and let $\xi : 0 \longrightarrow \tau M \longrightarrow E \longrightarrow M \longrightarrow 0$ be a nonsplit exact sequence. Show that the following conditions are equivalent:

(a) ξ is almost split.

(b) For every homomorphism $u : \tau M \longrightarrow U$ that is not a section, we have $\operatorname{Ext}_A^1(M, u)(\xi) = 0$.

(c) For every homomorphism $v : V \longrightarrow M$ that is not a retraction, we have $\operatorname{Ext}_A^1(v, \tau M)(\xi) = 0$.

6. Let M be an indecomposable nonprojective right A-module and let $\xi : 0 \xrightarrow{f} \tau M \xrightarrow{g} E \longrightarrow M \longrightarrow 0$ be a nonsplit exact sequence. Show that the following conditions are equivalent:

(a) The sequence ξ is almost split.

(b) For every indecomposable A-module U and every nonisomorphism $u : \tau M \longrightarrow U$, there exists $\overline{u} : E \longrightarrow U$ such that $\overline{u}f = u$.

(c) For every indecomposable A-module V and every nonisomorphism $v : V \longrightarrow M$, there exists $\overline{v} : V \longrightarrow E$ such that $g\overline{v} = v$.

7. Let $0 \longrightarrow L \xrightarrow{f} M \xrightarrow{g} N \longrightarrow 0$ be an almost split sequence in $\operatorname{mod} A$. Prove the following statements:

(a) If N' is a nonzero proper submodule of N, then the short exact sequence $0 \longrightarrow L \longrightarrow g^{-1}(N') \longrightarrow N' \longrightarrow 0$ is split.

(b) If L' is a nonzero submodule of L, then the short exact sequence $0 \longrightarrow L/L' \longrightarrow M/f(L') \longrightarrow N' \longrightarrow 0$ is split.

8. Let $0 \longrightarrow L \xrightarrow{f} M \xrightarrow{g} N \longrightarrow 0$ be an almost split sequence in mod A. Prove the following statements:

(a) For every nonsplit exact sequence $0 \longrightarrow X \xrightarrow{u} Y \xrightarrow{v} N \longrightarrow 0$ and every commutative diagram with exact rows

$$
\begin{array}{ccccccccc}
0 & \longrightarrow & L & \xrightarrow{f} & M & \xrightarrow{g} & N & \longrightarrow & 0 \\
& & \downarrow{\scriptstyle h} & & \downarrow{\scriptstyle k} & & \| & & \\
0 & \longrightarrow & X & \xrightarrow{u} & Y & \xrightarrow{v} & N & \longrightarrow & 0
\end{array}
$$

there exists a commutative diagram with exact rows

$$
\begin{array}{ccccccccc}
0 & \longrightarrow & X & \xrightarrow{u} & Y & \xrightarrow{v} & N & \longrightarrow & 0 \\
& & \downarrow{\scriptstyle h'} & & \downarrow{\scriptstyle k'} & & \| & & \\
0 & \longrightarrow & L & \xrightarrow{f} & M & \xrightarrow{g} & N & \longrightarrow & 0
\end{array}
$$

such that $h'h = 1_L$ and $k'k = 1_M$. In particular, h and k are sections.

(b) For every nonsplit exact sequence $0 \longrightarrow L \xrightarrow{u} X \xrightarrow{v} Y \longrightarrow 0$ and every commutative diagram with exact rows

$$
\begin{array}{ccccccccc}
0 & \longrightarrow & L & \xrightarrow{u} & X & \xrightarrow{v} & Y & \longrightarrow & 0 \\
& & \| & & \downarrow{\scriptstyle h} & & \downarrow{\scriptstyle k} & & \\
0 & \longrightarrow & L & \xrightarrow{f} & M & \xrightarrow{g} & N & \longrightarrow & 0
\end{array}
$$

there exists a commutative diagram with exact rows

$$
\begin{array}{ccccccccc}
0 & \longrightarrow & L & \xrightarrow{f} & M & \xrightarrow{g} & N & \longrightarrow & 0 \\
& & \| & & \downarrow{\scriptstyle h'} & & \downarrow{\scriptstyle k'} & & \\
0 & \longrightarrow & L & \xrightarrow{u} & X & \xrightarrow{v} & Y & \longrightarrow & 0
\end{array}
$$

such that $hh' = 1_M$ and $kk' = 1_N$. In particular, h and k are retractions.

9. Let $\xi : 0 \longrightarrow L \xrightarrow{f} M \xrightarrow{g} N \longrightarrow 0$ be a nonsplit short exact sequence in mod A. Prove the following statements:

(a) The homomorphism f is irreducible if and only if

(i) Im f is a direct summand of every proper submodule M' of M such that Im $f \subseteq M'$, and

(ii) if X is an A-module and $\eta \in \operatorname{Ext}^1_A(N, X)$, then either there exists an A-module homomorphism $u : X \longrightarrow L$ such that $\operatorname{Ext}^1_A(N, u)(\eta) = \xi$ or an A-module homomorphism $v : L \longrightarrow X$ such that $\operatorname{Ext}^1_A(N, v)(\xi) = \eta$.

(b) The homomorphism g is irreducible if and only if

(i) $g : M/L' \longrightarrow N$ is a retraction if L' is a nonzero submodule of $L = \operatorname{Ker} g$, and

(ii) if X is a module and $\eta \in \operatorname{Ext}^1_A(X, L)$, then either there exists a homomorphism $u : N \to X$ such that $\operatorname{Ext}^1_A(u, L)(\eta) = \xi$ or a homomorphism $v : X \to N$ such that $\operatorname{Ext}^1_A(v, L)(\xi) = \eta$.

10. (a) Let $f : L \longrightarrow M$ be an irreducible monomorphism in $\operatorname{mod} A$, with M indecomposable. Let $h : X \longrightarrow N$ be an irreducible morphism, where $N = \operatorname{Coker} f$. Show that h is an epimorphism.

(b) Let $g : M \longrightarrow N$ be an irreducible epimorphism in $\operatorname{mod} A$, with M indecomposable. Let $h : L \longrightarrow X$ be an irreducible morphism, where $L = \operatorname{Ker} g$. Show that h is a monomorphism.

11. Let $f : L \longrightarrow M$ be an irreducible morphism in $\operatorname{mod} A$, and X be a right A-module.

(a) Show that $\operatorname{Ext}^1_A(X, f) : \operatorname{Ext}^1_A(X, L) \to \operatorname{Ext}^1_A(X, M)$ is a monomorphism, if $\operatorname{Hom}_A(M, X) = 0$.

(b) Show that $\operatorname{Ext}^1_A(f, X) : \operatorname{Ext}^1_A(M, X) \to \operatorname{Ext}^1_A(L, X)$ is a monomorphism, if $\operatorname{Hom}_A(X, L) = 0$.

12. Let $g : M \longrightarrow N$ be a right almost split epimorphism. If $\operatorname{Ker} g$ is not indecomposable, show that there exists a right almost split morphism $g_1 : M_1 \longrightarrow N$ such that $\ell(M_1) < \ell(M)$. Deduce that if M is of minimal length such that there exists a right almost split epimorphism $g : M \longrightarrow N$, then the short exact sequence $0 \longrightarrow \operatorname{Ker} g \longrightarrow M \xrightarrow{g} N \longrightarrow 0$ is almost split.

13. State and prove the dual of Exercise 12.

14. Let $0 \longrightarrow \tau M \longrightarrow \bigoplus_{i=1}^{n} E_i \longrightarrow M \longrightarrow 0$ be an almost split sequence, with the E_i indecomposable. Show that, for every i, we have $\ell(E_i) \neq \ell(M)$ and $\ell(E_i) \neq \ell(\tau M)$ so that no E_i is isomorphic to M or τM.

15. Let X be a nonzero module in $\operatorname{mod} A$. Show that there exists at most finitely many nonisomorphic almost split sequences

$$0 \longrightarrow L_i \longrightarrow M_i \longrightarrow N_i \longrightarrow 0$$

with X isomorphic to a direct summand of M_i.

16. Let $0 \longrightarrow L \longrightarrow M \longrightarrow N \longrightarrow 0$ be an almost split sequence in the category $\operatorname{mod} A$ and suppose that M is not indecomposable. Show that $\operatorname{Hom}_A(L, N) \neq 0$.

17. Let $0 \longrightarrow L \longrightarrow M \longrightarrow N \longrightarrow 0$ be an almost split sequence in the category $\operatorname{mod} A$. Show that if P is a nonzero projective module, the following conditions are equivalent:

(a) P is isomorphic to a direct summand of M.

(b) There exists an irreducible morphism $P \longrightarrow N$.

(c) There exists an irreducible morphism $L \longrightarrow P$.

(d) L is isomorphic to a direct summand of $\operatorname{rad} P$.

(e) There is an indecomposable direct summand R of $\operatorname{rad} P$ such that $N \cong \tau^{-1} R$.

(f) If $f : X \longrightarrow N$ is an epimorphism in $\operatorname{mod} A$ that is not a retraction, then P is isomorphic to a direct summand of X.

18. Let $0 \longrightarrow L \longrightarrow M \longrightarrow N \longrightarrow 0$ be an almost split sequence in $\operatorname{mod} A$. Prove the following statements:

(a) If there exists an irreducible epimorphism $h : P \longrightarrow N$ with P indecomposable projective, then $N \cong P/S$, where S is a simple submodule of P.

(b) If $N/\operatorname{rad} N$ is simple and M has a nonzero projective direct summand, there exists an irreducible epimorphism $h : P \longrightarrow N$, with P indecomposable projective.

19. Let A be the K-algebra of Example 4.13. Let M and N be the simple A-modules such that $\operatorname{\mathbf{dim}} M = \begin{smallmatrix} & 0 & \\ 0 & 1 & 0 \\ & 0 & \end{smallmatrix}$ and $\operatorname{\mathbf{dim}} N = \begin{smallmatrix} & 0 & \\ 0 & 0 & 1 \\ & 0 & \end{smallmatrix}$. Show that $\operatorname{\mathbf{dim}} \tau M = \begin{smallmatrix} & 1 & \\ 1 & 1 & 0 \\ & 0 & \end{smallmatrix}$, and that $\operatorname{Hom}_A(DA, \tau M) = 0$.

20. Let A be given by the quiver
$$\underset{2'}{\circ} \overset{\alpha'}{\underset{\beta'}{\rightleftarrows}} \underset{1'}{\circ} \overset{\gamma}{\longleftarrow} \underset{1}{\circ} \overset{\alpha}{\underset{\beta}{\rightleftarrows}} \underset{2}{\circ}$$
bound by the relations $\beta\alpha = 0$, $\beta'\alpha' = 0$, and $\alpha\beta\gamma = \gamma\beta'\alpha'$. Show that $P(1) \cong I(1')$, $P(2) \cong I(2')$ and deduce the almost split sequences having as middle terms $P(1)$ and $P(2)$, respectively.

21. Construct the Auslander–Reiten quiver of the algebra defined by each of the following bound quivers:

(a) $\alpha\beta = \gamma\delta;$

(b) $\alpha\beta = 0,\ \gamma\delta = 0;$

(c) $\alpha\beta = 0.$

In each case describe the structure of each indecomposable module.

22. Construct the Auslander–Reiten quiver of the algebra defined by each of the following bound quivers:

(a) $\alpha\beta = 0,\ \gamma\delta = 0$;

(b) $\alpha\beta = 0,\ \gamma\lambda = 0,\ \beta\gamma = \delta\varepsilon$;

(c) $\mu\alpha = 0$;

(d) $\alpha\beta = 0,\ \gamma\delta = 0,\ \delta\varepsilon = 0$;

(e) $\xi\eta = 0,\ \mu\lambda = 0,\ \nu\mu = 0$;

(f) $\gamma\delta = 0,\ \alpha\beta = 0$;

(g) $\alpha\beta = \gamma\delta,\ \lambda\alpha\beta\varepsilon = \mu\nu$;

(h) $\alpha\eta = \lambda\mu,\ \beta\gamma = \eta\nu$;

(i) $\alpha\beta = \gamma\delta,\ \alpha\mu = 0,\ \mu\delta = 0;$

(j) $\alpha\beta = \gamma\delta,\ \alpha\beta\varepsilon = 0;$

(k) $\alpha\eta = \lambda\mu,\ \beta\gamma = \eta\nu,\ \alpha\beta = 0,\ \mu\nu = 0;$

(l) $\alpha\beta = 0;$

(m) $\alpha\beta = 0,\ \beta\alpha = 0.$

23. Let Q be either of the following quivers:

(a)   (b)

Construct the component of the Auslander–Reiten quiver of the path K-algebra $A = KQ$ containing the indecomposable projective modules, and show that it contains no injective modules.

24. Let A be a K-algebra such that $\operatorname{rad}_A^m = \operatorname{rad}_{\operatorname{mod} A}^m = 0$ for some $m \geq 1$. Prove that any nonzero nonisomorphism between indecomposable modules in $\operatorname{mod} A$ is a sum of compositions of irreducible morphisms.

Hint: Follow the proof of (5.6).

25. Complete the proof of Proposition 2.10.

26. Let $0 \longrightarrow L \xrightarrow{f} M \xrightarrow{g} N \longrightarrow 0$ be a nonsplit short exact sequence in $\operatorname{mod} A$. Prove the following statements:

(a) f is irreducible if and only if, for every subfunctor F of the functor $\operatorname{Hom}_A(-, N)$, F either contains or is contained in the image of the functorial morphism $\operatorname{Hom}_A(-, g) : \operatorname{Hom}_A(-, M) \longrightarrow \operatorname{Hom}_A(-, N)$.

(b) g is irreducible if and only if, for every subfunctor F of $\operatorname{Hom}_A(L, -)$, F either contains or is contained in the image of the functorial morphism $\operatorname{Hom}_A(f, -) : \operatorname{Hom}_A(M, -) \longrightarrow \operatorname{Hom}_A(L, -)$.

Chapter V

Nakayama algebras and representation–finite group algebras

In this chapter we describe the representation theory of one of the best understood classes of algebras, that of the Nakayama algebras (which some authors call generalised uniserial algebras, see [68]). These algebras are always representation–finite and, using only elementary methods, we are able to give a complete list of their nonisomorphic indecomposable modules. The latter turn out to have a particularly simple structure; indeed, Nakayama algebras are characterised by the fact that any indecomposable module is uniserial, that is, has a unique composition series. As a consequence, it is also easy to describe the homomorphisms between two indecomposable modules and to compute all almost split sequences. The understanding of the module category of Nakayama algebras is very useful in the sequel, for instance, when we study the regular modules over representation–infinite hereditary algebras.

The final section of this chapter is devoted to a criterion allowing us to verify whether a group algebra is representation–finite. It was obtained in 1954 by Higman [92].

Throughout this chapter, we let A denote a finite dimensional K-algebra and all A-modules are, unless otherwise specified, right finite dimensional A-modules.

V.1. The Loewy series and the Loewy length of a module

For an A-module M, we consider the decreasing sequence of submodules of M given by

$$M \supset \operatorname{rad} M \supset \operatorname{rad}^2 M \supset \ldots \operatorname{rad}^i M \supset \ldots \supset 0.$$

This sequence is called the **radical series**, or the **descending Loewy series** of M. Because M has finite dimension as a K-vector space, it has finite composition length. Hence there exists a least positive integer m such that $\mathrm{rad}^m M = 0$. It follows that the radical series is finite and has m nonzero terms. The integer m is called the **length of the radical series** and is denoted by $r\ell(M)$.

The dual notion is that of the **socle series** or **ascending Loewy series** of M. We recall that the socle of M, $\mathrm{soc}\,M$, is the sum of all the simple submodules of M. For an integer $i \geq 0$, we define $\mathrm{soc}^i M$ inductively as follows: $\mathrm{soc}^0 M = 0$ and, if $\mathrm{soc}^i M$ is already defined and $p : M \to M/\mathrm{soc}^i M$ denotes the canonical epimorphism, we set

$$\mathrm{soc}^{i+1} M = p^{-1}(\mathrm{soc}(M/\mathrm{soc}^i M)).$$

Thus, by definition, $\mathrm{soc}^{i+1} M \supset \mathrm{soc}^i M$, and we obtain an increasing sequence

$$0 = \mathrm{soc}^0 M \subset \mathrm{soc}\,M = \mathrm{soc}^1 M \subset \mathrm{soc}^2 M \subset \ldots \subset \mathrm{soc}^i M \subset \ldots M$$

of submodules of M. Because M has finite composition length, there exists a least positive integer m such that $\mathrm{soc}^m M = M$; it is called the **length of the socle series** and is denoted by $s\ell(M)$.

It follows directly from the definition that $r\ell(M)$ and $s\ell(M)$ are at most equal to the composition length $\ell(M)$ of M, that is, to the dimension of M as a K-vector space.

In general, the radical and the socle series of a module M do not coincide (see, for instance, Example 1.5). However, we prove that $r\ell(M) = s\ell(M)$.

1.1. Lemma. *Let $f : M_A \to N_A$ be an A-module epimorphism. Then $f(\mathrm{rad}^i M) = \mathrm{rad}^i N$ for every $i \geq 0$.*

Proof. It clearly suffices to show the result for $i = 1$. By (I.3.7), we have $f(\mathrm{rad}\,M) = f(M\,\mathrm{rad}\,A) = f(M)\mathrm{rad}\,A = N\mathrm{rad}\,A = \mathrm{rad}\,N$. □

1.2. Corollary. *Let $0 \to L_A \xrightarrow{f} M_A \xrightarrow{g} N_A \to 0$ be an exact sequence of A-modules. Then $r\ell(M) \geq \max\{r\ell(L), r\ell(N)\}$.*

Proof. Indeed, we have $f(\mathrm{rad}^i L) \subseteq \mathrm{rad}^i M$ and, by (1.1), $g(\mathrm{rad}^i M) = \mathrm{rad}^i N$. Hence $\mathrm{rad}^i M = 0$ implies $\mathrm{rad}^i L = 0$ and $\mathrm{rad}^i N = 0$. □

We now show that $s\ell(M) = r\ell(DM)$ for any module M. We start with some remarks on the construction of the socle series of a module. Let M_A be a module and let $i \geq 1$. Consider the exact sequence

$$0 \longrightarrow \mathrm{soc}^i M \longrightarrow M \xrightarrow{p} M/\mathrm{soc}^i M \longrightarrow 0$$

together with the inclusion $j : \mathrm{soc}(M/\mathrm{soc}^i M) \hookrightarrow M/\mathrm{soc}^i M$. It easily follows from (A.5.3) in the Appendix that $\mathrm{soc}^{i+1} M$ is the fibered product in the commutative diagram with exact rows

$$0 \longrightarrow \operatorname{soc}^i M \longrightarrow \operatorname{soc}^{i+1} M \longrightarrow \operatorname{soc}(M/\operatorname{soc}^i M) \longrightarrow 0$$

$$\downarrow 1 \qquad\qquad \downarrow \qquad\qquad \downarrow j$$

$$0 \longrightarrow \operatorname{soc}^i M \longrightarrow M \xrightarrow{p} M/\operatorname{soc}^i M \longrightarrow 0$$

where the homomorphisms in the upper sequence are induced from those in the lower one.

We now show by induction on i that, for any module M, there is an isomorphism $D(\operatorname{soc}^i M) \cong DM/\operatorname{rad}^i DM$. For $i = 1$, the isomorphism follows immediately from the properties of the duality D collected in (I.5.13); we leave it as an exercise. Assume $i \geq 2$. In view of (I.5.13), taking the dual of the diagram yields that $D(\operatorname{soc}^{i+1} M)$ is isomorphic to the amalgamated sum N in the commutative diagram with exact rows

$$0 \rightarrow \operatorname{rad}^i DM \longrightarrow DM \longrightarrow DM/\operatorname{rad}^i DM \rightarrow 0$$

$$\downarrow \qquad\qquad \downarrow \qquad\qquad \downarrow 1$$

$$0 \rightarrow \operatorname{rad}^i DM/\operatorname{rad}^{i+1} DM \longrightarrow N \xrightarrow{p} DM/\operatorname{rad}^i DM \rightarrow 0$$

because, by induction, $D(\operatorname{soc}^i M) \cong DM/\operatorname{rad}^i DM$ and hence $D(M/\operatorname{soc}^i M) \cong \operatorname{rad}^i DM$, so that, by applying the formula $D(\operatorname{soc} X) \cong DX/\operatorname{rad} DX$, we obtain the isomorphism $D(\operatorname{soc}(M/\operatorname{soc}^i M)) \cong \operatorname{rad}^i DM/\operatorname{rad}^{i+1} DM$. Because an obvious application of the Snake lemma yields an A-module isomorphism $N \cong DM/\operatorname{rad}^{i+1} DM$, the proof of the required isomorphism is complete.

As an easy consequence, we get $s\ell(M) = r\ell(DM)$.

1.3. Proposition. *For every A-module M, we have $r\ell(M) = s\ell(M)$.*

Proof. We first prove by induction on $s\ell(M)$ that $s\ell(M) \leq r\ell(M)$. Because $s\ell(M) = 0$ if and only if $M = 0$, if and only if $r\ell(M) = 0$, the statement holds whenever $s\ell(M) = 0$.

Assume that $s\ell(X) \leq r\ell(X)$ for every module X such that $s\ell(X) = i \geq 0$ and let M be such that $s\ell(M) = i + 1$. Put $r\ell(M) = j$. Then $j > 0$ and $\operatorname{rad}^{j-1} M$ is a semisimple submodule of M, because $\operatorname{rad}(\operatorname{rad}^{j-1} M) = 0$. Hence $\operatorname{rad}^{j-1} M \subseteq \operatorname{soc} M$. Thus there exists an A-module epimorphism $M/\operatorname{rad}^{j-1} M \rightarrow M/\operatorname{soc} M$. By (1.2), this implies that $r\ell(M/\operatorname{rad}^{j-1} M) \geq r\ell(M/\operatorname{soc} M)$. Because $r\ell(M) = j = 1 + r\ell(M/\operatorname{rad}^{j-1} M)$ we deduce that $r\ell(M) \geq 1 + r\ell(M/\operatorname{soc} M)$. On the other hand, $s\ell(M) = 1 + s\ell(M/\operatorname{soc} M)$. Hence, by the induction hypothesis, $r\ell(M/\operatorname{soc} M) \geq s\ell(M/\operatorname{soc} M)$. Consequently, $r\ell(M) \geq 1 + r\ell(M/\operatorname{soc} M) \geq 1 + s\ell(M/\operatorname{soc} M) = s\ell(M)$, which proves our claim.

By applying this inequality to the left A-module DM, and using the equality $s\ell(M) = r\ell(DM)$ proved earlier, we get $s\ell(M) = r\ell(DM) \geq s\ell(DM) = r\ell(D(DM)) = r\ell(M)$. This finishes the proof. □

1.4. Definition. The **Loewy length** $\ell\ell(M)$ of a module M_A is the common value of $r\ell(M)$ and $s\ell(M)$.

Again, it is clear that $\ell\ell(M) \leq \ell(M)$ for every module M. Also, it follows directly from the definition of a radical (or socle) series and (I.3.7), that a decomposition $M = M_1 \oplus \ldots \oplus M_m$ yields

$$\ell\ell(M) = \max\{\ell\ell(M_1), \ldots, \ell\ell(M_m)\}.$$

1.5. Example. Let A be the path K-algebra of the following quiver $\circ \underset{\delta}{\overset{\beta}{\rightleftarrows}} \circ \underset{\gamma}{\overset{\alpha}{\rightleftarrows}} \circ$ bound by two zero relations $\alpha\beta = 0$ and $\gamma\delta = 0$. Let M_A be the representation

$$K \underset{[0\,0\,1]}{\overset{[0\,1\,0]}{\rightleftarrows}} K^3 \underset{\left[\begin{smallmatrix}0\\1\\0\end{smallmatrix}\right]}{\overset{\left[\begin{smallmatrix}1\\0\\0\end{smallmatrix}\right]}{\rightleftarrows}} K.$$

The radical series of M is:

$$M \supset \left(K \underset{0}{\overset{[0\,1]}{\rightleftarrows}} K^2 \overset{0}{\rightleftarrows} 0\right) \supset \left(K \underset{0}{\overset{0}{\rightleftarrows}} K \overset{0}{\rightleftarrows} 0\right) \supset 0,$$

and its socle series is:

$$0 \subset \left(K \underset{0}{\overset{0}{\rightleftarrows}} K \overset{0}{\rightleftarrows} 0\right) \subset \left(K \underset{[0\,0\,1]}{\overset{[0\,1\,0]}{\rightleftarrows}} K^3 \overset{0}{\rightleftarrows} 0\right) \subset M.$$

They are clearly distinct. We have $\ell\ell(M) = 3$, while $\ell(M) = \dim_K M = 5$.

V.2. Uniserial modules and right serial algebras

One may ask which modules M have the property that $\ell\ell(M) = \ell(M)$. This leads to the following definition.

2.1. Definition. An A-module M_A is said to be **uniserial** if it has a unique composition series.

In other words, M is uniserial if and only if its submodule lattice is a chain. Clearly, if M is uniserial, then so is every submodule of M, and every quotient of M. Moreover, the dual DM of M is a uniserial left A-module. Because a uniserial module M necessarily has a simple top (and a simple socle), it must be indecomposable.

We also notice that uniserial modules are determined up to isomorphism by their composition series, that is, if M and N are uniserial modules and have the same composition factors in the same order, then they are isomorphic. An isomorphism is constructed by an obvious induction on the common composition length of M and N.

The following lemma characterises the uniseriality of a module by means of its Loewy series.

2.2. Lemma. *The following conditions are equivalent for a right A-module M:*

(a)   *M is uniserial.*

(b)   *The radical series $M \supset \operatorname{rad} M \supset \operatorname{rad}^2 M \supset \ldots \supset 0$ is a composition series.*

(c)   *The socle series $0 \subset \operatorname{soc} M \subset \operatorname{soc}^2 M \subset \ldots \subset M$ is a composition series.*

(d)   *$\ell(M) = \ell\ell(M)$.*

Proof. We first prove the equivalence of (a) and (b). The proof of the equivalence of (a) and (c) is similar. Then we prove the equivalence of these conditions with (d).

We show that (a) implies (b) by induction on the composition length $\ell(M)$ of M. If $\ell(M) = 1$, then M is simple and the statement is trivial. Assume the result holds for every uniserial module of composition length $< t$, and let M be uniserial of composition length t. Because M is uniserial, it has a unique maximal submodule, which is necessarily equal to $\operatorname{rad} M$. Because $\operatorname{rad} M \subset M$, the module $\operatorname{rad} M$ is also uniserial. By the induction hypothesis, $\operatorname{rad} M \supset \operatorname{rad}^2 M \supset \ldots \supset 0$ is a composition series for $\operatorname{rad} M$. Hence $M \supset \operatorname{rad} M \supset \operatorname{rad}^2 M \supset \ldots \supset 0$ is a composition series for M. Conversely, assume that

$$M = M_0 \supset M_1 \supset \ldots \supset M_t = 0 \quad \text{and} \quad M = N_0 \supset N_1 \supset \ldots \supset N_t = 0$$

are two composition series for M. We show by induction on i that $M_i = N_i = \operatorname{rad}^i M$ for every $0 \leq i \leq t$. This is trivial if $i = 0$. Assume the result holds for some $i \geq 0$. Because $\operatorname{rad}^i M / \operatorname{rad}^{i+1} M$ is simple, $\operatorname{rad}^i M$ has a unique maximal submodule, which is necessarily equal to $\operatorname{rad}^{i+1} M$. Hence $M_{i+1} = N_{i+1} = \operatorname{rad}^{i+1} M$, and we have established our claim.

It follows directly from (b) that $\ell(M) = \ell\ell(M)$, thus (b) implies (d). To prove that (d) implies (b), assume that $m = \ell(M) = \ell\ell(M)$. It follows from (I.3.11) that $m = \sum_{i=0}^{m-1} \ell(\operatorname{rad}^i M / \operatorname{rad}^{i+1} M)$ and therefore $\ell(\operatorname{rad}^i M / \operatorname{rad}^{i+1} M) = 1$ for $i = 0, \ldots, m-1$, because $\operatorname{rad}^i M / \operatorname{rad}^{i+1} M \neq 0$ for $i \leq m-1$. This shows that the radical series of M is a composition series.   □

We now describe those algebras that have the property that every indecomposable projective module is uniserial.

2.3. Definition. An algebra A is said to be **right serial** if every indecomposable projective right A-module is uniserial. An algebra A is called **left serial** if every indecomposable projective left A-module is uniserial.

Equivalently, A is right serial if every indecomposable injective left A-module is uniserial, and A is left serial if every indecomposable injective right A-module is uniserial. Thus, an algebra A is right serial if and only if its opposite algebra A^{op} is left serial.

2.4. Examples. (a) It follows from the results of (2.5) and (3.2) that the finite dimensional K-algebra $K[t]/(t^n)$, $n \geq 2$, and the algebra $\mathbb{T}_n(K)$ of lower triangular matrices are both left and right serial.

(b) If G is a cyclic group of order $m = p^n$ and K is a field of characteristic $p > 0$ then $KG \cong K[t]/(t^m - 1)$, as will be seen in (5.3), and therefore KG is a left and right serial algebra.

(c) Readers familiar with commutative algebra recall that those commutative discrete valuation domains that are also K-algebras are right and left serial. This is the case, for instance, of the infinite dimensional K-algebra $K[[t]]$ of formal power series in one indeterminate t, whose ideals form the infinite chain

$$K[[t]] \supset (t) \supset (t^2) \supset \ldots \supset (t^n) \supset (t^{n+1}) \supset \ldots \supset (0).$$

We will show later that there exist left serial algebras that are not right serial.

The shape of the ordinary quiver of a right serial algebra follows easily from the next lemma.

2.5. Lemma. *An algebra A is right serial if and only if for every indecomposable projective right module P the module $\operatorname{rad} P/\operatorname{rad}^2 P$ is simple or zero.*

Proof. If A is right serial and P is indecomposable projective, it follows from (2.2) that the radical series $P \supset \operatorname{rad} P \supset \operatorname{rad}^2 P \supset \ldots \supset 0$ is a composition series. In particular, $\operatorname{rad} P/\operatorname{rad}^2 P$ is simple or zero.

Conversely, assume that for every indecomposable projective right module P, $\operatorname{rad} P/\operatorname{rad}^2 P$ is simple or zero. By (2.2), we must show that the radical series $P \supset \operatorname{rad} P \supset \operatorname{rad}^2 P \supset \ldots \supset 0$ is a composition series. We know that $\operatorname{top} P = P/\operatorname{rad} P$ of P is simple. We prove by induction on $i \geq 1$ that $\operatorname{rad}^{i-1} P/\operatorname{rad}^i P$ is simple or zero, and this implies the wanted result.

By hypothesis, the statement holds for $i = 2$. Let $i \geq 2$, and assume that $\mathrm{rad}^{i-1}P/\mathrm{rad}^i P$ is simple. Let $f : P' \to \mathrm{rad}^{i-1}P$ be a projective cover, and $p : \mathrm{rad}^{i-1}P \to \mathrm{rad}^{i-1}P/\mathrm{rad}^i P$ be the canonical epimorphism. Then $pf : P' \to \mathrm{rad}^{i-1}P/\mathrm{rad}^i P$ is a projective cover: indeed, f is minimal by hypothesis, and p is minimal because $\mathrm{rad}^i P = \mathrm{rad}(\mathrm{rad}^{i-1}P)$, hence the composition pf is minimal, see (I.5.6). Because, by the induction hypothesis, $\mathrm{rad}^{i-1}P/\mathrm{rad}^i P$ is simple, P' is indecomposable. By (1.1), the epimorphism f restricts to epimorphisms $f_1 : \mathrm{rad}\,P' \to \mathrm{rad}^i P$ and $f_2 : \mathrm{rad}^2 P' \to \mathrm{rad}^{i+1}P$. By passing to the cokernels, we deduce the existence of a unique epimorphism $\overline{f} : \mathrm{rad}P'/\mathrm{rad}^2 P' \to \mathrm{rad}^i P/\mathrm{rad}^{i+1}P$ such that we have a commutative diagram with exact rows:

$$
\begin{array}{ccccccccc}
0 & \longrightarrow & \mathrm{rad}^2 P' & \longrightarrow & \mathrm{rad}\,P' & \longrightarrow & \mathrm{rad}\,P'/\mathrm{rad}^2 P' & \longrightarrow & 0 \\
& & \downarrow{\scriptstyle f_2} & & \downarrow{\scriptstyle f_1} & & \downarrow{\scriptstyle \overline{f}} & & \\
0 & \longrightarrow & \mathrm{rad}^{i+1}P & \longrightarrow & \mathrm{rad}^i P & \longrightarrow & \mathrm{rad}^i P/\mathrm{rad}^{i+1}P & \longrightarrow & 0
\end{array}
$$

Because P' is indecomposable projective, $\mathrm{rad}\,P'/\mathrm{rad}^2 P'$ is simple or zero, by hypothesis, hence so is $\mathrm{rad}^i P/\mathrm{rad}^{i+1}P$.   □

2.6. Theorem. *A basic K-algebra A is right serial if and only if, for every point a of its ordinary quiver Q_A, there exists at most one arrow of source a.*

Proof. It follows from (2.5) that the algebra A is right serial if and only if, for every $a \in (Q_A)_0$, the A-module

$$\mathrm{rad}\,P(a)/\mathrm{rad}^2 P(a) = e_a(\mathrm{rad}\,A/\mathrm{rad}^2 A)$$

is simple or zero, that is, is at most one dimensional as a K-vector space. This is the case if and only if there is at most one point $b \in (Q_A)_0$ such that the K-vector space $e_a(\mathrm{rad}\,A/\mathrm{rad}^2 A)e_b \neq 0$ and then, this vector space is at most one dimensional. By definition of Q_A, this happens if and only if there is at most one point $b \in (Q_A)_0$ such that there is an arrow $a \to b$, and then there is at most one such arrow.   □

Two examples of connected quivers satisfying the conditions of the theorem are:

In particular, the ordinary quiver Q_A of a connected right serial algebra A either is a tree with a unique sink or contains a unique (oriented) cycle towards which all other arrows are pointing. We also remark that if $A \cong KQ_A/\mathcal{I}$ is right serial, the theorem imposes a condition on the quiver Q_A of A, but the admissible ideal $\mathcal{I}$ is arbitrary.

2.7. Notation. The following notation is useful when dealing with uniserial modules. Let M_A be uniserial, with the radical series

$$M = M_0 \supset M_1 \supset \ldots \supset M_t = 0,$$

where $M_i/M_{i+1} \cong S(a_i)$ for some point a_i in Q_A, and $0 \le i < t$. Using the fact that uniserial modules are uniquely determined up to isomorphism by their composition series, the module M is written as

$$M = \begin{pmatrix} a_0 \\ a_1 \\ \vdots \\ a_{t-1} \end{pmatrix}.$$

Not only does this notation make the structure of M more apparent but, by exhibiting the composition factors of M, it allows us to compute more easily the homomorphisms. Indeed, it follows from Schur's lemma that if $f : M \to N$ is a homomorphism between uniserial modules M and N, the simple top of M maps into an isomorphic simple in the composition series of N.

2.8. Example. Let A be the right serial K-algebra given by the quiver

$$\overset{1}{\circ} \overset{\alpha}{\longrightarrow} \overset{2}{\circ} \, \Big(\Big) \beta$$

and bound by $\alpha\beta^2 = 0$ and $\beta^3 = 0$. Then, as representations of the bound quiver, the indecomposable projective A-modules are given by:

$$P(1)_A = \qquad K \xrightarrow{\left[\begin{smallmatrix}1\\0\end{smallmatrix}\right]} K^2 \, \Big(\Big) \left[\begin{smallmatrix}0 & 0\\1 & 0\end{smallmatrix}\right]$$

and

$$P(2)_A = \qquad 0 \longrightarrow K^3 \, \Big(\Big) \left[\begin{smallmatrix}0 & 0 & 0\\1 & 0 & 0\\0 & 1 & 0\end{smallmatrix}\right]$$

Using Notation 2.7, we can write them as $P(1)_A = \left(\begin{smallmatrix}1\\2\\2\end{smallmatrix}\right)$ and $P(2)_A = \left(\begin{smallmatrix}2\\2\\2\end{smallmatrix}\right)$. In particular, $\mathrm{Hom}_A(P(1), P(2)) = 0$, because the simple top $S(1)$ of $P(1)$ does not appear as a composition factor of $P(2)$, while there are two (linearly independent) homomorphisms from $P(2)$ to $P(1)$, namely having as respective images the radical $\left(\begin{smallmatrix}2\\2\end{smallmatrix}\right)$ of $P(1)$ and its socle (2).

V.3. Nakayama algebras

3.1. Definition. An algebra A is called a **Nakayama algebra** if it is both right and left serial.

That is, A is a Nakayama algebra if and only if every indecomposable projective A-module and every indecomposable injective A-module are uniserial. Clearly, A is a Nakayama algebra if and only if its opposite algebra A^{op} is also.

3.2. Theorem. *A basic and connected algebra A is a Nakayama algebra if and only if its ordinary quiver Q_A is one of the following two quivers:*

(a) $\underset{1}{\circ} \longleftarrow \underset{2}{\circ} \longleftarrow \underset{3}{\circ} \longleftarrow \cdots \longleftarrow \underset{n-1}{\circ} \longleftarrow \underset{n}{\circ}$

(b)

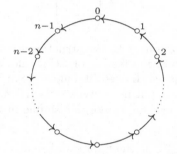

(with $n \geq 1$ points).

Proof. In view of (2.6), A is a Nakayama algebra if and only if every point of Q_A is the source of at most one arrow and the target of at most one arrow. □

Again, if $A \cong kQ_A/\mathcal{I}$ is a Nakayama algebra, the theorem imposes a condition on Q_A, but the admissible ideal $\mathcal{I}$ is arbitrary.

We now show that every indecomposable module over a Nakayama algebra is uniserial, and we give a concrete description of these indecomposables. We first need two easy lemmas.

3.3. Lemma. *Let A be an algebra, and J be a proper ideal of A.*
(a) *If A is right serial, then A/J is also right serial.*
(b) *If A is a Nakayama algebra, then A/J is also a Nakayama algebra.*

Proof. We only prove (a); (b) follows from (a) and its dual. If $A_A = \bigoplus_{i=1}^{n} P_i$ is a direct sum decomposition of A, with the P_i indecomposable, then

$A/J = \bigoplus_{i=1}^{n}(P_i/P_iJ)$ is a direct sum decomposition of A/J, with the P_i/P_iJ indecomposable or zero. In particular, every indecomposable projective A/J-module P' is isomorphic to P_i/P_iJ, for some i. Then the module P' is uniserial, because it is a quotient of the uniserial module P_i.

$\square$

3.4. Lemma. *Let A be a Nakayama algebra, and let P_A be an indecomposable projective A-module with $\ell\ell(P) = \ell\ell(A_A)$. Then P is also injective.*

Proof. Let $P \to E$ be an injective envelope in $\mathrm{mod}\, A$. Because P is uniserial, its socle is simple and hence so is that of E. Consequently, E is indecomposable. Because A is a Nakayama algebra, E is uniserial and we have

$$\ell\ell(A_A) = \ell\ell(P) = \ell(P) \le \ell(E) = \ell\ell(E) \le \ell\ell(A_A).$$

Therefore, $\ell(P) = \ell(E)$ and $P \cong E$ is injective. $\square$

3.5. Theorem. *Let A be a basic and connected Nakayama algebra, and let M be an indecomposable A-module. There exists an indecomposable projective A module P and an integer t with $1 \le t \le \ell\ell(P)$ such that $M \cong P/\mathrm{rad}^t P$. In particular, A is representation–finite.*

Proof. Observe that each of the A-modules $P/\mathrm{rad}^t P$ with P indecomposable projective and $1 \le t \le \ell\ell(P)$, is uniserial and hence indecomposable. Let now M_A be an arbitrary indecomposable A-module, and $t = \ell\ell(M)$ denote its Loewy length. In particular, $0 = \mathrm{rad}^t M = M\mathrm{rad}^t A$ shows that M is annihilated by $\mathrm{rad}^t A$ and hence M has a natural structure of $A/\mathrm{rad}^t A$-module. Also, $\mathrm{rad}^{t-1} M \ne 0$ implies that $\mathrm{rad}^{t-1} A \ne 0$ and so $\ell\ell(A/\mathrm{rad}^t A) = t$. On the other hand, by (3.3), $A/\mathrm{rad}^t A$ is itself a Nakayama algebra. Moreover, there is a direct sum decomposition

$$A/\mathrm{rad}^t A \cong \bigoplus_{i=1}^{n}(P_i/P_i\mathrm{rad}^t A) = \bigoplus_{i=1}^{n}(P_i/\mathrm{rad}^t P_i)$$

with the modules $P_i/\mathrm{rad}^t P_i$ indecomposable.

Let $f : \bigoplus_{j=1}^{r} P'_j \to M$ be a projective cover of M in $\mathrm{mod}(A/\mathrm{rad}^t A)$, with the P'_j indecomposable. Then

$$t = \ell\ell(A/\mathrm{rad}^t A) \ge \max\{\ell\ell(P'_1), \ldots, \ell\ell(P'_t)\} \ge \ell\ell(M) = t.$$

Hence there exists an index j, with $1 \le j \le r$, such that $\ell\ell(P'_j) = t$. We may assume that $\ell\ell(P'_j) = t$ whenever $1 \le j \le s$ and that $\ell\ell(P'_j) < t$ for all j such that $s < j \le r$. Let f_j denote the restriction $f|_{P'_j}$ of f to P'_j. If no f_j

with $j \leq s$ is a monomorphism, we would have $\ell\ell(\operatorname{Im} f_j) < t$ for all j, while the homomorphism $\bigoplus_{j=1}^{r} \operatorname{Im} f_j \to M$ induced by f is an epimorphism, and this would imply, by (1.2), that $\ell\ell(M) < t$, which is a contradiction. Hence there exists an index $q \leq s$ such that $f_q : P_q' \to M$ is a monomorphism. Because $\ell\ell(P_q') = t = \ell(A/\mathrm{rad}^t A)$, it follows from (3.4) that P_q' is injective as an $A/\mathrm{rad}^t A$-module. Consequently, $f_q : P_q' \to M$ is a section. Because M is indecomposable, f_q is an isomorphism. P_q' is an indecomposable projective $A/\mathrm{rad}^t A$-module. Hence there exists an index i with $1 \leq i \leq n$ such that $P_q' \cong P_i/\mathrm{rad}^t P_i$, and therefore there is an isomorphism $M \cong P_i/\mathrm{rad}^t P_i$. $\square$

A direct consequence of the theorem is that the number of nonisomorphic indecomposable A-modules is equal to

$$\sum_{i=1}^{n} \ell\ell(P_i) \leq n \cdot \ell\ell(A),$$

where n and the P_i are as in the proof. We also remark that if $M \cong P/\mathrm{rad}^t P$, for P indecomposable projective and $1 \leq t \leq \ell\ell(P)$, the canonical epimorphism $P \to M$ is a projective cover. Moreover, every indecomposable A-module is uniquely determined, up to isomorphism, by its simple top (or its simple socle) and its composition length. Indeed, let $S(a)$ be the simple top of an indecomposable A-module M, and $t \geq 1$ be its composition length. Because M is necessarily uniserial, $t = \ell\ell(M)$ and hence $M \cong P(a)/\mathrm{rad}^t P(a)$. We have the following useful fact.

3.6. Corollary. *A basic and connected algebra A is a Nakayama algebra if and only if every indecomposable A-module is uniserial.*

Proof. The sufficiency follows from the definition, the necessity from (3.5). $\square$

3.7. Example. Let A be given by the quiver $\underset{1}{\circ} \xleftarrow{\gamma} \underset{2}{\circ} \xleftarrow{\beta} \underset{3}{\circ} \xleftarrow{\alpha} \underset{4}{\circ}$ and bound by $\alpha\beta\gamma = 0$ (see (IV.4.11)). The indecomposable projective A-modules are listed as representations of the bound quiver in the notation of Section 2:

$$P(1) = (K \leftarrow 0 \leftarrow 0 \leftarrow 0) = (1),$$
$$P(2) = (K \xleftarrow{1} K \leftarrow 0 \leftarrow 0) = \left(\begin{smallmatrix}2\\1\end{smallmatrix}\right),$$
$$P(3) = (K \xleftarrow{1} K \xleftarrow{1} K \leftarrow 0) = \left(\begin{smallmatrix}3\\2\\1\end{smallmatrix}\right) = I(1),$$
$$P(4) = (0 \leftarrow K \xleftarrow{1} K \xleftarrow{1} K) = \left(\begin{smallmatrix}4\\3\\2\end{smallmatrix}\right) = I(2).$$

By (3.5), the remaining indecomposable A-modules are

$$
\begin{aligned}
P(2)/\operatorname{rad}P(2) &= (0\longleftarrow K\longleftarrow 0\longleftarrow 0) &=& (2), \\
P(3)/\operatorname{rad}P(3) &= (0\longleftarrow 0\longleftarrow K\longleftarrow 0) &=& (3), \\
P(3)/\operatorname{rad}^2 P(3) &= (0\longleftarrow K\overset{1}{\longleftarrow} K\longleftarrow 0) &=& \binom{3}{2}, \\
P(4)/\operatorname{rad}P(4) &= (0\longleftarrow 0\longleftarrow 0\longleftarrow K) &=& (4) = I(4), \\
P(4)/\operatorname{rad}^2 P(4) &= (0\longleftarrow 0\longleftarrow K\overset{1}{\longleftarrow} K) &=& \binom{4}{3} = I(3).
\end{aligned}
$$

The notation of Section 2 allows us to easily see the homomorphisms. For instance, there exists a homomorphism $P(3)/\operatorname{rad}^2 P(3) \to P(4)/\operatorname{rad}^2 P(4)$ of image $S(3)$ and a homomorphism $P(3) \to P(4)$ of image $P(3)/\operatorname{rad}^2 P(3)$. Neither of these homomorphisms is a monomorphism or an epimorphism. On the other hand, we have a monomorphism $P(2) \to P(3)$ of cokernel $S(3)$, and an epimorphism $P(4) \to P(4)/\operatorname{rad}^2 P(4)$ of kernel $S(2)$.

We now characterise the self-injective Nakayama algebras. We recall that an algebra is said to be **self-injective** (or a **quasi-Frobenius algebra**) if the right module A_A is an injective A-module, or, equivalently, if each projective right A_A-module is injective.

3.8. Proposition. *Let A be a basic and connected algebra, which is not isomorphic to K. Then A is a self-injective Nakayama algebra if and only if $A \cong KQ/I$, where Q is the quiver*

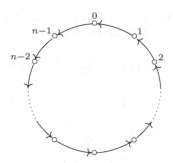

with $n \geq 1$ and $I = R^h$ for some $h \geq 2$, where R denotes the arrow ideal of KQ.

Proof. If A is of the given form, then it is a Nakayama algebra by (3.2) and it follows directly from the computation of the indecomposable projective and injective A-modules (see (III.2.4) and (III.2.6)) that A is self-injective.

Conversely, assume that A is a self-injective Nakayama algebra and $A \not\cong K$. The ordinary quiver $Q = Q_A$ of A cannot be of the form

$$\underset{1}{\circ} \longleftarrow \underset{2}{\circ} \longleftarrow \underset{3}{\circ} \longleftarrow \cdots \longleftarrow \underset{n-1}{\circ} \longleftarrow \underset{n}{\circ}$$

with $n > 1$, because then $P(1)_A$ would be a simple projective noninjective module. By (3.2), Q has the required form. If $n = 1$, the only admissible ideals of KQ are of the form $\mathcal{I} = R^h$ for some $h \geq 2$. We may thus suppose that $n > 1$.

For each i with $0 \leq i < n$, let t_i denote the length of the shortest path w_{i,t_i} of source i that belongs to $\mathcal{I}$, and let $h = \max\{t_i \mid 0 \leq i \leq n-1\}$. Because $\mathcal{I}$ is admissible, $h \geq 2$. Clearly, $\{w_{i,t_i} \mid 0 \leq i \leq n-1\}$ is a set of generators for $\mathcal{I}$, hence it suffices to show that $t_i = h$ for every i. Indeed, assume that this is not the case; then there exists an index i such that $t_i < h$. Let $s \in Q_0$ be the source of the unique arrow in Q with target i. We may clearly assume that $t_s = h$. Let now $j \in Q_0$ be such that $j + 1 \equiv i + t_i \pmod n$. Because $P(i)_A$ is injective, w_{i,t_i-1} is the longest path of target j that does not belong to $\mathcal{I}$. Hence $w_{s,t_i} \in \mathcal{I}$, because the target of w_{s,t_i} is j and it is longer than w_{i,t_i-1}. By definition of t_s, we have $h = t_s \leq t_i < h$, which is a contradiction. $\qquad\square$

3.9. Example. Let A be the K-algebra given by the quiver

and bound by $\alpha\beta\gamma = 0$, $\beta\gamma\alpha = 0$, $\gamma\alpha\beta = 0$. Then A is a self-injective Nakayama algebra. Its indecomposable projective (= injective) modules are given by:

$$P(1) = \begin{pmatrix} 1 \\ 3 \\ 2 \end{pmatrix} = \begin{smallmatrix} & K & \\ {}_K\nearrow & & \searrow^0 \\ K & \xrightarrow{1} & K \end{smallmatrix} = I(2),$$

$$P(2) = \begin{pmatrix} 2 \\ 1 \\ 3 \end{pmatrix} = \begin{smallmatrix} & K & \\ {}_K\nearrow & & \searrow^1 \\ K & \xrightarrow{0} & K \end{smallmatrix} = I(3),$$

$$P(3) = \begin{pmatrix} 3 \\ 2 \\ 1 \end{pmatrix} = \begin{smallmatrix} & K & \\ {}_K\nearrow^0 & & \searrow^1 \\ K & \xrightarrow{1} & K \end{smallmatrix} = I(1),$$

and the remaining indecomposable modules are given by:

$$P(1)/\operatorname{rad} P(1) = (1) = \begin{smallmatrix} & K & \\ \nearrow & & \searrow \\ 0 & \longrightarrow & 0 \end{smallmatrix},$$

$$P(1)/\operatorname{rad}^2 P = \begin{pmatrix} 1 \\ 3 \end{pmatrix} = \begin{smallmatrix} & K & \\ {}^1\nearrow & & \searrow \\ K & \longrightarrow & 0 \end{smallmatrix},$$

$$P(2)/\operatorname{rad} P(2) = (2) = \begin{smallmatrix} & K & \\ \nearrow & & \searrow \\ 0 & \longrightarrow & K \end{smallmatrix},$$

$$P(2)/\mathrm{rad}^2 P(2) \;=\; \binom{2}{1} \;=\; \begin{smallmatrix} & & K & \\ & \nearrow & & \searrow 1 \\ 0 & & & \\ & \xrightarrow{\;\;0\;\;} & & K \end{smallmatrix},$$

$$P(3)/\mathrm{rad}\, P(3) \;=\; (3) \;=\; \begin{smallmatrix} & & K & \\ & \nearrow & & \searrow \\ K & & & \\ & \xrightarrow{\;\;0\;\;} & & 0 \end{smallmatrix},$$

$$P(3)/\mathrm{rad}^2 P(3) \;=\; \binom{3}{2} \;=\; \begin{smallmatrix} & & K & \\ & \nearrow & & \searrow \\ K & & & \\ & \xrightarrow{\;\;1\;\;} & & K \end{smallmatrix}.$$

V.4. Almost split sequences for Nakayama algebras

We now show how to compute all almost split sequences in the module category of a Nakayama algebra A. We recall that if M is an indecomposable A-module of Loewy length t, then there exists, up to isomorphism, a unique indecomposable projective A-module P (the projective cover of M) such that $M \cong P/\mathrm{rad}^t P$. Moreover, M is nonprojective if and only if $t < \ell\ell(P)$.

4.1. Theorem. *Let $M \cong P/\mathrm{rad}^t P$ be an indecomposable nonprojective A-module. The sequence*

$$0 \longrightarrow \mathrm{rad}\, P/\mathrm{rad}^{t+1} P \xrightarrow{\left[\begin{smallmatrix} q \\ i \end{smallmatrix}\right]} (\mathrm{rad}\, P/\mathrm{rad}^t P) \oplus (P/\mathrm{rad}^{t+1} P) \xrightarrow{[-j\,p]} P/\mathrm{rad}^t P \longrightarrow 0$$

(where q and p are the canonical epimorphisms and i and j are the inclusion homomorphisms) is an almost split sequence.

Proof. The given sequence is easily seen to be exact. It is not split and has indecomposable end terms; hence, by (IV.1.13), it suffices to prove that the homomorphism $g = [-j\,p]$ is right almost split. It is clear that g is not a retraction. Let V be an indecomposable A-module and $v : V \to M$ be a nonisomorphism. We have two cases. If v is not surjective, $\mathrm{Im}\, v$ is contained in the unique maximal submodule $\mathrm{rad}\, M = \mathrm{rad}\, P/\mathrm{rad}^t P$ of $M = P/\mathrm{rad}^t P$. But then the homomorphism $\left[\begin{smallmatrix} -v \\ 0 \end{smallmatrix}\right] : V \longrightarrow (\mathrm{rad}\, P/\mathrm{rad}^t P) \oplus (P/\mathrm{rad}^{t+1} P)$ satisfies $g \cdot \left[\begin{smallmatrix} -v \\ 0 \end{smallmatrix}\right] = v$. If, on the other hand, v is surjective, because it is not an isomorphism, we must have $V \cong P/\mathrm{rad}^s P$ for some $s \geq t + 1$. Hence there exists an epimorphism $v' : V \to P/\mathrm{rad}^{t+1} P$ such that $v = pv'$. The homomorphism $\left[\begin{smallmatrix} 0 \\ v' \end{smallmatrix}\right] : V \to (\mathrm{rad}\, P/\mathrm{rad}^t P) \oplus (P/\mathrm{rad}^{t+1} P)$ satisfies $g \cdot \left[\begin{smallmatrix} 0 \\ v' \end{smallmatrix}\right] = v$. $\square$

It follows immediately that an almost split sequence in the module category of a Nakayama algebra has at most two indecomposable middle terms.

4.2. Corollary. *For every indecomposable nonprojective A-module M, we have $\ell(\tau M) = \ell(M)$. In particular, all the nonisomorphic simple*

A-modules belong to the same τ-orbit.

Proof. By (3.5), if t denotes the Loewy length of M and P is the projective cover of M, then $M \cong P/\mathrm{rad}^t P$. Hence, by (4.1), $\tau M \cong \mathrm{rad}P/\mathrm{rad}^{t+1}P$. Then, by (2.2),

$$\ell(\tau M) = \ell(\mathrm{rad}P/\mathrm{rad}^{t+1}P) = t = \ell(P/\mathrm{rad}^t P) = \ell(M).$$

This shows that all modules in the τ-orbit of M have the same length as M.
$\square$

4.3. Examples. We construct, with the help of (4.1), the Auslander–Reiten quivers of the algebras of the examples of Section 3.

(a) Let A be the K-algebra given by the quiver $\underset{1}{\circ} \xleftarrow{\gamma} \underset{2}{\circ} \xleftarrow{\beta} \underset{3}{\circ} \xleftarrow{\alpha} \underset{4}{\circ}$ and bound by $\alpha\beta\gamma = 0$. Then $\Gamma(\mathrm{mod}\,A)$ is given by:

$$
\begin{array}{ccccccc}
(1) & & (2) & & (3) & & (4) \\
& \searrow \quad \nearrow & & \searrow \quad \nearrow & & \searrow \quad \nearrow & \\
& \binom{2}{1} & & \binom{3}{2} \to \binom{4}{3}{2} \to \binom{4}{3} & & \\
& & \searrow \quad \nearrow & & & & \\
& & \binom{3}{2}{1} & & & &
\end{array}
$$

(compare with (IV.4.11)).

(b) Let A be the K-algebra given by the quiver

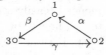

and bound by $\alpha\beta\gamma = 0$, $\beta\gamma\alpha = 0$, $\gamma\alpha\beta = 0$. Then $\Gamma(\mathrm{mod}\,A)$ is given by

$$
\begin{array}{ccccccccc}
(1) & & (2) & & (3) & & & & (1) \\
\vdots & \searrow \quad \nearrow & & \searrow \quad \nearrow & & \searrow \quad \nearrow & & \vdots \\
\vdots & \binom{2}{1} \to \binom{3}{2}{1} \to \binom{3}{2} & & & & \binom{1}{3} \to \binom{2}{1}{3} & \vdots \\
\binom{2}{1}{3} & \nearrow & & & \searrow & \nearrow & & \\
& & & & \binom{1}{3}{2} & & &
\end{array}
$$

Notice that the indecomposable modules (1) and $\binom{2}{1}{3}$ appear at both the extreme left and the extreme right of the quiver. One may thus think of $\Gamma(\mathrm{mod}\,A)$ as lying on a cylinder.

V.5. Representation–finite group algebras

The aim of this section is to prove Higman's characterisation [92] of the representation–finite group algebras. Throughout this section, we let K denote a commutative field (not necessarily algebraically closed) and G a finite group. By algebra A is meant, as usual, a finite dimensional K-algebra. We note that, if H is a subgroup of G, then the group algebra AH of H can be identified to a subalgebra of the group algebra AG of G. We thus have a restriction functor $\mod AG \to \mod AH$ defined in the obvious way. Given an AG-module M, we also denote by M the corresponding AH-module; it is always clear from the context which module structure is being considered.

5.1. Lemma. *Let A be an algebra, G be a finite group, and H be a subgroup of G.*

(a) *If AG is representation–finite, then AH is also representation–finite.*

(b) *If the index $[G : H]$ of H in G is invertible as an element of A then every right AG-module M is isomorphic to a direct summand of $M \otimes_{AH} AG$. Further, if AH is representation–finite, then AG is also representation–finite.*

Proof. (a) Let $\{M_1, \ldots, M_t\}$ be a complete set of representatives of the isomorphism classes of indecomposable AG-modules. Considering each M_i as an AH-module and applying the unique decomposition theorem (I.4.10) we have that $M_i \cong N_{i1} \oplus \cdots \oplus N_{it_i}$, where each N_{ij} is an indecomposable AH-module. We show that each indecomposable AH-module N is isomorphic to N_{ij} for some pair (i, j) with $1 \le i \le t$, $1 \le j \le t_i$. This clearly means that AH is representation–finite.

For this purpose, we first consider the K-linear map $p : AG \longrightarrow AH$ defined by the formula $\sum_{g\in G} a_g g \mapsto \sum_{h\in H} a_h h$. Then p is clearly an epimorphism of AH-AH-bimodules and actually a retraction of left and right AH-modules. Let N be an indecomposable AH-module. The composed epimorphism $N \otimes_{AH} AG_{AH} \xrightarrow{\ 1_N \otimes p\ } N \otimes_{AH} AH_{AH} \cong N_{AH}$ of AH-modules is a retraction, so that N is isomorphic to an indecomposable direct summand of $N \otimes_{AH} AG_{AH}$. The AG-module $N \otimes_{AH} AG_{AH}$ is isomorphic to the direct sum of the modules M_i, each of which is isomorphic as an AH-module to the direct sum of the modules N_{ij}, with $1 \le j \le t_i$. Another application of the unique decomposition theorem (I.4.10) yields that $N \cong N_{ij}$ for some pair (i, j).

(b) Let $s = [G : H]$ and $\{g_1, g_2, \ldots, g_s\}$ be a complete set of representatives of the left cosets of H in G, so that $G = Hg_1 \cup \cdots \cup Hg_s$. Given a right AG-module M_{AG} we define two homomorphisms of AG-modules by

$$M_{AG} \xrightarrow{\ f\ } M \otimes_{AH} AG_{AG}, \qquad x \mapsto \sum_{i=1}^{s} x g_i \otimes g_i^{-1}, \quad \text{and}$$

$$M \otimes_{AH} AG_{AG} \xrightarrow{\ f'\ } M_{AG}, \qquad x \otimes g \mapsto x g s^{-1},$$

where $x \in M$ and $g \in G$. It is easily verified that f and f' are indeed homomorphisms of AG-modules. Moreover, $f' \circ f = 1_M$; indeed, for any $x \in M$,

$$\begin{aligned}
(f' \circ f)(x) &= f'(\sum_{i=1}^{s} x g_i \otimes g_i^{-1}) \\
&= \sum_{i=1}^{s} x g_i g_i^{-1} s^{-1} = x s s^{-1} = x.
\end{aligned}$$

Thus, f is a section, that is, M_{AG} is isomorphic to a direct summand of $M \otimes_{AH} AG_{AG}$.

Assume now that AH is representation–finite and let $\{N_1, \dots, N_t\}$ be a complete set of representatives of the isomorphism classes of indecomposable AH-modules. Let M be an indecomposable AG-module. Then M_{AG} is isomorphic to a direct summand of $M \otimes_{AH} AG_{AG}$. On the other hand, the unique decomposition theorem allows us to write the AH-module M as $M_{AH} \cong N_1^{n_1} \oplus \cdots \oplus N_t^{n_t}$, where $n_i \geq 0$ for each $1 \leq i \leq t$. Hence M_{AG} is isomorphic to an indecomposable direct summand of $\bigoplus_{i=1}^{t} (N_i \otimes_{AH} AG_{AG})^{n_i}$. Applying the unique decomposition theorem (I.4.10) to the AG-modules $N_i \otimes_{AH} AG_{AG}$, where $1 \leq i \leq t$, we can write

$$N_i \otimes_{AH} AG_{AG} \cong M_{i1} \oplus \cdots \oplus M_{i t_i},$$

where each M_{ij} is an indecomposable AG-module. Consequently, $M \cong M_{ij}$ for some pair (i, j) with $1 \leq i \leq t_i$. This shows that the algebra AG is representation–finite. □

As an easy consequence of (5.1), we obtain Maschke's theorem (I.3.5).

5.2. Corollary. *If the characteristic p of K does not divide the order of the group G then the group algebra KG is semisimple.*

Proof. We apply (5.1) to $A = K$ and $H = \{e\}$; then $AH \cong K$. It follows from (5.1)(b) that every indecomposable KG-module is isomorphic to an indecomposable summand of $K \otimes_K KG_{KG} \cong KG_{KG}$ and thus is projective. Consequently, the algebra KG is semisimple. □

5.3. Lemma. *Let K be a field of characteristic $p > 0$ and C_{p^m} denote the cyclic group of order p^m with $m \geq 0$.*

(a) *There exists an isomorphism $K(C_{p^m} \oplus C_{p^n}) \cong K[t_1, t_2]/(t_1^{p^m}, t_2^{p^n})$ of K-algebras.*

(b) *There exists an isomorphism $KC_{p^m} \cong K[t]/(t^{p^m})$ of K-algebras.*
(c) *The group algebra $K(C_p \oplus C_p)$ is representation–infinite.*

Proof. (a) Let a and b denote, respectively, generators of the cyclic groups C_{p^m} and C_{p^n}, and consider the K-algebra homomorphism

$$f : K[T_1, T_2] \longrightarrow K(C_{p^m} \oplus C_{p^n})$$

defined by $\sum_{i,j} \lambda_{ij} T_1^i T_2^j \mapsto \sum_{i,j} \lambda_{ij}(a^i, b^j)$, where $\lambda_{ij} \in K$ for all i, j. Clearly, f is surjective and the ideal $(T_1^{p^m} - 1, T_2^{p^n} - 1)$ is contained in Ker f. Consequently f induces, by passing to the quotient, a surjective K-algebra homomorphism

$$\overline{f} : K[T_1, T_2]/(T_1^{p^m} - 1, T_2^{p^n} - 1) \longrightarrow K(C_{p^m} \oplus C_{p^n}).$$

We have now

$$\dim_K K[T_1, T_2]/(T_1^{p^m} - 1, T_2^{p^n} - 1) = p^{m+n} = \dim_K K(C_{p^m} \oplus C_{p^n}).$$

Therefore $\overline{f}$ is an isomorphism. Finally, let $t_1 = T_1 - 1$ and $t_2 = T_2 - 1$. Because p is the characteristic of the field K, $t_1^{p^m} = T_1^{p^m} - 1$ and $t_2^{p^m} = T_2^{p^m} - 1$ so that $K(C_{p^m} \oplus C_{p^n}) \cong K[t_1, t_2]/(t_1^{p^m}, t_2^{p^n})$, as required.

(b) The required isomorphism follows from the isomorphism in (a) after setting $n = 0$.

(c) Let $A = K[t_1, t_2]/(t_1, t_2)^2$. Because $(t_1^p, t_2^p) \subseteq (t_1, t_2)^2$, we have a surjective K-algebra homomorphism given by the composition

$$K(C_p \oplus C_p) \cong K[t_1, t_2]/(t_1^p, t_2^p) \longrightarrow K[t_1, t_2]/(t_1, t_2)^2 = A,$$

which induces a full and faithful embedding mod $A \to \operatorname{mod} K(C_p \oplus C_p)$. Hence it suffices to show that mod A is representation–infinite. For this purpose, we construct an infinite family $\{M_d\}_{d \geq 1}$ of pairwise nonisomorphic indecomposable A-modules.

Let $d \geq 1$ be an arbitrary natural number. Consider the $K[t]$-module $N_d = K[t]/(t^d)$ of dimension d. It is well-known and easy to check that N_d is indecomposable as a $K[t]$-module and that $\operatorname{End}_{K[t]} N_d \cong K[t]/(t^d)$.

We define a $K[t_1, t_2]$-module structure on the K-vector space $M_d = N_d \oplus N_d$ by the formulas $(r, q) \cdot t_1 = (0, r \cdot t)$ and $(r, q) \cdot t_2 = (0, r)$, for $r, q \in N_d$. Because $(r, q) \cdot t_1^2 = 0$, $(r, q) \cdot t_2^2 = 0$, and $(r, q) \cdot t_1 t_2 = 0$ for any $r, q \in N_d$, we see that M_d is annihilated by the ideal $(t_1, t_2)^2$ and thus has a natural A-module structure. Moreover, $\dim_K M_d = 2d$; hence the modules M_d are pairwise nonisomorphic. To complete the proof, we show that for any $d \geq 1$ the endomorphism algebra $\operatorname{End}_A M_d$ is local, so that M_d is indecomposable as an A-module. Let

$$f = \begin{bmatrix} f_{11} & f_{12} \\ f_{21} & f_{22} \end{bmatrix} : N_d \oplus N_d \longrightarrow N_d \oplus N_d$$

be a K-linear endomorphism of the A-module $M_d = N_d \oplus N_d$, where $f_{ij} :$ $N_d \to N_d$ are K-linear endomorphisms. Clearly, f is a homomorphism of A-modules if and only if $f((r,q) \cdot t_1) = (f(r,q)) \cdot t_1$ and

$$f((r,q) \cdot t_2) = (f(r,q)) \cdot t_2$$

for all $r, q \in N_d$. An immediate calculation shows that this is the case if and only if $f_{12} = 0$, $f_{11} = f_{22}$ and f_{11} is an endomorphism of N_d viewed as a $K[t]$-module. Consider the K-algebra homomorphism

$$\varphi : \operatorname{End}_A M_d \to \operatorname{End}_{K[t]} N_d \cong K[t]/(t^d)$$

defined by $f \mapsto f_{11}$. Clearly, φ is surjective and $\operatorname{Ker}\varphi$ consists of those $f \in \operatorname{End}_A M_d$ such that $f_{11} = f_{12} = f_{22} = 0$ (thus $f \in \operatorname{Ker}\varphi$ implies $f^2 = 0$). To show that $\operatorname{End}_A M_d$ is local, it suffices, by (I.4.6), to show that any idempotent $e \in \operatorname{End}_A M_d$ equals either zero or the identity. Because $\varphi(e)$ is an idempotent of the local algebra

$$\operatorname{End}_{K[t]} N_d \cong K[t]/(t^d),$$

$\varphi(e)$ is either zero or the identity. In the former case, $e \in \operatorname{Ker}\varphi$, hence $e^2 = 0$ so that $e = e^2 = 0$. In the latter, $1_{M_d} - e \in \operatorname{Ker}\varphi$ yields $(1_{M_d} - e) = (1_{M_d} - e)^2 = 0$, hence $e = 1_{M_d}$. This completes the proof. □

We note that the proof of (c) shows in fact that $K(C_{p^m} \oplus C_{p^n})$ is representation–infinite for all $m, n \geq 1$. Moreover, the isomorphisms of the lemma allow us to construct bound quivers representing the group algebras arising from groups of the form $C_{p^m} \oplus C_{p^n}$. For instance, over a field K of characteristic 2, the group algebra of the Klein four group $C_2 \oplus C_2$ is given by the quiver

$$\alpha \,\circlearrowleft\, \underset{1}{\overset{\circ}{\cdot}} \,\circlearrowright\, \beta$$

and bound by $\alpha^2 = 0$, $\beta^2 = 0$, $\alpha\beta = \beta\alpha$. Moreover, this algebra is representation–infinite (by (c)). On the other hand, over a field of characteristic $p > 0$, the group algebra of the cyclic group C_{p^m} is given by the quiver

$$1 \,\circ\, \circlearrowright\, \alpha$$

and bound by $\alpha^{p^m} = 0$. Such an algebra is a Nakayama algebra and thus is representation–finite (by (3.7) and (3.5)).

We now need to recall a few facts from elementary group theory. Let G be a finite group acting on itself by conjugation. To determine the number of elements in the conjugacy class of an element $x \in G$, we consider the **centraliser**

$$Z_x = \{y \in G \mid yxy^{-1} = x\}$$

of x: this is a subgroup of G containing x. Clearly, $yxy^{-1} = zxz^{-1}$ if and only if $yZ_x = zZ_x$ so that the number of distinct conjugates of x is the same as the number $[G : Z_x]$ of left cosets of Z_x in G. In particular, x coincides with all its conjugates if and only if x belongs to the centre $Z(G)$ of G. Because every element of G belongs to exactly one conjugacy class, we deduce the so-called **class equation**

$$|G| = |Z(G)| + \sum [G : Z_x],$$

where the sum is taken over a set of representatives $\{x\}$ of those conjugacy classes of G such that $[G : Z_x] \neq 1$. Let p be a prime number. A finite group G is called a p-**group** if $|G| = p^m$ for some $m > 0$.

We need the following lemma.

Lemma 5.4. *Let G be a p-group; then the centre $Z(G)$ of G is nontrivial. If, moreover, G is not abelian, then $G/Z(G)$ is a nontrivial noncyclic group.*

Proof. The first assertion follows from the class equation. Indeed, if the conjugacy class of $x \in G$ contains more than one element, then $Z_x \neq G$. By Lagrange's theorem, p divides $[G : Z_x]$. The class equation then implies that p divides $|Z(G)|$. In particular, $Z(G)$ is nontrivial.

Because G is not abelian, $G/Z(G)$ is not trivial. Assume that $G/Z(G)$ is cyclic and is generated, say, by a coset $\overline{x}$ for some $x \in G$. Then any element $y \in G$ is of the form $y = x^s z$, where $s \geq 0$ and $z \in Z(G)$. But this implies that G is abelian, which is a contradiction. Hence $G/Z(G)$ is not cyclic. $\square$

Corollary 5.5. *If $|G| = p^2$, then G is abelian.*

Proof. It suffices to show that $G/Z(G)$ is cyclic, and this follows from the fact that $Z(G)$ is not trivial, so that $|G/Z(G)|$ equals 1 or p. $\square$

We are now able to prove Higman's characterisation of the representation–finite group algebras. We recall that if G is a finite group of order $p^m n$, where p is a prime that does not divide n, a **Sylow p-subgroup** G_p of G is a subgroup of order p^m. The celebrated Sylow theorems assert that G contains a p-Sylow subgroup and that all Sylow p-subgroups are conjugate (and, in particular, are isomorphic).

5.6. Theorem. *Let G be a finite group and let K be a field of characteristic p dividing the order of G. The group algebra KG is representation–*

finite if and only if the Sylow p-subgroups G_p of G are cyclic.

Proof. By definition of G_p, the integer p does not divide $s = [G : G_p]$ and therefore s is invertible in K. By (5.1)(b), it suffices to prove the theorem in case $G = G_p$ is a p-group. Assume that $|G| = p^m$.

One implication is trivial: indeed, assume that G is cyclic, that is, $G \cong C_{p^m}$. Then, by (5.3)(b), the group algebra KG is a Nakayama algebra; hence it is representation–finite (by (3.5)).

Conversely, assume that G is not cyclic. We must prove that KG is representation–infinite. For this purpose, we first show by induction on m that there exists a group epimorphism $G \to C_p \oplus C_p$.

If $m = 2$, then G is of order p^2, hence is abelian, by (5.5), so that $G \cong C_p \oplus C_p$.

Assume that $m > 2$. Clearly, the statement holds if G is abelian. If this is not the case, then, by (5.4), $\overline{G} = G/Z(G)$ is a nontrivial noncyclic group, of order p^k with $k < m$, because $Z(G)$ is nontrivial. The inductive hypothesis implies the existence of a group epimorphism $\overline{G} \to C_p \oplus C_p$, and the required epimorphism follows after composing with the canonical epimorphism $G \to \overline{G}$. This finishes the proof of our claim.

The group epimorphism $G \to C_p \oplus C_p$ obviously induces a surjective algebra homomorphism $KG \to K(C_p \oplus C_p)$ and consequently a full and faithful K-linear functor mod $K(C_p \oplus C_p) \to \bmod KG$. By (5.3), the algebra $K(C_p \oplus C_p)$ is representation–infinite. Hence KG is also representation–infinite. $\square$

5.7. Example. Let A_4 denote the alternating group on four objects. Then KA_4 is representation–finite if K is a field of characteristic 3 and representation–infinite if K is a field of characteristic 2. Indeed, a straight-forward calculation, left as an exercise to the reader, shows that the Sylow 3-subgroup of A_4 is isomorphic to the cyclic group C_3, while the Sylow 2-subgroup of A_4 is isomorphic to the Klein four group $C_2 \oplus C_2$.

V.6. Exercises

1. A module M over an arbitrary algebra A is called a **Nakayama module** if M is the direct sum of uniserial modules. Let A be a right (or left, respectively) serial algebra. Show that every submodule (or quotient module, respectively) of a Nakayama module is a Nakayama module.

2. Show that A is a Nakayama algebra if and only if $A/\mathrm{rad}^2 A$ is a Nakayama algebra.

3. For each of the following bound quivers $(Q, \mathcal{I})$

(a) $\underset{1}{\circ} \xleftarrow{\gamma} \underset{2}{\circ} \xleftarrow{\beta} \underset{3}{\circ} \xleftarrow{\alpha} \underset{4}{\circ}$ $\alpha\beta = 0, \beta\gamma = 0;$

(b) $\underset{1}{\circ} \xleftarrow{\delta} \underset{2}{\circ} \xleftarrow{\gamma} \underset{3}{\circ} \xleftarrow{\beta} \underset{4}{\circ} \xleftarrow{\alpha} \underset{5}{\circ}$ $\alpha\beta = 0, \beta\gamma\delta = 0;$

(c) $1 \circ \underset{\beta}{\overset{\alpha}{\rightleftarrows}} \circ 2$ $\alpha\beta = 0, \beta\alpha = 0;$

(d)
$$\begin{array}{ccc}
1 \circ & \xrightarrow{\alpha} & \circ 2 \\
\delta \uparrow & & \downarrow \beta \\
4 \circ & \xleftarrow{\gamma} & \circ 3
\end{array}$$
$\alpha\beta = 0, \beta\gamma = 0, \gamma\delta = 0$

describe the path algebra $A = KQ/\mathcal{I}$, all the indecomposable A-modules, and the homomorphisms between them.

4. Let $0 \longrightarrow L \longrightarrow M \longrightarrow N \longrightarrow 0$ be a short exact sequence. Show that $\ell\ell(M) \le \ell\ell(L) + \ell\ell(N)$.

5. For each of the following bound quivers $(Q, \mathcal{I})$

(a) $\circ \xleftarrow{\beta} \circ \xleftarrow{\alpha} \circ$ $\alpha\beta = 0;$

(b)
$$\begin{array}{c}
\circ \\
{}^{\beta}\swarrow \quad \nwarrow^{\alpha} \\
\circ \xrightarrow{\gamma} \circ
\end{array}$$
$\alpha\beta\gamma\alpha = 0, \gamma\alpha\beta = 0$

describe the Nakayama algebra $A = KQ/\mathcal{I}$ and compute all the indecomposable A-modules. Then, for each pair (M, N) of indecomposable modules, compute the vector spaces $\mathrm{Hom}_A(M, N)$, $\underline{\mathrm{Hom}}_A(M, N)$, and $\overline{\mathrm{Hom}}_A(M, N)$.

6. Construct the Auslander–Reiten quiver of the Nakayama algebras defined by each of the following bound quivers:

(a) $\circ \,\alpha$ (loop) $\beta^2 = 0;$

(b) $\circ \xleftarrow{\alpha} \circ$

(c) $\circ \xleftarrow{\beta} \circ \xleftarrow{\alpha} \circ$ $\alpha\beta = 0;$

(d)
$$\begin{array}{c}
\circ \\
{}^{\beta}\swarrow \quad \nwarrow^{\alpha} \\
\circ \xrightarrow{\gamma} \circ
\end{array}$$
$\gamma\alpha\beta = 0, \alpha\beta\gamma\alpha = 0;$

(e) $\circ \xleftarrow{\ \delta\ } \circ \xleftarrow{\ \gamma\ } \circ \xleftarrow{\ \beta\ } \circ \xleftarrow{\ \alpha\ } \circ$ $\alpha\beta = \beta\gamma = \gamma\delta = 0;$

(f) $\circ \xleftarrow{\ \delta\ } \circ \xleftarrow{\ \gamma\ } \circ \xleftarrow{\ \beta\ } \circ \xleftarrow{\ \alpha\ } \circ$ $\alpha\beta\gamma = 0,\ \gamma\delta = 0;$

(g) $\circ \underset{\beta}{\overset{\alpha}{\rightleftarrows}} \circ$ $\alpha\beta = 0;$

(h) $\circ \underset{\beta}{\overset{\alpha}{\rightleftarrows}} \circ$ $\alpha\beta\alpha = 0;$

(i)   triangle with top vertex, β down-left, α down-right, γ along the bottom $\alpha\beta = \beta\gamma = \gamma\alpha = 0;$

(j)   square with α top, β right down, γ bottom, δ left up $\alpha\beta = \beta\gamma = \gamma\delta = 0.$

7. Let A be a Nakayama algebra and P be indecomposable projective with $P/\operatorname{rad} P = S$. Show that

$$\operatorname{rad}^i P/\operatorname{rad}^{i+1} P \cong \tau^i S$$

for every $0 \leq i < \ell(P)$ (so that all the composition factors of P belongs to the same τ-orbit).

8. Let A be a connected Nakayama algebra. Show that there exists an ordering $\{P_1, P_2, \ldots, P_n\}$ of the nonisomorphic indecomposable projective A-modules such that
  (a) $P_{i+1}/\operatorname{rad} P_{i+1} \cong \tau^{-1}(P_i/\operatorname{rad} P_i)$ for $1 \leq i \leq n-1$, and if $\ell(P_1) \neq 1$ then $P_1/\operatorname{rad} P_1 \cong (P_n/\operatorname{rad} P_n)$;
  (b) $\ell(P_i) \geq 2$ for $i = 2, \ldots, n$; and
  (c) $\ell(P_{i+1}) \leq \ell(P_i) + 1$ for every $i = 1, \ldots, n-1$ and $\ell(P_1) \leq \ell(P_n) + 1$.
Such an ordering, called a **Kupisch series** for A, is unique up to a cyclic permutation (or simply unique if $\ell(P_1) = 1$).

9. Assume that A is a connected Nakayama K-algebra with Kupisch series $\{P_1, \ldots, P_n\}$. Show that $\ell(P_{i+1}) = \ell(P_i) + 1$ if and only if P_i is not injective for $i = 1, \ldots, n-1$ and $\ell(P_1) = \ell(P_n) + 1$ if and only if P_n is not injective.

10. Compute a Kupisch series for each of the Nakayama algebras of Exercise 6.

11. Let A be a self-injective connected Nakayama algebra. Show that every indecomposable projective A-module has the same length $\ell\ell(A)$.

12. Let $A = KQ/R^2$, where Q is the quiver

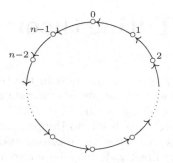

$(n \geq 3)$ and R is the two-sided ideal of KQ generated by the arrows. Show that A is self-injective, but that eAe, where $e = e_0 + e_1 + \ldots + e_k$ $(k < n-1)$, is not.

13. Let $(a_1, \ldots, a_n)$ be a sequence of integers such that $a_j \geq 2$ for all $j \geq 2$, $a_{j+1} \leq 1 + a_j$ for $j \leq n-1$, and $a_1 \leq 1 + a_n$. Construct a Nakayama K-algebra having the sequence $(a_1, \ldots, a_n)$ as a Kupisch series.

14. Construct the Auslander–Reiten quiver of the K-algebra A defined by the following bound quiver:

$$\gamma\alpha\beta = 0, \quad \beta\gamma = 0.$$

Compute the global dimension gl.dim A of A.

15. Let A be the K-algebra given by the quiver $1 \circ \underset{\beta}{\overset{\alpha}{\rightleftarrows}} \circ 2$ and bound by the relation $\alpha\beta\alpha\beta = 0$. Using the notation (2.7), show that

$$P(2)_A = \begin{pmatrix} 2 \\ 1 \\ 2 \\ 1 \\ 2 \end{pmatrix} \qquad \text{and} \qquad P(1)_A = \begin{pmatrix} 1 \\ 2 \\ 1 \\ 2 \end{pmatrix}.$$

Prove that the K-vector space $\mathrm{Hom}_A(P(2), P(1))$ is of dimension two.

Chapter VI

Tilting theory

Tilting theory is one of the main tools in the representation theory of algebras. It originated with the study of reflection functors [32], [18]. The first set of axioms for a tilting module is due to Brenner and Butler [46]; the one generally accepted now is due to Happel and Ringel [89]. The main idea of tilting theory is that when the representation theory of an algebra A is difficult to study directly, it may be convenient to replace A with another simpler algebra B and to reduce the problem on A to a problem on B. We then construct an A-module T, called a tilting module, which can be thought of as being close to the Morita progenerators such that, if $B = \operatorname{End} T_A$, then the categories $\operatorname{mod} A$ and $\operatorname{mod} B$ are reasonably close to each other (but generally not equivalent). As will be seen, the knowledge of one of these module categories implies the knowledge of two distinguished full subcategories of the other, which form a torsion pair and thus determine up to extensions the whole module category. Because this procedure can be seen as generalising Morita theory, it is reasonable to give special attention to the full subcategory $\operatorname{Gen} T_A$ of all A-modules generated by T and to use the adjoint functors $\operatorname{Hom}_A(T, -)$ and $- \otimes_B T$ to compare $\operatorname{mod} A$ and $\operatorname{mod} B$.

Some notation is useful. Throughout this chapter, we let A denote an algebra, by which is meant, as usual, a finite dimensional, basic, and connected algebra over a fixed algebraically closed field K. For an A-module M, we denote by $\operatorname{add} M$ the smallest additive full subcategory of $\operatorname{mod} A$ containing M, that is, the full subcategory of $\operatorname{mod} A$ whose objects are the direct sums of direct summands of the module M. In many places, we consider the restriction to a subcategory $\mathcal{C}$ of a functor F defined originally on a module category, and we denote it by $F|_{\mathcal{C}}$.

VI.1. Torsion pairs

It is a well-known fact from elementary abelian group theory that there exists no nonzero homomorphism from a torsion group to a torsion-free one and that these two classes of abelian groups are maximal for this property. Generalising this situation, we obtain the concept of a torsion pair, valid in any abelian category, but which we need only for module categories. The following definition is due to Dickson [53].

1.1. Definition. A pair $(\mathcal{T}, \mathcal{F})$ of full subcategories of mod A is called a **torsion pair** (or a **torsion theory**) if the following conditions are satisfied:

(a) $\operatorname{Hom}_A(M, N) = 0$ for all $M \in \mathcal{T}$, $N \in \mathcal{F}$.
(b) $\operatorname{Hom}_A(M, -)|_{\mathcal{F}} = 0$ implies $M \in \mathcal{T}$.
(c) $\operatorname{Hom}_A(-, N)|_{\mathcal{T}} = 0$ implies $N \in \mathcal{F}$.

The first condition of the definition says that there is no nonzero homomorphism from an object in $\mathcal{T}$ to one in $\mathcal{F}$, and the other two conditions say that these two subcategories are maximal for this property. In analogy with the situation for abelian groups, the subcategory $\mathcal{T}$ is called the **torsion class**, and its objects are called **torsion objects**, while the subcategory $\mathcal{F}$ is called the **torsion-free class**, and its objects are called **torsion-free objects**. It follows directly from the definition that the torsion class and the torsion-free class determine uniquely each other.

1.2. Examples. (a) An arbitrary class $\mathcal{C}$ of A-modules induces a torsion pair as follows: let $\mathcal{F} = \{N \mid \operatorname{Hom}_A(-, N)|_{\mathcal{C}} = 0\}$ and $\mathcal{T} = \{M \mid \operatorname{Hom}_A(M, -)|_{\mathcal{F}} = 0\}$. Then $(\mathcal{T}, \mathcal{F})$ is a torsion pair, and $\mathcal{T}$ is in fact the smallest torsion class containing $\mathcal{C}$. The dual construction yields the smallest torsion-free class containing $\mathcal{C}$.

(b) If $(\mathcal{T}, \mathcal{F})$ is a torsion pair in the category mod A of all finite dimensional right A-modules, and $D : \operatorname{mod} A \to \operatorname{mod} A^{\mathrm{op}}$ denotes the standard duality, then $(D\mathcal{F}, D\mathcal{T})$ is a torsion pair in mod A^{op}.

(c) Let A be the path algebra of the quiver

$$\overset{1}{\circ} \longleftarrow \overset{2}{\circ} \longleftarrow \overset{3}{\circ}$$

and let $\mathcal{T} = \operatorname{add}\{010 \oplus 011 \oplus 001\}$, $\mathcal{F} = \operatorname{add}\{100 \oplus 110 \oplus 111\}$ (where the indecomposable A-modules are represented by their dimension vectors). Then $(\mathcal{T}, \mathcal{F})$ is a torsion pair. We may illustrate $(\mathcal{T}, \mathcal{F})$ in the Auslander–Reiten quiver $\Gamma(\operatorname{mod} A)$ of A, adopting the convention (which we keep throughout this chapter and the next) to shade the class $\mathcal{T}$ as ⟋ and the class $\mathcal{F}$ as ⬭ :

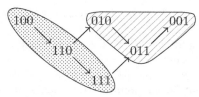

(d) Let A be as in (c). Then we have another torsion pair $(\mathcal{T}, \mathcal{F})$, illustrated as follows in $\Gamma(\operatorname{mod} A)$:

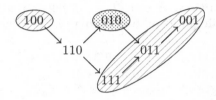

Our first objective is to give an intrinsic characterisation of torsion (or torsion-free) classes. For this purpose, we need one further definition.

1.3. Definition. A subfunctor t of the identity functor on $\operatorname{mod} A$ is called an **idempotent radical** if, for every module M_A, we have $t(tM) = tM$ and $t(M/tM) = 0$.

We recall that a subfunctor of the identity functor on $\operatorname{mod} A$ is a functor $t : \operatorname{mod} A \longrightarrow \operatorname{mod} A$ that assigns to each module M a submodule $tM \subseteq M$ such that each homomorphism $M \longrightarrow N$ restricts to a homomorphism $tM \longrightarrow tN$. As we now show, each torsion pair induces an idempotent radical and conversely.

1.4. Proposition. (a) *Let* $\mathcal{T}$ *be a full subcategory of* $\operatorname{mod} A$. *The following conditions are equivalent:*

(i) $\mathcal{T}$ *is the torsion class of some torsion pair* $(\mathcal{T}, \mathcal{F})$ *in* $\operatorname{mod} A$.

(ii) $\mathcal{T}$ *is closed under images, direct sums, and extensions.*

(iii) *There exists an idempotent radical* t *such that* $\mathcal{T} = \{M \mid tM = M\}$.

(b) *Let* $\mathcal{F}$ *be a full subcategory of* $\operatorname{mod} A$. *The following conditions are equivalent:*

(i) $\mathcal{F}$ *is the torsion-free class of some torsion pair* $(\mathcal{T}, \mathcal{F})$ *in* $\operatorname{mod} A$.

(ii) $\mathcal{F}$ *is closed under submodules, direct products, and extensions.*

(iii) *There exists an idempotent radical* t *such that* $\mathcal{F} = \{N \mid tN = 0\}$.

Proof. We only prove (a); the proof of (b) is similar.

(i) implies (ii). A short exact sequence $0 \to M' \to M \to M'' \to 0$ of A-modules induces a left exact sequence of functors

$$0 \longrightarrow \operatorname{Hom}_A(M'', -)|_{\mathcal{F}} \longrightarrow \operatorname{Hom}_A(M, -)|_{\mathcal{F}} \longrightarrow \operatorname{Hom}_A(M', -)|_{\mathcal{F}}.$$

Hence $M \in \mathcal{T}$ implies $M'' \in \mathcal{T}$ and, similarly, $M', M'' \in \mathcal{T}$ imply $M \in \mathcal{T}$. The statement follows.

(ii) implies (iii). Let M be any A-module and tM denote the **trace** of $\mathcal{T}$ in M, that is, the sum of the images of all A–homomorphisms from modules in $\mathcal{T}$ to M. Because $\mathcal{T}$ is closed under images and direct (hence arbitrary) sums, tM is the largest submodule of M that lies in $\mathcal{T}$. The trace defines a subfunctor of the identity: if $f : M \to N$ is a homomorphism, then $f(tM) \subseteq tN$ for, if $g : X \to M$ is a homomorphism with $X \in \mathcal{T}$, then $fg : X \to N$ has its image lying in tN. Moreover, we clearly have $t(tM) = tM$ and $M \in \mathcal{T}$ if and only if $tM = M$. Finally, let M be arbitrary and assume that $t(M/tM) = M'/tM$ with $tM \subseteq M' \subseteq M$. Because $\mathcal{T}$ is closed under extensions, tM, $M'/tM \in \mathcal{T}$ yield $M' \in \mathcal{T}$. Hence $M' \subseteq tM$ and $t(M/tM) = 0$.

(iii) implies (i). Let $\mathcal{F} = \{N \mid tN = 0\}$. Clearly, $\mathrm{Hom}_A(M, -)|_{\mathcal{F}} = 0$ for all $M \in \mathcal{T}$. We claim that, conversely, $\mathrm{Hom}_A(M, -)|_{\mathcal{F}} = 0$ implies $M \in \mathcal{T}$. Indeed, $t(M/tM) = 0$ gives $M/tM \in \mathcal{F}$. The canonical surjection $M \to M/tM$ being zero, we have $M/tM = 0$ so that $M = tM \in \mathcal{T}$. Similarly, $\mathrm{Hom}_A(-, N)|_{\mathcal{T}} = 0$ implies that $N \in \mathcal{F}$. □

An immediate consequence is that a torsion (or a torsion-free) class is an additive, hence K-linear, subcategory of $\mathrm{mod}\, A$, closed under isomorphic images, extensions, and direct summands.

The idempotent radical t attached to a given torsion pair is called the **torsion radical**. It follows from its definition that, for any module M_A, we have $tM \in \mathcal{T}$ and $M/tM \in \mathcal{F}$. The uniqueness follows from the next proposition, which also says that any module can be written in a unique way as the extension of a torsion-free module by a torsion module.

1.5. Proposition. *Let $(\mathcal{T}, \mathcal{F})$ be a torsion pair in $\mathrm{mod}\, A$ and M be an A-module. There exists a short exact sequence*

$$0 \longrightarrow tM \longrightarrow M \longrightarrow M/tM \longrightarrow 0$$

with $tM \in \mathcal{T}$ and $M/tM \in \mathcal{F}$. This sequence is unique in the sense that, if $0 \to M' \to M \to M'' \to 0$ is exact with $M' \in \mathcal{T}$, $M'' \in \mathcal{F}$, then the two sequences are isomorphic.

Proof. Only the second statement needs a proof. Because $M' \in \mathcal{T}$ and tM is the largest torsion submodule of M, there exists a commutative diagram with exact rows

$$
\begin{array}{ccccccccc}
0 & \longrightarrow & M' & \longrightarrow & M & \longrightarrow & M'' & \longrightarrow & 0 \\
 & & {\scriptstyle j}\downarrow & & {\scriptstyle 1_M}\downarrow & & {\scriptstyle f}\downarrow & & \\
0 & \longrightarrow & tM & \longrightarrow & M & \longrightarrow & M/tM & \longrightarrow & 0
\end{array}
$$

where j denotes the inclusion and f is obtained by passing to the cokernels. The Snake lemma (I.5.1) yields $tM/M' \cong \mathrm{Ker}\, f$. Because $tM/M' \in \mathcal{T}$ and $\mathrm{Ker}\, f \in \mathcal{F}$, we get $M'' \cong M/tM$ and $tM/M' = 0$. □

A short exact sequence as in the proposition is called the **canonical sequence** for M. For instance, in Example 1.2 (d), the canonical sequence for the indecomposable module $M = 110$ (which is neither torsion nor torsion-free) is $0 \longrightarrow 100 \longrightarrow 110 \longrightarrow 010 \longrightarrow 0$. The following obvious corollary is sometimes useful.

1.6. Corollary. *Every simple module is either torsion or torsion-free.*

$\square$

A torsion pair $(\mathcal{T}, \mathcal{F})$ such that each indecomposable A-module lies either in $\mathcal{T}$ or in $\mathcal{F}$ is called **splitting**. This is the case in example (1.2)(c) (but not in (1.2)(d)). Splitting torsion pairs are characterised as follows.

1.7. Proposition. *Let $(\mathcal{T}, \mathcal{F})$ be a torsion pair in* mod A. *The following conditions are equivalent:*

   (a) *$(\mathcal{T}, \mathcal{F})$ is splitting.*
   (b) *For each A-module M, the canonical sequence for M splits.*
   (c) $\operatorname{Ext}_A^1(N, M) = 0$ *for all $M \in \mathcal{T}$, $N \in \mathcal{F}$.*
   (d) *If $M \in \mathcal{T}$, then $\tau^{-1} M \in \mathcal{T}$.*
   (e) *If $N \in \mathcal{F}$, then $\tau N \in \mathcal{F}$.*

Proof. (a) implies (b). Let M_A be any module and M' (or M'') denote the direct sum of all the indecomposable summands of M that belong to $\mathcal{T}$ (or $\mathcal{F}$, respectively). We have a split short exact sequence $0 \to M' \to M \to M'' \to 0$ with $M' \in \mathcal{T}$, $M'' \in \mathcal{F}$, which is, by (1.5), isomorphic to the canonical sequence.

(b) implies (c). Any short exact sequence $0 \to M \to E \to N \to 0$ with $M \in \mathcal{T}$ and $N \in \mathcal{F}$ is a canonical sequence, by (1.5).

(c) implies (a). Let M be indecomposable. The hypothesis implies that the canonical sequence for M splits. Hence $M \cong tM \oplus (M/tM)$ so that either $M \cong tM$ or $M \cong M/tM$.

(a) implies (d). Let $0 \to M \to \bigoplus_{i=1}^n E_i \to \tau^{-1} M \to 0$ be the almost split sequence starting with M, where the modules $E_1, \dots, E_n$ are indecomposable. Because $\operatorname{Hom}_A(M, E_i) \neq 0$ for all i, the hypothesis implies that $E_i \in \mathcal{T}$ for all i. Hence $\bigoplus_{i=1}^n E_i \in \mathcal{T}$ so that $\tau^{-1} M \in \mathcal{T}$. We prove similarly that (a) implies (e).

(d) implies (c). Let $M \in \mathcal{T}$ and $N \in \mathcal{F}$. By the Auslander–Reiten formulas (IV.2.13), $\operatorname{Ext}_A^1(N, M) \cong D\underline{\operatorname{Hom}}_A(\tau^{-1} M, N)$. Because $\tau^{-1} M \in \mathcal{T}$ and $N \in \mathcal{F}$, we have $\operatorname{Hom}_A(\tau^{-1} M, N) = 0$. Hence $\operatorname{Ext}_A^1(N, M) = 0$. We prove similarly that (e) implies (c). $\square$

Let T be an arbitrary A-module. We define $\operatorname{Gen} T$ to be the class of all modules M in mod A generated by T, that is, the modules M such that there exist an integer $d \geq 0$ and an epimorphism $T^d \to M$ of A-modules. Dually, we define $\operatorname{Cogen} T$ to be the class of all modules N in mod A cogenerated by T, that is, the modules N such that there exist an integer $d \geq 0$ and a monomorphism $N \to T^d$ of A-modules.

We ask when the class $\operatorname{Gen} T$ is a torsion class and when the class $\operatorname{Cogen} T$ is a torsion-free class. It is clear that $\operatorname{Gen} T$ is closed under images, $\operatorname{Cogen} T$ is closed under submodules, and both classes are closed under direct sums. There remains thus, by (1.4), to see when they are closed under extensions. This is generally not the case: let A be an algebra having two nonisomorphic simple modules S, S' such that $\operatorname{Ext}_A^1(S, S') \neq 0$; then neither $\operatorname{Gen}(S \oplus S')$ nor $\operatorname{Cogen}(S \oplus S')$ is closed under extensions.

Before answering these questions, we derive a necessary and sufficient condition for an A-module to belong to $\operatorname{Gen} T$ (or to $\operatorname{Cogen} T$). We write $B = \operatorname{End} T_A$ so that T is endowed with a natural left B-module structure, compatible with the action of A, making it a B–A-bimodule.

1.8. Lemma. *Let M be an A-module.*

(a) $M \in \operatorname{Gen} T$ *if and only if the canonical homomorphism*

$$\varepsilon_M : \operatorname{Hom}_A(T, M) \otimes_B T \longrightarrow M$$

defined by $f \otimes t \mapsto f(t)$ is surjective, where $B = \operatorname{End} T_A$.
(b) $M \in \operatorname{Cogen} T$ *if and only if the canonical homomorphism*

$$\eta_M : M \longrightarrow \operatorname{Hom}_B(\operatorname{Hom}_A(M, T), T)$$

defined by $x \mapsto (g \mapsto g(x))$ is injective.

Proof. We only prove (a); the proof of (b) is similar. Assume $M \in \operatorname{Gen} T$ and let $f_1, \ldots, f_d$ be a basis of the K-vector space $\operatorname{Hom}_A(T, M)$. Then $f = [f_1 \ldots f_d] : T^d \to M$ is an epimorphism. Indeed, there exist $m > 0$ and an epimorphism $g : T^m \to M$. It follows from the definition of f that there exists $h : T^m \to T^d$ such that $g = fh$, so that f is surjective. Let $L = \operatorname{Ker} f$, and apply $\operatorname{Hom}_A(T, -)$ to the short exact sequence

$$0 \longrightarrow L \longrightarrow T^d \overset{f}{\longrightarrow} M \longrightarrow 0.$$

Because $\operatorname{Hom}_A(T, f)$ is an epimorphism by the definition of f, this yields a short exact sequence

$$0 \longrightarrow \operatorname{Hom}_A(T, L) \longrightarrow \operatorname{Hom}_A(T, T^d) \overset{\operatorname{Hom}_A(T,f)}{\longrightarrow} \operatorname{Hom}_A(T, M) \longrightarrow 0.$$

Applying $- \otimes_B T$, we obtain the upper row in the commutative diagram with exact rows

$$
\begin{array}{ccccccc}
\operatorname{Hom}_A(T, L) \otimes_B T & \to & \operatorname{Hom}_A(T, T^d) \otimes_B T & \to & \operatorname{Hom}_A(T, M) \otimes_B T & \to & 0 \\
\varepsilon_L \downarrow & & \varepsilon_{T^d} \downarrow & & \varepsilon_M \downarrow & & \\
0 \longrightarrow L & \longrightarrow & T^d & \longrightarrow & M & \longrightarrow & 0
\end{array}
$$

The composite homomorphism

$$\varepsilon_{T^d} : \mathrm{Hom}_A(T, T^d) \otimes_B T \cong B^d \otimes_B T \cong T^d$$

is an isomorphism. By the commutativity of the right square, the homomorphism ε_M is surjective.

Conversely, because $\mathrm{Hom}_A(T, M)$ is a finitely generated B-module, there exist $m > 0$ and an epimorphism $g : B^m \to \mathrm{Hom}_A(T, M)$, hence an epimorphism

$$T^m \cong B^m \otimes_B T \xrightarrow{\ g \otimes T\ } \mathrm{Hom}_A(T, M) \otimes_B T \xrightarrow{\ \varepsilon_M\ } M,$$

so $M \in \mathrm{Gen}\, T$. $\square$

The following lemma answers our questions.

1.9. Lemma. (a) *Assume that* $\mathrm{Ext}^1_A(T, -)|_{\mathrm{Gen}\, T} = 0$; *then* $\mathrm{Gen}\, T$ *is a torsion class. If this is the case, then the corresponding torsion-free class is the class* $\{M \mid \mathrm{Hom}_A(T, M) = 0\}$.

(b) *Assume that* $\mathrm{Ext}^1_A(-, T)|_{\mathrm{Cogen}\, T} = 0$; *then* $\mathrm{Cogen}\, T$ *is a torsion-free class. If this is the case, then the corresponding torsion class is the class* $\{M \mid \mathrm{Hom}_A(M, T) = 0\}$.

Proof. We only prove (a); the proof of (b) is similar. Assume that

$$0 \longrightarrow M' \longrightarrow M \longrightarrow M'' \longrightarrow 0$$

is a short exact sequence with $M', M'' \in \mathrm{Gen}\, T$. Because $\mathrm{Ext}^1_A(T, M') = 0$, we have a short exact sequence

$$0 \longrightarrow \mathrm{Hom}_A(T, M') \longrightarrow \mathrm{Hom}_A(T, M) \longrightarrow \mathrm{Hom}_A(T, M'') \longrightarrow 0,$$

which yields, after applying $- \otimes_B T$, the upper row in the commutative diagram with exact rows

$$
\begin{array}{ccccccc}
\mathrm{Hom}_A(T, M') \otimes_B T & \to & \mathrm{Hom}_A(T, M) \otimes_B T & \to & \mathrm{Hom}_A(T, M'') \otimes_B T & \to & 0 \\
\varepsilon_{M'} \downarrow & & \varepsilon_M \downarrow & & \varepsilon_{M''} \downarrow & & \\
0 \longrightarrow \quad M' & \longrightarrow & M & \longrightarrow & M'' & \longrightarrow & 0
\end{array}
$$

Because, by (1.8), $\varepsilon_{M'}$ and $\varepsilon_{M''}$ are epimorphisms, so is ε_M. A further application of (1.8) yields that $M \in \mathrm{Gen}\, T$ so that $\mathrm{Gen}\, T$ is indeed closed under extensions.

For the second statement, we notice that every torsion-free module M satisfies $\mathrm{Hom}_A(T, M) = 0$. Conversely, if $\mathrm{Hom}_A(T, M) = 0$ and $X \in \mathrm{Gen}\, T$, there exist $m > 0$ and an epimorphism $T^m \to X$. But this implies that $\mathrm{Hom}_A(X, M) = 0$. $\square$

1.10. Definition. Let $\mathcal{C}$ be a full K-subcategory of $\mathrm{mod}\,A$. An A-module $M \in \mathcal{C}$ is called **Ext-projective** in $\mathcal{C}$ if $\mathrm{Ext}^1_A(M, -)|_{\mathcal{C}} = 0$. Dually, it is called **Ext-injective** in $\mathcal{C}$ if $\mathrm{Ext}^1_A(-, M)|_{\mathcal{C}} = 0$.

This definition, due to Auslander and Smalø [22], is clearly motivated by Lemma 1.9. Thus $\mathrm{Gen}\,T$ is a torsion class if T is Ext-projective in $\mathrm{Gen}\,T$ and, dually, $\mathrm{Cogen}\,T$ is a torsion-free class if T is Ext-injective in $\mathrm{Cogen}\,T$. The following proposition characterises completely Ext-projectives and Ext-injectives in torsion or torsion-free classes.

1.11. Proposition. *Let $(\mathcal{T}, \mathcal{F})$ be a torsion pair in $\mathrm{mod}\,A$ and M be an indecomposable A-module.*

(a) *Assume that M lies in $\mathcal{T}$.*

    (i) *M is Ext-projective in $\mathcal{T}$ if and only if $\tau M \in \mathcal{F}$.*

    (ii) *M is Ext-injective in $\mathcal{T}$ if and only if there exist an injective module $E \notin \mathcal{F}$ and an isomorphism $M \cong tE$.*

(b) *Assume that M lies in $\mathcal{F}$.*

    (i) *M is Ext-injective in $\mathcal{F}$ if and only if $\tau^{-1}M \in \mathcal{T}$.*

    (ii) *M is Ext-projective in $\mathcal{F}$ if and only if there exist a projective module $P \notin \mathcal{T}$ and an isomorphism $M \cong P/tP$.*

Proof. We only prove (a); the proof of (b) is similar. Suppose $\tau M \in \mathcal{F}$. Then, for any $X \in \mathcal{T}$, we have

$$\mathrm{Ext}^1_A(M, X) \cong D\overline{\mathrm{Hom}}_A(X, \tau M) \subseteq D\mathrm{Hom}_A(X, \tau M) = 0.$$

Thus, M is Ext-projective in $\mathcal{T}$. Conversely, if $\tau M \notin \mathcal{F}$, then, in the canonical sequence

$$0 \longrightarrow t(\tau M) \overset{u}{\longrightarrow} \tau M \overset{v}{\longrightarrow} \tau M/t(\tau M) \longrightarrow 0,$$

the epimorphism v is not an isomorphism and, in particular, is not a retraction. Considering the almost split sequence

$$0 \longrightarrow \tau M \overset{f}{\longrightarrow} N \overset{g}{\longrightarrow} M \longrightarrow 0,$$

we deduce the existence of a homomorphism $h : N \to \tau M/t(\tau M)$ such that $hf = v$. Because v is surjective, so is h, and we have a commutative diagram with exact rows and columns:

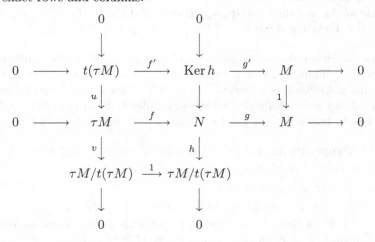

The first row is not split (for if g' were a retraction, so would be g) and consequently $\operatorname{Ext}^1_A(M, t(\tau M)) \neq 0$. Thus, M is not Ext-projective in $\mathcal{T}$.

Let $E \notin \mathcal{F}$ be injective and $X \in \mathcal{T}$. The functor $\operatorname{Hom}_A(X, -)$ applied to the short exact sequence $0 \to tE \to E \to E/tE \to 0$ yields

$$0 = \operatorname{Hom}_A(X, E/tE) \longrightarrow \operatorname{Ext}^1_A(X, tE) \longrightarrow \operatorname{Ext}^1_A(X, E) = 0.$$

Thus tE is Ext-injective in $\mathcal{T}$. Conversely, let $M \in \mathcal{T}$ be Ext-injective and E be its injective envelope. Because $M \subseteq E$, we have $M \subseteq tE$. Consider the short exact sequence $0 \to M \to tE \to tE/M \to 0$. Because $tE \in \mathcal{T}$, we have $tE/M \in \mathcal{T}$. The Ext-injectivity of M in $\mathcal{T}$ implies that the sequence splits. Hence M is a direct summand of tE. The statement follows.   □

In example (1.2)(c), $\mathcal{T} = \operatorname{Gen}(010 \oplus 011)$, the indecomposable Ext-projectives in $\mathcal{T}$ are 010 and 011, and the indecomposable Ext-injectives are 001 and 011, whereas $\mathcal{F} = \operatorname{Cogen}(111)$ and every indecomposable in $\mathcal{F}$ is both Ext-injective and Ext-projective.

VI.2. Partial tilting modules and tilting modules

We now introduce a class of modules that induce torsion pairs in a natural way.

2.1. Definition. Let A be an algebra. An A-module T is called a **partial tilting module** if the following two conditions are satisfied:

(T1) $\operatorname{pd} T_A \leq 1$,
(T2) $\operatorname{Ext}^1_A(T, T) = 0$.

A partial tilting module T is called a **tilting module** if it also satisfies the following additional condition:

(T3) There exists a short exact sequence $0 \to A_A \to T'_A \to T''_A \to 0$ with T', T'' in add T.

Thus, any projective A-module is trivially a partial tilting module, and any Morita progenerator is a tilting module. In fact, the axioms can be understood to mean that a partial tilting module is a module that is "close enough" to a projective module, and a tilting module is a module that is "close enough" to a Morita progenerator. The third condition (T3) may be reformulated to say that a partial tilting module T_A is a tilting module if and only if, for any indecomposable projective A-module P, there exists a short exact sequence

$$0 \longrightarrow P_A \longrightarrow T'_A \longrightarrow T''_A \longrightarrow 0$$

with T', T'' in add T.

One easy consequence of (T3) is that every tilting module is faithful. We recall that an A-module is **faithful** if its **right annihilator**

$$\operatorname{Ann} M = \{a \in A \mid Ma = 0\}$$

vanishes. We need the following characterisation of faithful modules.

2.2. Lemma. *Let A be an algebra and M be an A-module. The following conditions are equivalent:*

(a) *M_A is faithful.*

(b) *For any basis $\{f_1, \ldots, f_d\}$ of the K-vector space $\operatorname{Hom}_A(A, M)$, the K-linear map $f = [f_1 \ldots f_d]^t : A_A \longrightarrow M^d$ is injective.*

(c) *A_A is cogenerated by M_A.*

(d) *DA_A is generated by M_A.*

Proof. Let $\{f_1, \ldots, f_d\}$ be a basis of the K-vector space $\operatorname{Hom}_A(A, M)$. Then M is faithful if and only if

$$f = [f_1 \ldots f_d]^t : A_A \longrightarrow M^d$$

is a monomorphism; indeed, $f(a) = 0$ for some $a \in A$ if and only if $g(a) = 0$ for some $a \in A$ and any $g \in \operatorname{Hom}_A(A, M)$. Using the canonical isomorphism $M_A \cong \operatorname{Hom}_A(A, M)$, this is equivalent to saying that $Ma = 0$ for some $a \in A$. This implies the equivalence of (a), (b), and (c).

The right annihilator $\{a \in A \mid Ma = 0\}$ of M_A coincides with the left annihilator $\{a \in A \mid aDM = 0\}$ of $_ADM$. Therefore, M_A is faithful if and only if $_AA$ is cogenerated by $_ADM$ or, equivalently, DA_A is generated by $D(DM) \cong M$. $\qquad \square$

Applying the equivalence of (a) and (c), the monomorphism $A_A \to T'_A$ of (T3) shows that every tilting module is faithful.

Given a partial tilting module T_A, we ask whether the class $\operatorname{Gen} T$ is a torsion class. We also consider the full subcategory $\mathcal{T}(T)$ of mod A defined by $\mathcal{T}(T) = \{M_A \mid \operatorname{Ext}^1_A(T, M) = 0\}$.

2.3. Lemma. *Let T be a partial tilting module. Then*

(a) $\operatorname{Gen} T$ *is a torsion class in which T is Ext-projective, and the corresponding torsion-free class is $\mathcal{F}(T) = \{M_A \mid \operatorname{Hom}_A(T, M) = 0\}$;*

(b) $\mathcal{T}(T)$ *is a torsion class in which T is Ext-projective; and and the corresponding torsion-free class is $\operatorname{Cogen} \tau T$; and*

(c) $\operatorname{Gen} T \subseteq \mathcal{T}(T)$.

Proof. Assume that $M \in \operatorname{Gen} T$. There exist $m > 0$ and an epimorphism $T^m \to M$. Because $\operatorname{pd} T \le 1$, this epimorphism induces an epimorphism $0 = \operatorname{Ext}^1_A(T, T^m) \to \operatorname{Ext}^1_A(T, M)$. Hence $\operatorname{Ext}^1_A(T, M) = 0$. Thus the functor $\operatorname{Ext}^1_A(T, -)|_{\operatorname{Gen} T}$ equals zero and, by (1.9)(a), $\operatorname{Gen} T$ is a torsion class in which T is Ext-projective. Moreover, we have shown that $\operatorname{Gen} T \subseteq \mathcal{T}(T)$ and (1.9)(a) implies that the torsion-free class corresponding to $\operatorname{Gen} T$ is $\mathcal{F}(T)$. This shows (a) and (c).

To prove (b), let $0 \to M' \to M \to M'' \to 0$ be a short exact sequence. Applying $\operatorname{Hom}_A(T, -)$ yields a right exact sequence

$$\operatorname{Ext}^1_A(T, M') \longrightarrow \operatorname{Ext}^1_A(T, M) \longrightarrow \operatorname{Ext}^1_A(T, M'') \longrightarrow 0;$$

hence $M', M'' \in \mathcal{T}(T)$ imply $M \in \mathcal{T}(T)$ and $M \in \mathcal{T}(T)$ implies $M'' \in \mathcal{T}(T)$. Because $\mathcal{T}(T)$ is closed under direct sums, it is a torsion class, in which T is clearly Ext-projective. For the corresponding torsion-free class, we observe that, because $\operatorname{pd} T \le 1$, we have, by (IV.2.14), that $\operatorname{Ext}^1_A(T, M) \cong D\operatorname{Hom}_A(M, \tau T)$ and thus $M \in \mathcal{T}(T)$ if and only if $\operatorname{Hom}_A(M, \tau T) = 0$. Moreover, for each X in $\operatorname{Cogen} \tau T$, we have

$$\operatorname{Ext}^1_A(X, \tau T) \cong D\underline{\operatorname{Hom}}_A(T, X) \subseteq D\operatorname{Hom}_A(T, X) = 0,$$

because $\operatorname{Hom}_A(T, \tau T) = 0$. It follows that the restriction of $\operatorname{Ext}^1_A(-, \tau T)$ to $\operatorname{Cogen} \tau T$ is zero. Hence, by (1.9)(b), $\operatorname{Cogen} \tau T$ is a torsion-free class whose corresponding torsion class is $\{M \mid \operatorname{Hom}_A(M, \tau T) = 0\} = \mathcal{T}(T)$. $\square$

It is easy to see that every injective A-module is torsion in the torsion pair $(\mathcal{T}(T), \operatorname{Cogen} \tau T)$. Also, if a projective module P lies in $\operatorname{Gen} T$, then $P \in \operatorname{add} T$. Indeed, if $P \in \operatorname{Gen} T$, there exist $m > 0$ and an epimorphism $T^m \to P$ that must split, because P is projective.

In Example 1.2 (c), the module $T = 010 \oplus 011$ is a partial tilting module. Indeed, $\mathrm{pd}\, T \le 1$, as seen from the projective resolutions

$$0 \longrightarrow P(1) \longrightarrow P(2) \longrightarrow 010 \longrightarrow 0,$$
$$0 \longrightarrow P(1) \longrightarrow P(3) \longrightarrow 011 \longrightarrow 0.$$

In fact, it is easy to see that in this example, we have $\mathrm{gl.dim}\, A = 1$. Algebras with global dimension one are called hereditary and are studied in detail in the following chapters. Because 011 is injective,

$$\mathrm{Ext}^1_A(T, T) \cong \mathrm{Ext}^1_A(010 \oplus 011, 010) \cong D\mathrm{Hom}_A(010, \tau(010 \oplus 011))$$
$$\cong D\mathrm{Hom}_A(010, 100 \oplus 110) = 0.$$

The torsion pair illustrated in Example 1.2 (c) is the pair $(\mathrm{Gen}\, T, \mathcal{F}(T))$; the pair $(\mathcal{T}(T), \mathrm{Cogen}\, \tau T)$ is illustrated as follows:

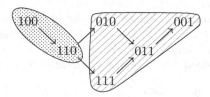

In this case, the inclusion of $(2.3)(c)$ is proper.

In Example 1.2 (d), the module $T = 100 \oplus 111 \oplus 001$ is a partial tilting module. Indeed, $\mathrm{pd}\, T \le 1$ because $\mathrm{gl.dim}\, A = 1$. Because $100 \oplus 111$ is projective, whereas $001 \oplus 111$ is injective, we have

$$\mathrm{Ext}^1_A(T, T) \cong \mathrm{Ext}^1_A(001, 100) \cong D\mathrm{Hom}_A(100, \tau(001))$$
$$\cong D\mathrm{Hom}_A(100, 010) = 0.$$

In fact, T is even a tilting module: because $P(1), P(3) \in \mathrm{add}\, T$, the short exact sequence

$$0 \longrightarrow P(2) \longrightarrow 111 \longrightarrow 001 \longrightarrow 0$$

shows that (T3) is satisfied. In this case, the classes $(\mathrm{Gen}\, T, \mathcal{F}(T))$ and $(\mathcal{T}(T), \mathrm{Cogen}\, \tau T)$ coincide and are illustrated in Example 1.2 (d).

As the reader may have noticed, the formula of (IV.2.14), asserting that $\mathrm{Ext}^1_A(T, M) \cong D\mathrm{Hom}_A(M, \tau T)$ whenever $\mathrm{pd}\, T \le 1$, is extremely useful in these computations.

The following lemma, known as **Bongartz's lemma** [33], justifies the name of partial tilting module; it asserts that a partial tilting module may always be completed to a tilting module.

2.4. Lemma. *Let T_A be a partial tilting module. There exists an A-module E such that $T \oplus E$ is a tilting module.*

Proof. Let $e_1, \dots, e_d$ be a basis of the K-vector space $\operatorname{Ext}_A^1(T, A)$. Represent each e_i by a short exact sequence $0 \longrightarrow A \xrightarrow{f_i} E_i \xrightarrow{g_i} T \longrightarrow 0$. Consider the commutative diagram with exact rows

$$
\begin{array}{ccccccccc}
0 & \longrightarrow & A^d & \xrightarrow{f} & \displaystyle\bigoplus_{i=1}^{d} E_i & \xrightarrow{g} & T^d & \longrightarrow & 0 \\
 & & {\scriptstyle k}\downarrow & & {\scriptstyle u}\downarrow & & {\scriptstyle 1}\downarrow & & \\
(*) \qquad 0 & \longrightarrow & A & \xrightarrow{v} & E & \xrightarrow{w} & T^d & \longrightarrow & 0
\end{array}
$$

where $f = \begin{bmatrix} f_1 & & 0 \\ & \ddots & \\ 0 & & f_d \end{bmatrix}$, $g = \begin{bmatrix} g_1 & & 0 \\ & \ddots & \\ 0 & & g_d \end{bmatrix}$ and $k = [1, \dots, 1]$ is the codiagonal homomorphism. We denote by $\mathbf{e}$ the element of $\operatorname{Ext}_A^1(T^d, A)$ represented by the lower sequence $(*)$. Let $u_i : T \to T^d$ be the inclusion homomorphism in the ith coordinate. We claim that $\mathbf{e}_i = \operatorname{Ext}_A^1(u_i, A)\mathbf{e}$ for each i with $1 \le i \le d$. Indeed, consider the commutative diagram with exact rows

$$
\begin{array}{ccccccccc}
0 & \longrightarrow & A & \xrightarrow{f_i} & E_i & \xrightarrow{g_i} & T & \longrightarrow & 0 \\
 & & {\scriptstyle u_i''}\downarrow & & {\scriptstyle u_i'}\downarrow & & {\scriptstyle u_i}\downarrow & & \\
0 & \longrightarrow & A^d & \xrightarrow{f} & \displaystyle\bigoplus_{i=1}^{d} E_i & \xrightarrow{g} & T^d & \longrightarrow & 0 \\
 & & {\scriptstyle k}\downarrow & & {\scriptstyle u}\downarrow & & {\scriptstyle 1}\downarrow & & \\
(*) \qquad 0 & \longrightarrow & A & \xrightarrow{v} & E & \xrightarrow{w} & T^d & \longrightarrow & 0
\end{array}
$$

where u_i', u_i'' denote the respective inclusion homomorphisms in the ith coordinate. Because $ku_i'' = 1_A$, we deduce a commutative diagram with exact rows

$$
\begin{array}{ccccccccc}
0 & \longrightarrow & A & \xrightarrow{f_i} & E_i & \xrightarrow{g_i} & T & \longrightarrow & 0 \\
 & & {\scriptstyle 1}\downarrow & & {\scriptstyle uu_i'}\downarrow & & {\scriptstyle u_i}\downarrow & & \\
(*) \qquad 0 & \longrightarrow & A & \xrightarrow{v} & E & \xrightarrow{w} & T^d & \longrightarrow & 0
\end{array}
$$

hence our claim. Applying $\operatorname{Hom}_A(T, -)$ to $(*)$ yields an exact sequence

$$
\cdots \to \operatorname{Hom}_A(T, T^d) \xrightarrow{\delta} \operatorname{Ext}_A^1(T, A) \longrightarrow \operatorname{Ext}_A^1(T, E) \longrightarrow \operatorname{Ext}_A^1(T, T^d) = 0.
$$

Because $\mathbf{e}_i = \mathrm{Ext}_A^1(u_i, A)\mathbf{e} = \delta(u_i)$, each basis element of $\mathrm{Ext}_A^1(T, A)$ lies in the image of the connecting homomorphism δ, which is therefore surjective. Hence $\mathrm{Ext}_A^1(T, E) = 0$. Applying now $\mathrm{Hom}_A(-, T)$ and $\mathrm{Hom}_A(-, E)$ to $(*)$ yields respectively

$$0 = \mathrm{Ext}_A^1(T^d, T) \longrightarrow \mathrm{Ext}_A^1(E, T) \longrightarrow \mathrm{Ext}^1(A, T) = 0,$$
$$0 = \mathrm{Ext}_A^1(T^d, E) \longrightarrow \mathrm{Ext}_A^1(E, E) \longrightarrow \mathrm{Ext}^1(A, E) = 0;$$

hence $\mathrm{Ext}_A^1(E \oplus T, E \oplus T) = 0$. It follows from the short exact sequence $(*)$ that $\mathrm{pd}\, E \leq 1$, hence that $\mathrm{pd}\,(T \oplus E) \leq 1$ and the module $T \oplus E$ satisfies the axiom (T3). $\qquad\square$

The short exact sequence $(*)$ constructed in the proof of the lemma is referred to as **Bongartz's exact sequence**. As a first consequence, we obtain the following characterisation of tilting modules.

2.5. Theorem. *Let T_A be a partial tilting module. The following conditions are equivalent:*

(a) *T_A is a tilting module.*
(b) *$\mathrm{Gen}\, T = \mathcal{T}(T)$.*
(c) *For every module $M \in \mathcal{T}(T)$, there exists a short exact sequence $0 \to L \to T_0 \to M \to 0$ with $T_0 \in \mathrm{add}\, T$ and $L \in \mathcal{T}(T)$.*
(d) *Let X be an A-module. Then $X \in \mathrm{add}\, T$ if and only if X is Ext-projective in $\mathcal{T}(T)$.*
(e) *$\mathcal{F}(T) = \mathrm{Cogen}\, \tau T$.*

Proof. Because (b) and (e) are clearly equivalent (by (2.3)), it suffices to establish the equivalence of the first four conditions.

(a) implies (b). Assume that T is a tilting module and let $M \in \mathcal{T}(T)$. We must show that $M \in \mathrm{Gen}\, T$ or, equivalently, that $M \cong tM$, where t is the torsion radical associated to the torsion pair $(\mathrm{Gen}\, T, \mathcal{F}(T))$. Applying $\mathrm{Hom}_A(T, -)$ to the canonical sequence $0 \to tM \to M \to M/tM \to 0$ yields an epimorphism $\mathrm{Ext}_A^1(T, M) \to \mathrm{Ext}_A^1(T, M/tM)$. Because $\mathrm{Ext}_A^1(T, M) = 0$, we have $\mathrm{Ext}_A^1(T, M/tM) = 0$. Further, because $M/tM \in \mathcal{F}(T)$, we have $\mathrm{Hom}_A(T, M/tM) = 0$. On the other hand, because T is a tilting module, there exists a short exact sequence $0 \to A \to T' \to T'' \to 0$ with $T', T'' \in \mathrm{add}\, T$. Applying the functor $\mathrm{Hom}_A(-, M/tM)$ to this sequence yields an exact sequence $0 = \mathrm{Hom}_A(T', M/tM) \to \mathrm{Hom}_A(A, M/tM) \to \mathrm{Ext}_A^1(T'', M/tM) = 0$ so that $M/tM \cong \mathrm{Hom}_A(A, M/tM) = 0$ and $M = tM \in \mathrm{Gen}\, T$.

(b) implies (c). Let $M \in \mathcal{T}(T)$ and $f_1, \ldots, f_d$ be a basis of the K-vector space $\mathrm{Hom}_A(T, M)$. Because $M \in \mathrm{Gen}\, T$, the homomorphism $f = [f_1 \ldots f_d] : T^d \to M$ is surjective (see the proof of (1.8)). Letting $L = \mathrm{Ker}\, f$

and applying $\operatorname{Hom}_A(T, -)$ to the short exact sequence $0 \to L \to T^d \xrightarrow{f} M \to$
0 yields an exact sequence

$$\cdots \longrightarrow \operatorname{Hom}_A(T, T^d) \xrightarrow{\operatorname{Hom}_A(T,f)} \operatorname{Hom}_A(T, M) \longrightarrow \operatorname{Ext}_A^1(T, L) \longrightarrow 0.$$

By construction, $\operatorname{Hom}_A(T, f)$ is an epimorphism. Hence $\operatorname{Ext}_A^1(T, L) = 0$
and $L \in \mathcal{T}(T)$.

(c) implies (d). Let $X \in \operatorname{add} T$; then X is clearly Ext-projective in
$\mathcal{T}(T) = \{M \mid \operatorname{Ext}_A^1(T, M) = 0\}$. Conversely, let X be Ext-projective in
$\mathcal{T}(T)$, and consider the exact sequence $0 \to L \to T_0 \to X \to 0$ with
$T_0 \in \operatorname{add} T$ and $L \in \mathcal{T}(T)$. Because X is Ext-projective in $\mathcal{T}(T)$, this
sequence splits and $X \in \operatorname{add} T$.

(d) implies (a). Let $0 \to A \to E \to T^d \to 0$ be Bongartz's exact sequence
corresponding to the partial tilting module T. To show that T is a tilting
module, it suffices to show that $E \in \operatorname{add} T$ or, equivalently, that E is Ext-
projective in $\mathcal{T}(T)$. First, we observe that, because $T \oplus E$ is a tilting module
by (2.4), we have $\operatorname{Ext}_A^1(T, E) = 0$ so that $E \in \mathcal{T}(T)$. Letting $M \in \mathcal{T}(T)$ and
applying $\operatorname{Hom}_A(-, M)$ to the previous Bongartz sequence yields an exact
sequence

$$0 = \operatorname{Ext}_A^1(T^d, M) \longrightarrow \operatorname{Ext}_A^1(E, M) \longrightarrow \operatorname{Ext}_A^1(A, M) = 0.$$

Hence $\operatorname{Ext}_A^1(E, M) = 0$. □

2.6. Corollary. *Let T_A be a tilting module and $M \in \mathcal{T}(T)$. Then there
exists an exact sequence*

$$\cdots \to T_2 \longrightarrow T_1 \longrightarrow T_0 \longrightarrow M \longrightarrow 0$$

with all T_i in $\operatorname{add} T$.

Proof. This follows from (2.5)(c) and an obvious induction. □

In the sequel, if T_A is a tilting module, we refer to the torsion pair
$(\operatorname{Gen} T, \mathcal{F}(T_A)) = (\mathcal{T}(T_A), \operatorname{Cogen} \tau T)$ as the **torsion pair induced by** T
in $\operatorname{mod} A$, and we usually denote it by $(\mathcal{T}(T_A), \mathcal{F}(T_A))$.

As another consequence of (2.5), we can refine the result of (1.8)(a), in
the case where T is a tilting module.

2.7. Corollary. *Let T_A be a tilting module, and $B = \operatorname{End} T_A$. Then
$M \in \mathcal{T}(T)$ if and only if the canonical A-module homomorphism $\varepsilon_M :
\operatorname{Hom}_A(T, M) \otimes_B T \to M$ is bijective.*

Proof. The sufficiency follows from (1.8) and (2.5). For the necessity,
we apply twice (2.5)(c) and find short exact sequences

$$0 \longrightarrow L_0 \longrightarrow T_0 \longrightarrow M \longrightarrow 0,$$
$$0 \longrightarrow L_1 \longrightarrow T_1 \longrightarrow L_0 \longrightarrow 0,$$

with $T_0, T_1 \in \operatorname{add} T$ and $L_0, L_1 \in \mathcal{T}(T)$. Applying $\operatorname{Hom}_A(T, -)$ yields short exact sequences

$$0 \longrightarrow \operatorname{Hom}_A(T, L_0) \longrightarrow \operatorname{Hom}_A(T, T_0) \longrightarrow \operatorname{Hom}_A(T, M) \longrightarrow 0,$$
$$0 \longrightarrow \operatorname{Hom}_A(T, L_1) \longrightarrow \operatorname{Hom}_A(T, T_1) \longrightarrow \operatorname{Hom}_A(T, L_0) \longrightarrow 0,$$

because $\operatorname{Ext}_A^1(T, L_0) = 0$ and $\operatorname{Ext}_A^1(T, L_1) = 0$. Applying the right exact functor $\operatorname{Hom}_A(T, -) \otimes_B T$ to the exact sequence $T_1 \longrightarrow T_0 \longrightarrow M \longrightarrow 0$ we get the commutative diagram

$$\operatorname{Hom}_A(T, T_1) \otimes_B T \;\to\; \operatorname{Hom}_A(T, T_0) \otimes_B T \;\to\; \operatorname{Hom}_A(T, M) \otimes_B T \to 0$$

$$\varepsilon_{T_1} \downarrow \qquad\qquad \varepsilon_{T_0} \downarrow \qquad\qquad \varepsilon_M \downarrow$$

$$T_1 \xrightarrow{\hspace{2cm}} T_0 \xrightarrow{\hspace{2cm}} M \longrightarrow 0$$

with exact rows. Because ε_T is just the canonical A-module isomorphism $\operatorname{Hom}_A(T, T) \otimes_B T \cong B \otimes_B T_A \cong T_A$, it follows that $\varepsilon_{T_0}, \varepsilon_{T_1}$ are isomorphisms. Hence so is ε_M. $\qquad\qquad\qquad\qquad\qquad\qquad\qquad$ $\square$

2.8. Examples. (a) Let A be given by the quiver

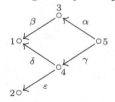

bound by $\alpha\beta = \gamma\delta$, $\gamma\varepsilon = 0$. Representing the indecomposable A-modules by their dimension vectors, we consider the module

$$T_A = \begin{smallmatrix} & 0 & \\ 1 & & 0 \\ & 0 & \end{smallmatrix} \oplus \begin{smallmatrix} & 0 & \\ 1 & & 0 \\ & 1 & \end{smallmatrix} \oplus \begin{smallmatrix} & 0 & \\ 1 & & 0 \\ & 1 & \end{smallmatrix} \oplus \begin{smallmatrix} & 1 & \\ 1 & & 1 \\ & 1 & \end{smallmatrix} \oplus \begin{smallmatrix} & 0 & \\ 0 & & 1 \\ & 1 & \end{smallmatrix}.$$

Then T_A is a tilting module. Indeed, we have the following

(T1) $\operatorname{pd} T_A \leq 1$, because the modules $\begin{smallmatrix} & 0 & \\ 1 & & 0 \\ & 0 & \end{smallmatrix} = P(1)$, $\begin{smallmatrix} & 0 & \\ 1 & & 0 \\ & 1 & \end{smallmatrix} = P(4)$, $\begin{smallmatrix} & 1 & \\ 1 & & 1 \\ & 0 & \end{smallmatrix} = P(5)$ are projective, and we have projective resolutions for the other two summands of T

$$0 \longrightarrow P(2) \longrightarrow P(4) \longrightarrow \begin{smallmatrix} & 0 & \\ 1 & & 0 \\ & 1 & \end{smallmatrix} \longrightarrow 0,$$

$$0 \longrightarrow P(3) \longrightarrow P(5) \longrightarrow \begin{smallmatrix} & 0 & \\ 0 & & 1 \\ & 1 & \end{smallmatrix} \longrightarrow 0.$$

(T2) $\mathrm{Ext}^1_A(T,T) = 0$. Because $1\,{}^0_0\,0 \oplus 1\,{}^0_1\,0 \oplus 1\,{}^1_0\,1$ is projective and $1\,{}^1_0\,1 \oplus 0\,{}^0_1\,1$ is injective, this follows from

$$\mathrm{Ext}^1_A(T,T) \cong \mathrm{Ext}^1_A\left(1\,{}^0_0\,0 \oplus 0\,{}^0_1\,1,\ 1\,{}^0_0\,0 \oplus 1\,{}^0_1\,0 \oplus 1\,{}^0_0\,0 \right)$$

$$\cong D\mathrm{Hom}_A\left(1\,{}^0_0\,0 \oplus 1\,{}^0_1\,0 \oplus 1\,{}^0_0\,0,\ \tau\left(1\,{}^0_1\,0 \oplus 0\,{}^0_1\,1 \right) \right)$$

$$\cong D\mathrm{Hom}_A\left(1\,{}^0_0\,0 \oplus 1\,{}^0_1\,0 \oplus 1\,{}^0_0\,0,\ 0\,{}^0_1\,0 \oplus 0\,{}^1_0\,0 \right) = 0.$$

(T3) There exists, for each point a in the quiver of A, a short exact sequence $0 \to P(a) \to T' \to T'' \to 0$ with $T', T'' \in \mathrm{add}\,T$. Because $P(1), P(4), P(5) \in \mathrm{add}\,T$, it suffices to consider the two short exact sequences presented in (T1).

The torsion pair $(\mathcal{T}(T), \mathcal{F}(T))$ induced by T in $\mathrm{mod}\,A$ is illustrated as follows in $\Gamma(\mathrm{mod}\,A)$, where we represent the indecomposable summands of T by squares:

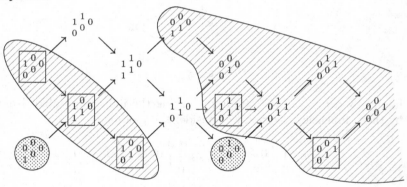

(b) Let A be given by the quiver

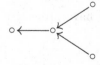

and consider the module $T_A = 1\,1\,{}^1_0 \oplus 1\,1\,{}^1_1 \oplus 0\,1\,{}^1_0 \oplus 0\,0\,{}^1_0$. We leave it to the reader to verify that T is a tilting module and that the torsion pair $(\mathcal{T}(T), \mathcal{F}(T))$ induced by T in $\mathrm{mod}\,A$ is as illustrated here:

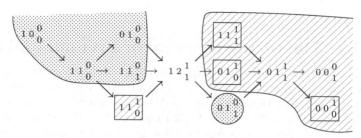

(c) The following class of tilting modules, whose construction is due to Auslander, Platzeck, and Reiten [18] (and, accordingly, are called APR-tilting modules), were at the origins of the theory. Let A be an algebra and $S(a)_A$ be a simple projective that is not injective (thus, the corresponding point a is a sink in the quiver of A and there exists at least one arrow having a as a target). We claim that

$$T_A = T[a] = \tau^{-1}S(a) \oplus (\bigoplus_{b \neq a} P(b))$$

is a tilting module.

First, we note that, according to (IV.3.9) and (IV.4.4), the almost split sequence in $\operatorname{mod} A$ starting from the simple projective module $S(a) = P(a)$ has the form

$$0 \longrightarrow S(a) \longrightarrow \bigoplus_{c \neq a} P(c)^{m_c} \longrightarrow \tau^{-1}S(a) \longrightarrow 0,$$

where $m_c = \dim_K \operatorname{Irr}(S(a), P(c))$. This immediately yields (T1) and (T3). The statement (T2) is a consequence of $\operatorname{Ext}_A^1(T, T) \cong D\operatorname{Hom}_A(T, \tau T) = 0$, because $\tau T = S(a)$ is simple projective. In this case, the only indecomposable A-module lying in $\mathcal{F}(T_A)$ is $S(a)$, whereas $\mathcal{T}(T_A)$ is the additive subcategory generated by all remaining indecomposables. Indeed, if M_A is indecomposable, then $M \in \mathcal{T}(T)$ if and only if $0 = \operatorname{Ext}_A^1(T, M) \cong D\operatorname{Hom}_A(M, S(a))$ if and only if $M \not\cong S(a)$. In particular, $(\mathcal{T}(T), \mathcal{F}(T))$ is splitting.

For instance, if A is as in (a), then there exist two APR-tilting modules $T[1]$ and $T[2]$, whereas, if A is as in (b), then there exists a unique APR-tilting module corresponding to the only sink in the quiver of A.

The reader may have observed that in all of the examples, the number of indecomposable nonisomorphic summands of a tilting A-module is equal to the number of nonisomorphic simple A-modules (that is, to the rank of the Grothendieck group $K_0(A)$ of A). This is no accident, as will be shown in (4.4).

VI.3. The tilting theorem of Brenner and Butler

Tilting theory aims at comparing the module categories of two finite dimensional algebras. Namely, let A be an algebra, T_A be a tilting module, and $B = \operatorname{End} T_A$. Because T_A is, by definition, a module "close to" a Morita progenerator, thus "close to" A_A, it turns out that $B = \operatorname{End} T_A$ is "close to" $\operatorname{End} A_A \cong A$. An obvious functor allowing to pass from $\operatorname{mod} A$ to $\operatorname{mod} B$ is the functor $\operatorname{Hom}_A(T, -)$. The following easy lemma shows that this functor maps the objects in $\operatorname{add} T$ onto the projective B-modules. For this reason, the procedure of passing from an algebra to the endomorphism algebra of one of its modules is sometimes called **projectivisation**; see [21].

    3.1. Lemma. *Let A be an algebra, T be any A-module, and $B = \operatorname{End} T_A$.*
  (a) *For each module $T_0 \in \operatorname{add} T$ and each A-module M, the K-linear map $f \mapsto \operatorname{Hom}_A(T, f)$ induces a functorial isomorphism*

$$\operatorname{Hom}_A(T_0, M) \cong \operatorname{Hom}_B(\operatorname{Hom}_A(T, T_0), \operatorname{Hom}_A(T, M)).$$

  (b) *The functor $\operatorname{Hom}_A(T, -)$ induces an equivalence of categories between $\operatorname{add} T$ and the subcategory $\operatorname{proj} B$ of $\operatorname{mod} B$ consisting of the projective modules.*

    Proof. (a) This follows from the additivity of the functors and from the fact that the defined map is an isomorphism when $T_0 = T$.
    (b) Clearly, P_B is an indecomposable projective B-module if and only if P is an indecomposable summand of

$$B_B = (\operatorname{End} T_A)_B = \operatorname{Hom}_A({}_B T_A, T_A),$$

if and only if $P_B \cong \operatorname{Hom}_A({}_B T_A, T_0)$ for some indecomposable summand T_0 of T. Thus the functor $\operatorname{Hom}_A(T, -)|_{\operatorname{add} T}$ maps into $\operatorname{proj} B$ and is dense. Also, (a) shows that it is full and faithful.   □

    As an obvious consequence of (3.1)(b) we get that B is a basic algebra if and only if two distinct indecomposable summands of T are not isomorphic (we then say that T is **multiplicity-free**).
    In (3.1), no assumption on T was necessary. Until the end of this section, we assume that T is a tilting A-module and

$$B = \operatorname{End} T_A.$$

We consider the functor

$$\operatorname{Hom}_A(T, -) : \mathcal{T}(T_A) \longrightarrow \operatorname{mod} B.$$

The following lemma ensures that this functor embeds $\mathcal{T}(T)$ as a full subcategory of $\operatorname{mod} B$, closed under extensions.

3.2. Lemma. *Let $M, N \in \mathcal{T}(T)$; then we have functorial isomorphisms:*
(a) $\operatorname{Hom}_A(M, N) \cong \operatorname{Hom}_B(\operatorname{Hom}_A(T, M), \operatorname{Hom}_A(T, N))$.
(b) $\operatorname{Ext}^1_A(M, N) \cong \operatorname{Ext}^1_B(\operatorname{Hom}_A(T, M), \operatorname{Hom}_A(T, N))$.

Proof. By (2.6), there exists an exact sequence

$$T_* : \quad \cdots \to T_2 \xrightarrow{d_2} T_1 \xrightarrow{d_1} T_0 \xrightarrow{d_0} M \longrightarrow 0$$

with $T_i \in \operatorname{add} T$ for all i. Applying $\operatorname{Hom}_A(-, N)$ to the right exact sequence $T_1 \to T_0 \to M \to 0$ yields a left exact sequence

$$0 \longrightarrow \operatorname{Hom}_A(M, N) \longrightarrow \operatorname{Hom}_A(T_0, N) \longrightarrow \operatorname{Hom}(T_1, N).$$

By (3.1)(a), we have a commutative diagram with exact columns

$$
\begin{array}{ccc}
0 & & 0 \\
\downarrow & & \downarrow \\
\operatorname{Hom}_A(M, N) & \dashrightarrow & \operatorname{Hom}_B(\operatorname{Hom}_A(T, M), \operatorname{Hom}_A(T, N)) \\
\downarrow & & \downarrow \\
\operatorname{Hom}_A(T_0, N) & \xrightarrow{\cong} & \operatorname{Hom}_B(\operatorname{Hom}_A(T, T_0), \operatorname{Hom}_A(T, N)) \\
\downarrow & & \downarrow \\
\operatorname{Hom}_A(T_1, N) & \xrightarrow{\cong} & \operatorname{Hom}_B(\operatorname{Hom}_A(T, T_1), \operatorname{Hom}_A(T, N))
\end{array}
$$

where the dotted arrow is induced by the others. This shows (a) by passing to the kernels. For (b), let $L = \operatorname{Im} d_1$; we have a short exact sequence

$$0 \longrightarrow L \xrightarrow{j} T_0 \xrightarrow{d_0} M \longrightarrow 0,$$

to which we apply $\operatorname{Hom}_A(-, N)$, thus obtaining an exact sequence

$$0 \longrightarrow \operatorname{Hom}_A(M, N) \longrightarrow \operatorname{Hom}_A(T_0, N) \xrightarrow{\operatorname{Hom}_A(j, N)} \operatorname{Hom}_A(L, N)$$
$$\longrightarrow \operatorname{Ext}^1_A(M, N) \longrightarrow 0$$

so that $\operatorname{Ext}^1_A(M, N) \cong \operatorname{Coker} \operatorname{Hom}_A(j, N)$ is isomorphic to the first cohomology group of the complex $\operatorname{Hom}_A(T_*, N)$. On the other hand, if we apply $\operatorname{Hom}_A(T, -)$ to the complex T_*, we obtain, by (3.1)(b), a projective resolution $\operatorname{Hom}_A(T, T_*)$ of $\operatorname{Hom}_A(T, M)$ in $\operatorname{mod} B$, because $\operatorname{Ker} d_i \in \mathcal{T}(T)$ and hence $\operatorname{Ext}^1_A(T, \operatorname{Ker} d_i) = 0$ for any $i \geq 1$. Therefore $\operatorname{Ext}^1_B(\operatorname{Hom}_A(T, M), \operatorname{Hom}_A(T, N))$ is isomorphic to the first cohomology group of the complex $\operatorname{Hom}_B(\operatorname{Hom}_A(T, T_*), \operatorname{Hom}_A(T, N))$, which is, by (3.1)(a), isomorphic (as a complex) to $\operatorname{Hom}_A(T_*, N)$. This completes the proof of (b). $\qquad\square$

The key observation of tilting theory is that the tilting module T_A induces a tilting B-module, which is the left B-module ${}_B T$. Moreover, the algebra A can be recovered from B and ${}_B T$.

3.3. Lemma. *Let* T_A *be a tilting* A-*module and* $B = \operatorname{End} T_A$.

(a) $D(_BT) \cong \operatorname{Hom}_A(T, DA)$.

(b) $_BT$ *is a tilting left* B-*module*.

(c) *The canonical* K-*algebra homomorphism* $A \to \operatorname{End}(_BT)^{\mathrm{op}}$, *given by* $a \mapsto (t \mapsto ta)$, *is an isomorphism*.

Proof. (a) $D(_BT) \cong D(_BT_A \otimes_A A) \cong \operatorname{Hom}_A(T, DA)$.

(b) We verify the axioms of tilting module:

(T1) pd $_BT \leq 1$. Indeed, because T_A is a tilting module, there exists a short exact sequence $0 \to A_A \to T' \to T'' \to 0$ with $T', T'' \in \operatorname{add} T$. Applying $\operatorname{Hom}_A(-, _BT_A)$, we get a short exact sequence

$$0 \longrightarrow \operatorname{Hom}_A(T'', _BT_A) \longrightarrow \operatorname{Hom}_A(T', _BT_A) \longrightarrow \operatorname{Hom}_A(A, _BT_A) \longrightarrow 0.$$

Because

$\operatorname{Hom}_A(A, _BT_A) \cong {}_BT$ and $\operatorname{Hom}_A(T', _BT_A), \operatorname{Hom}_A(T'', _BT_A) \in \operatorname{add}(_BB)$,

we are done.

(T2) $\operatorname{Ext}_B^1(T, T) = 0$. Indeed, using (a) and the fact that $DA \in \mathcal{T}(T)$, we get, by (3.2)(b),

$$\operatorname{Ext}_B^1(DT, DT) \cong \operatorname{Ext}_B^1(\operatorname{Hom}_A(T, DA), \operatorname{Hom}_A(T, DA))$$
$$\cong \operatorname{Ext}_A^1(DA, DA) = 0,$$

hence the result.

(T3) Let $0 \to P_1 \to P_0 \to T_A \to 0$ be a projective resolution. Applying $\operatorname{Hom}_A(-, _BT_A)$, we get a short exact sequence

$$0 \longrightarrow \operatorname{Hom}_A(T, _BT_A) \longrightarrow \operatorname{Hom}_A(P_0, _BT_A) \longrightarrow \operatorname{Hom}_A(P_1, _BT_A) \longrightarrow 0.$$

Because

$\operatorname{Hom}_A(T, _BT_A) \cong {}_BB$ and $\operatorname{Hom}_A(P_0, _BT_A), \operatorname{Hom}_A(P_1, _BT_A) \in \operatorname{add}(_BT)$,

we are done.

(c) Let $a \in A$ belong to the kernel of this homomorphism. Then $Ta = 0$. But every tilting module is faithful, hence $a = 0$. Thus the given homomorphism is injective. By (a) and the fact that $DA \in \mathcal{T}(T)$, (3.2)(a) yields vector space isomorphisms

$$A \cong \operatorname{End} DA \cong \operatorname{End} \operatorname{Hom}_A(T, DA) \cong \operatorname{End} DT,$$

so that $\dim_K A = \dim_K \operatorname{End}(_BT)$ and the canonical homomorphism is an isomorphism. $\qquad\square$

A first consequence of (3.3) is that B is a connected algebra. In fact, we show more, namely that the centre is preserved under the tilting process.

3.4. Lemma. *Let A be an algebra and T_A be a tilting A-module. Then the centre $Z(A)$ of A is isomorphic to the centre $Z(B)$ of $B = \operatorname{End} T_A$.*

Proof. We define $\varphi : Z(A) \to Z(B)$ by $a \mapsto (\rho_a : t \mapsto ta)$. Indeed, let $a \in Z(A)$; then ρ_a is an endomorphism of T_A for, if $t_1, t_2 \in T$ and $a_1, a_2 \in A$, then we have
$$\rho_a(t_1a_1 + t_2a_2) = t_1a_1a + t_2a_2a = t_1aa_1 + t_2aa_2 = \rho_a(t_1)a_1 + \rho_a(t_2)a_2.$$
Also, ρ_a is central for, if $f \in \operatorname{End} T_A = B$ and $t \in T$, we have $(\rho_a f)(t) = f(t)a = f(ta) = (f\rho_a)(t)$. Finally, φ is an algebra homomorphism for, if $a_1, a_2 \in Z(A)$ then $\varphi(a_1a_2) = \rho_{a_1a_2} = \rho_{a_2a_1} = \varphi(a_1)\varphi(a_2)$ and, clearly, $\varphi(a_1 + a_2) = \varphi(a_1) + \varphi(a_2)$ and $\varphi(1) = 1$.

To show that φ is an isomorphism, we construct its inverse. Following (3.3)(c), we identify the algebra A with $\operatorname{End}(_BT)^{\mathrm{op}}$ via $a \mapsto \rho_a$, then we define $\psi : Z(B) \to Z(A)$ by $b \mapsto (\lambda_b : t \mapsto bt)$. By (3.3)(b) and the first part, ψ is an algebra homomorphism. Let $a \in Z(A)$ and consider $\psi\varphi(a) = \lambda_{\rho_a}$; it is given by $\lambda_{\rho_a} : t \mapsto \rho_a(t) = ta$, that is, by the element $a \in A$ as identified to the endomorphism $\rho_a \in \operatorname{End}(_BT)$. Thus $\psi\varphi(a) = a$ for every $a \in Z(A)$ and $\psi\varphi = 1_{Z(A)}$. By symmetry, we have $\varphi\psi = 1_{Z(B)}$. $\quad\square$

3.5. Corollary. *Let A be an algebra. If T_A is a tilting A-module, then the algebra $B = \operatorname{End} T_A$ is connected.*

Proof. Note that an algebra is connected if and only if its centre is (see Exercise 8.8 in Chapter I), and then apply (3.4). $\quad\square$

Another consequence of (3.3) and the considerations in Section 2 is that $_BT$ induces a torsion pair $(\mathcal{T}(_BT), \mathcal{F}(_BT))$ in the category of left B-modules, where, as before,
$$\mathcal{T}(_BT) = \operatorname{Gen}(_BT) = \{_BU \mid \operatorname{Ext}_B^1(T, U) = 0\},$$
$$\mathcal{F}(_BT) = \operatorname{Cogen}\tau(_BT) = \{_BV \mid \operatorname{Hom}_B(T, V) = 0\}.$$
Because we are interested in the category $\operatorname{mod} B$ of right B-modules, we must rather consider the torsion pair (see Example 1.2 (b))
$$(\mathcal{X}(T_A), \mathcal{Y}(T_A)) = (D\mathcal{F}(_BT), D\mathcal{T}(_BT)).$$

3.6. Corollary. *Let A be an algebra. Any tilting A-module T_A induces a torsion pair $(\mathcal{X}(T_A), \mathcal{Y}(T_A))$ in the category $\operatorname{mod} B$, where $B = \operatorname{End} T_A$ and*
$$\mathcal{X}(T_A) = \{X_B \mid \operatorname{Hom}_B(X, DT) = 0\} = \{X_B \mid X \otimes_B T = 0\},$$
$$\mathcal{Y}(T_A) = \{Y_B \mid \operatorname{Ext}_B^1(Y, DT) = 0\} = \{Y_B \mid \operatorname{Tor}_1^B(Y, T) = 0\}.$$

Proof. This follows from the remark and the functorial isomorphisms $\operatorname{Hom}_B(X, DT) \cong D(X \otimes_B T)$ and $\operatorname{Ext}^1_B(Y, DT) \cong D\operatorname{Tor}^B_1(Y, T)$. The first is the adjoint isomorphism. The second is a consequence of (A.4.11) in the Appendix. $\square$

Note that $\mathcal{Y}(T_A)$ contains all the projective B-modules. This subcategory of mod B plays a rôle fairly similar to that of $\mathcal{T}(T_A)$ in mod A. In fact, we have the following analogue of (2.5)(c) and (2.7).

3.7. Lemma. *Let A be an algebra, T_A be a tilting A-module, $B = \operatorname{End} T_A$, and $Y_B \in \mathcal{Y}(T_A)$.*

(a) *There exists a short exact sequence $0 \to Y \to T^* \to Z \to 0$ with T^* in add DT and Z in $\mathcal{Y}(T_A)$.*

(b) *The canonical homomorphism $\delta_Y : Y_B \to \operatorname{Hom}_A(T, Y \otimes_B T)$ defined by $y \mapsto (t \mapsto y \otimes t)$ is an isomorphism.*

Proof. (a) Because $_BT$ is a tilting module and $D(Y_B) \in \mathcal{T}(_BT)$, there exists a short exact sequence $0 \to {}_BY' \to {}_BT' \to {}_B(DY) \to 0$ with $T' \in \operatorname{add}(_BT)$, $Y' \in \mathcal{T}(_BT)$. Taking $T^* = DT'$ and $Z = DY'$ completes the proof.

(b) The duality isomorphism $\operatorname{Hom}_B(X, DT) \cong D(X \otimes_B T)$ yields $DA \cong D\operatorname{Hom}_B(T, T) \cong DT \otimes_B T$, so that $\delta_{DT} : D(_BT) \to \operatorname{Hom}_A(T, DA) \cong \operatorname{Hom}_A(T, DT \otimes_B T)$ is an isomorphism. Therefore, so is δ_{T^*}, for any $T^* \in \operatorname{add} DT$. Applying (a) twice to $Y \in \mathcal{Y}(T_A)$, we obtain short exact sequences $0 \to Y \to T^*_0 \to Y_0 \to 0$ and $0 \to Y_0 \to T^*_1 \to Y_1 \to 0$ with $T^*_0, T^*_1 \in \operatorname{add} DT$ and $Y_0, Y_1 \in \mathcal{Y}(T_A)$, and so $\operatorname{Tor}^B_1(Y_0, T) = 0$ and $\operatorname{Tor}^B_1(Y_1, T) = 0$. Applying $- \otimes_B T$ yields short exact sequences
$$0 \to Y \otimes_B T \to T^*_0 \otimes_B T \to Y_0 \otimes_B T \to 0 \text{ and}$$
$$0 \to Y_0 \otimes_B T \to T^*_1 \otimes_B T \to Y_1 \otimes_B T \to 0.$$
These combine to a left exact sequence
$$0 \longrightarrow Y \otimes_B T \longrightarrow T^*_0 \otimes_B T \longrightarrow T^*_1 \otimes_B T$$
to which we apply $\operatorname{Hom}_A(T, -)$, thus obtaining the lower row of the commutative diagram with exact rows

$$
\begin{array}{ccccc}
0 & \longrightarrow & Y & \longrightarrow & T^*_0 & \longrightarrow & T^*_1 \\
& & \delta_Y \downarrow & & \delta_{T^*_0} \downarrow \cong & & \delta_{T^*_1} \downarrow \cong \\
0 & \longrightarrow & \operatorname{Hom}_A(T, Y \otimes_B T) & \longrightarrow & \operatorname{Hom}_A(T, T^*_0 \otimes_B T) & \longrightarrow & \operatorname{Hom}_A(T, T^*_1 \otimes_B T)
\end{array}
$$

Because $\delta_{T^*_0}$ and $\delta_{T^*_1}$ are isomorphisms, so is δ_Y. $\square$

We are now able to prove the main result of this section, which is known as the **Brenner–Butler theorem** or the **tilting theorem**.

3.8. Theorem. *Let A be an algebra, T_A be a tilting module, $B =$ $\operatorname{End} T_A$, and $(\mathcal{T}(T_A), \mathcal{F}(T_A))$, $(\mathcal{X}(T_A), \mathcal{Y}(T_A))$ be the induced torsion pairs in $\operatorname{mod} A$ and $\operatorname{mod} B$, respectively. Then T has the following properties:*

(a) *$_BT$ is a tilting module, and the canonical K-algebra homomorphism $A \to \operatorname{End}(_BT)^{\mathrm{op}}$ defined by $a \mapsto (t \mapsto ta)$ is an isomorphism.*

(b) *The functors $\operatorname{Hom}_A(T, -)$ and $- \otimes_B T$ induce quasi-inverse equivalences between $\mathcal{T}(T_A)$ and $\mathcal{Y}(T_A)$.*

(c) *The functors $\operatorname{Ext}^1_A(T, -)$ and $\operatorname{Tor}^B_1(-, T)$ induce quasi-inverse equivalences between $\mathcal{F}(T_A)$ and $\mathcal{X}(T_A)$.*

Proof. Because (a) is (3.3)(b) and (3.3)(c), we prove (b). Let $M \in \mathcal{T}(T_A)$. The duality isomorphism established in (3.6) yields

$$D\operatorname{Hom}_A(T, M) \cong {}_BT_A \otimes DM \in \operatorname{Gen}(_BT),$$

and therefore $\operatorname{Hom}_A(T, M) \in \operatorname{Cogen} DT = \mathcal{Y}(T)$. By (2.7), we have $M \cong \operatorname{Hom}_A(T, M) \otimes_B T$. Conversely, if $Y \in \mathcal{Y}(T_A)$, then $Y \otimes_B T_A \in \operatorname{Gen} T_A = \mathcal{T}(T_A)$ and, by (3.7), we have $Y \cong \operatorname{Hom}_A(T, Y \otimes_B T)$.

To show (c), we take $N \in \mathcal{F}(T_A)$. There is a short exact sequence $0 \longrightarrow N \longrightarrow E \longrightarrow L \longrightarrow 0$ with E injective. In particular, $E \in \mathcal{T}(T_A)$ and hence $L \in \mathcal{T}(T_A)$. Applying $\operatorname{Hom}_A(T, -)$, we get a short exact sequence $0 \longrightarrow \operatorname{Hom}_A(T, E) \longrightarrow \operatorname{Hom}_A(T, L) \longrightarrow \operatorname{Ext}^1_A(T, N) \longrightarrow 0$. Applying $- \otimes_B T$, we get the left column in the commutative diagram

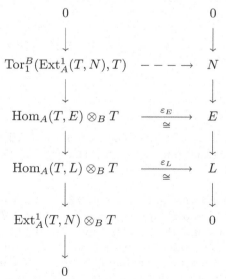

with exact columns, because $L \in \mathcal{T}(T)$ implies $\operatorname{Tor}^B_1(\operatorname{Hom}_A(T, L), T) = 0$, by (b). Therefore we get $\operatorname{Ext}^1_A(T, N) \otimes_B T = 0$ (hence $\operatorname{Ext}^1_A(T, N) \in \mathcal{X}(T_A)$)

and $N \cong \mathrm{Tor}_1^B(\mathrm{Ext}_A^1(T, N), T)$. Dually, let $X_B \in \mathcal{X}(T)$ and consider the short exact sequence

$$0 \to Y \to P \to X \to 0$$

with P projective. Then $P \in \mathcal{Y}(T)$ and $Y \in \mathcal{Y}(T)$. Applying $- \otimes_B T$, we get a short exact sequence

$$0 \longrightarrow \mathrm{Tor}_1^B(X, T) \longrightarrow Y \otimes_B T \longrightarrow P \otimes_B T \longrightarrow 0.$$

Applying $\mathrm{Hom}_A(T, -)$, we get the right column in the commutative diagram with exact columns

$$
\begin{array}{ccc}
0 & & \mathrm{Hom}_A(T, \mathrm{Tor}_1^B(X, T)) \\
\downarrow & & \downarrow \\
Y & \xrightarrow[\cong]{\delta_Y} & \mathrm{Hom}_A(T, Y \otimes_B T) \\
\downarrow & & \downarrow \\
P & \xrightarrow[\cong]{\delta_P} & \mathrm{Hom}_A(T, P \otimes_B T) \\
\downarrow & & \downarrow \\
X & \dashrightarrow & \mathrm{Ext}_A^1(T, \mathrm{Tor}_1^B(X, T)) \\
\downarrow & & \downarrow \\
0 & & 0
\end{array}
$$

because $\mathrm{Ext}_A^1(T, Y \otimes_B T) = 0$ by (b). Therefore $\mathrm{Hom}_A(T, \mathrm{Tor}_1^B(X, T)) = 0$ (hence $\mathrm{Tor}_1^B(X, T) \in \mathcal{F}(T_A)$) and $X \cong \mathrm{Ext}_A^1(T, \mathrm{Tor}_1^B(X, T))$. $\qquad\square$

It is possible to visualise the equivalence of (3.8) in the Auslander–Reiten quivers of the algebras A and B. If one keeps in mind that $\mathcal{T}(T_A)$ contains the injective A-modules and thus lies (roughly speaking) "at the right" of $\Gamma(\mathrm{mod}\, A)$, while $\mathcal{F}(T_A)$ lies "on the left" of $\mathcal{T}(T_A)$ (because there is no homomorphism from a torsion module to a torsion-free one) and, similarly, $\mathcal{Y}(T_A)$ contains the projective B-modules and thus lies "at the left" of $\Gamma(\mathrm{mod}\, B)$, while $\mathcal{X}(T_A)$ lies "on its right," one obtains the following picture, which also shows the quasi-inverse equivalences:

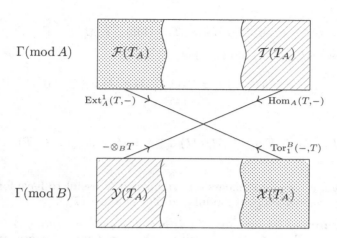

Here, and in the sequel, the equivalent subcategories $\mathcal{T}(T_A)$ and $\mathcal{Y}(T_A)$ are shaded as ⬭ and the equivalent subcategories $\mathcal{F}(T_A)$ and $\mathcal{X}(T_A)$ are shaded as ⬭.

The following corollary asserts that the composition of any two of the four functors $\operatorname{Hom}_A(T, -)$, $\operatorname{Ext}_A^1(T, -)$, $- \otimes_B T$, and $\operatorname{Tor}_A^B(-, T)$, which are not quasi-inverse to each other on one of the shaded subcategories, vanishes.

3.9. Corollary. (a) *Let M be an arbitrary A-module. Then*

(i) $\operatorname{Tor}_1^B(\operatorname{Hom}_A(T, M), T) = 0$;

(ii) $\operatorname{Ext}_A^1(T, M) \otimes_B T = 0$; *and*

(iii) *the canonical sequence of M in $(\mathcal{T}(T_A), \mathcal{F}(T_A))$ is*

$$0 \longrightarrow \operatorname{Hom}_A(T, M) \otimes_B T \xrightarrow{\varepsilon_M} M \longrightarrow \operatorname{Tor}_1^B(\operatorname{Ext}_A^1(T, M), T)) \longrightarrow 0.$$

(b) *Let X be an arbitrary B-module. Then*

(i) $\operatorname{Hom}_A(T, \operatorname{Tor}_1^B(X, T)) = 0$;

(ii) $\operatorname{Ext}_A^1(T, X \otimes_B T) = 0$; *and*

(iii) *the canonical sequence of X in $(\mathcal{X}(T_A), \mathcal{Y}(T_A))$ is*

$$0 \longrightarrow \operatorname{Ext}_A^1(T, \operatorname{Tor}_1^B(X, T)) \longrightarrow X \xrightarrow{\delta_X} \operatorname{Hom}_A(T, X \otimes_B T) \longrightarrow 0.$$

Proof. We only prove (a); the proof of (b) is similar. Indeed, let

$$0 \to tM \to M \to M/tM \to 0$$

be the canonical sequence of M in $(\mathcal{T}(T), \mathcal{F}(M))$. Applying the functor $\operatorname{Hom}_A(T, -)$, we obtain isomorphisms $\operatorname{Hom}_A(T, M) \cong \operatorname{Hom}_A(T, tM)$ and $\operatorname{Ext}_A^1(T, M) \cong \operatorname{Ext}_A^1(T, M/tM)$. Therefore $tM \in \mathcal{T}(T)$ implies that

$$\operatorname{Tor}_1^B(\operatorname{Hom}_A(T, M), T) \cong \operatorname{Tor}_1^B(\operatorname{Hom}_A(T, tM), T) = 0$$

and
$$tM \cong \mathrm{Hom}_A(T, tM) \otimes_B T \cong \mathrm{Hom}_A(T, M) \otimes_B T.$$
Similarly, $M/tM \in \mathcal{F}(T)$ implies that
$$\mathrm{Ext}^1_A(T, M) \otimes_B T \cong \mathrm{Ext}^1_A(T, M/tM) \otimes_B T = 0$$
and
$$M/tM \cong \mathrm{Tor}_1^B((\mathrm{Ext}^1_A(T, M/tM)), T) \cong \mathrm{Tor}_1^B(\mathrm{Ext}^1_A(T, M), T). \qquad \square$$

To illustrate these statements on examples it is useful to have formulas for the dimension vectors of modules in $\mathcal{X}(T_A)$ and $\mathcal{Y}(T_A)$.

3.10. Lemma. *Assume that T_A is a multiplicity-free tilting A-module, $T_A = T_1 \oplus \ldots \oplus T_n$ is its decomposition into a direct sum of indecomposable modules, and $B = \mathrm{End}\, T_A$. Let $e_i \in \mathrm{End}\, T_A$ be the composition of the canonical projection $p_i : T \to T_i$ with the canonical injection $u_i : T_i \to T$.*

(a) *The elements $e_1, \ldots, e_n$ are primitive orthogonal idempotents of B such that $1 = e_1 + \ldots + e_n$; there is a B-module isomorphism $e_a B \cong \mathrm{Hom}_A(T, T_a)$, for all a; and there exist K-linear isomorphisms*

$$e_a B e_b \cong \mathrm{Hom}_A(T_b, T_a) \text{ and } \mathrm{Ext}^1_A(e_a T, N) \cong \mathrm{Ext}^1_A(T, N)e_a$$

for all a, b and for any A-module N.

(b) *For any pair of A-modules $M \in \mathcal{T}(T_A)$ and $N \in \mathcal{F}(T_A)$, we have*

$$\mathbf{dim}\, \mathrm{Hom}_A(T, M) = [\dim_K \mathrm{Hom}_A(T_1, M) \ldots \dim_K \mathrm{Hom}_A(T_n, M)]^t \text{ and}$$
$$\mathbf{dim}\, \mathrm{Ext}^1_A(T, N) = [\dim_K \mathrm{Hom}_A(N, \tau T_1) \ldots \dim_K \mathrm{Hom}_A(N, \tau T_n)]^t.$$

Proof. We recall that, for any L in mod A, the vector space $\mathrm{Hom}_A(T, L)$ has a right B-module structure defined by $fb = f \circ b$ for $f \in \mathrm{Hom}_A(T, L)$ and $b \in B$, where $f \circ b$ means the composition of $b : T \to T$ with $f : T \to L$. It follows from (3.1)(b) and from the assumption that T_A is multiplicity-free that the B-modules $\mathrm{Hom}_A(T, T_1), \ldots, \mathrm{Hom}_A(T, T_n)$ form a complete set of pairwise nonisomorphic indecomposable projective B-modules and, obviously, there is a B-module isomorphism

$$B \cong \mathrm{Hom}_A(T, T_1) \oplus \ldots \oplus \mathrm{Hom}_A(T, T_n).$$

It is easy to see that for any j the B-module homomorphism $\mathrm{Hom}_A(T, T_j) \to e_j B$, defined by $f \mapsto u_j f = e_j u_j f$, is an isomorphism, and the first part of

(a) follows. The isomorphism $\operatorname{Hom}_B(e_bB, e_aB) \cong e_aBe_b$, defined by $h \mapsto h(e_b)$ (see (I.4.2)), together with (3.8)(b) yields $e_aBe_b \cong \operatorname{Hom}_B(e_bB, e_aB) \cong \operatorname{Hom}_B(\operatorname{Hom}_A(T, T_b), \operatorname{Hom}_A(T, T_a)) \cong \operatorname{Hom}_A(T_b, T_a)$.

Because $p_i = p_i \circ e_i$, for each A-module L, the K-linear map

$$\operatorname{Hom}_A(e_iT, L) \longrightarrow \operatorname{Hom}_A(T, L)e_i$$

$g \mapsto g \circ p_i = (g \circ p_i)e_i$ is a K-linear isomorphism, which is functorial in L. Hence, if $I^\bullet$ is an injective resolution of an A-module N, there is an isomorphism $\operatorname{Hom}_A(e_iT, I^\bullet) \cong \operatorname{Hom}_A(T, I^\bullet)e_i$ of complexes and it induces K-linear isomorphisms of the cohomology spaces. In view of (A.4.1) in the Appendix, this yields the isomorphisms $\operatorname{Ext}_A^1(e_iT, N) \cong H^1(\operatorname{Hom}_A(e_iT, I^\bullet)) \cong H^1(\operatorname{Hom}_A(T, I^\bullet)e_i) \cong H^1(\operatorname{Hom}_A(T, I^\bullet))e_i \cong \operatorname{Ext}_A^1(T, N)e_i$. It follows that the ith coordinates of the vectors $\mathbf{dim}\operatorname{Hom}_A(T, M)$ and $\mathbf{dim}\operatorname{Ext}_A^1(T, N)$ are as follows:

$$
\begin{aligned}
(\mathbf{dim}\operatorname{Hom}_A(T, M))_i &= \dim_K \operatorname{Hom}_A(T, M)e_i = \dim_K \operatorname{Hom}_A(e_iT, M) \\
&= \dim_K \operatorname{Hom}_A(e_i(T), M) = \dim_K \operatorname{Hom}_A(T_i, M), \\
(\mathbf{dim}\operatorname{Ext}_A^1(T, N))_i &= \dim_K \operatorname{Ext}_A^1(T, N)e_i = \dim_K \operatorname{Ext}_A^1(e_iT, N) \\
&= \dim_K \operatorname{Ext}_A^1(e_i(T), N) = \dim_K \operatorname{Ext}_A^1(T_i, N) \\
&= \dim_K D\operatorname{Hom}_A(N, \tau T_i) = \dim_K \operatorname{Hom}_A(N, \tau T_i),
\end{aligned}
$$

because $\operatorname{pd} T_i \leq 1$ yields $\operatorname{Ext}_A^1(T_i, N) \cong D\operatorname{Hom}_A(N, \tau T_i)$, by (IV.2.14). $\square$

3.11. Examples. (a) Consider, as in Example 1.2 (d), the algebra A given by the quiver $\circ\!\longleftarrow\!\circ\!\longrightarrow\!\circ$. The tilting module $T_A = 100 \oplus 111 \oplus 001$ induces a torsion pair $(\mathcal{T}(T), \mathcal{F}(T))$ in $\operatorname{mod} A$ illustrated as follows:

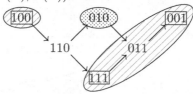

Hence, $B = \operatorname{End} T_A$ is given by the quiver $\circ\!\overset{\mu}{\longleftarrow}\!\circ\!\overset{\lambda}{\longleftarrow}\!\circ$ bound by $\lambda\mu = 0$. The induced torsion pair $(\mathcal{X}(T), \mathcal{Y}(T))$ in $\operatorname{mod} B$ is illustrated in $\Gamma(\operatorname{mod} B)$ as follows:

The effect of the functors $\operatorname{Hom}_A(T,-)$ and $\operatorname{Ext}^1_A(T,-)$ can easily be computed. We have

$$\operatorname{Hom}_A(T,100) \cong 100, \qquad \operatorname{Hom}_A(T,111) \cong 110,$$
$$\operatorname{Hom}_A(T,011) \cong 010, \qquad \operatorname{Hom}_A(T,001) \cong 011,$$

and finally $\operatorname{Ext}^1_A(T,010) \cong 001$.

(b) Consider, as in Example 2.8 (a), the algebra A given by the quiver

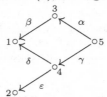

bound by $\alpha\beta = \gamma\delta$ and $\gamma\varepsilon = 0$. The tilting module

$$T_A = \begin{smallmatrix} &0& \\ 1&&0 \\ &0& \end{smallmatrix} \ \oplus\ \begin{smallmatrix} &0& \\ 1&&0 \\ &1& \end{smallmatrix} \ \oplus\ \begin{smallmatrix} &0& \\ 1&&0 \\ &0&1 \end{smallmatrix} \ \oplus\ \begin{smallmatrix} &1& \\ 1&&1 \\ &0& \end{smallmatrix} \ \oplus\ \begin{smallmatrix} &0& \\ 0&&1 \\ &1& \end{smallmatrix}$$

induces a torsion pair $(\mathcal{T}(T), \mathcal{F}(T))$ in $\operatorname{mod} A$ illustrated as follows:

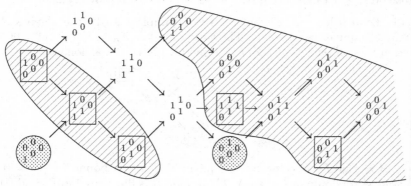

Hence, $B = \operatorname{End} T_A$ is given by the quiver

$$\circ \xleftarrow{\ \eta\ } \circ \xleftarrow{\ \nu\ } \circ \xleftarrow{\ \mu\ } \circ \xleftarrow{\ \lambda\ } \circ$$

bound by $\lambda\mu\nu\eta = 0$. The induced torsion pair $(\mathcal{X}(T), \mathcal{Y}(T))$ in $\operatorname{mod} B$ is illustrated as follows:

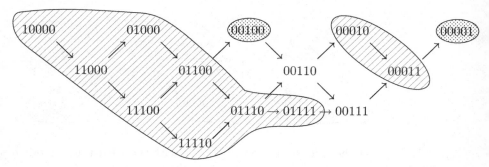

Here, we have

$$\text{Hom}_A\left(T, {}^{\ }_1{}^0_0{}^0_{\ }\right) = 10000, \ \text{Hom}_A\left(T, {}^{\ }_1{}^0_1{}^0_{\ }\right) = 11000, \ \text{Hom}_A\left(T, {}^{\ }_0{}^0_1{}^0_{\ }\right) = 11100,$$

$$\text{Hom}_A\left(T, {}^{\ }_0{}^1_1{}^1_{\ }\right) = 11110, \ \text{Hom}_A\left(T, {}^{\ }_1{}^0_0{}^0_{\ }\right) = 01000, \ \text{Hom}_A\left(T, {}^{\ }_1{}^0_0{}^0_{\ }\right) = 01100,$$

$$\text{Hom}_A\left(T, {}^{\ }_1{}^1_0{}^1_{\ }\right) = 01110, \ \text{Hom}_A\left(T, {}^{\ }_0{}^0_0{}^1_{\ }\right) = 01111, \ \text{Hom}_A\left(T, {}^{\ }_0{}^1_0{}^1_{\ }\right) = 00010,$$

$$\text{Hom}_A\left(T, {}^{\ }_0{}^0_0{}^1_{\ }\right) = 00011, \ \ \text{Ext}^1_A\left(T, {}^{\ }_1{}^0_0{}^0_{\ }\right) = 00100, \ \ \text{Ext}^1_A\left(T, {}^{\ }_0{}^1_0{}^0_{\ }\right) = 00001.$$

Observe that

$$(DT)_B = \text{Hom}_A(T, DA) = 11110 \oplus 01000 \oplus 01111 \oplus 00010 \oplus 00011.$$

(c) Consider, as in Example 2.8 (b), the algebra A given by the quiver

and the tilting module $T_A = 11 {}^1_0 \oplus 11 {}^1_1 \oplus 01 {}^1_0 \oplus 00 {}^1_0$. Here, $B = \text{End}\, T_A$ is given by the quiver

bound by $\alpha\beta = \gamma\delta$. The induced torsion pair $(\mathcal{X}(T), \mathcal{Y}(T))$ in $\text{mod}\, B$ is illustrated as:

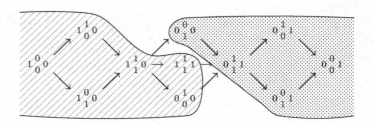

(d) Consider the algebra A of Example (b), with the APR-tilting module $T[2]$. Here, $B = \operatorname{End} T[2]_A$ is given by the quiver

bound by $\lambda\mu = \nu\eta\sigma$. The induced torsion pair $(\mathcal{X}(T[2]), \mathcal{Y}(T[2]))$ in $\operatorname{mod} B$ is illustrated as:

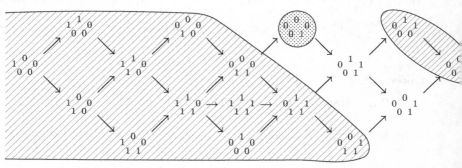

If, on the other hand, one considers the APR-tilting module $T[1]$, one obtains the algebra $\operatorname{End} T[1]_A$ given by the quiver

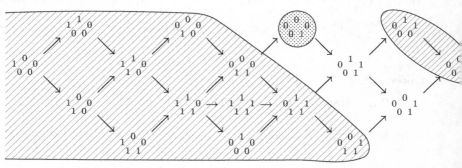

bound by the relation $\lambda\mu\nu = 0$. We leave to the reader the calculation of $(\mathcal{X}(T[1]), \mathcal{Y}(T[1]))$.

VI.4. Consequences of the tilting theorem

In this section, we investigate the connection between an algebra A and the endomorphism algebras of its tilting modules, using the tilting theorem of Brenner and Butler. Throughout, we keep the notation used in Section 3.

Our first result says that, under tilting, the global dimension of an algebra changes by at most one. As a consequence, this entails that the class of algebras of finite global dimension is closed under the tilting process. We need one lemma.

4.1. Lemma. *Let A be an algebra, T_A be a tilting module, and $B = \operatorname{End} T_A$. If $M \in \mathcal{T}(T)$, then $\operatorname{pd} \operatorname{Hom}_A(T, M) \leq \operatorname{pd} M$.*

Proof. We use induction on $n = \operatorname{pd} M$. If $n = 0$, then M is projective. Because $M \in \mathcal{T}(T) = \operatorname{Gen} T$, this implies that $M \in \operatorname{add} T$. Therefore $\operatorname{Hom}_A(T, M)$ is projective (by (3.1)(b)), and we are done.

Now, assume $n \geq 1$. By $(2.5)(c)$, there exists a short exact sequence

$$0 \longrightarrow L \longrightarrow T_0 \longrightarrow M \longrightarrow 0$$

with $T_0 \in \operatorname{add} T$ and $L \in \mathcal{T}(T)$. Therefore we have a short exact sequence

$$0 \longrightarrow \operatorname{Hom}_A(T, L) \longrightarrow \operatorname{Hom}_A(T, T_0) \longrightarrow \operatorname{Hom}_A(T, M) \longrightarrow 0.$$

Assume $n = 1$. Then the first short exact sequence yields an exact sequence of functors

$$0 = \operatorname{Ext}_A^1(T_0, -)|_{\mathcal{T}(T)} \longrightarrow \operatorname{Ext}_A^1(L, -)|_{\mathcal{T}(T)} \longrightarrow \operatorname{Ext}_A^2(M, -)|_{\mathcal{T}(T)} = 0;$$

therefore $\operatorname{Ext}_A^1(L, -)|_{\mathcal{T}(T)} = 0$, that is, L is Ext-projective in $\mathcal{T}(T)$. By $(2.5)(d)$, $L \in \operatorname{add} T$, so that $\operatorname{Hom}_A(T, L)$ is projective and the second exact sequence implies that $\operatorname{pd} \operatorname{Hom}_A(T, M) \leq 1$. Finally, assume $n \geq 2$. Then, according to (A.4.7) of the Appendix, the first short exact sequence yields $\operatorname{pd} L \leq n - 1$, because $\operatorname{pd} T_0 \leq 1$. By the induction hypothesis, this implies that $\operatorname{pd} \operatorname{Hom}_A(T, L) \leq n - 1$. Hence the second short exact sequence gives

$$\operatorname{pd} \operatorname{Hom}_A(T, M) \leq 1 + \operatorname{pd} \operatorname{Hom}_A(T, L) \leq 1 + (n - 1) = n. \qquad \square$$

4.2. Theorem. *Let A be an algebra, T_A be a tilting module, and $B = \operatorname{End} T_A$. Then $|\operatorname{gl.dim} A - \operatorname{gl.dim} B| \leq 1$.*

Proof. Let X be any B-module. There exists a short exact sequence

$$0 \longrightarrow Y \longrightarrow P \longrightarrow X \longrightarrow 0$$

with P projective. Because $P \in \mathcal{Y}(T)$, we have $Y \in \mathcal{Y}(T)$ as well. By the tilting theorem (3.8), there exists $M \in \mathcal{T}(T)$ such that $Y = \operatorname{Hom}_A(T, M)$. By (4.1), we have $\operatorname{pd} Y \le \operatorname{pd} M$. Hence $\operatorname{pd} X \le 1 + \operatorname{pd} Y \le 1 + \operatorname{pd} M \le 1 + \operatorname{gl.dim} A$, and consequently $\operatorname{gl.dim} B \le 1 + \operatorname{gl.dim} A$. Because, again by the tilting theorem, $_BT$ is also a tilting module, we have $\operatorname{gl.dim} A \le 1 + \operatorname{gl.dim} B$. □

In Example 3.11 (a), we have $\operatorname{gl.dim} B = 2$, whereas $\operatorname{gl.dim} A = 1$ (hence the bound of (4.2) is sharp). In Example 3.11 (b), we have $\operatorname{gl.dim} A = \operatorname{gl.dim} B = 2$.

There are the following other relations between the homological dimensions in $\operatorname{mod} A$ and $\operatorname{mod} B$ (see Exercise 20):

    (a) If $N \in \mathcal{F}(T)$, then $\operatorname{pd} \operatorname{Ext}_A^1(T, N) \le 1 + \max(1, \operatorname{pd} N)$.
    (b) If $M \in \mathcal{T}(T)$, then $\operatorname{id} \operatorname{Hom}_A(T, M) \le 1 + \operatorname{id} M$.
    (c) If $N \in \mathcal{F}(T)$, then $\operatorname{id} \operatorname{Ext}_A^1(T, N) \le \operatorname{id} N$.

In our next application, we show that the number of simple modules is preserved under the tilting process. For this purpose we recall from (III.3.5) that the Grothendieck group $K_0(A)$ of A is free abelian and that the elements $[S]$, where S ranges over a complete set of representatives of the isomorphism classes of simple A-modules, constitute a basis of $K_0(A)$. The map $[X] \mapsto \mathbf{dim}\, X$ defines a group isomorphism

$$\mathbf{dim} \, : K_0(A) \xrightarrow{\;\cong\;} \mathbb{Z}^n,$$

where n is the number of the isomorphism classes of simple A-modules. Throughout, we identify the group $K_0(A)$ with $\mathbb{Z}^n$ and the element $[X]$ of $K_0(A)$ with the dimension vector $\mathbf{dim}\, X$ in $\mathbb{Z}^n$, for any module X in $\operatorname{mod} A$.

4.3. Theorem. *Let A be an algebra, T_A be a tilting module, and $B = \operatorname{End} T_A$. Then the correspondence*

$$\mathbf{dim}\, M \mapsto \mathbf{dim} \operatorname{Hom}_A(T, M) - \mathbf{dim} \operatorname{Ext}_A^1(T, M),$$

where M is an A-module, induces an isomorphism $f : K_0(A) \to K_0(B)$ of the Grothendieck groups of A and B.

Proof. Because $\operatorname{pd} T_A \le 1$, any short exact sequence $0 \to L_A \to M_A \to N_A \to 0$ in $\operatorname{mod} A$ induces an exact cohomology sequence

$$0 \longrightarrow \operatorname{Hom}_A(T, L) \longrightarrow \operatorname{Hom}_A(T, M) \longrightarrow \operatorname{Hom}_A(T, N)$$
$$\longrightarrow \operatorname{Ext}_A^1(T, L) \longrightarrow \operatorname{Ext}_A^1(T, M) \longrightarrow \operatorname{Ext}_A^1(T, N) \longrightarrow 0$$

in $\operatorname{mod} B$, from which we deduce the equality

$$\mathbf{dim} \operatorname{Hom}_A(T, M) - \mathbf{dim} \operatorname{Ext}_A^1(T, M) =$$
$$= [\mathbf{dim} \operatorname{Hom}_A(T, L) - \mathbf{dim} \operatorname{Ext}^1(T, L)] +$$
$$+ [\mathbf{dim} \operatorname{Hom}_A(T, N) - \mathbf{dim} \operatorname{Ext}^1(T, N)]$$

in $K_0(B)$ (see (III.3.3) and (III.3.5)). Hence the given correspondence defines indeed a group homomorphism $f : K_0(A) \to K_0(B)$.

Let S be a simple B-module. Because $(\mathcal{X}(T), \mathcal{Y}(T))$ is a torsion pair, we have $S \in \mathcal{X}(T)$ or $S \in \mathcal{Y}(T)$ (by (1.6)). In the latter case, we have $S \cong \operatorname{Hom}_A(T, S \otimes_B T)$ while $\operatorname{Ext}^1_A(T, S \otimes_B T) = 0$, so that $\dim S = f(\dim S \otimes_B T)$. In the former case, we have $S \cong \operatorname{Ext}^1_A(T, \operatorname{Tor}^B_1(S, T))$ while $\operatorname{Hom}_A(T, \operatorname{Tor}^B_1(S, T)) = 0$, so that $\dim S = f(-\dim \operatorname{Tor}^B_1(S, T))$. In either case, $\dim S$ lies in the image of f. Because, according to (III.3.5), the vectors of the form $\dim S$, where S ranges over a complete set of representatives of the isomorphism classes of simple B-modules, constitute a basis of $K_0(B)$, this shows that f is surjective. Consequently, the rank of $K_0(A)$ is greater than or equal to that of $K_0(B)$. Because $_BT$ is also a tilting module and $A \cong \operatorname{End}(_BT)^{\operatorname{op}}$, we have, by symmetry, that the rank of $K_0(B)$ is greater than or equal to that of $K_0(A)$. Therefore these ranks are equal, and the group epimorphism f is an isomorphism. □

For instance, in Example 3.11 (a), it is easily seen that $f(100) = (100)$, $f(010) = -(001)$, and $f(001) = (011)$. Hence the matrix $\mathbf{F}$ of f in the canonical bases of $K_0(A)$ and $K_0(B)$ is of the form

$$\mathbf{F} = \begin{bmatrix} 1 & 0 & 0 \\ 0 & 0 & 1 \\ 0 & -1 & 1 \end{bmatrix}$$

(where the elements of $K_0(A)$ and $K_0(B)$ are considered as column vectors). Thus, the image of the dimension vector of the torsion module $I(2) = 011$ is given by

$$\begin{bmatrix} 1 & 0 & 0 \\ 0 & 0 & 1 \\ 0 & -1 & 1 \end{bmatrix} \begin{bmatrix} 0 \\ 1 \\ 1 \end{bmatrix} = \begin{bmatrix} 0 \\ 1 \\ 0 \end{bmatrix},$$

that is, is the dimension vector of the B-module 010.

We deduce from (4.3) and Bongartz's lemma (2.4) a very useful criterion for deciding whether a partial tilting module is a tilting module or not.

4.4. Corollary. *Let T_A be a partial tilting module. Then T_A is a tilting module if and only if the number of pairwise nonisomorphic indecomposable summands of T equals the number of pairwise nonisomorphic simple modules (that is, the rank of $K_0(A)$).*

Proof. If T_A is a tilting module, and $B = \operatorname{End}T_A$, then by (3.1)(b), the number t of pairwise nonisomorphic indecomposable summands of T equals the rank of $K_0(B)$. Hence, by (4.3), t equals the rank of $K_0(A)$.

Conversely, assume that T_A is a partial tilting module satisfying the stated condition. By Bongartz's lemma (2.3), there exists an A-module E such that $T \oplus E$ is a tilting module. The necessity part says that the number of pairwise nonisomorphic indecomposable summands of $T \oplus E$ equals the rank of $K_0(A)$, hence, by hypothesis, equals the number of pairwise nonisomorphic indecomposable summands of T. Therefore $E \in \operatorname{add} T$ and T is indeed a tilting module. □

Assume now that A is an algebra of finite global dimension. We recall from (III.3.11) and (III.3.13) that the Euler characteristic of A is the bilinear form on $K_0(A)$ defined by

$$\langle \operatorname{\mathbf{dim}} M, \operatorname{\mathbf{dim}} N \rangle_A = \sum_{s=0}^{\infty} (-1)^s \dim_K \operatorname{Ext}_A^s(M, N),$$

where M, N are modules in $\operatorname{mod} A$. The preceding sum is finite due to our hypothesis on A. We next show that the Euler characteristic of A is preserved under tilting; namely, that the isomorphism between the Grothendieck groups of A and B defined in (4.3) is an isometry of the Euler characteristics of A and B.

4.5. Proposition. *Let A be an algebra of finite global dimension, T_A be a tilting module, $B = \operatorname{End} T_A$, and $f : K_0(A) \to K_0(B)$ be the isomorphism of (4.3). Then for any A-modules M and N we have*

$$\langle \operatorname{\mathbf{dim}} M, \operatorname{\mathbf{dim}} N \rangle_A = \langle f(\operatorname{\mathbf{dim}} M), f(\operatorname{\mathbf{dim}} N) \rangle_B.$$

Proof. Let $T_1, \ldots, T_n$ denote the pairwise nonisomorphic indecomposable summands of T. We claim that the vectors $\operatorname{\mathbf{dim}} T_i$, where $1 \le i \le n$, constitute a basis of $K_0(A)$. Indeed, by (3.1)(b), the B-modules

$$\operatorname{Hom}_A(T, T_1), \ldots, \operatorname{Hom}_A(T, T_n)$$

form a complete set of representatives of the isomorphism classes of indecomposable projective modules. Because, by (4.2), B also has finite global dimension, the vectors $f(\operatorname{\mathbf{dim}} T_i) = \operatorname{\mathbf{dim}} \operatorname{Hom}_A(T, T_i)$, where $1 \le i \le n$, constitute a basis of $K_0(B)$. Because, by (4.3), f is an isomorphism, this implies our claim.

Also, the projectivity of the B-modules $\operatorname{Hom}_A(T, T_i)$ and the tilting theorem imply that, for any i, j such that $1 \le i, j \le n$,

$$\begin{aligned}
\langle f(\operatorname{\mathbf{dim}} T_i), f(\operatorname{\mathbf{dim}} T_j) \rangle_B &= \langle \operatorname{\mathbf{dim}} \operatorname{Hom}_A(T, T_i), \operatorname{\mathbf{dim}} \operatorname{Hom}_A(T, T_j) \rangle_B \\
&= \dim_K \operatorname{Hom}_B(\operatorname{Hom}_A(T, T_i), \operatorname{Hom}_A(T, T_j)) \\
&= \dim_K \operatorname{Hom}_A(T_i, T_j) = \langle \operatorname{\mathbf{dim}} T_i, \operatorname{\mathbf{dim}} T_j \rangle_A,
\end{aligned}$$

because $\text{Ext}_A^1(T_i, T_j) = 0$. The conclusion follows from our claim. □

Let $\mathbf{A}$ and $\mathbf{B}$ be the matrices defining the Euler characteristics of the algebras A and B, respectively, and let $\mathbf{F}$ denote the matrix defining the isomorphism f of (4.3). It follows from (4.3) that $\mathbf{A}$, $\mathbf{B}$, and $\mathbf{F}$ are all square matrices of the same size, and from the explicit expression of f that the matrix $\mathbf{F}$ has integral coefficients. Because for $\mathbf{x}, \mathbf{y} \in K_0(A)$, we have

$$\langle \mathbf{x}, \mathbf{y} \rangle_A = \mathbf{x}^t \mathbf{A} \mathbf{y} \quad \text{and} \quad \langle f(\mathbf{x}), f(\mathbf{y}) \rangle_B = (\mathbf{F}\mathbf{x})^t \mathbf{B}(\mathbf{F}\mathbf{y}) = \mathbf{x}^t(\mathbf{F}^t \mathbf{B} \mathbf{F})\mathbf{y},$$

we infer from (4.5) that $\mathbf{x}^t \mathbf{A} \mathbf{y} = \mathbf{x}^t(\mathbf{F}^t \mathbf{B} \mathbf{F})\mathbf{y}$ for all $\mathbf{x}, \mathbf{y} \in K_0(A)$. That is, $\mathbf{A} = \mathbf{F}^t \mathbf{B} \mathbf{F}$; the matrices $\mathbf{A}$ and $\mathbf{B}$ are $\mathbb{Z}$-congruent.

We deduce the following corollary.

4.6. Corollary. *Let A be an algebra of finite global dimension, T_A be a tilting module, and $B = \text{End}\, T_A$. Then the Cartan matrices $\mathbf{C}_A$ of A and $\mathbf{C}_B$ of B are $\mathbb{Z}$-congruent.*

Proof. By (III.3.11) and the preceding discussion, we have $\mathbf{A} = (\mathbf{C}_A^{-1})^t$ and $\mathbf{B} = (\mathbf{C}_B^{-1})^t$. Thus, the equality $\mathbf{A} = \mathbf{F}^t \mathbf{B} \mathbf{F}$ can be written as $(\mathbf{C}_A^{-1})^t = \mathbf{F}^t(\mathbf{C}_B^{-1})^t\mathbf{F}$, or, equivalently, as $\mathbf{C}_B = \mathbf{F}\mathbf{C}_A\mathbf{F}^t$. □

These considerations also apply to the integral Euler quadratic form $q_A : K_0(A) \to \mathbb{Z}$ attached to the Euler characteristic of A by the formula

$$q_A(\dim M) = \langle \dim M, \dim M \rangle_A,$$

where M is an A-module; see (III.3.11). The equality $\mathbf{A} = \mathbf{F}^t \mathbf{B} \mathbf{F}$ yields the following corollary.

4.7. Corollary. *Let A be an algebra of finite global dimension, T_A be a tilting module, and $B = \text{End}\, T_A$. Then the Euler quadratic forms q_A and q_B are $\mathbb{Z}$-congruent.* □

Let, for instance, A be as in Example 3.11 (a), that is, A is given by the quiver

$$\overset{1}{\circ} \longleftarrow \overset{2}{\circ} \longleftarrow \overset{3}{\circ}$$

Then

$$\mathbf{C}_A = \begin{bmatrix} 1 & 1 & 1 \\ 0 & 1 & 1 \\ 0 & 0 & 1 \end{bmatrix},$$

and consequently

$$\mathbf{A} = (\mathbf{C}_A^{-1})^t = \begin{bmatrix} 1 & 0 & 0 \\ -1 & 1 & 0 \\ 0 & -1 & 1 \end{bmatrix},$$

so that

$$q_A(\mathbf{x}) = \mathbf{x}^t \mathbf{A}\mathbf{x} = x_1^2 + x_2^2 + x_3^2 - x_1x_2 - x_2x_3, \text{ for } \mathbf{x} = \begin{bmatrix} x_1 \\ x_2 \\ x_3 \end{bmatrix} \in K_0(A).$$

We tilt A to B, where B is given by the quiver

$$\underset{1}{\circ} \xleftarrow{\;\;\mu\;\;} \underset{2}{\circ} \xleftarrow{\;\;\lambda\;\;} \underset{3}{\circ}$$

bound by $\lambda\mu = 0$. Then

$$\mathbf{C}_B = \begin{bmatrix} 1 & 1 & 0 \\ 0 & 1 & 1 \\ 0 & 0 & 1 \end{bmatrix},$$

and consequently

$$\mathbf{B} = (\mathbf{C}_B^{-1})^t = \begin{bmatrix} 1 & 0 & 0 \\ -1 & 1 & 0 \\ 1 & -1 & 1 \end{bmatrix},$$

so that

$$q_B(\mathbf{x}) = \mathbf{x}^t \mathbf{B}\mathbf{x} = x_1^2 + x_2^2 + x_3^2 - x_1x_2 - x_2x_3 + x_1x_3, \text{ for } \mathbf{x} = \begin{bmatrix} x_1 \\ x_2 \\ x_3 \end{bmatrix} \in K_0(B).$$

We have already observed that the matrix $\mathbf{F}$ defining the group isomorphism $f : K_0(A) \xrightarrow{\cong} K_0(B)$ is of the form

$$\mathbf{F} = \begin{bmatrix} 1 & 0 & 0 \\ 0 & 0 & 1 \\ 0 & -1 & 1 \end{bmatrix}.$$

Finally, it is easily verified that

$$\mathbf{F}^t\mathbf{B}\mathbf{F} = \begin{bmatrix} 1 & 0 & 0 \\ 0 & 0 & -1 \\ 0 & 1 & 1 \end{bmatrix} \begin{bmatrix} 1 & 0 & 0 \\ -1 & 1 & 0 \\ 1 & -1 & 1 \end{bmatrix} \begin{bmatrix} 1 & 0 & 0 \\ 0 & 0 & 1 \\ 0 & -1 & 1 \end{bmatrix}$$

$$= \begin{bmatrix} 1 & 0 & 0 \\ -1 & 1 & 0 \\ 0 & -1 & 1 \end{bmatrix} = \mathbf{A}.$$

As a third and final application of the tilting theorem, we consider those almost split sequences in mod B whose left term lies in $\mathcal{Y}(T)$ and whose right term lies in $\mathcal{X}(T)$; such sequences are called **connecting sequences**. The following easy lemma shows that there are only finitely many connecting sequences.

4.8. Lemma. *If $0 \to Y_B \to E_B \to X_B \to 0$ is a connecting sequence, then there exists an indecomposable injective A-module $I(a)$ such that $Y \cong \mathrm{Hom}_A(T, I(a))$.*

Proof. Because $Y \in \mathcal{Y}(T)$, according to (3.8), there exists $M \in \mathcal{T}(T)$ such that $Y \cong \mathrm{Hom}_A(T, M)$. Let $f : M \to N$ be an injective envelope in $\mathrm{mod}\,A$ and consider the short exact sequence

$$0 \longrightarrow M \xrightarrow{\ f\ } N \longrightarrow N/M \longrightarrow 0.$$

Because $N \in \mathcal{T}(T)$, this sequence lies entirely in $\mathcal{T}(T)$. Applying the functor $\mathrm{Hom}_A(T, -)$ yields a short exact sequence in $\mathcal{Y}(T)$

$$0 \longrightarrow Y \xrightarrow{\ \mathrm{Hom}_A(T,f)\ } \mathrm{Hom}_A(T, N) \longrightarrow \mathrm{Hom}_A(T, N/M) \longrightarrow 0.$$

Since $\tau^{-1}Y = X \in \mathcal{X}(T)$, we deduce from (1.11)(b) that Y is Ext-injective in $\mathcal{Y}(T)$. Therefore the preceding short exact sequence splits, that is, $\mathrm{Hom}_A(T, f)$ is a section. Applying $- \otimes_B T$ shows that f is a section. We have thus shown that M is injective. Its indecomposability follows from the indecomposability of Y. Hence M is isomorphic to an indecomposable injective module $I(a)$. $\qquad\square$

Of course, not all indecomposable injective A-modules correspond to connecting sequences. The next lemma, known as the **connecting lemma**, characterises those that do and gives the right term of such a sequence. More precisely, one can show, exactly as in (4.8), that the right term X of a connecting sequence $0 \to Y \to E \to X \to 0$ satisfies $X \cong \mathrm{Ext}_A^1(T, P)$ for some indecomposable projective A-module P. The connecting lemma says that the top of P is isomorphic to the socle of I, and that $P \notin \mathrm{add}\,T$.

4.9. Connecting lemma. *Let A be an algebra, T_A be a tilting module, and $B = \mathrm{End}\,T_A$. Let $P(a)$ be the projective cover of a simple module $S(a)_A$ and $I(a)$ be its injective envelope. Then*

$$\tau^{-1}\mathrm{Hom}_A(T, I(a)) \cong \mathrm{Ext}_A^1(T, P(a)).$$

In particular, $P(a) \in \mathrm{add}\,T$ if and only if $\mathrm{Hom}_A(T, I(a))$ is an injective B-module.

Proof. Let $P = P(a) = e_a A$ and $I = I(a) = D(Ae_a)$, where $e_a \in A$ is a primitive idempotent. By (III.2.11), there is a functorial isomorphism $D\mathrm{Hom}_A(T, I) \cong \mathrm{Hom}_A(P, T)$. We need to show that the transpose Tr of $\mathrm{Hom}_A(P, T)$ is isomorphic to $\mathrm{Ext}_A^1(T, P)$. For this purpose, we use the definition of the transpose (IV.2).

Because T_A is a tilting module, there exists a short exact sequence

$$0 \longrightarrow P_A \longrightarrow T'_A \overset{f}{\longrightarrow} T''_A \longrightarrow 0$$

with $T', T'' \in \operatorname{add} T$. Applying $\operatorname{Hom}_A(-, {}_B T_A)$ yields a short exact sequence

$$0 \rightarrow \operatorname{Hom}_A(T'', {}_B T_A) \xrightarrow{\operatorname{Hom}_A(f,T)} \operatorname{Hom}_A(T', {}_B T_A) \rightarrow \operatorname{Hom}_A(P(a), {}_B T_A) \rightarrow 0,$$

which is a projective resolution for the left B-module $\operatorname{Hom}_A(P,T)$. The transpose (in $\operatorname{mod} B$) of $\operatorname{Hom}_A(P,T)$ is obtained by applying to the previous sequence the functor $(-)^t = \operatorname{Hom}_B(-,B) = \operatorname{Hom}_B(-, \operatorname{Hom}_A(T,T))$. If $T_0 \in \operatorname{add} T$, we have a functorial isomorphism in $\operatorname{add} T$ given by

$$\operatorname{Hom}_A(T, T_0) \cong \operatorname{Hom}_B(\operatorname{Hom}_A(T_0, T), \operatorname{Hom}_A(T, T)).$$

Indeed, such an isomorphism exists when $T_0 = T$ and the functors are additive. Hence the commutative square

$$
\begin{array}{ccc}
\operatorname{Hom}_B(\operatorname{Hom}_A(T', T), \operatorname{Hom}_A(T, T)) & \overset{\cong}{\longrightarrow} & \operatorname{Hom}_A(T, T') \\
{\scriptstyle \operatorname{Hom}_B(\operatorname{Hom}_A(f,T), \operatorname{Hom}_A(T,T))} \Big\downarrow & & \Big\downarrow {\scriptstyle \operatorname{Hom}_A(T,f)} \\
\operatorname{Hom}_B(\operatorname{Hom}_A(T'', T), \operatorname{Hom}_A(T, T)) & \underset{\cong}{\longrightarrow} & \operatorname{Hom}_A(T, T'')
\end{array}
$$

shows that $\operatorname{Hom}_A(f,T)^t \cong \operatorname{Hom}_A(T,f)$. On the other hand, applying $\operatorname{Hom}_A(T, -)$ to the first short exact sequence yields an exact sequence

$$0 \longrightarrow \operatorname{Hom}_A(T, P) \longrightarrow \operatorname{Hom}_A(T, T') \xrightarrow{\operatorname{Hom}_A(T,f)} \operatorname{Hom}_A(T, T'')$$
$$\longrightarrow \operatorname{Ext}^1_A(T, P) \longrightarrow 0.$$

By definition of the transpose, we deduce, as required

$$\operatorname{Ext}^1_A(T, P) \cong \operatorname{Tr} \operatorname{Hom}_A(P, T) \cong \operatorname{Tr} D\operatorname{Hom}_A(T, I) = \tau^{-1}\operatorname{Hom}_A(T, I).$$

The second statement follows from the fact that a projective module P lies in $\operatorname{add} T$ if and only if it lies in $\mathcal{T}(T) = \operatorname{Gen} T$, that is, if and only if $\operatorname{Ext}^1_A(T, P) = 0$. $\qquad\qquad \square$

The middle term of a connecting sequence, on the other hand, can only be approximated by means of its canonical sequence.

4.10. Corollary. *Let* $P(a)$, $I(a)$, *and* $S(a)$ *be as in* (4.9), *with* $P(a) \notin$ add T. *Consider the connecting sequence*

$$0 \longrightarrow \mathrm{Hom}_A(T, I(a)) \xrightarrow{u} E_B \xrightarrow{v} \mathrm{Ext}^1_A(T, P(a)) \longrightarrow 0.$$

The canonical sequence of E_B *in the torsion pair* $(\mathcal{X}(T), \mathcal{Y}(T))$ *is*

$$0 \longrightarrow \mathrm{Ext}^1_A(T, \mathrm{rad}\, P(a)) \longrightarrow E_B \longrightarrow \mathrm{Hom}_A(T, I(a)/S(a)) \longrightarrow 0.$$

Proof. Because $(\mathcal{T}(T), \mathcal{F}(T))$ is a torsion pair, the simple module $S(a)$ lies in either $\mathcal{T}(T)$ or $\mathcal{F}(T)$ (by (1.6)).

(a) Assume that $S(a) \in \mathcal{T}(T)$; then $\mathrm{Ext}^1_A(T, S(a)) = 0$. Hence the short exact sequence

$$0 \to S(a) \to I(a) \to I(a)/S(a) \to 0$$

induces a short exact sequence

$$0 \longrightarrow \mathrm{Hom}_A(T, S(a)) \xrightarrow{f} \mathrm{Hom}_A(T, I(a)) \xrightarrow{g} \mathrm{Hom}_A(T, I(a)/S(a)) \longrightarrow 0.$$

On the other hand, $P(a) \notin$ add T implies $P(a) \notin \mathcal{T}(T)$ so that $P(a) \neq tP(a)$ and hence $tP(a) \subseteq \mathrm{rad}\, P(a)$, which yields a K-linear isomorphism $\mathrm{Hom}_A(T, P(a)) \cong \mathrm{Hom}_A(T, \mathrm{rad}\, P(a))$ and the exact sequence in mod A

$$0 \to \mathrm{rad}\, P(a) \to P(a) \to S(a) \to 0$$

induces a short exact sequence

$$0 \longrightarrow \mathrm{Hom}_A(T, S(a)) \xrightarrow{h} \mathrm{Ext}^1_A(T, \mathrm{rad}\, P(a)) \xrightarrow{k} \mathrm{Ext}^1_A(T, P(a)) \longrightarrow 0.$$

This sequence does not split; otherwise, there would exist a nonzero homomorphism from the torsion B-module

$$\mathrm{Ext}^1_A(T, \mathrm{rad}\, P(a)) \cong \mathrm{Ext}^1_A(T, \mathrm{rad}\, P(a)/t\mathrm{rad}\, P(a))$$

to the torsion-free module $\mathrm{Hom}_A(T, S(a))$ (see (3.9)), a contradiction. In particular, k is not a retraction. Because the given connecting sequence is almost split, there exists a homomorphism $f' : \mathrm{Ext}^1_A(T, \mathrm{rad}\, P(a)) \to E$ such that $k = vf'$. By passing to the kernels, there exists a homomorphism $\mathrm{Hom}_A(T, S(a)) \to \mathrm{Hom}_A(T, I(a))$ whose composition with u equals $f'h$. But the K-vector space $\mathrm{Hom}_B(\mathrm{Hom}_A(T, S(a)), \mathrm{Hom}_A(T, I(a))) \cong \mathrm{Hom}_A(S, I(a))$ is one-dimensional. Hence this homomorphism can be taken equal to f, after replacing h, if necessary, by one of its scalar multiples, so that we have a commutative diagram with exact rows and columns

$$
\begin{array}{ccccc}
0 & & 0 & & \\
\downarrow & & \downarrow & & \\
0\to\operatorname{Hom}_A(T,S(a)) & \xrightarrow{\ h\ } & \operatorname{Ext}^1_A(T,\operatorname{rad}P(a)) & \xrightarrow{\ k\ } & \operatorname{Ext}^1_A(T,P(a))\to 0 \\
f\downarrow & & f'\downarrow & & 1\downarrow \\
0\to\operatorname{Hom}_A(T,I(a)) & \xrightarrow{\ u\ } & E & \xrightarrow{\ v\ } & \operatorname{Ext}^1_A(T,P(a))\to 0 \\
g\downarrow & & g'\downarrow & & \\
\operatorname{Hom}_A(T,I(a)/S(a)) & \xrightarrow{\ 1\ } & \operatorname{Hom}_A(T,I(a)/S(a)) & & \\
\downarrow & & \downarrow & & \\
0 & & 0 & &
\end{array}
$$

The middle column yields the result.

(b) Assume that $S(a)\in\mathcal{F}(T)$; then $\operatorname{Hom}_A(T,S(a))=0$ and hence we have short exact sequences

$$0\longrightarrow \operatorname{Ext}^1(T,\operatorname{rad}P(a))\longrightarrow \operatorname{Ext}^1_A(T,P(a))\longrightarrow \operatorname{Ext}^1_A(T,S(a))\longrightarrow 0,$$
$$0\longrightarrow \operatorname{Hom}_A(T,I(a))\longrightarrow \operatorname{Hom}_A(T,I(a)/S(a))\longrightarrow \operatorname{Ext}^1_A(T,S(a))\longrightarrow 0.$$

The second sequence does not split and we deduce, exactly as in (a), a commutative diagram with exact rows and columns

$$
\begin{array}{ccccc}
 & 0 & & 0 & \\
 & \downarrow & & \downarrow & \\
 & \operatorname{Ext}^1_A(T,\operatorname{rad}P(a)) & \xrightarrow{\ 1\ } & \operatorname{Ext}^1_A(T,\operatorname{rad}P(a)) & \\
 & \downarrow & & \downarrow & \\
0\to\operatorname{Hom}_A(T,I(a)) & \xrightarrow{\ u\ } & E & \xrightarrow{\ v\ } & \operatorname{Ext}^1_A(T,P(a))\to 0 \\
1\downarrow & & \downarrow & & \downarrow \\
0\to\operatorname{Hom}_A(T,I(a)) & \longrightarrow & \operatorname{Hom}_A(T,I(a)/S(a)) & \longrightarrow & \operatorname{Ext}^1_A(T,S(a))\to 0 \\
 & & \downarrow & & \downarrow \\
 & & 0 & & 0
\end{array}
$$

Again the middle column yields the result. $\square$

For instance, in Example 3.11 (a), the only connecting sequence is the sequence

$$0 \longrightarrow 010 \longrightarrow 011 \longrightarrow 001 \longrightarrow 0.$$

Here, $S_A = 010$, $I_A = 011$, $P_A = 110$ and we have $\mathrm{Hom}_A(T, I) = 010$ and $\mathrm{Ext}_A^1(T, P) = 001$. The middle term E lies entirely in $\mathcal{Y}(T)$, hence

$$E \cong \mathrm{Hom}_A(T, I/S) = \mathrm{Hom}_A(T, 001) = 011.$$

In Example 3.11 (c), the connecting sequence

$$0 \longrightarrow 1\,{\textstyle\frac{1}{1}}\,0 \longrightarrow 0\,{\textstyle\frac{0}{1}}\,0 \oplus 1\,{\textstyle\frac{1}{1}}\,1 \oplus 0\,{\textstyle\frac{1}{0}}\,0 \longrightarrow 0\,{\textstyle\frac{1}{1}}\,1 \longrightarrow 0$$

corresponds to the simple A-module $S = 0\,1\,{\textstyle\frac{0}{0}}$. Here, $I_A = 0\,1\,{\textstyle\frac{1}{1}}$, $P_A = 1\,1\,{\textstyle\frac{0}{0}}$, $\mathrm{Hom}_A(T, I) = 1\,{\textstyle\frac{1}{1}}\,0$, $\mathrm{Ext}_A^1(T, P) = 0\,{\textstyle\frac{1}{1}}\,1$. The middle term E is a direct sum of three indecomposable modules. Indeed, $I/S = 0\,0\,{\textstyle\frac{1}{0}} \oplus 0\,0\,{\textstyle\frac{0}{1}}$ so that

$$\mathrm{Hom}_A(T, I/S) = \mathrm{Hom}_A\left(T,\ 0\,0\,{\textstyle\frac{1}{0}}\right) \oplus \mathrm{Hom}_A\left(T,\ 0\,0\,{\textstyle\frac{0}{1}}\right) = 1\,{\textstyle\frac{1}{1}}\,1 \oplus 0\,{\textstyle\frac{1}{0}}\,0,$$

whereas $\mathrm{rad}\, P = 1\,0\,{\textstyle\frac{0}{0}}$, so that $\mathrm{Ext}_A^1(T, \mathrm{rad}\, P) = 0\,{\textstyle\frac{0}{1}}\,0$.

The reader may have noticed that in Examples (3.11), it turns out that the indecomposable summands of E are either torsion or torsion-free (that is, the corresponding canonical sequence splits). This is generally not the case, as will be shown in Exercise 14.

VI.5. Separating and splitting tilting modules

It is reasonable to consider those tilting modules that induce splitting torsion pairs, one in $\mathrm{mod}\,A$ and the other in $\mathrm{mod}\,B$, where $B = \mathrm{End}\,T_A$. This leads to the following definition.

5.1. Definition. Let A be an algebra, T_A be a tilting module, and $B = \mathrm{End}\,T_A$. Then

(a) T_A is said to be **separating** if the induced torsion pair $(\mathcal{T}(T), \mathcal{F}(T))$ in $\mathrm{mod}\,A$ is splitting, and

(b) T_A is said to be **splitting** if the induced torsion pair $(\mathcal{X}(T), \mathcal{Y}(T))$ in $\mathrm{mod}\,B$ is splitting.

For instance, let, as in Example 3.11 (a), A be given by the quiver

$$\overset{1}{\circ}\longleftarrow\overset{2}{\circ}\longleftarrow\overset{3}{\circ}$$

Then the shown tilting module $T_A = 100 \oplus 111 \oplus 001$ is splitting but not separating. On the other hand, it is easily seen that, over the same algebra A, the APR-tilting module $T[1]_A$ is both splitting and separating. In general, however, an APR-tilting module is necessarily separating, as we showed in Example 2.8 (c), but it is not always splitting, as was seen in (3.11)(d). Finally, Example 3.11 (b) showed a tilting module that is neither separating nor splitting.

Clearly, if T_A is a splitting tilting module, then every indecomposable B-module is the image of an indecomposable A-module via one of the functors $\operatorname{Hom}_A(T, -)$ or $\operatorname{Ext}^1_A(T, -)$, so that B has fewer indecomposable modules than A (in particular, if A is representation–finite, then so is B). Moreover, the almost split sequences in $\operatorname{mod} B$ are easily characterised.

5.2. Proposition. *Let A be an algebra, T_A be a splitting tilting module, and $B = \operatorname{End} T_A$. Then any almost split sequence in $\operatorname{mod} B$ lies entirely in either $\mathcal{X}(T)$ or $\mathcal{Y}(T)$, or else it is of the form*

$$0 \to \operatorname{Hom}_A(T, I) \to \operatorname{Hom}_A(T, I/\operatorname{soc} I) \oplus \operatorname{Ext}^1_A(T, \operatorname{rad} P) \to \operatorname{Ext}^1_A(T, P) \to 0,$$

where P is an indecomposable projective module not lying in $\operatorname{add} T$ and I is the indecomposable injective module such that $P/\operatorname{rad} P \cong \operatorname{soc} I$.

Proof. Let $0 \to E' \to E \to E'' \to 0$ be an almost split sequence in $\operatorname{mod} B$. Because $(\mathcal{X}(T), \mathcal{Y}(T))$ is a splitting torsion pair, either this sequence lies entirely in one of the subcategories $\mathcal{X}(T)$ and $\mathcal{Y}(T)$ or we have $E' \in \mathcal{Y}(T)$ and $E'' \in \mathcal{X}(T)$; that is, it is a connecting sequence. In this last case, it follows from (4.8) and (4.9) that it is of the form

$$0 \longrightarrow \operatorname{Hom}_A(T, I) \longrightarrow E_B \longrightarrow \operatorname{Ext}^1_A(T, P) \longrightarrow 0,$$

where P and I are as required. Further, it follows from (4.10) that the canonical sequence for E in $(\mathcal{X}(T), \mathcal{Y}(T))$ is of the form

$$0 \longrightarrow \operatorname{Ext}^1_A(T, \operatorname{rad} P) \longrightarrow E_B \longrightarrow \operatorname{Hom}_A(T, I/\operatorname{soc} I) \longrightarrow 0.$$

Because $(\mathcal{X}(T), \mathcal{Y}(T))$ is splitting, this canonical sequence splits (1.7) so that $E \cong \operatorname{Ext}^1_A(T, \operatorname{rad} P) \oplus \operatorname{Hom}_A(T, I/\operatorname{soc} I)$. $\qquad\square$

The following lemma shows that the almost split sequences in $\operatorname{mod} A$ lying entirely inside one of the classes $\mathcal{T}(T)$ and $\mathcal{F}(T)$ give rise to almost split sequences in $\operatorname{mod} B$.

5.3. Lemma. *Let A be an algebra, T_A be a splitting tilting module, and $B = \operatorname{End} T_A$. Let $0 \longrightarrow L \xrightarrow{f} M \xrightarrow{g} N \to 0$ be an almost split sequence in $\operatorname{mod} A$.*

(a) *If the modules L, M, and N lie in $\mathcal{T}(T)$, then*

$$0 \to \operatorname{Hom}_A(T, L) \xrightarrow{\operatorname{Hom}_A(T,f)} \operatorname{Hom}_A(T, M) \xrightarrow{\operatorname{Hom}_A(T,g)} \operatorname{Hom}_A(T, N) \to 0$$

is an almost split sequence in $\operatorname{mod} B$, all of whose terms lie in $\mathcal{Y}(T)$.

(b) *If the modules L, M, and N lie in $\mathcal{F}(T)$, then*

$$0 \to \operatorname{Ext}^1_A(T, L) \xrightarrow{\operatorname{Ext}^1_A(T,f)} \operatorname{Ext}^1_A(T, M) \xrightarrow{\operatorname{Ext}^1_A(T,g)} \operatorname{Ext}^1_A(T, N) \to 0$$

is an almost split sequence in $\operatorname{mod} B$, all of whose terms lie in $\mathcal{X}(T)$.

Proof. We only prove (a); the proof of (b) is similar. Because the modules L, M, and N lie in $\mathcal{T}(T) = \operatorname{Gen} T_A$, $\operatorname{Ext}^1_A(T, L) = 0$ and the sequence of B-modules

$$0 \to \operatorname{Hom}_A(T, L) \xrightarrow{\operatorname{Hom}_A(T,f)} \operatorname{Hom}_A(T, M) \xrightarrow{\operatorname{Hom}_A(T,g)} \operatorname{Hom}_A(T, N) \to 0$$

is exact. Moreover, the B-modules $\operatorname{Hom}_A(T, L)$ and $\operatorname{Hom}_A(T, N)$ are indecomposable, because N and L are. By (IV.1.13), it suffices to show that $\operatorname{Hom}_A(T, f)$ and $\operatorname{Hom}_A(T, g)$ are irreducible. By (3.8), the functor $\operatorname{Hom}_A(T, -)$ induces an equivalence of categories $\mathcal{Y}(T) \xrightarrow{\cong} \mathcal{T}(T)$, and therefore the homomorphism $\operatorname{Hom}_A(T, f)$ is neither a section nor a retraction. Assume that there exist $u : \operatorname{Hom}_A(T, L) \to Y$ and $v : Y \to \operatorname{Hom}_A(T, M)$ in $\operatorname{mod} B$ such that $\operatorname{Hom}_A(T, f) = vu$. Because $u \neq 0$ (because $f \neq 0$), $Y \in \mathcal{Y}(T)$ and there exists $E \in \mathcal{T}(T)$ such that $Y \cong \operatorname{Hom}_A(T, E)$. Moreover, there exist homomorphisms of A-modules $u' : L \to E$ and $v' : E \to M$ such that $u = \operatorname{Hom}_A(T, u')$ and $v = \operatorname{Hom}_A(T, v')$. It follows that $f = v'u'$, and therefore u' is a retraction or v' is a section. Hence u is a retraction, or v is a section. This shows that $\operatorname{Hom}_A(T, f)$ is an irreducible morphism. The proof that $\operatorname{Hom}_A(T, g)$ is an irreducible morphism is similar.   □

The following technical property will be needed in Chapter VIII.

5.4. Lemma. *Let A be an algebra, I be an indecomposable injective A-module, T_A be a splitting tilting module, and $B = \operatorname{End} T_A$.*

(a) *If $Y_B \in \mathcal{Y}(T)$ is indecomposable, then there exists an irreducible morphism $\operatorname{Hom}_A(T, I) \to Y$ in $\operatorname{mod} B$ if and only if there exists an indecomposable A-module J such that $Y \cong \operatorname{Hom}_A(T, J)$ and J is isomorphic to a direct summand of $I/\operatorname{soc} I$.*

(b) *If $X_B \in \mathcal{X}(T)$ is indecomposable, then there exists an irreducible morphism $\operatorname{Hom}_A(T, I) \to X$ in $\operatorname{mod} B$ if and only if there exists an indecomposable injective A-module J such that $\tau X \cong \operatorname{Hom}_A(T, J)$ and I is a direct summand of $J/\operatorname{soc} J$. Further, in this case, $X \cong \operatorname{Ext}^1_A(T, P)$, where P is the projective cover of $\operatorname{soc} J$.*

Proof. Let $p : I \rightarrow I/\mathrm{soc}\, I$ be the canonical surjection. We claim that the homomorphism $f = \mathrm{Hom}_A(T, p)$ is irreducible in $\mathrm{mod}\, B$. By (3.8), the functor $\mathrm{Hom}_A(T, -)$ induces an equivalence of categories $\mathcal{Y}(T) \xrightarrow{\cong} \mathcal{T}(T)$, and therefore f is neither a section nor a retraction. Assume that $f = hg$, where $g : \mathrm{Hom}_A(T, I) \rightarrow Z$ and $h : Z \rightarrow \mathrm{Hom}_A(T, I/\mathrm{soc}\,I)$ are in $\mathrm{mod}\, B$. Because $h \neq 0$ (because $f \neq 0$), $Z \notin \mathcal{X}(T)$ and therefore $Z \in \mathcal{Y}(T)$, because T_A is a splitting tilting module. By (3.8)(b), there exists $M \in \mathcal{T}(T)$ such that $Z \cong \mathrm{Hom}_A(T, M)$. Moreover, there exist homomorphisms of A-modules $g' : I \rightarrow M$ and $h' : M \rightarrow I/\mathrm{soc}\, I$ such that $g = \mathrm{Hom}_A(T, g')$ and $h = \mathrm{Hom}_A(T, h')$. It follows that $p = h'g'$, and therefore h' is a retraction or g' is a section. Hence h is a retraction or g is a section. This shows that $\mathrm{Hom}_A(T, p)$ is an irreducible morphism. The sufficiency follows from (IV.1.10) and (IV.4.2).

For the necessity, let $Y_B \in \mathcal{Y}(T)$ be an indecomposable module and $f : \mathrm{Hom}_A(T, I) \rightarrow Y$ be an irreducible morphism in $\mathrm{mod}\, B$. Then there exists an indecomposable A-module J such that $Y \cong \mathrm{Hom}_A(T, J)$ and a homomorphism of B-modules $f' : I \rightarrow J$ such that $f = \mathrm{Hom}_A(T, f')$. Because, according to (IV.3.5)(b), $p : I \longrightarrow I/\mathrm{soc}\, I$ is left minimal almost split, there exists $g' : I/\mathrm{soc}\, I \longrightarrow J$ such that $f' = g'p$. Moreover, because f is irreducible, so is f' (by the equivalence $\mathcal{Y}(T) \xrightarrow{\cong} \mathcal{T}(T)$). Therefore g' is a retraction and so J is isomorphic to a direct summand of $I/\mathrm{soc}\, I$.

(b) Let $f : \mathrm{Hom}_A(T, I) \rightarrow X_B$ be irreducible with $X_B \in \mathcal{X}(T)$ indecomposable. Because all the projective B-modules lie in $\mathcal{Y}(T)$, the module X is not projective, hence there exists an irreducible morphism $\tau X \rightarrow \mathrm{Hom}_A(T, I)$. Because $\mathrm{Hom}_A(T, I) \in \mathcal{Y}(T)$, we deduce that $\tau X \in \mathcal{Y}(T)$. By (5.2), the almost split sequence ending with X is a connecting sequence, so that there exists an indecomposable injective A-module J such that $\tau X \cong \mathrm{Hom}_A(T, J)$. If P denotes the projective cover of $\mathrm{soc}\, J$, then $X \cong \mathrm{Ext}^1_A(T, P)$. By (a), the existence of an irreducible morphism $g : \mathrm{Hom}_A(T, J) \rightarrow \mathrm{Hom}_A(T, I)$ implies that I is isomorphic to a direct summand of $J/\mathrm{soc}\, J$. This shows the necessity.

Conversely, assume that J_A is an indecomposable injective module such that $\tau X \cong \mathrm{Hom}_A(T, J)$ and I a direct summand of $J/\mathrm{soc}\, J$. Then (a) yields an irreducible morphism $\tau X \rightarrow \mathrm{Hom}_A(T, I)$. Hence, in view of (IV.3.8), there exists an irreducible morphism $\mathrm{Hom}_A(T, I) \rightarrow X$. $\square$

There exists a characterisation of separating and splitting tilting modules, due to Hoshino [94]. To prove it, we need the following lemma.

5.5. Lemma. *Let A be an algebra, T_A be a tilting module, and $B = \mathrm{End}\, T_A$. If $M \in \mathcal{T}(T)$ and $N \in \mathcal{F}(T)$, then, for any $j \geq 1$, there is an* *isomorphism*

$$\mathrm{Ext}_A^j(M, N) \cong \mathrm{Ext}_B^{j-1}(\mathrm{Hom}_A(T, M), \mathrm{Ext}_A^1(T, N)).$$

Proof. Let $0 \to N \to I \to N' \to 0$ be a short exact sequence, with I injective. Thus I and N' belong to $\mathcal{T}(T)$. Applying $\mathrm{Hom}_A(T, -)$ yields a short exact sequence in mod B

$$0 \longrightarrow \mathrm{Hom}_A(T, I) \longrightarrow \mathrm{Hom}_A(T, N') \longrightarrow \mathrm{Ext}_A^1(T, N) \longrightarrow 0.$$

Applying the functor $\mathrm{Hom}_B(\mathrm{Hom}_A(T, M), -)$, we obtain the long exact cohomology sequence

$$0 \to \mathrm{Hom}_B(\mathrm{Hom}_A(T, M), \mathrm{Hom}_A(T, I)) \to \mathrm{Hom}_B(\mathrm{Hom}_A(T, M), \mathrm{Hom}_A(T, N'))$$
$$\to \mathrm{Hom}_B(\mathrm{Hom}_A(T, M), \mathrm{Ext}_A^1(T, N)) \to \mathrm{Ext}_B^1(\mathrm{Hom}_A(T, M), \mathrm{Hom}_A(T, I))$$
$$\to \ldots$$

$$\ldots \to \mathrm{Ext}_B^j(\mathrm{Hom}_A(T, M), \mathrm{Hom}_A(T, I)) \to \mathrm{Ext}_B^j(\mathrm{Hom}_A(T, M), \mathrm{Hom}_A(T, N'))$$
$$\to \mathrm{Ext}_B^j(\mathrm{Hom}_A(T, M), \mathrm{Ext}_A^1(T, N)) \to \mathrm{Ext}_B^{j+1}(\mathrm{Hom}_A(T, M), \mathrm{Hom}_A(T, I))$$
$$\to \ldots.$$

By the tilting theorem (3.8), we have

$$\mathrm{Ext}_B^j(\mathrm{Hom}_A(T, M), \mathrm{Hom}_A(T, I)) \cong \mathrm{Ext}_A^j(M, I) = 0,$$

for all $j \geq 1$, because I is injective. Then the sequence

$$0 \to \mathrm{Hom}_B(\mathrm{Hom}_A(T, M), \mathrm{Hom}_A(T, I)) \to \mathrm{Hom}_B(\mathrm{Hom}_A(T, M), \mathrm{Hom}_A(T, N'))$$
$$\to \mathrm{Hom}_B(\mathrm{Hom}_A(T, M), \mathrm{Ext}_A^1(T, N)) \to 0$$

is exact, and there is an isomorphism

$$\mathrm{Ext}_B^j(\mathrm{Hom}_A(T, M), \mathrm{Ext}_A^1(T, N)) \cong \mathrm{Ext}_B^j(\mathrm{Hom}_A(T, M), \mathrm{Hom}_A(T, N'))$$

for all $j \geq 1$. Compare this exact sequence with the short exact sequence

$$0 \longrightarrow \mathrm{Hom}_A(M, I) \longrightarrow \mathrm{Hom}_A(M, N') \longrightarrow \mathrm{Ext}_A^1(M, N) \longrightarrow 0$$

obtained by applying the functor $\mathrm{Hom}_A(M, -)$ to the short exact sequence $0 \to N \to I \to N' \to 0$, using the injectivity of I and the fact that $N \in \mathcal{F}(T)$. Because, by the tilting theorem (3.8), there are isomorphisms

$$\mathrm{Hom}_B(\mathrm{Hom}_A(T, M), \mathrm{Hom}_A(T, E)) \cong \mathrm{Hom}_A(M, E),$$
$$\mathrm{Hom}_B(\mathrm{Hom}_A(T, M), \mathrm{Hom}_A(T, N')) \cong \mathrm{Hom}_A(M, N'),$$

by passing to the cokernels, we obtain an isomorphism

$$\operatorname{Hom}_B(\operatorname{Hom}_A(T, M), \operatorname{Ext}_A^1(T, N)) \cong \operatorname{Ext}_A^1(M, N),$$

which is the required statement whenever $j = 1$. Assume now $j \geq 1$. Then the tilting theorem (3.8) again gives

$$\operatorname{Ext}_B^j(\operatorname{Hom}_A(T, M), \operatorname{Ext}_A^1(T, N)) \cong \operatorname{Ext}_B^j(\operatorname{Hom}_A(T, M), \operatorname{Hom}_A(T, N'))$$

$$\cong \operatorname{Ext}_A^j(M, N') \cong \operatorname{Ext}_A^{j+1}(M, N). \quad \square$$

5.6. Theorem. *Let A be an algebra, T_A be a tilting A-module, and $B = \operatorname{End} T_A$.*

(a) *T_A is separating if and only if $\operatorname{pd} X = 1$ for every $X_B \in \mathcal{X}(T)$.*
(b) *T_A is splitting if and only if $\operatorname{id} N = 1$ for every $N_A \in \mathcal{F}(T)$.*

Proof. We only prove (b); (a) follows using that $_BT$ is a tilting module. We first show the sufficiency of the condition. Assume that, for every $N \in \mathcal{F}(T)$, we have $\operatorname{id} N = 1$. Let $X \in \mathcal{X}(T)$ and $Y \in \mathcal{Y}(T)$. Then there exist $M \in \mathcal{T}(T)$ and $N \in \mathcal{F}(T)$ such that $X \cong \operatorname{Ext}_A^1(T, N)$ and $Y \cong \operatorname{Hom}_A(T, M)$. Hence, by (5.5),

$$\operatorname{Ext}_B^1(Y, X) \cong \operatorname{Ext}_B^1(\operatorname{Hom}_A(T, M), \operatorname{Ext}_A^1(T, N)) \cong \operatorname{Ext}_A^2(M, N) = 0,$$

because $\operatorname{id} N = 1$. Therefore, by (1.7), the pair $(\mathcal{X}(T), \mathcal{Y}(T))$ is splitting.

Conversely, assume that $(\mathcal{X}(T), \mathcal{Y}(T))$ is splitting and let $N \in \mathcal{F}(T)$. Take an injective resolution of N

$$0 \longrightarrow N \xrightarrow{d^0} I^0 \xrightarrow{d^1} I^1 \xrightarrow{d^2} I^2 \longrightarrow \cdots .$$

Let $L^0 = \operatorname{Im} d^1$ and $L^1 = \operatorname{Im} d^2$. Then, by (5.5), because $L^1 \in \mathcal{T}(T)$ and $N \in \mathcal{F}(T)$, we have

$$\operatorname{Ext}_A^1(L^1, L^0) \cong \operatorname{Ext}_A^2(L^1, N) \cong \operatorname{Ext}_B^1(\operatorname{Hom}_A(T, L^1), \operatorname{Ext}_A^1(T, N)) = 0,$$

because $\operatorname{Hom}_A(T, L^1) \in \mathcal{Y}(T)$ and $\operatorname{Ext}_A^1(T, N) \in \mathcal{X}(T)$, and $(\mathcal{X}(T), \mathcal{Y}(T))$ is splitting (see (1.7)). This implies that the short exact sequence $0 \to L^0 \to I^1 \to L^1 \to 0$ splits. Therefore, L^0 is injective and consequently $\operatorname{id} N \leq 1$. Finally, because $N \in \mathcal{F}(T)$, N cannot be injective so that $\operatorname{id} N = 1$. $\quad \square$

If A is an algebra and $P(a)$ is simple projective noninjective, then the APR-tilting module $T[a]$ (which is always separating, by (2.8)(c)) is splitting if and only if $\operatorname{id} P(a) = 1$. Moreover, we have the following corollary.

5.7. Corollary. *If* gl.dim $A \leq 1$, *then every tilting A-module is splitting.*

This is the case for the algebras of Examples 3.11 (a) and (c). These algebras are studied in detail in future chapters.

Let T_A be a tilting A-module and let $T_1, \ldots, T_n$ denote the pairwise nonisomorphic indecomposable summands of T. By (3.1), the modules $\mathrm{Hom}_A(T, T_1), \ldots, \mathrm{Hom}_A(T, T_n)$ form a complete set of pairwise nonisomorphic indecomposable projective modules over the algebra $B = \mathrm{End}\, T_A$. It is less easy in general to describe the indecomposable injective B-modules. In the splitting case, however, we have the following result.

5.8. Proposition. *Let A be an algebra, T_A be a splitting tilting module, $B = \mathrm{End}\, T_A$, and $T_1, \ldots, T_n$ be a complete set of pairwise nonisomorphic indecomposable direct summands of T. Assume that the modules $T_1, \ldots, T_m$ are projective, the remaining modules $T_{m+1}, \ldots, T_n$ are not projective and $I_1, \ldots, I_m$ are indecomposable injective A-modules with* soc $I_j \cong T_j/\mathrm{rad}\, T_j$, *for $j = 1, \ldots, m$. Then the right B-modules*

$$\mathrm{Hom}_A(T, I_1), \ldots, \mathrm{Hom}_A(T, I_m), \mathrm{Ext}_A^1(T, \tau T_{m+1}), \ldots, \mathrm{Ext}_A^1(T, \tau T_n)$$

form a complete set of pairwise nonisomorphic indecomposable injective modules.

Proof. It follows from (4.9) that $\mathrm{Hom}_A(T, I_1), \ldots, \mathrm{Hom}_A(T, I_m)$ are paiwise non-isomorphic indecomposable injective B-modules, and belong to $\mathcal{Y}(T)$. If $m = n$, they form a complete set of pairwise nonisomorphic indecomposable injective B-modules.

Assume that $m < n$. Clearly, $\mathrm{Ext}_A^1(T, \tau T_{m+1}), \ldots, \mathrm{Ext}_A^1(T, \tau T_n)$ are pairwise nonisomorphic objects of the torsion class $\mathcal{X}(T_A)$ of $\mathrm{mod}\, B$. It then suffices to show that, for each i such that $m + 1 \leq i \leq n$, the B-module $\mathrm{Ext}_A^1(T, \tau T_i)$ is injective. Indeed, if this is not the case, then there exists an almost split sequence $0 \longrightarrow \mathrm{Ext}_A^1(T, \tau T_i) \longrightarrow F_B \longrightarrow X_B \longrightarrow 0$ in $\mathrm{mod}\, B$. Because, by our assumption, the torsion pair $(\mathcal{X}(T), \mathcal{Y}(T))$ in $\mathrm{mod}\, B$ is splitting and $\mathrm{Ext}_A^1(T, \tau T_i)$ maps to no module from $\mathcal{Y}(T)$, we deduce that $F_B \in \mathcal{X}(T)$, and similarly $X_B \in \mathcal{X}(T)$. Thus, there exist an A-module E and an indecomposable A-module N in $\mathcal{F}(T)$ such that $F_B \cong \mathrm{Ext}_A^1(T, E)$ and $X_B \cong \mathrm{Ext}_A^1(T, N)$, and the almost split exact sequence becomes

$$0 \longrightarrow \mathrm{Ext}_A^1(T, \tau T_i) \longrightarrow \mathrm{Ext}_A^1(T, E) \longrightarrow \mathrm{Ext}_A^1(T, N) \longrightarrow 0.$$

The equivalence $\mathcal{X}(T) \cong \mathcal{F}(T)$ yields a short exact sequence in $\mathcal{F}(T)$

$$0 \longrightarrow \tau T_i \longrightarrow E \longrightarrow N \longrightarrow 0.$$

Because $T_i = \tau^{-1}(\tau T_i) \in \mathcal{T}(T)$, by (1.11)(b), the A-module τT_i is Ext-injective in $\mathcal{F}(T)$. Therefore, the short exact sequence splits, and applying $\mathrm{Ext}_A^1(T, -)$ to it yields a split–almost split sequence, a contradiction. $\square$

VI.6. Torsion pairs induced by tilting modules

It is natural to ask which torsion pairs $(\mathcal{T}, \mathcal{F})$ in a module category mod A are in fact induced by tilting modules, that is, are such that there exists a tilting module T_A such that $\mathcal{T} = \mathcal{T}(T_A)$ and $\mathcal{F} = \mathcal{F}(T_A)$. This is useful in practice, because in many applications it is easier to start by constructing the torsion pairs and then finding the corresponding tilting module. Clearly, because a torsion class induced by a tilting module T is of the form $\operatorname{Gen} T$, we may start our investigation by asking what the properties of a module U are so that the class $\operatorname{Gen} U$ is a torsion class. We need one definition.

An A-module U will be called Gen-**minimal** if, whenever $U = U' \oplus U''$, $U' \notin \operatorname{Gen} U''$. We define dually Cogen-**minimal** modules.

Our first lemma is a partial converse of (1.9).

6.1. Lemma. *Let A be an algebra.*

(a) *Let U be a* Gen-*minimal A-module such that* $\operatorname{Gen} U$ *is a torsion class. Then U is* Ext-*projective in* $\operatorname{Gen} U$.

(b) *Let V be a* Cogen-*minimal A-module such that* $\operatorname{Cogen} V$ *is a torsion-free class. Then V is* Ext-*injective in* $\operatorname{Cogen} V$.

Proof. We only prove (a); the proof of (b) is similar. Under the stated assumptions, let $M \in \operatorname{Gen} U$ be such that $\operatorname{Ext}_A^1(U, M) \neq 0$. Then there exists an indecomposable summand U_0 of U such that $\operatorname{Ext}_A^1(U_0, M) \neq 0$, and hence a nonsplit extension

$$0 \longrightarrow M \overset{u}{\longrightarrow} E \overset{v}{\longrightarrow} U_0 \longrightarrow 0.$$

Because $M, U_0 \in \operatorname{Gen} U$, and $\operatorname{Gen} U$ is a torsion class, we have $E \in \operatorname{Gen} U$, and thus there exists an epimorphism $p : U^m \to E$ for some $m > 0$. Let $U^m = R \oplus U_0^m$; then the composition $f = vp : U^m \to U_0$ can be written as $f = [g, f_1, \dots, f_m]$ with $g \in \operatorname{Hom}_A(R, U_0)$ and $f_i \in \operatorname{End} U_0$ for each i.

The surjectivity of f means that $U_0 = g(R) + \sum_{i=1}^m f_i(U_0)$. Because v is not a retraction, no f_i is an isomorphism, and consequently, $f_i(U_0) \subseteq (\operatorname{rad} \operatorname{End} U_0) \cdot U_0$ (because the indecomposability of U_0 implies that $\operatorname{End} U_0$ is local) for any i such that $1 \leq i \leq m$. So $U_0 = g(R) + (\operatorname{rad} \operatorname{End} U_0) \cdot U_0$. Applying Nakayama's lemma (I.2.2) to the left $\operatorname{End} U_0$-module U_0, we get that $U_0 = g(R)$ so that g is an epimorphism. This, however, contradicts the Gen-minimality of U. Thus $\operatorname{Ext}_A^1(U, M) = 0$ for all M in $\operatorname{Gen} U$. $\square$

6.2. Corollary. *Let A be an algebra.*

(a) *Let U be a* Gen-*minimal A-module. Then* $\operatorname{Gen} U$ *is a torsion class if and only if U is* Ext-*projective in* $\operatorname{Gen} U$.

(b) *Let V be a* Cogen-*minimal A-module. Then* $\operatorname{Cogen} V$ *is a torsion-free class if and only if V is* Ext-*injective in* $\operatorname{Cogen} V$.

Proof. This follows from (1.9) and (6.1). □

6.3. Corollary. *Let A be an algebra and let U be a* Gen-*minimal faithful A-module such that* Gen U *is a torsion class. Then U is a partial tilting module.*

Proof. Because $U \in \operatorname{Gen} U$, (6.1) yields $\operatorname{Ext}_A^1(U, U) = 0$. On the other hand, because U is faithful, by (2.2), we have $DA \in \operatorname{Gen} U$, whereas the Ext-projectivity of U in the torsion class $\operatorname{Gen} U$ implies, by (1.11), that τU lies in the corresponding torsion-free class. Thus, we have $\operatorname{Hom}_A(DA, \tau U) = 0$. Therefore, by (IV.2.7), we have $\operatorname{pd} U \leq 1$. □

6.4. Lemma. *Let A be an algebra.*

(a) *If $\mathcal{T} = \operatorname{Gen} U$ is a torsion class, then the numbers of isomorphism classes of indecomposable* Ext-*projectives in $\mathcal{T}$ and of indecomposable* Ext-*injectives in $\mathcal{T}$ are finite and equal.*

(b) *If $\mathcal{F} = \operatorname{Cogen} V$ is a torsion-free class, then the number of isomorphism classes of indecomposable* Ext-*projectives in $\mathcal{F}$ and of indecomposable* Ext-*injectives in $\mathcal{F}$ are finite and equal.*

Proof. We only prove (a); the proof of (b) is similar. Because there clearly exists a direct summand U_0 of U that is Gen-minimal and such that $\operatorname{Gen} U = \operatorname{Gen} U_0$, we may assume from the start that U is Gen-minimal. Because, on the other hand, U is clearly faithful as an $A/\operatorname{Ann} U$-module and we have embeddings

$$\mathcal{T} \hookrightarrow \operatorname{mod}(A/\operatorname{Ann} U) \hookrightarrow \operatorname{mod} A,$$

we may also assume that U is faithful.

By (6.3) and (6.1), U is a partial tilting module and is Ext-projective in $\mathcal{T}$. Because $DA \in \operatorname{Gen} U$ (by (2.2)), all the indecomposable injective A-modules are torsion and so, by (1.11), they coincide with the indecomposable Ext-injectives in $\mathcal{T}$.

Let $u_1, \ldots, u_d$ be a basis of the K-vector space $\operatorname{Hom}_A(A, U)$ and consider the homomorphism $u = \begin{bmatrix} u_1 \\ \vdots \\ u_d \end{bmatrix} : A_A \longrightarrow U_A^d$. Because U is faithful, according to (2.2), the map u is injective. We thus have a short exact sequence

$$0 \longrightarrow A \xrightarrow{u} U^d \longrightarrow U' \longrightarrow 0,$$

where $U' = \operatorname{Coker} u$. Notice that $U' \in \mathcal{T}$. Also, because $\operatorname{pd} U \leq 1$, we have $\operatorname{pd} U' \leq 1$. We now show that U' is Ext-projective in $\mathcal{T}$. Let $M \in \mathcal{T}$ and apply $\operatorname{Hom}_A(-, M)$ to the preceding sequence. This yields an exact sequence

$$0 \longrightarrow \operatorname{Hom}_A(U', M) \longrightarrow \operatorname{Hom}_A(U^d, M) \xrightarrow{\operatorname{Hom}_A(u,M)} \operatorname{Hom}_A(A, M)$$
$$\longrightarrow \operatorname{Ext}^1_A(U', M) \longrightarrow 0,$$

because $\operatorname{Ext}^1_A(U^d, M) = 0$ due to the Ext-projectivity of U in $\mathcal{T}$. We claim that $\operatorname{Hom}_A(u, M)$ is surjective. Because $M \in \mathcal{T}$, there exists an epimorphism $p : U^m \to M$ for some $m > 0$. Because A_A is a projective module, the homomorphism $\operatorname{Hom}_A(A, p) : \operatorname{Hom}_A(A, U^m) \to \operatorname{Hom}_A(A, M)$ is surjective. On the other hand, it follows from the definition of u that $\operatorname{Hom}_A(u, U^m) : \operatorname{Hom}_A(U^d, U^m) \to \operatorname{Hom}_A(A, U^m)$ is surjective. Therefore the composition $\operatorname{Hom}_A(u, p) : \operatorname{Hom}_A(U^d, U^m) \to \operatorname{Hom}_A(A, M)$ is surjective. Because $\operatorname{Hom}_A(u, p) = \operatorname{Hom}_A(u, M) \circ \operatorname{Hom}_A(U^d, p)$, this shows that $\operatorname{Hom}_A(u, M)$ is surjective. Therefore $\operatorname{Ext}^1_A(U, M) = 0$, and hence U' is Ext-projective in $\mathcal{T}$.

We deduce that $T_A = U \oplus U'$ is a tilting module. Indeed, $\operatorname{pd} T \le 1$ and the Ext-projectivity of both U and U' implies that $\operatorname{Ext}^1_A(T, T) = 0$. Finally, the short exact sequence $0 \longrightarrow A \xrightarrow{u} U^d \longrightarrow U' \longrightarrow 0$ shows that T is indeed a tilting module. It follows from (2.5) that $\mathcal{T}(T) = \operatorname{Gen} T = \operatorname{Gen} U = \mathcal{T}$. By (2.5)(d), the pairwise nonisomorphic indecomposable Ext-projectives in $\mathcal{T}$ coincide with the pairwise nonisomorphic indecomposable direct summands of T. Therefore, by (4.4), their number equals the rank of $K_0(A)$ and thus equals the number of pairwise nonisomorphic indecomposable Ext-injectives in $\mathcal{T} = \mathcal{T}(T)$.   □

6.5. Theorem. *Let A be an algebra and let $(\mathcal{T}, \mathcal{F})$ be a torsion pair in* $\operatorname{mod} A$. *Then there exists a tilting module T_A such that $\mathcal{T} = \mathcal{T}(T_A)$ if and only if $\mathcal{T} = \operatorname{Gen} M$ for some A-module M, and $\mathcal{T}$ contains the injectives.*

Proof. Because the necessity is obvious, we only show the sufficiency. Let $\mathcal{T}$ be a torsion class containing all the injectives such that $\mathcal{T} = \operatorname{Gen} M$ for some A-module M. Let $T_1, \dots, T_t$ be a complete set of pairwise nonisomorphic indecomposable Ext-projectives in $\mathcal{T}$, and let $T_A = \bigoplus_{i=1}^t T_i$. We claim that T_A is a tilting module. Indeed, the Ext-projectivity of T_A in $\mathcal{T}$ implies that $\operatorname{Ext}^1_A(T, T) = 0$. On the other hand,

$$\operatorname{Hom}_A(DA, \tau T) = \bigoplus_{i=1}^t \operatorname{Hom}_A(DA, \tau T_i) = 0$$

(because τT_i is zero or torsion-free, by (1.11)(a), whereas $DA \in \mathcal{T}$ by hypothesis). Hence, by (IV.2.7), $\operatorname{pd} T \le 1$. Also, by (6.4), t equals the number of pairwise nonisomorphic indecomposable injective A-modules. Therefore t equals the rank of $K_0(A)$ and so T is a tilting module, by (4.4).

Because M is itself Ext-projective in $\mathcal{T}$, its indecomposable direct summands are also summands of T. Therefore $\mathcal{T} \subseteq \mathcal{T}(T)$. Because $T \in \operatorname{Gen} M$, we also have $\mathcal{T}(T) \subseteq \mathcal{T}$ so that $\mathcal{T}(T) = \mathcal{T}$.   □

We give an application of this theorem, but first we prove two important corollaries. The first is obvious.

6.6. Corollary. *Let A be a representation–finite algebra and $(\mathcal{T}, \mathcal{F})$ be a torsion pair in $\operatorname{mod} A$. Then there exists a tilting module T_A such that $\mathcal{T} = \mathcal{T}(T_A)$ and $\mathcal{F} = \mathcal{F}(T_A)$ if and only if $\mathcal{T}$ contains the injectives.*

Proof. Let $\{M_1, \dots, M_r\}$ be a complete set of pairwise nonisomorphic indecomposable modules in $\mathcal{T}$ (such a set is finite, because A is representation–finite), and let $M = M_1 \oplus \dots \oplus M_r$. Then $\mathcal{T} = \operatorname{Gen} M$, and the required equivalence is a direct consequence of (2.5) and (6.5).   □

6.7. Corollary. *Let B be an algebra and $(\mathcal{X}, \mathcal{Y})$ be a torsion pair in $\operatorname{mod} B$. Then there exists an algebra A and a tilting module T_A such that $B = \operatorname{End} T_A$, $\mathcal{X} = \mathcal{X}(T_A)$ and $\mathcal{Y} = \mathcal{Y}(T_A)$ if and only if $\mathcal{Y} = \operatorname{Cogen} Y$ for some B-module Y, and $\mathcal{Y}$ contains the projectives.*

Proof. We first show the necessity. Let A be an algebra and T_A be a tilting module such that $B = \operatorname{End} T_A$. It follows from (3.1)(b) that $\mathcal{Y}(T_A)$ contains the projective B-modules. We claim that $\mathcal{Y}(T_A)$ is the class cogenerated by the B-module $D(_BT) = \operatorname{Hom}_A(T, DA) \in \mathcal{Y}(T_A)$. Let $Y \in \mathcal{Y}(T_A)$; there exists an A-module $M \in \mathcal{T}(T)$ such that $Y = \operatorname{Hom}_A(T, M)$. There exists an injective A-module U and a monomorphism $M \to U$ and hence a monomorphism $Y = \operatorname{Hom}_A(T, M) \to \operatorname{Hom}_A(T, U)$. Because $\operatorname{Hom}_A(T, U) \in \operatorname{add} D(_BT)$, we deduce from (3.3)(a) that $\mathcal{Y}(T) \subseteq \operatorname{Cogen} D(_BT)$. Because, on the other hand, $D(_BT) \in \mathcal{Y}(T)$, we have established our claim.

To prove the sufficiency, we notice that, by (6.4), the torsion class of left B-modules $D\mathcal{Y}$ is induced by a tilting module, that is, there exists a left B-module $_BT$ such that $D\mathcal{Y} = \mathcal{T}(_BT)$ and $D\mathcal{X} = \mathcal{F}(_BT)$. Letting $A = \operatorname{End}(_BT)^{\operatorname{op}}$, we deduce from (3.3) that T_A is a tilting A-module and $B = \operatorname{End} T_A$. Moreover, by (3.6), $\mathcal{Y}(T_A) = D\mathcal{T}(_BT) = \mathcal{Y}$ and $\mathcal{X}(T_A) = D\mathcal{F}(_BT) = \mathcal{X}$.   □

To apply Corollary 6.7 in examples, we need the following easy computational lemma.

6.8. Lemma. *Assume that the torsion pair $(\mathcal{X}, \mathcal{Y})$ in $\operatorname{mod} B$ satisfies the equivalent conditions of (6.7). Then $D(_BT)$ equals the direct sum of a complete set of pairwise nonisomorphic indecomposable Ext-injectives in $\mathcal{Y}$.*

Proof. We recall that $D(_BT) = \operatorname{Hom}_A(T, DA)$ equals the direct sum of modules of the form $\operatorname{Hom}_A(T, I(a))$, where $I(a)$ runs over a complete set of indecomposable injective A-modules. Let $I(a)_A$ be indecomposable injective. By the connecting lemma (4.9), either $\mathrm{Hom}_A(T, I(a))$ is injective in mod B (if the corresponding indecomposable projective lies in add T_A) or $\tau^{-1}\mathrm{Hom}_A(T, I(a)) \in \mathcal{X}$. By (1.11), $\mathrm{Hom}_A(T, I(a))$ is Ext-projective in $\mathcal{Y}$.

Conversely, let Y be indecomposable Ext-injective in $\mathcal{Y}$; then $\tau^{-1}Y \in \mathcal{X}$. If $\tau^{-1}Y \neq 0$; then, by (4.8), there exists an indecomposable injective A-module $I(a)$ such that $Y \cong \mathrm{Hom}_A(T, I(a))$. Assume now that $\tau^{-1}Y = 0$, that is, Y is injective. Because $Y \in \mathcal{Y}$, there exists an indecomposable A-module $M \in \mathcal{T}(T_A)$ such that $Y \cong \mathrm{Hom}_A(T, M)$. Let $M \to E$ be an injective envelope of M in mod A. Applying $\mathrm{Hom}_A(T, -)$ to the short exact sequence

$$0 \longrightarrow M \longrightarrow E \longrightarrow E/M \longrightarrow 0$$

yields an exact sequence in mod B

$$0 \longrightarrow Y \longrightarrow \mathrm{Hom}_A(T, E) \longrightarrow \mathrm{Hom}_A(T, E/M) \longrightarrow 0,$$

because $\mathrm{Ext}^1_A(T, M) = 0$. Because, by hypothesis, Y is Ext-injective in $\mathcal{Y}$ and the previous sequence lies in $\mathcal{Y}$, it splits. Hence Y is isomorphic to a direct summand of $\mathrm{Hom}_A(T, E)$, that is, there exists an indecomposable summand $I(a)$ of E such that $Y \cong \mathrm{Hom}_A(T, I(a))$. $\qquad\square$

Assume thus that $(\mathcal{X}, \mathcal{Y})$ satisfies the conditions of (6.7). We indicate how to find an algebra A and a tilting module T_A from which $(\mathcal{X}, \mathcal{Y})$ arises. We first compute $D(_BT)$ using (6.8): Let $Y_1, \ldots, Y_n$ be a complete set of pairwise nonisomorphic indecomposable Ext-injectives in $\mathcal{Y}$, then $D(_BT) = \bigoplus_{i=1}^n Y_i$. We next find

$$A = \mathrm{End}_{B^{\mathrm{op}}}(_BT) = \mathrm{End}_B(D(_BT)) = \mathrm{End}_B\left(\bigoplus_{i=1}^n Y_i\right).$$

In doing the last calculation, we associate each of the Y_i to a point in the quiver of A. Thus, without loss of generality, we may assume that $Y_i = \mathrm{Hom}_A(T, I(i))$ for each i such that $1 \leq i \leq n$. Letting $T = \bigoplus_{j=1}^n T_j$, we have

$$\begin{aligned}
(T_j)_i &= \mathrm{Hom}_A(P(i)_A, T_j) \\
&\cong D\mathrm{Hom}_A(T_j, I(i)) \\
&\cong D\mathrm{Hom}_B(\mathrm{Hom}_A(T, T_j), \mathrm{Hom}_A(T, I(i))) \\
&\cong D\mathrm{Hom}_B(P(j)_B, Y_i).
\end{aligned}$$

Thus, in particular, $\dim_K(T_j)_i$ is the jth coordinate of Y_i. This gives $\dim T_j$. The method is explained in the following example.

6.9. Examples. (a) Let B be given by the quiver

$$\underset{1}{\circ}\xleftarrow{\;\mu\;}\underset{2}{\circ}\xleftarrow{\;\lambda\;}\underset{3}{\circ}$$

bound by $\lambda\mu = 0$ and $(\mathcal{X}, \mathcal{Y})$ be the shown torsion pair in $\operatorname{mod} B$ (compare with (3.11)(a))

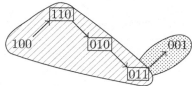

where $\mathcal{Y}$ is shaded as ⬭ and $\mathcal{X}$ as ⬬. Clearly, $(\mathcal{X}, \mathcal{Y})$ satisfies the conditions of (6.7). To find an algebra A and a tilting module T_A from which $(\mathcal{X}, \mathcal{Y})$ arises, we consider the indecomposable Ext-injectives in $\mathcal{Y}$; these are $Y_1 = 110$, $Y_2 = 010$, $Y_3 = 011$. Thus $D(_BT) = 110 \oplus 010 \oplus 011$. Hence $A = \operatorname{End}_{B^{\mathrm{op}}}(_BT) = \operatorname{End}_B(D(_BT)) = \operatorname{End}_B(\bigoplus_{i=1}^{3} Y_i)$ is given by the quiver

$$\underset{1}{\circ}\xleftarrow{\hspace{1cm}}\underset{2}{\circ}\xleftarrow{\hspace{1cm}}\underset{3}{\circ}$$

where the point i corresponds to Y_i (for each i with $1 \le i \le 3$). To recover T_A, we notice that, in the preceding notation,

$$\operatorname{Hom}_A(T, I(1)) = 110, \quad \operatorname{Hom}_A(T, I(2)) = 010, \quad \operatorname{Hom}_A(T, I(3)) = 011.$$

Thus, if one writes $T = T_1 \oplus T_2 \oplus T_3$, with T_1, T_2, T_3 indecomposable, one gets

$$T_1 = 100, \qquad T_2 = 111, \qquad T_3 = 001.$$

(b) Let B be given by the quiver

$$\underset{1}{\circ}\xleftarrow{\;\eta\;}\underset{2}{\circ}\xleftarrow{\;\nu\;}\underset{3}{\circ}\xleftarrow{\;\mu\;}\underset{4}{\circ}\xleftarrow{\;\lambda\;}\underset{5}{\circ}$$

bound by $\lambda\mu\nu\eta = 0$ and $(\mathcal{X}, \mathcal{Y})$ be the shown torsion pair in $\operatorname{mod} B$ (compare with (3.11)(b))

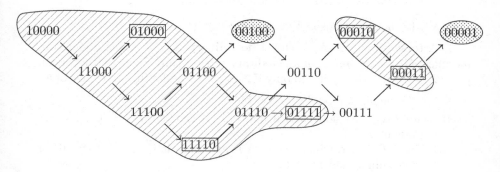

where $\mathcal{Y}$ is shaded as 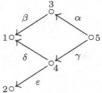 and $\mathcal{X}$ as ⬤. Clearly, $(\mathcal{X}, \mathcal{Y})$ satisfies the conditions of (6.7). The indecomposable Ext-injective modules in $\mathcal{Y}$ are $Y_1 = 11110$, $Y_2 = 01000$, $Y_3 = 00010$, $Y_4 = 01111$, and $Y_5 = 00011$. Thus, $A = \operatorname{End}(\bigoplus_{i=1}^{5} Y_i)$ is given by the quiver

bound by $\alpha\beta = \gamma\delta$ and $\gamma\varepsilon = 0$, where the point i corresponds to Y_i (for each i with $1 \le i \le 5$). To recover T_A, we notice that

$$\operatorname{Hom}_A(T, I(1)) = 11110, \operatorname{Hom}_A(T, I(2)) = 01000, \operatorname{Hom}_A(T, I(3)) = 00010,$$
$$\operatorname{Hom}_A(T, I(4)) = 01111, \operatorname{Hom}_A(T, I(5)) = 00011.$$

Thus if one writes $T = T_1 \oplus T_2 \oplus T_3 \oplus T_4 \oplus T_5$, with T_1, T_2, T_3, T_4, T_5 indecomposable, one gets

$$T_1 = \begin{smallmatrix} & 0 & \\ 1 & & 0 \\ & 0 & \end{smallmatrix}, \quad T_2 = \begin{smallmatrix} & 0 & \\ 1 & & 0 \\ & 1 & \end{smallmatrix}, \quad T_3 = \begin{smallmatrix} & 0 & \\ 1 & & 0 \\ & 1 & \end{smallmatrix}, \quad T_4 = \begin{smallmatrix} & 1 & \\ 1 & & 1 \\ & 1 & \end{smallmatrix}, \quad T_5 = \begin{smallmatrix} & 0 & \\ 0 & & 1 \\ & 1 & \end{smallmatrix}$$

VI.7. Exercises

1. Show that a pair $(\mathcal{T}, \mathcal{F})$ of full subcategories of mod A is a torsion pair if and only if it satisfies the following four conditions:

(a) $\mathcal{T} \cap \mathcal{F} = \{0\}$;

(b) $\mathcal{T}$ is closed under images;

(c) $\mathcal{F}$ is closed under submodules; and

(d) for every module M, there exists a short exact sequence $0 \to M' \to M \to M'' \to 0$ with $M' \in \mathcal{T}$ and $M'' \in \mathcal{F}$.

2. Verify the assertions in Example 1.2 (a).

3. A torsion pair $(\mathcal{T}, \mathcal{F})$ is called **hereditary** if $\mathcal{T}$ is closed under submodules. Give an example of a hereditary torsion pair. Show that a torsion pair $(\mathcal{T}, \mathcal{F})$ is hereditary if and only if $\mathcal{F}$ is closed under injective envelopes.

4. Let T_A be an A-module. Show that:

(a) $\operatorname{Gen} T$ is a torsion class if and only if $\operatorname{Ext}_A^1(T, T'') = 0$ for every quotient T'' of T.

(b) $\operatorname{Cogen} T$ is a torsion-free class if and only if $\operatorname{Ext}_A^1(T', T) = 0$ for every submodule T' of T.

5. Assume that Gen T is a torsion class for some module T_A. Show that τT belongs to the corresponding torsion-free class.

6. Assume that Gen T is a torsion class for some module T_A.

(a) Show that if T_A is faithful, then T_A is a partial tilting module.
(b) Give an example showing that if T_A is not faithful, then T_A is generally not a partial tilting module.

7. Let T_A be a partial tilting module. Show that:

(a) If $\mathcal{T}$ is a torsion class such that T_A is Ext-projective in $\mathcal{T}$, then Gen $T \subseteq \mathcal{T} \subseteq \mathcal{T}(T)$.
(b) $\mathcal{T}(T)$ is induced by a tilting module having T as a summand.

8. Let T_A be a partial tilting module and E be the middle term of Bongartz's exact sequence. Show that any indecomposable direct summand E' of E is projective or satisfies $\operatorname{Hom}_A(E', T) \neq 0$.

9. An A-module M is called **sincere** if $\operatorname{Hom}_A(P, M) \neq 0$ for any projective A-module P. Show that any faithful module is sincere (consequently, any tilting module is sincere).

10. Let T_A be a tilting module. Show that any indecomposable projective-injective A-module is a direct summand of T.

11. Let T_A be a tilting module and $(\mathcal{T}(T), \mathcal{F}(T))$ be the induced torsion pair in mod A. Show that if $M_3 \to M_2 \to M_1 \to M_0$ is exact with $M_i \in \mathcal{T}(T)$ for all i, then the induced sequence

$$\operatorname{Hom}_A(T, M_2) \longrightarrow \operatorname{Hom}_A(T, M_1) \longrightarrow \operatorname{Hom}_A(T, M_0)$$

is exact.

12. Let T_A be a tilting module and $\mathcal{X}(T)$ be the induced torsion class in mod B. Show that $\mathcal{X}(T) = \operatorname{Gen} \operatorname{Ext}_A^1(T, A)$.

13. Let T_A be a tilting module and E_A be injective. Show that if $N \in \mathcal{F}(T)$, then we have a functorial isomorphism

$$\operatorname{Hom}_A(N, E) \cong \operatorname{Ext}_B^1(\operatorname{Ext}_A^1(T, N), \operatorname{Hom}_A(T, E)).$$

14. Let A be a K-algebra given by each of the bound quivers (i)–(iv).

(a) Verify that the given module T_A is a tilting module.
(b) Compute the bound quiver of $B = \operatorname{End} T_A$.
(c) Illustrate in $\Gamma(\operatorname{mod} A)$ and $\Gamma(\operatorname{mod} B)$ the classes $\mathcal{T}(T)$, $\mathcal{F}(T)$, $\mathcal{X}(T)$, and $\mathcal{Y}(T)$.

(d) Describe explicitly the equivalences $\mathcal{T}(T) \cong \mathcal{Y}(T)$, $\mathcal{X}(T) \cong \mathcal{F}(T)$.

(e) Compute the global dimensions of A and B.

(f) Describe all connecting sequences in $\operatorname{mod} A$ and $\operatorname{mod} B$. For which ones is the canonical sequence of the middle term not split?

(g) Find the matrix $\mathbf{F}$ of the isomorphism $K_0(A) \to K_0(B)$, the matrices $\mathbf{A}$ and $\mathbf{B}$ of the Euler characteristics for A and B, respectively, and verify the relation $\mathbf{A} = \mathbf{F}^t \mathbf{B} \mathbf{F}$.

(i)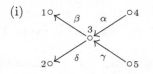

bound by $\alpha\beta = 0$, $\gamma\delta = 0$,

$$T_A = {}^1_0{}^0_0{}^0_0 \oplus {}^1_0{}^1_1{}^0_1 \oplus {}^0_0{}^1_1{}^1_1 \oplus {}^0_1{}^1_1{}^1_0 \oplus {}^0_1{}^1_1{}^1_0$$

(ii)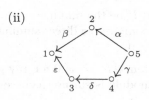

bound by $\alpha\beta = \gamma\delta\varepsilon$,

$$T_A = 1{}^0_1{}^0_1 0 \oplus 1{}^1_1{}^1_1 1 \oplus 0{}^0_1{}^0_0 0 \oplus 0{}^0_1{}^0_1 0 \oplus 0{}^0_1{}^0_1$$

(iii)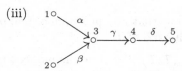

bound by $\gamma\delta = 0$,

$$T_A = {}^0_0 001 \oplus {}^0_0 011 \oplus {}^1_0 110 \oplus {}^0_1 110 \oplus {}^1_1 110$$

(iv)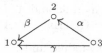

bound by $\beta\alpha = 0$,

$$T_A = \begin{pmatrix} 1 \\ 2 \\ 1 \end{pmatrix} \oplus (1) \quad \text{(in the notation of (V.2}$$

15. Let A be given by the quiver

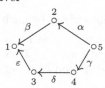

bound by $\alpha\beta = 0$. Find all (nontrivial, multiplicity-free) tilting A-modules and compute the bound quiver of the endomorphism algebra of each.

16. Let A be given by the quiver

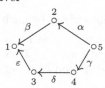

bound by $\alpha\beta = 0$, $\gamma\delta = 0$, and $\delta\varepsilon = 0$. Compute the bound quiver of the endomorphism algebra B of the unique APR-tilting module and the Auslander–Reiten quivers of each of A and B and then describe the equivalences $\mathcal{T}(T) \cong \mathcal{Y}(T)$, $\mathcal{F}(T) \cong \mathcal{X}(T)$.

17. Repeat Exercise 16 with A given by the quiver

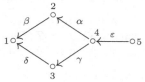

bound by $\alpha\beta = \gamma\delta$, $\varepsilon\alpha = 0$, and $\varepsilon\gamma = 0$.

18. Let T_A be a tilting module and $B = \operatorname{End} T_A$. Show that if $J_B \in \mathcal{Y}(T)$ is an indecomposable injective B-module, then there exists an indecomposable injective A-module E_A such that $J \cong \operatorname{Hom}_A(T, E)$ and the indecomposable projective P_A such that $P/\operatorname{rad} P \cong \operatorname{soc} I$ and P_A are not in $\operatorname{add} T$.

19. Let T_A be a tilting module and $B = \operatorname{End} T_A$. If, for a point a of Q_A, both $P(a)$ and $I(a)$ are in $\operatorname{add} T$, then show that $\operatorname{Hom}_A(T, I(a))$ is a projective-injective B-module and, conversely, show that every indecomposable projective-injective B-module is of this form.

20. Let T_A be a tilting module. Prove the following implications:
  (a) If $N \in \mathcal{F}(T)$, then $\operatorname{pd} \operatorname{Ext}^1_A(T, N) \leq 1 + \max(1, \operatorname{pd} N)$.
  (b) If $M \in \mathcal{T}(T)$, then $\operatorname{id} \operatorname{Hom}_A(T, M) \leq 1 + \operatorname{id} M$.
  (c) If $N \in \mathcal{F}(T)$, then $\operatorname{id} \operatorname{Ext}^1_A(T, N) \leq \operatorname{id} N$.
Hint: See the remark following (4.2).

21. The following construction, due to Brenner and Butler, generalises that of the APR-tilting modules. Let A be an algebra and $S(a)$ be a simple A-module such that: (i) $\operatorname{pd} \tau^{-1}S(a) \leq 1$ and (ii) $\operatorname{Ext}^1_A(S(a), S(a)) = 0$. Show that
  (a) $T = \tau^{-1}S(a) \oplus (\bigoplus_{b \neq a} P(b))$ is a tilting module,
  (b) $\mathcal{F}(T) = \operatorname{add} S(a)$.
Let A be as in Exercise 14 (ii). Find a simple A-module $S(a)$ satisfying (i) and (ii), construct the corresponding tilting module T as in (a), compute the bound quiver and the Auslander–Reiten quiver of $B = \operatorname{End} T$, and describe the equivalences $\mathcal{T}(T) \cong \mathcal{Y}(T)$, $\mathcal{F}(T) \cong \mathcal{X}(T)$.

22. An A-module T_A is called a **partial cotilting module** if T satisfies
(CT1) $\operatorname{id} T \leq 1$ and
(CT2) $\operatorname{Ext}^1_A(T, T) = 0$

and a **cotilting module** if it also satisfies

(CT3) the number of pairwise nonisomorphic indecomposable summands of T equals the rank of $K_0(A)$.

Show that T_A is a (partial) cotilting module if and only if $_A DT$ is a (partial) tilting module. Then state and prove the analogues for (partial) cotilting modules of the results of Sections 2 and 3.

23. Let $(\mathcal{T}, \mathcal{F})$ be a torsion pair in $\mathrm{mod}\,A$. Show that there exists a tilting module T_A such that $\mathcal{T} = \mathcal{T}(T_A)$, $\mathcal{F} = \mathcal{F}(T_A)$ if and only if $\mathcal{F}$ is cogenerated by a module N such that $\mathrm{pd}\,(\tau^{-1}N) \leq 1$.

24. Let A be given by the quiver

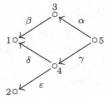

bound by $\alpha\beta = \gamma\delta$.

(a) Show that $\mathcal{X} = \mathrm{add}\left\{{}^0_1{}^0_{1}\oplus{}^0_1{}^0_{}{}_1\oplus{}^0_{}{}^1_1{}_1\oplus{}^0_1{}^1_{0}\oplus{}^0_0{}^0_{0}{}_1\right\}$ is a torsion-free class in $\mathrm{mod}\,A$.

(b) Find a class $\mathcal{Y}$ such that $(\mathcal{X}, \mathcal{Y})$ is a torsion pair in $\mathrm{mod}\,A$.

(c) Show that there exists an algebra C and a tilting module T_C such that $A = \mathrm{End}\,T_C$, $\mathcal{X} = \mathcal{X}(T_C)$, and $\mathcal{Y} = \mathcal{Y}(T_C)$. Compute the algebra C and the module T_C.

Chapter VII

Representation–finite hereditary algebras

As we saw in Chapter II, any basic and connected finite dimensional algebra A over an algebraically closed field K admits a presentation as a bound quiver algebra $A \cong KQ/\mathcal{I}$, where Q is a finite connected quiver and $\mathcal{I}$ is an admissible ideal of KQ. It is thus natural to study the representation theory of the algebras of the form $A \cong KQ$, that is, of the path algebras of finite, connected, and acyclic quivers. It turns out that an algebra A is of this form if and only if it is hereditary, that is, every submodule of a projective A-module is projective. We are thus interested in the representation theory of hereditary algebras. In [72], Gabriel showed that a connected hereditary algebra is representation–finite if and only if the underlying graph of its quiver is one of the Dynkin diagrams $\mathbb{A}_m$ with $m \geq 1$; $\mathbb{D}_n$ with $n \geq 4$; and $\mathbb{E}_6$, $\mathbb{E}_7$, $\mathbb{E}_8$, that appear also in Lie theory (see, for instance, [41]). Later, Bernstein, Gelfand, and Ponomarev [32] gave a very elegant and conceptual proof underlining the links between the two theories, by applying the nice concept of reflection functors. In this chapter, using reflection functors (which may now be thought of as tilting functors), we prove Gabriel's theorem and show how to compute all the (isomorphism classes of) indecomposable modules over a representation–finite hereditary algebra.

VII.1. Hereditary algebras

This introductory section is devoted to defining and giving various characterisations of hereditary algebras. In particular, we show that the hereditary algebras coincide with the path algebras of finite, connected, and acyclic quivers. Throughout, we let A denote a basic and connected finite dimensional algebra over an algebraically closed field K.

1.1. Definition. An algebra A is said to be **right hereditary** if any right ideal of A is projective as an A-module.

Left hereditary algebras are defined dually. It is not clear a priori whether a right hereditary algebra is also left hereditary, though we show in (1.4) that this is the case. The most obvious example of a right (and left) hereditary algebra is provided by the class of semisimple algebras; because any right (or left) module over a semisimple algebra is projective, then so is any right (or left, respectively) ideal of the algebra. On the other hand, let A be the full 2×2 lower triangular matrix algebra $A = \begin{bmatrix} K & 0 \\ K & K \end{bmatrix}$; see (I.2.4). Then, denoting by $e_1 = \left(\begin{smallmatrix} 1 & 0 \\ 0 & 0 \end{smallmatrix}\right)$ and $e_2 = \left(\begin{smallmatrix} 0 & 0 \\ 0 & 1 \end{smallmatrix}\right)$ the matrix idempotents, an immediate calculation shows that the only proper right ideals are e_1A, e_2A, and $e_{21}K = \left(\begin{smallmatrix} 0 & 0 \\ K & 0 \end{smallmatrix}\right) \cong e_1A$, where $e_{21} = \left(\begin{smallmatrix} 0 & 0 \\ 1 & 0 \end{smallmatrix}\right)$. Because e_1A and e_2A are direct summands of A_A, all these are projective A-modules and A is right hereditary.

The following theorem, due to Kaplansky [100], is fundamental. We warn the reader that, contrary to our custom, the modules we consider in (1.2)–(1.4) are not necessarily finitely generated.

1.2. Theorem. *Let A be a right hereditary algebra. Every submodule of a free A-module is isomorphic to a direct sum of right ideals of A.*

Proof. Let L be a free A-module with basis $(e_\lambda)_{\lambda \in \Lambda}$ and M be a submodule of L. We wish to show that M is isomorphic to a direct sum of right ideals of A. Without loss of generality, we may assume the index set Λ to be well-ordered. For each $\lambda \in \Lambda$, let $L_\lambda = \bigoplus_{\mu < \lambda}(e_\mu A)$. Then $L_0 = 0$ and $L_{\lambda+1} = \bigoplus_{\mu \leq \lambda}(e_\mu A) = L_\lambda \oplus (e_\lambda A)$. An element $x \in M \cap L_{\lambda+1}$ has a unique expression of the form $x = y + e_\lambda a$ with $y \in L_\lambda$ and $a \in A$. We may thus define an A-module homomorphism $f_\lambda : M \cap L_{\lambda+1} \to A$ by $x \mapsto a$, and hence we have a short exact sequence

$$0 \longrightarrow M \cap L_\lambda \longrightarrow M \cap L_{\lambda+1} \xrightarrow{f_\lambda} \operatorname{Im} f_\lambda \longrightarrow 0.$$

Because $\operatorname{Im} f_\lambda$ is a right ideal of the right hereditary algebra A, it is projective and the sequence splits. Hence there exists a submodule N_λ of $M \cap L_{\lambda+1}$, isomorphic to $\operatorname{Im} f_\lambda$ and such that $M \cap L_{\lambda+1} = (M \cap L_\lambda) \oplus N_\lambda$. To complete the proof, it suffices to show that $M \cong \bigoplus_{\lambda \in \Lambda} N_\lambda$.

First, we show that M is equal to its submodule $N = \sum_{\lambda \in \Lambda} N_\lambda$. Because L equals the union of the increasing chain of submodules $(L_\lambda)_{\lambda \in \Lambda}$, for each $x \in L$, there exists a least index $\lambda \in \Lambda$ such that $x \in L_{\lambda+1}$. Denote this index by μ_x. If $N \subsetneq M$, there exists $x \in M$ such that $x \notin N$. Let μ denote the least μ_x with $x \in M$, $x \notin N$ and take $y \in M$ such that $y \notin N$ and $\mu = \mu_y$. We have $y \in M \cap L_{\mu+1}$ hence $y = u + v$ with $u \in M \cap L_\mu$ and $v \in N_\mu$. Therefore $u = y - v \in M$ and $u \notin N$ (otherwise, $y \in N$, which is a contradiction). But, on the other hand, $u \in M \cap L_\mu$ gives $\mu_u < \mu$, and this contradicts the minimality of μ. Hence $M = \sum_{\lambda \in \Lambda} N_\lambda$.

There remains to show that the sum $\sum_{\lambda \in \Lambda} N_\lambda$ is direct. Assume that $x_1 + \ldots + x_n = 0$ with $x_i \in N_{\lambda_i}$, where we can suppose that $\lambda_1 < \ldots < \lambda_n$. Then $x_1 + \ldots + x_{n-1} = -x_n \in (M \cap L_{\lambda_n}) \cap N_{\lambda_n} = 0$ gives $x_n = 0$. By descending induction, $x_i = 0$ for each i. $\qquad\square$

1.3. Corollary. *Let A be a right hereditary algebra. Every submodule of a projective A-module is projective.*

Proof. Indeed, any projective module is isomorphic to a direct summand of a free module. $\qquad\square$

We are now able to state and prove our first characterisation of right hereditary algebras.

1.4. Theorem. *Let A be an algebra. The following conditions are equivalent:*

(a) *A is right hereditary.*

(b) *The global dimension of A is at most one.*

(c) *Every submodule of a projective right A-module is projective.*

(d) *Every quotient of an injective right A-module is injective.*

(e) *Every submodule of a finitely generated projective right A-module is projective.*

(f) *Every quotient of a finitely generated injective right A-module is injective.*

(g) *The radical of any indecomposable finitely generated projective right A-module is projective.*

(h) *The quotient of any indecomposable finitely generated injective right A-module by its socle is injective.*

Proof. (a) is equivalent to (c). Indeed, it follows from (1.3) that (a) implies (c). The converse is obvious.

(b) is equivalent to (c). If gl.dim $A \leq 1$ and M_A is a submodule of a projective module P_A then, in the short exact sequence

$$0 \longrightarrow M \longrightarrow P \longrightarrow P/M \longrightarrow 0,$$

we have pd $(P/M) \leq 1$; hence, by (A.4.7) of the Appendix, M is projective. Conversely, if every submodule of a projective module is projective, let N be an arbitrary A-module. Then there exists a projective module P_A and an epimorphism $f : P \to N$. Because Ker f is a submodule of P, it is projective. Hence the exact sequence $0 \longrightarrow \text{Ker } f \longrightarrow P \xrightarrow{f} N \longrightarrow 0$ gives pd $N \leq 1$. Consequently, gl.dim $A \leq 1$.

Obviously, (c) implies (e) and (e) implies (a), because A_A is finitely generated as an A-module.

(e) is equivalent to (g). The necessity being obvious, let us show the sufficiency. Let P be a finitely generated projective A-module and M be a submodule of P. We prove that M is projective by induction on $d = \dim_K P$. If $d = 1$, there is nothing to show. Assume $d > 1$ and that the statement holds for every finitely generated projective A-module of dimension $< d$. The module P can be written in the form $P = P_1 \oplus P_2$, where P_1 is indecomposable and P_2 may be zero. Let $p : P \to P_1$ denote the canonical projection. If $p(M) = P_1$, then the composition of the injection $j : M \to P$ with $p : P \to P_1$ is an epimorphism and hence splits, because P_1 is projective. Therefore $M \cong P_1 \oplus M'$, where $M' \cong M \cap P_2 \subseteq P_2$. Because $\dim_K P_2 < d$, the induction hypothesis yields that M' is projective. Hence M is also projective. If $p(M) \neq P_1$, then $M \subseteq (\operatorname{rad} P_1) \oplus P_2$, where $\operatorname{rad} P_1$ is projective by hypothesis. Now $\dim_K[(\operatorname{rad} P_1) \oplus P_2] = d - 1$, because $\operatorname{rad} P_1$ is a maximal submodule of P_1. The induction hypothesis again implies that M is projective. The equivalence with the remaining conditions is proven similarly and left to the reader. $\qquad\square$

Because condition (b) of the theorem is right-left symmetric (see (A.4.9) of the Appendix), it follows that a finite dimensional algebra is right hereditary if and only if it is left hereditary. Thus, from now on, we speak about hereditary algebras without further specification, and hereditary algebras also satisfy the "left-hand" analogues of the equivalent conditions of the theorem. On the other hand, conditions (e) to (h) show that we may revert to our custom of considering only finitely generated modules. From now on, the term *module* means, as usual, a finitely generated module.

1.5. Corollary. *Let A be a hereditary algebra.*

(a) *Any nonzero A-homomorphism between indecomposable projective A-modules is a monomorphism.*

(b) *If P is an indecomposable projective A-module, then $\operatorname{End} P \cong K$.*

Proof. Let $f : P \to P'$ be a nonzero homomorphism, with P and P' indecomposable projective. Because $\operatorname{Im} f \subseteq P'$ is projective, the short exact sequence $0 \to \operatorname{Ker} f \to P \to \operatorname{Im} f \to 0$ splits and $P \cong \operatorname{Im} f \oplus \operatorname{Ker} f$. Because the module P is indecomposable and $\operatorname{Im} f \neq 0$, $\operatorname{Ker} f = 0$ and f is a monomorphism, hence (a) follows. The statement (b) is an immediate consequence of (a). $\qquad\square$

The following lemma is used repeatedly in the sequel. We first recall that if A is a K-algebra and M, N are indecomposable modules in $\mathrm{mod}\,A$, then $\mathrm{rad}_A(M,N)$ is the subspace of $\mathrm{Hom}_A(M,N)$ consisting of all nonisomorphisms, and the subspace $\mathrm{rad}_A^2(M,N)$ of $\mathrm{rad}_A(M,N)$ consists of the sums $f_1 f_1' + \ldots + f_t f_t'$, where for each $i \in \{1, \ldots, ,t\}$, $f_i' \in \mathrm{rad}_A(M, L_i)$ and $f_i \in \mathrm{rad}_A(L_i, N)$ for some indecomposable module L_i. The space of irreducible morphisms from M to N is then the K-vector space $\mathrm{Irr}(M,N) = \mathrm{rad}_A(M,N)/\mathrm{rad}_A^2(M,N)$. We use essentially the functorial isomorphism $\theta : \mathrm{Hom}_A(eA, M) \xrightarrow{\sim} Me$, $f \mapsto f(e)$, established in (I.4.2).

1.6. Lemma. *Let A be a basic hereditary K-algebra and e, e' primitive idempotents of A. There exists an isomorphism of K-vector spaces*

$$\mathrm{Irr}(e'A, eA) \cong e(\mathrm{rad}\,A/\mathrm{rad}^2 A)e'.$$

Proof. First we note that, because the canonical A-module projection $e(\mathrm{rad}\,A)e' \longrightarrow e(\mathrm{rad}\,A/\mathrm{rad}^2 A)e'$ has kernel $e(\mathrm{rad}^2 A)e'$, it induces a K-linear isomorphism $e(\mathrm{rad}\,A/\mathrm{rad}^2 A)e' \cong e(\mathrm{rad}\,A)e'/e(\mathrm{rad}^2 A)e'$.

We split the proof into two cases. Assume first that $e = e'$. By (1.5), any nonzero A-homomorphism $eA \to eA$ is injective, and hence is an isomorphism. Consequently, $\mathrm{Hom}_A(eA, eA) \cong K$ and $\mathrm{rad}_A(eA, eA) = 0$ (so that $\mathrm{Irr}(eA, eA) = 0$). On the other hand, $e(\mathrm{rad}\,A)e = \mathrm{rad}(eAe) = 0$. This establishes the statement in this case.

Assume next that $e \neq e'$. Because A is basic, $eA \not\cong e'A$ and therefore $\mathrm{rad}_A(e'A, eA) = \mathrm{Hom}_A(e'A, eA) \cong \mathrm{Hom}_A(e'A, \mathrm{rad}\,eA)$, because the idempotent e is primitive and $\mathrm{rad}\,eA$ is the unique maximal submodule of eA (by (I.4.5)). Because $\mathrm{rad}\,eA = eA(\mathrm{rad}\,A)$, it follows that the functorial isomorphism θ induces an isomorphism $\theta_1 : \mathrm{rad}_A(e'A, eA) \longrightarrow (\mathrm{rad}\,eA)e' = e(\mathrm{rad}\,A)e'$. Similarly, the isomorphism θ induces another A-module isomorphism $\theta_1' : \mathrm{Hom}_A(e'A, e(\mathrm{rad}^2 A)) \longrightarrow e(\mathrm{rad}^2 A)e'$. Denote by

$$e(\mathrm{rad}^2 A) \xrightarrow{u} e(\mathrm{rad}\,A) \xrightarrow{v} eA$$

the inclusion homomorphisms. Then the functoriality of θ implies the commutativity of the following square

$$
\begin{array}{ccc}
\mathrm{rad}_A(e'A, eA) & \xrightarrow[\simeq]{\theta_1} & e(\mathrm{rad}\,A)e' \\
{\scriptstyle j}\big\uparrow & & \big\uparrow{\scriptstyle j'} \\
\mathrm{Hom}_A(e'A, e(\mathrm{rad}^2 A)) & \xrightarrow[\simeq]{\theta_1'} & e(\mathrm{rad}^2 A)e'
\end{array}
$$

where $j = \mathrm{Hom}_A(e'A, vu)$ and j' is the restriction of u to $e(\mathrm{rad}^2 A)e'$.

We claim that the image of j is contained in $\mathrm{rad}_A^2(e'A, eA)$. Indeed, because A is hereditary, $\mathrm{rad}\,eA$ is projective. Because, clearly, no indecomposable summand of $\mathrm{rad}\,eA$ is isomorphic to eA, we have $v \in \mathrm{rad}_A(\mathrm{rad}\,eA, eA)$.

Similarly, $u \in \mathrm{rad}_A(\mathrm{rad}^2 eA, \mathrm{rad}\,eA)$. Consequently, $vu \in \mathrm{rad}_A^2(\mathrm{rad}^2 eA, eA)$ and, therefore, for any homomorphism $f \in \mathrm{Hom}_A(e'A, e(\mathrm{rad}^2 A))$ we have $vuf \in \mathrm{rad}\,{}_A^2(e'A, eA)$, because rad_A^2 defines a two-sided ideal in the category mod A.

Next we claim that θ_1 maps the space $\mathrm{rad}_A^2(e'A, eA)$ into $e(\mathrm{rad}^2 A)e'$. Let $f \in \mathrm{rad}_A^2(e'A, eA)$. Then there exist indecomposable modules $L_1, \ldots, L_t$ in mod A and, for each $s \in \{1, \ldots, , t\}$, homomorphisms $f'_s \in \mathrm{rad}_A(e'A, L_s)$ and $f_s \in \mathrm{rad}_A(L_s, eA)$ such that $f = f_1 f'_1 + \ldots + f_t f'_t$. For any $s \in \{1, \ldots, , t\}$, the submodule $\mathrm{Im}\, f_s$ of the projective module eA is itself projective, because A is hereditary. Hence $\mathrm{Im}\, f_s$ is isomorphic to a direct summand of the indecomposable module L_s, so that $L_s \cong \mathrm{Im}\, f_s$ is projective. Therefore there exists a primitive idempotent e_s of A such that $L_s \cong e_s A$. Because θ induces isomorphisms $\mathrm{Hom}_A(e'A, e_s A) \cong e_s(\mathrm{rad}\,A)e'$ and $\mathrm{rad}_A(e'A, e_s A) \cong e_s(\mathrm{rad}\,A)e'$, we deduce that

$$\theta_1(f_s f'_s) \in e(\mathrm{rad}\,A)e_s \cdot e_s(\mathrm{rad}\,A)e' \subseteq e(\mathrm{rad}^2 A)e'.$$

This shows that $\theta(f) \in e(\mathrm{rad}^2 A)e'$ and, consequently, that θ_1 restricts to a linear map $\theta_2 : \mathrm{rad}_A^2(e'A, eA) \longrightarrow e(\mathrm{rad}^2 A)e'$. Therefore the previous square induces the following commutative diagram:

$$
\begin{array}{ccc}
\mathrm{rad}_A(e'A, eA) & \xrightarrow[\simeq]{\theta_1} & e(\mathrm{rad}\,A)e' \\[4pt]
\big\uparrow & & \big\uparrow {\scriptstyle j'} \\[4pt]
\mathrm{rad}_A^2(e'A, eA) & \xrightarrow{\theta_2} & e(\mathrm{rad}^2 A)e' \\[4pt]
{\scriptstyle j}\big\uparrow & & \big\uparrow {\scriptstyle 1} \\[4pt]
\mathrm{Hom}_A(e'A, e(\mathrm{rad}^2 A)) & \xrightarrow[\simeq]{\theta'_1} & e(\mathrm{rad}^2 A)e'
\end{array}
$$

It follows that θ_2 is bijective. Passing to the quotients yields

$$\mathrm{Irr}(e'A, eA) = \frac{\mathrm{rad}_A(e'A, eA)}{\mathrm{rad}_A^2(e'A, eA)} \cong \frac{e(\mathrm{rad}\,A)e'}{e(\mathrm{rad}^2 A)e'} \cong e\left(\frac{\mathrm{rad}\,A}{\mathrm{rad}^2 A}\right)e'.$$

The lemma is proved. □

Our next objective is to prove that an algebra is hereditary if and only if it is the path algebra of a finite, connected, and acyclic quiver.

1.7. Theorem. (a) *If Q is a finite, connected, and acyclic quiver, then the algebra $A = KQ$ is hereditary and $Q_A = Q$.*

(b) *If A is a basic, connected, hereditary algebra and $\{e_1, \ldots, e_n\}$ is a complete set of primitive orthogonal idempotents of A, then*

(i) *the quiver Q_A of A is finite, connected, and acyclic; and*

(ii) *there exists a K-algebra isomorphism $A \cong KQ_A$.*

Proof. (a) Let Q be a finite, connected, and acyclic quiver and let ε_a be the stationary path at $a \in Q_0$. To show that $A = KQ$ is hereditary, it suffices, by (1.4), to show that the radical $\operatorname{rad} P(a)$ of each indecomposable projective KQ-module $P(a) = \varepsilon_a KQ$ is itself projective. In view of (III.1.6), we identify modules X in $\operatorname{mod} KQ$ with K-linear representations $(X_b, \varphi_\beta)_{b \in Q_0, \beta \in Q_1}$ of Q.

Let $a \in Q_0$. By (III.2.4)(a), we have $P(a) = (P(a)_b, \varphi_\beta)$, where $P(a)_b = \varepsilon_a(KQ)\varepsilon_b$ has as a basis the set of all the paths from a to b, and for an arrow $\beta : b \to c$ in Q, the K-linear map $\varphi_\beta : P(a)_b \to P(a)_c$ is given by the right multiplication by β, hence it is injective. For $x, y \in Q_0$, let $w(x, y)$ denote the number of paths from x to y. We thus have $\dim_K P(a)_b = w(a, b)$. By (III.2.4)(b), $\operatorname{rad} P(a) = (J_b, \gamma_\beta)$ is a representation of Q with $J_b = P(a)_b$ for $b \neq a$, $J_a = 0$ and $\gamma_\beta = \varphi_\beta$ for any arrow β of source $b \neq a$.

Let $\{b_1, \dots, b_t\}$ be the set of all direct successors of a in Q, and n_i be the number of arrows from a to b_i (for $1 \leq i \leq t$). By (III.2.2)(d), the top of $\operatorname{rad} P(a)$ is isomorphic to $\bigoplus_{i=1}^t S(b_i)^{n_i}$; hence we have a projective cover $f : \bigoplus_{i=1}^t P(b_i)^{n_i} \longrightarrow \operatorname{rad} P(a)$. On the other hand, for $b \neq a$, there are K-linear isomorphisms

$$J_b = (\operatorname{rad} \varepsilon_a(KQ))\varepsilon_b \cong \operatorname{Hom}_{KQ}(\varepsilon_b(KQ), \operatorname{rad} \varepsilon_a(KQ))$$
$$\cong \operatorname{Hom}_{KQ}(\varepsilon_b(KQ), \varepsilon_a(KQ)) \cong \varepsilon_a(KQ)\varepsilon_b = P(a)_b.$$

Note that the existence of the isomorphism

$$\operatorname{Hom}_{KQ}(\varepsilon_b(KQ), \operatorname{rad} \varepsilon_a(KQ)) \cong \operatorname{Hom}_{KQ}(\varepsilon_b(KQ), \varepsilon_a(KQ))$$

is a consequence of the facts that $\varepsilon_a(KQ) \not\cong \varepsilon_b(KQ)$ and $\operatorname{rad} \varepsilon_a(KQ)$ is the unique maximal submodule of the right ideal $\varepsilon_a(KQ)$. Consequently, for any $b \neq a$ in Q, we have

$$\dim_K[\operatorname{rad} P(a)]_b = \dim_K J_b = \dim_K P(a)_b = w(a, b) = \sum_{i=1}^t n_i w(b_i, b)$$
$$= \sum_{i=1}^t n_i \dim_K P(b_i)_b = \dim_K \left[\bigoplus_{i=1}^t P(b_i)^{n_i} \right]_b.$$

It follows that f is an isomorphism, and we are done.

Now we prove the statement (b).

(i) Because A is connected, its quiver Q_A of A is connected, by (II.3.4). We notice that to each arrow $\alpha : a \to b$ in Q_A corresponds an irreducible morphism $f_\alpha : e_b A \to e_a A$. By (1.5), f_α is a monomorphism and obviously $\operatorname{Im} f_\alpha \subseteq \operatorname{rad} e_a A$. To show that Q_A is acyclic, assume to the contrary that it is not and let $\alpha_1 \dots \alpha_t$ be a cycle in Q_A passing through a point a. Then $f = f_{\alpha_t} \cdots f_{\alpha_1} : e_a A \to e_a A$ is a monomorphism, because each f_{α_i} is. But also $\operatorname{Im} f \subseteq \operatorname{rad} e_a A$. Hence $\dim_K e_a A = \dim_K \operatorname{Im} f \leq \dim_K \operatorname{rad} e_a A < \dim_K e_a A$, which is a contradiction.

(ii) By (II.3.7), there exists an admissible ideal $\mathcal{I}$ of KQ_A such that $A \cong KQ_A/\mathcal{I}$. We identify A with $KQ_A/\mathcal{I}$ and the idempotent $e_a \in A$ with the class $\overline{\varepsilon}_a = \varepsilon_a + \mathcal{I}$ of the stationary path ε_a at $a \in (Q_A)_0$. By (III.2.4), for each $a \in Q_0$, the corresponding indecomposable projective module $P(a) = e_a A$ is viewed as a representation of Q_A as follows: $P(a) = (P(a)_b, \varphi_\beta)$, $P(a)_b = P(a)e_b = e_a A e_b = e_a(KQ)e_b/e_a \mathcal{I} e_b$ is the K-vector space with basis the set of all $\overline{w} = w + \mathcal{I}$, where w is a path from a to b, and, for an arrow $\beta : b \to c$, the K-linear map $\varphi_\beta : P(a)_b \to P(a)_c$ is given by the right multiplication by $\overline{\beta} = \beta + \mathcal{I}$. Note that, because $\dim_K(\varepsilon_a KQ\varepsilon_b)$ equals the number $w(a, b)$ of paths from a to b in Q_A, $\dim_K P(a)e_b = w(a, b) - \dim_K \varepsilon_a \mathcal{I} \varepsilon_b$.

We show that $\mathcal{I} = 0$. Assume that this is not the case. Because, according to (i), the quiver Q_A is acyclic, we may number its points so that the existence of a path from x to y implies $x > y$. Then there is a least a such that there exists $b \in (Q_A)_0$ with $\varepsilon_a \mathcal{I} \varepsilon_b \neq 0$. In particular, a is not a sink, and so $\operatorname{rad} P(a) \neq 0$, by (III.2.4). Because A is hereditary, the nonzero module $\operatorname{rad} P(a)$ is projective, and therefore there exist $t \geq 1$, vertices $b_1, \ldots, b_t \in (Q_A)_0$, and positive integers $n_1, \ldots, n_t$ such that

$$\operatorname{rad} P(a) \cong P(b_1)^{n_1} \oplus \cdots \oplus P(b_t)^{n_t}.$$

It follows from (III.2.4), (IV.4.3), and (1.6) that $\{b_1, \ldots, b_t\}$ is the set of direct successors of a in Q_A and

$$n_i = \dim_K \operatorname{Irr}(P(b_i), P(a)) = \dim_K \varepsilon_a (\operatorname{rad} A/\operatorname{rad}^2 A)\varepsilon_b,$$

that is, n_i is the number of arrows from a to b_i in Q_A for i such that $1 \leq i \leq t$. The minimality of a implies that $\varepsilon_{b_i} \mathcal{I} \varepsilon_b = 0$ and $\dim_K P(b_i)\varepsilon_b = \dim_K \varepsilon_{b_i} A \varepsilon_b = w(b_i, b)$ for each b and each i. It follows that

$$\dim_K(\operatorname{rad} P(a))\varepsilon_b = \sum_{i=1}^{t} n_i \dim_K P(b_i)\varepsilon_b = \sum_{i=1}^{t} n_i w(b_i, b) = w(a, b)$$
$$> w(a, b) - \dim_K \varepsilon_a \mathcal{I} \varepsilon_b = \dim_K P(a)\varepsilon_b,$$

and this is clearly a contradiction. The proof is complete. $\qquad\square$

We end this section with some remarks on the Auslander–Reiten translation and the Auslander–Reiten quiver of a hereditary algebra.

1.8. Lemma. *Let* A *be a hereditary algebra and* M *be an* A-*module. There exists a functorial isomorphism* $\mathrm{Tr}\, M \cong \mathrm{Ext}_A^1(M, A)$.

Proof. Because $\mathrm{gl.dim}\, A \leq 1$, a minimal projective resolution of the A-module M is of the form $0 \longrightarrow P_1 \xrightarrow{f} P_0 \longrightarrow M \longrightarrow 0$. Applying the functor $(-)^t = \mathrm{Hom}_A(-, A)$, we obtain an exact sequence of left A-modules

$$0 \longrightarrow M^t \longrightarrow P_0^t \xrightarrow{f^t} P_1^t \longrightarrow \mathrm{Ext}_A^1(M, A) \longrightarrow 0.$$

The statement follows at once. □

Actually, the proof shows that the isomorphism $\mathrm{Tr}\, M \cong \mathrm{Ext}_A^1(M, A)$ holds whenever $\mathrm{pd}\, M \leq 1$. One consequence of this lemma is that the Auslander–Reiten translations $\tau = D\mathrm{Tr}$ and $\tau^{-1} = \mathrm{Tr}\, D$ are endofunctors of the module category $\mathrm{mod}\, A$ of a hereditary algebra A.

1.9. Corollary. *Let* A *be a hereditary algebra, and* M *be an* A-*module. There exist functorial isomorphisms*

$$\tau M \cong D\mathrm{Ext}_A^1(M, A) \quad and \quad \tau^{-1} M \cong \mathrm{Ext}_A^1(DM, A). \qquad \square$$

We also have the following easy characterisation of hereditary algebras by means of the Auslander–Reiten quiver.

1.10. Proposition. *Let* A *be an algebra and* $\Gamma(\mathrm{mod}\, A)$ *be its Auslander–Reiten quiver. The following conditions are equivalent:*

  (a) *A is hereditary.*

  (b) *The predecessors of the points in $\Gamma(\mathrm{mod}\, A)$ corresponding to the indecomposable projective modules correspond to indecomposable projective modules.*

  (c) *The successors of the points in $\Gamma(\mathrm{mod}\, A)$ corresponding to the indecomposable injective modules correspond to indecomposable injective modules.*

Proof. We prove the equivalence of (a) and (b); the proof of the equivalence of (a) and (c) is similar.

For the necessity, let M be an immediate predecessor of an indecomposable projective P in $\Gamma(\mathrm{mod}\, A)$. Then there exists an irreducible morphism $f : M \longrightarrow P$. By (IV.1.10) and (IV.3.5), there exist a module N, an A-module isomorphism $h : M \oplus N \xrightarrow{\simeq} \mathrm{rad}\, P$, and a homomorphism $f' : N \longrightarrow P$ such that $[f\ f'] = jh$, where $j : \mathrm{rad}\, P \to P$ denotes the inclusion. Because A is hereditary, $\mathrm{rad}\, P$ is projective, hence the module M is projective. Consequently, every immediate predecessor of an indecomposable projective is an indecomposable projective module. The statement follows from an obvious induction. Note that, because $\Gamma(\mathrm{mod}\, A)$ contains only finitely many projectives, any indecomposable projective has only finitely many predecessors.

The sufficiency follows from the fact that the given condition implies that the radical of any indecomposable projective module is projective. □

VII.2. The Dynkin and Euclidean graphs

Certain graphs are of particular interest in this chapter (and the following ones).

(a) The Dynkin graphs

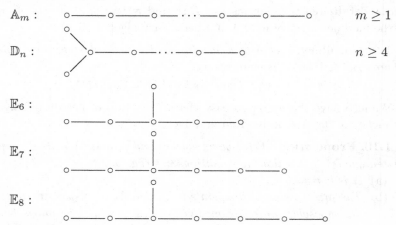

$\mathbb{A}_m$: $m \geq 1$

$\mathbb{D}_n$: $n \geq 4$

$\mathbb{E}_6$:

$\mathbb{E}_7$:

$\mathbb{E}_8$:

(b) The Euclidean graphs

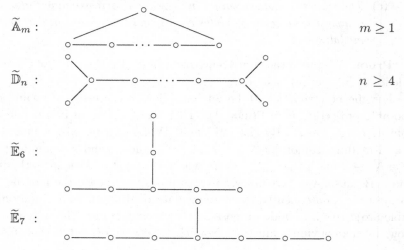

$\widetilde{\mathbb{A}}_m$: $m \geq 1$

$\widetilde{\mathbb{D}}_n$: $n \geq 4$

$\widetilde{\mathbb{E}}_6$:

$\widetilde{\mathbb{E}}_7$:

$\widetilde{\mathbb{E}}_8$:

The index in the Dynkin graphs always refers to the number of points in the graph, whereas in the Euclidean, it refers to the number of points minus one (thus, $\mathbb{A}_m$ has m points while $\widetilde{\mathbb{A}}_m$ has $m+1$ points). In fact, a Euclidean graph can be constructed from the corresponding Dynkin graph by adding one point. Dynkin graphs and Euclidean graphs are also called *Dynkin diagrams* and *Euclidean diagrams*, respectively (see [41] and [72]).

We are interested in the path algebras of quivers having one of the preceding as underlying graph, that is, of quivers arising from arbitrary orientations of these graphs (excluding the orientation making $\widetilde{\mathbb{A}}_m$ an oriented cycle; this orientation gives an infinite dimensional path algebra). As pointed out in the introduction, the main result of this chapter says that the path algebra of a quiver Q is representation–finite if and only if the underlying graph $\overline{Q}$ of Q is a Dynkin graph.

We start with a purely combinatorial lemma.

2.1. Lemma. *Let Q be a finite, connected, and acyclic quiver. If the underlying graph $\overline{Q}$ of Q is not a Dynkin graph, then $\overline{Q}$ contains a Euclidean graph as a subgraph.*

Proof. We show that if $\overline{Q}$ contains no Euclidean subgraph, then $\overline{Q}$ is a Dynkin graph. The exclusion of $\widetilde{\mathbb{A}}_m$ implies that $\overline{Q}$ is a tree. The exclusion of $\widetilde{\mathbb{D}}_4$ implies that no point in $\overline{Q}$ has more than three neighbours, and the exclusion of $\widetilde{\mathbb{D}}_n$ with $n \geq 5$ implies that at most one point has three neighbours. Hence $\overline{Q}$ is of the following form

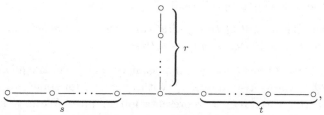

where we may assume without loss of generality that $r \leq s \leq t$. The exclusion of $\widetilde{\mathbb{E}}_6$ gives $r \leq 1$. If $r = 0$, then $\overline{Q} = \mathbb{A}_{s+t+1}$. If $r = 1$, the exclusion of $\widetilde{\mathbb{E}}_7$ gives $1 \leq s \leq 2$. If $s = 1$, then $\overline{Q} = \mathbb{D}_{t+3}$. Finally, if $s = 2$, the exclusion of $\widetilde{\mathbb{E}}_8$ gives $2 \leq t \leq 4$, so that $\overline{Q}$ is equal to $\mathbb{E}_6$, $\mathbb{E}_7$ or $\mathbb{E}_8$. $\square$

We use this lemma to show that if $A \cong KQ$ is representation–finite, then $\overline{Q}$ is a Dynkin graph. To do so, we start by showing that if Q' is a subquiver of Q such that KQ' is representation–infinite, then KQ itself is representation–infinite. It will then remain to show that if $\overline{Q'}$ is Euclidean, then KQ' is representation–infinite.

2.2. Lemma. *Let Q be a finite, connected, and acyclic quiver. If Q' is a subquiver of Q such that KQ' is representation–infinite, then KQ is representation–infinite.*

Proof. We must show that Q has at least as many nonisomorphic indecomposable representations as Q'. Let $M' = (M'_a, \varphi'_\alpha)$ be a representation of Q'. We define its extension $E(M')$ to be the representation (M_a, φ_α) of Q defined by

$$M_a = \begin{cases} M'_a & \text{if } a \in Q'_0, \\ 0 & \text{if } a \notin Q'_0, \end{cases} \quad \text{and} \quad \varphi_\alpha = \begin{cases} \varphi'_\alpha & \text{if } \alpha \in Q'_1, \\ 0 & \text{if } \alpha \notin Q'_1. \end{cases}$$

Given a morphism $f' : M' \to N'$ of representations of Q', where $M' = (M'_a, \varphi'_\alpha)$ and $N' = (N'_a, \psi'_\alpha)$, we define $f = E(f') : E(M') \to E(N')$ to be the morphism of representations of Q given by

$$f_a = \begin{cases} f'_a & \text{if } a \in Q'_0, \\ 0 & \text{if } a \notin Q'_0. \end{cases}$$

Clearly, E induces a full and faithful functor $\operatorname{mod} KQ' \to \operatorname{mod} KQ$ so that $\operatorname{End}_{KQ} E(M') \cong \operatorname{End}_{KQ'} M'$. In particular, $E(M')$ is indecomposable if and only if M' is indecomposable (see (I.4.8)), and we have $M' \cong N'$ if and only if $E(M') \cong E(N')$. □

We now want to show that if Q is a quiver whose underlying graph is Euclidean, then KQ is representation–infinite. The first step in this direction is the following proposition.

2.3. Proposition. *Let Q be a finite, connected, and acyclic quiver. If KQ is representation–finite, then Q is a tree.*

Proof. Because Q has no loops, that is, it is not a tree, is equivalent to saying that Q contains a subquiver Q' with $\overline{Q'} = \widetilde{\mathbb{A}}_m$ for some $m \geq 1$. We show that, in this case, KQ' is representation–infinite. We may suppose that the points of Q' are numbered from 1 to $m + 1$ and that there exists an arrow $\alpha : 1 \to 2$. For each scalar $\lambda \in K$, let $M(\lambda) = (M_i^{(\lambda)}, \varphi_\beta^{(\lambda)})$ be the representation of Q' defined as follows

$$M_i^{(\lambda)} = K \quad \text{for each } 1 \leq i \leq m + 1$$

and

$$\varphi_\beta^{(\lambda)}(x) = \begin{cases} \lambda x & \text{if } \beta = \alpha, \\ x & \text{if } \beta \neq \alpha, \end{cases}$$

(that is, $\varphi_\beta^{(\lambda)}$ is the identity map for each arrow $\beta \neq \alpha$, and $\varphi_\alpha^{(\lambda)}$ is the multiplication by λ). Let $\lambda, \mu \in K$. We claim that each nonzero homomorphism $f : M(\lambda) \to M(\mu)$ is an isomorphism and, if this is the case, then $\lambda = \mu$ and $\operatorname{End} M(\lambda) \cong K$. Indeed, if $f : M(\lambda) \to M(\mu)$ is a nonzero homomorphism, then the commutativity relations

$$\begin{array}{ccc} M_i^{(\lambda)} & \xrightarrow{\varphi_\beta^{(\lambda)}} & M_j^{(\lambda)} \\ {\scriptstyle f_i}\downarrow & & \downarrow{\scriptstyle f_j} \\ M_i^{(\mu)} & \xrightarrow{\varphi_\beta^{(\mu)}} & M_j^{(\mu)} \end{array}$$

corresponding to all arrows $\beta : i \to j$ with $\beta \neq \alpha$ give $f_1 = \ldots = f_{m+1}$. In particular, $f \neq 0$ implies $f_i \neq 0$ for each i. Therefore, the map f_i, being a nonzero K-linear endomorphism of K is an isomorphism (and actually is the multiplication by a nonzero scalar). Finally, the commutativity condition corresponding to $\alpha : 1 \to 2$ gives

$$\mu f_1(1) = \varphi_\alpha^{(\mu)} f_1(1) = f_2 \varphi_\alpha^{(\lambda)}(1) = f_2(\lambda) = \lambda f_2(1).$$

Because $f_1 = f_2$ and both are nonzero, we have $\lambda = \mu$. On the other hand, f is entirely determined by $f_1(1)$. Because f_1 is the multiplication by a nonzero scalar ν (say), we deduce that $f : M(\lambda) \to M(\mu)$ is the map $\nu 1_{M(\lambda)}$. Thus $\operatorname{End} M(\lambda) \cong K$ and $M(\lambda)$ is indecomposable.

We have shown that the family $(M(\lambda))_{\lambda \in K}$ consists of pairwise nonisomorphic indecomposable representations. Because K is an algebraically closed (hence infinite) field, this gives an infinite family of pairwise nonisomorphic indecomposable representations of Q'. Therefore KQ' is representation–infinite. By (2.2), KQ is also representation–infinite.   □

We have considered, in the preceding proof, representations M having the property that $\operatorname{End} M \cong K$. Such a representation carries a name.

2.4. Definition. Let A be a finite dimensional K-algebra. An A-module M such that $\operatorname{End} M \cong K$ is called a **brick**.

Clearly, each brick is an indecomposable module. On the other hand, there exist indecomposables that are not bricks. Let, for instance, A be a nonsimple local algebra (we may, for example, take $A = K[t]/\langle t^n \rangle$, with $n \geq 2$); then A_A is an indecomposable module that is not a brick, because $\operatorname{End} A_A \cong A \ncong K$. We showed in the proof of (2.3) that if Q' is a quiver with underlying graph $\widetilde{\mathbb{A}}_m$, with $m \geq 1$, then KQ' admits an infinite family of pairwise nonisomorphic bricks.

2.5. Proposition. *Let Q be a finite, connected, and acyclic quiver and M_{KQ} be a brick such that there exists $a \in Q_0$ with $\dim_K M_a > 1$. Let Q' be the quiver defined as follows: $Q' = (Q'_0, Q'_1)$, where $Q'_0 = Q_0 \cup \{b\}$; $Q'_1 = Q_1 \cup \{\alpha\}$; and $\alpha : b \to a$. Then KQ' is representation–infinite.*

Proof. Let $\psi : K \to M_a$ be a nonzero K-linear map. We define $M(\psi)$ to be the representation (M'_c, φ'_γ) of Q' given by the formulas:

$$M'_c = \begin{cases} M_c & \text{if } c \in Q_0, \\ K & \text{if } c = b \end{cases} \quad \text{and} \quad \varphi'_\gamma = \begin{cases} \varphi_\gamma & \text{if } \gamma \in Q_1, \\ \psi & \text{if } \gamma = \alpha. \end{cases}$$

Let $\psi, \eta : K \to M_a$ be nonzero K-linear maps and $f : M(\psi) \to M(\eta)$ be a nonzero morphism. Because the restriction $f|_M$ of f to M is an endomorphism of the brick M, $f|_M$ equals the multiplication by some scalar $\lambda \in K$. On the other hand, $f_b : M(\psi)_b \to M(\eta)_b$ is a K-linear endomorphism of K and hence it equals the multiplication by a scalar $\mu \in K$. Note that, because $f \neq 0$ and $\psi, \eta \neq 0$, we have $\lambda, \mu \neq 0$. Consider $x \in M(\eta)_b$ and the commutativity condition corresponding to the arrow α

$$\eta(x) = \eta f_b(x\mu^{-1}) = f_a \psi(x\mu^{-1}) = \psi(x) \cdot (\mu^{-1}\lambda)$$

$$\begin{array}{ccc} M(\psi)_b & \xrightarrow{\psi} & M(\psi)_a \\ f_b \downarrow & & \downarrow f_a \\ M(\eta)_b & \xrightarrow{\eta} & M(\eta)_a \end{array}$$

Thus $\eta = \psi \cdot (\mu^{-1}\lambda)$.

This relation implies that each $M(\psi)$ is a brick. Indeed, setting $\psi = \eta$, we see that each endomorphism f of $M(\psi)$ equals the multiplication by a scalar: the preceding relation gives $\mu^{-1}\lambda = 1$; hence $\lambda = \mu$ and f is the multiplication by λ (or μ).

Assume now $f : M(\psi) \to M(\eta)$ is an isomorphism. The maps ψ and η are given by column matrices with $d = \dim_K M_a$ coefficients (and $d \geq 2$ by hypothesis), that is, $\psi = [\psi_1 \ldots \psi_d]^t$ and $\eta = [\eta_1 \ldots \eta_d]^t$. Hence $\eta = \psi \cdot (\mu^{-1}\lambda)$ yields $\eta_i = \psi_i \cdot (\mu^{-1}\lambda)$ for each $1 \leq i \leq d$. This can be expressed by saying that $(\psi_1, \ldots, \psi_d)$ and $(\eta_1, \ldots, \eta_d)$ correspond to the same point of the projective space $\mathbb{P}_{d-1}(K)$. Because K is an algebraically closed (hence infinite) field, $\mathbb{P}_{d-1}(K)$ has infinitely many points. We have thus shown the existence of infinitely many pairwise nonisomorphic bricks of the form $M(\psi)$. $\qquad\square$

We apply this proposition as follows: For each of the Dynkin graphs $\mathbb{D}_n$, $\mathbb{E}_6$, $\mathbb{E}_7$, and $\mathbb{E}_8$, we consider a quiver Q having it as underlying graph, and we show that there exists a brick M over KQ and a point $a \in Q_0$ such that $\dim_K M_a > 1$; applying the construction of the proposition yields that the path algebra of the corresponding enlarged quiver (whose underlying graph is Euclidean) is representation–infinite.

2.6. Lemma. *Let Q be one of the following quivers with underlying graph a Dynkin diagram:*

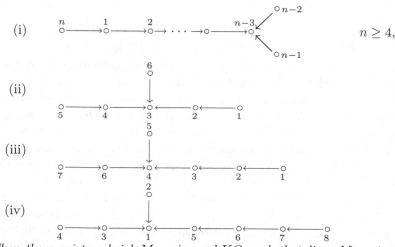

Then there exists a brick M_{KQ} in $\mathrm{mod}\,KQ$ such that $\dim_K M_a > 1$, where $a \in Q_0$ is the point 1, 6, 7, and 8 in cases (i), (ii), (iii), *and* (iv), *respectively.*

Proof. We exhibit in each case the wanted brick $M = (M_b, \varphi_\beta)$ such that $\dim_K M_a > 1$.

(i) $M_1 = \ldots = M_{n-3} = K^2$, where K^2 is given its canonical basis $\{\mathbf{e}_1, \mathbf{e}_2\}$, $M_{n-1} = \mathbf{e}_1 K$, $M_{n-2} = \mathbf{e}_2 K$, and $M_n = (\mathbf{e}_1 + \mathbf{e}_2)K$. All the φ_β are taken to be the canonical inclusions. We claim that M is a brick with $\dim_K M_1 > 1$. Let $\overline{f} \in \mathrm{End}\,M_{KQ}$. The commutativity conditions give $\overline{f}_1 = \ldots = \overline{f}_{n-3} = f$ (say) and $\overline{f}_i = f|_{M_i}$ for $i = n-2, n-1, n$. Therefore $f(\mathbf{e}_1) \in \mathbf{e}_1 K$, $f(\mathbf{e}_2) \in \mathbf{e}_2 K$, and $f(\mathbf{e}_1 + \mathbf{e}_2) \in (\mathbf{e}_1 + \mathbf{e}_2)K$. Letting $f(\mathbf{e}_1) = \mathbf{e}_1 \lambda_1$, $f(\mathbf{e}_2) = \mathbf{e}_2 \lambda_2$ where $\lambda_1, \lambda_2 \in K$, we have

$$f(\mathbf{e}_1 + \mathbf{e}_2) = f(\mathbf{e}_1) + f(\mathbf{e}_2) = \mathbf{e}_1 \lambda_1 + \mathbf{e}_2 \lambda_2 \in (\mathbf{e}_1 + \mathbf{e}_2)K;$$

hence $\lambda_1 = \lambda_2$ and therefore f is a multiplication by the scalar λ_1. This shows that M is indeed a brick with $\dim_K M_1 \geq 2$.

(ii) $M_3 = K^3$, where K^3 is given its canonical basis $\{\mathbf{e}_1, \mathbf{e}_2, \mathbf{e}_3\}$, $M_1 = \mathbf{e}_1 K$, $M_2 = \mathbf{e}_1 K \oplus \mathbf{e}_2 K$, $M_4 = \mathbf{e}_2 K \oplus \mathbf{e}_3 K$, $M_6 = (\mathbf{e}_1 + \mathbf{e}_2)K \oplus (\mathbf{e}_2 + \mathbf{e}_3)K$,

$M_5 = e_3 K$. All the φ_β are taken to be the canonical inclusions. We observe that $M_2 \cap M_4 = e_2 K$, $M_4 \cap M_6 = (e_2 + e_3)K$, $M_2 \cap M_6 = (e_1 + e_2)K$. We claim that M is a brick with $\dim_K M_6 > 1$. Let $\overline{f} \in \operatorname{End} M_{KQ}$. Then $\overline{f}_i = f|_{M_i}$ where $f = \overline{f}_3 \in \operatorname{End}_K M_3$. Because $f(M_i) \subseteq M_i$ for $1 \leq i \leq 6$, $f(M_2 \cap M_4) \subseteq f(M_2) \cap f(M_4) \subseteq M_2 \cap M_4$. Similarly, $f(M_4 \cap M_6) \subseteq M_4 \cap M_6$ and $f(M_2 \cap M_6) \subseteq M_2 \cap M_6$. Thus, there exist $\lambda_1, \lambda_2, \lambda_3, \mu, \nu \in K$ such that

$$f(e_1) = e_1 \lambda_1, \quad f(e_2) = e_2 \lambda_2, \quad f(e_3) = e_3 \lambda_3,$$
$$f(e_1 + e_2) = (e_1 + e_2)\mu, \quad f(e_2 + e_3) = (e_2 + e_3)\nu.$$

Hence $\lambda_1 = \mu = \lambda_2 = \nu = \lambda_3$ and f equals the multiplication by their common value. This shows that M is a brick such that $\dim_K M_6 \geq 2$.

(iii) $M_4 = K^4$, where K^4 is given its canonical basis $\{e_1, e_2, e_3, e_4\}$, $M_1 = e_1 K$, $M_2 = e_1 K \oplus e_2 K$, $M_3 = e_1 K \oplus e_2 K \oplus e_3 K$, $M_7 = (e_2 - e_3)K \oplus (e_1 + e_4)K$, $M_6 = (e_1 + e_2)K \oplus (e_1 + e_3)K \oplus (e_1 + e_4)K$, $M_5 = e_3 K \oplus e_4 K$. All the φ_β are taken to be the canonical inclusions. We observe that $M_3 \cap M_5 = e_3 K$, $M_2 \cap M_6 = (e_1 + e_2)K$, $M_5 \cap M_6 = (e_3 - e_4)K$, $M_3 \cap M_7 = (e_2 - e_3)K$, $M_7 \cap (M_1 + M_5) = (e_1 + e_4)K$, $M_6 \cap [M_1 + (M_3 \cap M_5)] = (e_1 + e_3)K$. We claim that M is a brick with $\dim_K M_7 > 1$. Let $\overline{f} \in \operatorname{End} M_{KQ}$. As earlier, we show that $\overline{f}_i = f|_{M_i}$ for $1 \leq i \leq 7$, where $f \in \operatorname{End}_K M_4$ is such that there exist $\lambda_1, \lambda_3, \mu_1, \mu_2, \mu_3, \mu_4, \mu_5 \in K$ satisfying the following conditions:

$$f(e_1) = e_1 \lambda_1, \qquad\qquad f(e_3) = e_3 \lambda_3,$$
$$f(e_3 - e_4) = (e_3 - e_4)\mu_1, \qquad f(e_1 + e_2) = (e_1 + e_2)\mu_2,$$
$$f(e_2 - e_3) = (e_2 - e_3)\mu_3, \qquad f(e_1 + e_4) = (e_1 + e_4)\mu_4,$$
$$f(e_1 + e_3) = (e_1 + e_3)\mu_5.$$

A straightforward calculation shows that f is indeed the multiplication by a scalar. Hence M is a brick such that $\dim_K M_7 \geq 2$.

(iv) $M_1 = K^6$, where K^6 is given its canonical basis $\{e_1, \ldots, e_6\}$, $M_2 = (e_4 + e_6)K \oplus (e_1 + e_3 + e_5)K \oplus (e_1 + e_2 + e_4)K$, $M_3 = e_1 K \oplus e_2 K \oplus e_3 K \oplus e_6 K$, $M_4 = e_1 K \oplus e_6 K$, $M_5 = e_1 K \oplus e_2 K \oplus e_3 K \oplus e_4 K \oplus e_5 K$, $M_6 = e_2 K \oplus e_3 K \oplus e_4 K \oplus e_5 K$, $M_7 = e_3 K \oplus e_4 K \oplus e_5 K$, $M_8 = e_4 K \oplus e_5 K$. All the φ_β are taken to be the canonical inclusions. We observe that $M_4 \cap M_5 = e_1 K$, $M_3 \cap M_7 = e_3 K$, $M_3 \cap M_6 = e_2 K \oplus e_3 K$, $M_2 \cap M_3 = (e_1 + e_2 - e_6)K$, $(M_4 + M_8) \cap M_2 = (e_4 + e_6)K$ and $M_2 \cap M_6 = (e_2 - e_3 + e_4 - e_5)K$. Let $\overline{f} \in \operatorname{End} M_{KQ}$. As earlier, we show that $\overline{f}_i = f|_{M_i}$ for $1 \leq i \leq 8$, where $f \in \operatorname{End}_K M_1$. Moreover, the subspaces $M_4 \cap M_5$, $M_3 \cap M_7$, $M_3 \cap M_6$, $(M_4 + M_8) \cap M_2$, and $M_2 \cap M_6$ of K^6 are invariant under f. A straightforward calculation shows that if f is given in the canonical basis $e_1, \ldots, e_6$ by a

6×6 matrix $[a_{ij}]$, then $a_{11} = a_{22} = a_{33} = a_{44} = a_{55} = a_{66}$ and $a_{ij} = 0$ for any $i \neq j$, and so f is a multiplication by the scalar a_{11}. Therefore M is a brick with $\dim_K M_8 \geq 2$. ☐

2.7. Corollary. *The path algebra of each of the following quivers is representation-infinite:*

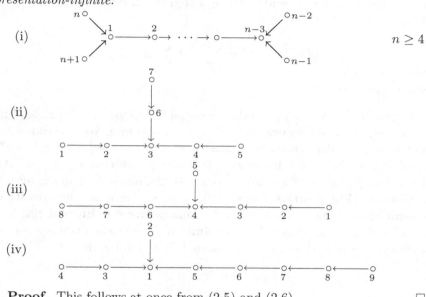

Proof. This follows at once from (2.5) and (2.6). ☐

We have shown in this section that if KQ is representation–finite, then Q is a tree (that is, $\overline{Q}$ contains no subgraph of the form $\widetilde{\mathbb{A}}_m$, for some $m \geq 1$) and contains no subquiver of one of the forms listed in (2.7). This does **not** yet imply that Q contains no subquivers whose underlying graph is Euclidean. Indeed, there remains to show that if Q is a tree, KQ is representation–infinite and Q' is a quiver such that $\overline{Q'} = \overline{Q}$ (that is, Q' has the same underlying graph as Q, but perhaps a different orientation), then KQ' is also representation–infinite. To prove this, we need to develop some new concepts.

VII.3. Integral quadratic forms

When studying hereditary algebras, it turns out that the Euler quadratic form, that is, the quadratic form arising from the Euler characteristic (see (III.3.11)) plays a prominent rôle. This quadratic form is an integral quadratic form, and this section is devoted to studying integral quadratic forms in general. Throughout, we denote by $\{e_1, \ldots, e_n\}$ the canonical basis of the free abelian group $\mathbb{Z}^n$ on n generators. As usual, elements in $\mathbb{Z}^n$ are written as column vectors.

3.1. Definition. A quadratic form $q = q(x_1, \ldots, x_n)$ on $\mathbb{Z}^n$ in n indeterminates $x_1, \ldots, x_n$ is said to be an **integral quadratic form** if it is of the form

$$q(x_1, \ldots, x_n) = \sum_{i=1}^n x_i^2 + \sum_{i<j} a_{ij} x_i x_j$$

where $a_{ij} \in \mathbb{Z}$ for all i, j.

Evaluating an integral quadratic form q on the vectors $\mathbf{x} = [x_1 \ldots x_n]^t$ in $\mathbb{Z}^n$, we obtain a mapping from $\mathbb{Z}^n$ to $\mathbb{Z}$, also denoted by q. We may endow $\mathbb{Z}^n$ with a partial order defined componentwise: a vector $\mathbf{x} = [x_1 \ldots x_n]^t \in \mathbb{Z}^n$ is called **positive** if $\mathbf{x} \neq 0$ and $x_j \geq 0$, for all j such that $1 \leq j \leq n$. We denote the positivity of a vector $\mathbf{x}$ as $\mathbf{x} > 0$. An integral quadratic form q is called **weakly positive** if $q(\mathbf{x}) > 0$ for all $\mathbf{x} > 0$; it is called **positive semidefinite** if $q(\mathbf{x}) \geq 0$ for all $\mathbf{x} \in \mathbb{Z}^n$, and **positive definite** if $q(\mathbf{x}) > 0$ for all $\mathbf{x} \neq 0$; finally, it is called **indefinite** if there exists a nonzero vector $\mathbf{x}$ such that $q(\mathbf{x}) < 0$. For a positive semidefinite form q, the set

$$\operatorname{rad} q = \{\mathbf{x} \in \mathbb{Z}^n \mid q(\mathbf{x}) = 0\}$$

is called the **radical** of q, and its elements are called **radical vectors**. It is a subgroup of $\mathbb{Z}^n$. Indeed, if $q(\mathbf{x}) = 0 = q(\mathbf{y})$, then

$$q(\mathbf{x} + \mathbf{y}) + q(\mathbf{x} - \mathbf{y}) = 2[q(\mathbf{x}) + q(\mathbf{y})] = 0 \text{ gives } q(\mathbf{x} + \mathbf{y}) = q(\mathbf{x} - \mathbf{y}) = 0,$$

by the positive semidefiniteness of q, and hence $\mathbf{x} + \mathbf{y}, \mathbf{x} - \mathbf{y} \in \operatorname{rad} q$.

The rank of the subgroup $\operatorname{rad} q$ is called the **corank** of q. Clearly, q is positive definite if and only if its corank is zero.

3.2. Examples. (a) The integral quadratic form

$$q(\mathbf{x}) = x_1^2 + x_2^2 + x_3^2 - x_1 x_2 + x_1 x_3 + x_2 x_3$$

on $\mathbb{Z}^3$ is weakly positive, positive semidefinite of corank 1 (hence is not positive definite). Indeed, $q(\mathbf{x}) = (x_1 - \frac{1}{2}x_2 + \frac{1}{2}x_3)^2 + \frac{3}{4}(x_2 + x_3)^2$ so that $\operatorname{rad} q$ is generated by the vector $[1 \ 1 \ -1]^t$. This implies our claim.

(b) The integral quadratic form $q(\mathbf{x}) = x_1^2 + x_2^2 - 2x_1 x_2 = (x_1 - x_2)^2$ on $\mathbb{Z}^2$ is positive semidefinite of corank 1 and $\operatorname{rad} q$ is generated by the vector $[1 \ 1]^t$. In particular, q is not weakly positive.

We denote by $(-, -)$ the **symmetric bilinear form** on $\mathbb{Z}^n$ corresponding to q, that is, for $\mathbf{x}, \mathbf{y} \in \mathbb{Z}^n$, we have

$$(\mathbf{x}, \mathbf{y}) = \frac{1}{4}[q(\mathbf{x} + \mathbf{y}) - q(\mathbf{x} - \mathbf{y})].$$

For instance, if q is as in Example 3.2 (b), we have

$$(\mathbf{x}, \mathbf{y}) = x_1 y_1 + x_2 y_2 - x_1 y_2 - x_2 y_1.$$

It is easily seen that the following relations hold:

(a) $q(\mathbf{x}) = (\mathbf{x}, \mathbf{x})$ for all $\mathbf{x} \in \mathbb{Z}^n$;

(b) $a_{ij} = 2(\mathbf{e}_i, \mathbf{e}_j)$ for all i, j such that $1 \le i < j \le n$; and $a_{ji} = 2(\mathbf{e}_i, \mathbf{e}_j)$ for all i, j such that $1 \le j < i \le n$;

(c) $q(\mathbf{x} + \mathbf{y}) = q(\mathbf{x}) + q(\mathbf{y}) + 2(\mathbf{x}, \mathbf{y})$ for all $\mathbf{x}, \mathbf{y} \in \mathbb{Z}^n$.

We also define the n **partial derivatives** of the quadratic form q to be the group homomorphisms from $\mathbb{Z}^n$ to $\mathbb{Z}$ defined by:

$$D_i q(x) = \frac{\partial q}{\partial x_i}(x) = 2(\mathbf{e}_i, \mathbf{x}) = 2x_i + \sum_{i<t} a_{it} x_t + \sum_{t<i} a_{ti} x_t$$

for each i such that $1 \le i \le n$.

3.3. Lemma. *Let q be a positive semidefinite quadratic form on $\mathbb{Z}^n$. Then $q(\mathbf{x}) = 0$ if and only if $D_i q(\mathbf{x}) = 0$ for all i such that $1 \le i \le n$.*

Proof. If $D_i q(\mathbf{x}) = 0$ for all i, then $(\mathbf{e}_i, \mathbf{x}) = 0$ for all i. Consequently, $q(\mathbf{x}) = (\mathbf{x}, \mathbf{x}) = \sum_{i=1}^{n} x_i (\mathbf{e}_i, \mathbf{x}) = 0$.

Conversely, assume that $q(\mathbf{x}) = 0$. For all $\lambda \in \mathbb{R}$ and $\mathbf{y} \in \mathbb{R}^n$, we have $q(\lambda \mathbf{y}) = \lambda^2 q(\mathbf{y})$. Because, by hypothesis, $q(\mathbf{y}) \ge 0$ for all $\mathbf{y} \in \mathbb{Z}^n$, we have $q(\mathbf{y}) \ge 0$ for all $\mathbf{y} \in \mathbb{Q}^n$. The continuity of q and the density of $\mathbb{Q}^n$ in $\mathbb{R}^n$ imply that $q(\mathbf{y}) \ge 0$ for all $\mathbf{y} \in \mathbb{R}^n$. Thus $q(\mathbf{x}) = 0$ if and only if the function $q : \mathbb{R}^n \to \mathbb{R}$ admits a global minimum at $\mathbf{x}$: the partial derivatives must then vanish at this point. $\square$

Let q be an integral quadratic form on $\mathbb{Z}^n$. A vector $\mathbf{x} \in \mathbb{Z}^n$ such that $q(\mathbf{x}) = 1$ is called a **root** of q. All the vectors of the canonical basis $\{\mathbf{e}_1, \ldots, \mathbf{e}_n\}$ of $\mathbb{Z}^n$ are clearly roots of q. The reason for studying roots is that, as we shall see, over a representation–finite hereditary algebra, there exists a bijection between the positive roots of the Euler quadratic form and the isomorphism classes of indecomposable modules. The following fundamental result, due to Drozd [59], shows that weakly positive quadratic forms have only finitely many roots that are positive vectors of $\mathbb{Z}^n$.

3.4. Proposition. *Let q be a weakly positive integral quadratic form on $\mathbb{Z}^n$. Then q has only finitely many positive roots.*

Proof. We consider q as a function from $\mathbb{R}^n$ to $\mathbb{R}$. As in the proof of (3.3), we see that $q(\mathbf{x}) > 0$ for all $\mathbf{x} > 0$ in $\mathbb{Q}^n$ and hence $q(\mathbf{x}) \geq 0$ for all $\mathbf{x} > 0$ in $\mathbb{R}^n$. We show by induction on n that in fact $q(\mathbf{x}) > 0$ for all $\mathbf{x} > 0$ in $\mathbb{R}^n$.

This is trivial if $n = 1$ because if $\lambda \in \mathbb{R}$, $\lambda \neq 0$, then $q(\lambda) = \lambda^2 q(1) > 0$. Assume that there exists a weakly positive quadratic form q in n indeterminates (with $n \geq 2$) and a positive vector $\mathbf{x} \in \mathbb{R}^n$ such that $q(\mathbf{x}) = 0$. It follows from the induction hypothesis that we can assume all the components x_i of $\mathbf{x}$ to be strictly positive. Then $\mathbf{x}$ lies in the positive cone of $\mathbb{R}^n$ and q attains a local minimum at $\mathbf{x}$. Consequently, we have $D_1 q(\mathbf{x}) = \ldots = D_n q(\mathbf{x}) = 0$. The linear forms $D_i q$ have integral, hence rational, coefficients, and $\mathbf{x} \in \bigcap_{i=1}^{n} \operatorname{Ker} D_i q$ implies that the real vector space

$$V = \{\mathbf{z} \in \mathbb{R}^n \mid D_1 q(\mathbf{z}) = \ldots = D_n q(\mathbf{z}) = 0\}$$

is nonzero . Hence the rank of the $n \times n$ matrix (with rational coefficients) determining this system of linear equations is smaller than n. Thus the rational vector space

$$U = \{\mathbf{y} \in \mathbb{Q}^n \mid D_1 q(\mathbf{y}) = \ldots = D_n q(\mathbf{y}) = 0\}$$

is nonzero, and V has a basis contained in U. In particular, V is the closure of U, because $\mathbb{Q}$ is dense in $\mathbb{R}$. Therefore, there exists a positive vector $\mathbf{x}'$ with rational coefficients lying in $\bigcap_{i=1}^{n} \operatorname{Ker} D_i q$. But then $q(\mathbf{x}') = 0$ because of (3.3) and the fact that $D_i q(\mathbf{x}') = 0$ for all $1 \leq i \leq n$, and this contradicts the fact that $q(\mathbf{x}') > 0$ because $\mathbf{x}' \in \mathbb{Q}^n$ is a positive vector. This completes the proof of our claim that $q(\mathbf{x}) > 0$ for all $\mathbf{x} > 0$ in $\mathbb{R}^n$.

Let now $\| - \| : \mathbb{R}^n \to \mathbb{R}$ denote the Euclidean norm. Because the set $C = \{x \in \mathbb{R}^n \mid \mathbf{x} > 0, \|\mathbf{x}\| = 1\}$ is compact in $\mathbb{R}^n$, $q|_C$ attains its minimum μ on a point of C. It follows from the preceding discussion that $\mu > 0$. For each $\mathbf{x} > 0$ in $\mathbb{R}^n$, we have

$$\mu \leq q\left(\frac{\mathbf{x}}{\|\mathbf{x}\|}\right) = \frac{1}{\|\mathbf{x}\|^2} q(\mathbf{x}).$$

Consequently, $\|\mathbf{y}\| \leq \frac{1}{\sqrt{\mu}}$ for each positive root $\mathbf{y}$ of q. Thus, q has only finitely many positive roots. $\qquad\square$

3.5. Corollary. *A weakly positive integral quadratic form always admits maximal positive roots.* □

Let $\mathbf{x} = \sum_{i=1}^{n} x_i e_i$ be a vector in $\mathbb{Z}^n$. Its **support** is the subset of $\{1, \ldots, n\}$ defined by $\operatorname{supp} \mathbf{x} = \{i \mid 1 \le i \le n,\ x_i \ne 0\}$.

3.6. Lemma. *Let q be a weakly positive integral quadratic form on $\mathbb{Z}^n$ and $\mathbf{x}$ be a positive root of q such that $\mathbf{x} \ne e_i$ for all i. Then there exists $i \in \operatorname{supp} \mathbf{x}$ such that $D_i q(\mathbf{x}) = 1$.*

Proof. We have $\sum_{i=1}^{n} x_i D_i q(\mathbf{x}) = 2 \sum_{i=1}^{n} x_i (e_i, \mathbf{x}) = 2(\mathbf{x}, \mathbf{x}) = 2$; hence there exists i such that $x_i D_i q(\mathbf{x}) \ge 1$. Because $\mathbf{x} > 0$, we have $x_i \ge 1$ and $D_i q(\mathbf{x}) \ge 1$. Therefore, $i \in \operatorname{supp} \mathbf{x}$. Because $\mathbf{x} \ne e_i$ by hypothesis, $\mathbf{x} - e_i > 0$ and

$$0 < q(\mathbf{x} - e_i) = q(\mathbf{x}) + q(e_i) - 2(e_i, \mathbf{x}) = 2 - D_i q(\mathbf{x})$$

gives $D_i q(\mathbf{x}) < 2$. Consequently, $D_i q(\mathbf{x}) = 1$. □

Let q be an integral quadratic form on $\mathbb{Z}^n$ and let $(-, -)$ be the corresponding symmetric bilinear form on $\mathbb{Z}^n$. For each i with $1 \le i \le n$, we define a mapping $s_i : \mathbb{Z}^n \to \mathbb{Z}^n$ by

$$s_i(\mathbf{x}) = \mathbf{x} - 2(\mathbf{x}, e_i) e_i.$$

Such a mapping is called a **reflection** at i. Note that $s_i(e_i) = -e_i$: that is, s_i transforms e_i to its negative. The properties of reflections are summarised in the following lemma.

3.7. Lemma. *Let $s_i : \mathbb{Z}^n \to \mathbb{Z}^n$ be a reflection. Then*

(a) *s_i is a group homomorphism;*
(b) *$(s_i(\mathbf{x}), s_i(\mathbf{y})) = (\mathbf{x}, \mathbf{y})$ for all $\mathbf{x}, \mathbf{y} \in \mathbb{Z}^n$; and*
(c) *$s_i^2 = 1$, thus s_i is an automorphism of $\mathbb{Z}^n$.*

Proof. (a) This is evident.
(b) $(s_i(\mathbf{x}), s_i(\mathbf{y})) = (\mathbf{x}, \mathbf{y}) - 2(\mathbf{x}, e_i)(\mathbf{y}, e_i) - 2(\mathbf{y}, e_i)(\mathbf{x}, e_i) + 4(\mathbf{x}, e_i)(\mathbf{y}, e_i)$
$\qquad = (\mathbf{x}, \mathbf{y})$.
(c) $s_i(s_i(\mathbf{x})) = s_i(\mathbf{x} - 2(\mathbf{x}, e_i) e_i) = \mathbf{x} - 2(\mathbf{x}, e_i) e_i + 2(\mathbf{x}, e_i) e_i = \mathbf{x}$. □

3.8. Lemma. *Let q be a weakly positive integral quadratic form on $\mathbb{Z}^n$ and $\mathbf{x}$ be a positive root of q such that $\mathbf{x} \ne e_i$ for all i. Then there exists $i \in \operatorname{supp} \mathbf{x}$ such that $s_i(\mathbf{x}) = \mathbf{x} - e_i$ is still a positive root.*

Proof. By (3.6), there exists $i \in \operatorname{supp} \mathbf{x}$ such that $D_i q(\mathbf{x}) = 1$. Now $D_i q(\mathbf{x}) = 2(\mathbf{x}, e_i)$ so that $s_i(\mathbf{x}) = \mathbf{x} - e_i > 0$. □

3.9. Corollary. *Let q be a weakly positive integral quadratic form on $\mathbb{Z}^n$ and $\mathbf{x}$ be a positive root of q. There exists a sequence $i_1, \ldots, i_t$, j of elements of $\{1, \ldots, n\}$ such that*

$$\mathbf{x} > s_{i_1}(\mathbf{x}) > s_{i_2} s_{i_1}(\mathbf{x}) > \ldots > s_{i_t} \ldots s_{i_1}(\mathbf{x}) = \mathbf{e}_j.$$

Proof. This follows at once from (3.8) and induction. $\qquad\square$

3.10. Definition. Let q be a weakly positive integral quadratic form on $\mathbb{Z}^n$. The subgroup W_q of the automorphism group of $\mathbb{Z}^n$ generated by the reflections $s_1, \ldots, s_n$ is called the **Weyl group** of q. A root $\mathbf{x}$ of q is called a **Weyl root** if there exist $w \in W_q$ and i with $1 \leq i \leq n$ such that $\mathbf{x} = w\mathbf{e}_i$.

It follows from (3.9) and (3.7)(c) that every positive root $\mathbf{x}$ of a weakly positive integral quadratic form q can be written as $\mathbf{x} = s_{i_1} \ldots s_{i_t} \mathbf{e}_j$: that is, every positive root of a weakly positive form is a Weyl root.

As is shown later, this applies to the Euler quadratic form for the representation–finite hereditary algebras; in this case, the form is positive definite, hence weakly positive, and therefore all positive roots are Weyl roots.

We end this section with an observation due to Happel [86] showing that the converse to (3.4) also holds.

3.11. Proposition. *Let q be an integral quadratic form having only finitely many positive roots. Then q is weakly positive.*

Proof. Let q be an integral quadratic form on $\mathbb{Z}^n$. Suppose that q is not weakly positive. Then $n \geq 2$ and there exists a positive vector $\mathbf{x} = [x_1 \ldots x_n]^t \in \mathbb{Z}^n$ such that $q(\mathbf{x}) \leq 0$. Because any restriction of q to a smaller number of indeterminates has also finitely many positive roots, we may assume that $x_i > 0$ for all i with $1 \leq i \leq n$. Clearly, we may also assume that $q(\mathbf{x}') > 0$ for any vector $\mathbf{x}' \in \mathbb{Z}^n$ with $0 < \mathbf{x}' < \mathbf{x}$. By our assumption on q, we may also choose a maximal positive root $\mathbf{y}$ of q. Then $(\mathbf{y}, \mathbf{e}_i) \geq 0$ for all i with $1 \leq i \leq n$, because, by (3.7), the reflections $s_i(\mathbf{y}) = \mathbf{y} - 2(\mathbf{y}, \mathbf{e}_i)\mathbf{e}_i$ are also roots of q. We claim that $(\mathbf{x}, \mathbf{y}) > 0$. Indeed, if $(\mathbf{x}, \mathbf{y}) \leq 0$ then $\sum_{i=1}^{n} x_i(\mathbf{e}_i, \mathbf{y}) \leq 0$, and hence $(\mathbf{e}_i, \mathbf{y}) = (\mathbf{y}, \mathbf{e}_i) = 0$ for all i with $1 \leq i \leq n$. But then we get $1 = q(\mathbf{y}) = (\mathbf{y}, \mathbf{y}) = \sum_{i=1}^{n} y_i(\mathbf{e}_i, \mathbf{y}) = 0$, a contradiction. Therefore, $\sum_{i=1}^{n} y_i(\mathbf{x}, \mathbf{e}_i) = (\mathbf{x}, \mathbf{y}) > 0$ and there exists i with $1 \leq i \leq n$ such that $(\mathbf{x}, \mathbf{e}_i) > 0$, because $\mathbf{y} > 0$. Take now $\mathbf{z} = \mathbf{x} - \mathbf{e}_i$. Then $\mathbf{z} > 0$ and $q(\mathbf{z}) = q(\mathbf{x} - \mathbf{e}_i) = q(\mathbf{x}) + q(\mathbf{e}_i) - 2(\mathbf{x}, \mathbf{e}_i) = 2 - 2(\mathbf{x}, \mathbf{e}_i) \leq 0$. This contradicts our choice of $\mathbf{x}$. Thus, q is weakly positive. $\qquad\square$

VII.4. The quadratic form of a quiver

Throughout this section, we let Q denote a finite, connected, and acyclic quiver. If we let $n = |Q_0|$ denote the number of points in Q, it follows from (III.3.5) that the Grothendieck group $K_0(KQ)$ of the path algebra KQ is isomorphic to $\mathbb{Z}^n$. We denote, as usual, by $\{e_1, \dots, e_n\}$ the canonical basis of $\mathbb{Z}^n$. It is sometimes convenient to work in a $\mathbb{Q}$-vector space rather than in the abelian group $\mathbb{Z}^n$. For this purpose, we denote by E the $\mathbb{Q}$-vector space

$$E = K_0(KQ) \otimes_{\mathbb{Z}} \mathbb{Q} \cong \mathbb{Q}^n$$

and by F the subgroup of E consisting of the vectors having only integral coordinates, that is,

$$F = \bigoplus_{i=1}^{n} e_i \mathbb{Z} \cong \mathbb{Z}^n \cong K_0(KQ).$$

The **quadratic form of a quiver** Q is defined to be the form

$$q_Q(\mathbf{x}) = \sum_{i \in Q_0} x_i^2 - \sum_{\alpha \in Q_1} x_{s(\alpha)} x_{t(\alpha)},$$

where $\mathbf{x} = [x_1 \dots x_n]^t \in \mathbb{Z}^n$.

Our first objective is to describe the Euler quadratic form of KQ by means of the quadratic form q_Q.

A first, but important, observation is that q_Q depends only on the underlying graph $\overline{Q}$ of Q, not on the particular orientation of the arrows in Q.

4.1. Lemma. *Let Q be a finite, connected, and acyclic quiver. Then the Euler quadratic form q_A of the path algebra $A = KQ$ and the quadratic form q_Q of the quiver Q coincide. Moreover,*

$$q_A(\mathbf{x}) = \sum_{i \in Q_0} x_i^2 - \sum_{i,j \in Q_0} a_{ij} x_i x_j,$$

where $a_{ij} = \dim_K \mathrm{Ext}_A^1(S(i), S(j))$.

Proof. By (III.3.13), the Euler characteristic is the bilinear form defined on the dimension vectors of the simple KQ-modules $S(i)$ by:

$$\langle \mathbf{dim}\, S(i), \mathbf{dim}\, S(j) \rangle = \sum_{l \geq 0} (-1)^l \dim_K \mathrm{Ext}_{KQ}^l(S(i), S(j))$$

$$= \dim_K \mathrm{Hom}_{KQ}(S(i), S(j)) - \dim_K \mathrm{Ext}_{KQ}^1(S(i), S(j)),$$

because, by (1.4) and (1.7), $\mathrm{gl.dim}\, KQ \leq 1$. Because there are no loops in Q at i, by (III.2.12), $\dim_K \mathrm{Ext}^1_{KQ}(S(i), S(j))$ equals the number a_{ij} of arrows from i to j.

Taking $i = j$, we get $\langle \mathbf{e}_i, \mathbf{e}_i \rangle = \langle \mathbf{dim}\, S(i), \mathbf{dim}\, S(i) \rangle = 1$. On the other hand, if $i \neq j$, we get

$$\langle \mathbf{e}_i, \mathbf{e}_j \rangle = \langle \mathbf{dim}\, S(i), \mathbf{dim}\, S(j) \rangle = -\dim_K \mathrm{Ext}^1_{KQ}(S(i), S(j)) = -a_{ij}.$$

Hence, for two arbitrary vectors $\mathbf{x} = \sum_{i=1}^n x_i \mathbf{e}_i$ and $\mathbf{y} = \sum_{i=1}^n y_i \mathbf{e}_i$, we get

$$\begin{aligned}
\langle \mathbf{x}, \mathbf{y} \rangle &= \sum_{i,j=1}^n x_i y_j \langle \mathbf{e}_i, \mathbf{e}_j \rangle = \sum_{i \in Q_0} x_i y_i - \sum_{i,j \in Q_0} a_{ij} x_i y_j \\
&= \sum_{i \in Q_0} x_i y_i - \sum_{\alpha \in Q_1} x_{s(\alpha)} y_{t(\alpha)}.
\end{aligned}$$

The result follows at once. $\qquad\qquad\square$

The Euler quadratic form of the algebra KQ will be simply referred to as the quadratic form of the quiver Q.

We denote by $(-, -)$ the symmetric bilinear form corresponding to q_Q, that is, the symmetrisation of the Euler characteristic. Thus,

$$(\mathbf{x}, \mathbf{y}) = \sum_{i \in Q_0} x_i y_i - \frac{1}{2} \sum_{\alpha \in Q_1} \{ x_{s(\alpha)} y_{t(\alpha)} + x_{t(\alpha)} y_{s(\alpha)} \}.$$

This can also be expressed in terms of the Cartan matrix $\mathbf{C}_{KQ}$; indeed, $\langle \mathbf{x}, \mathbf{y} \rangle = \mathbf{x}^t (\mathbf{C}_{KQ}^{-1})^t \mathbf{y}$, hence

$$(\mathbf{x}, \mathbf{y}) = \mathbf{x}^t \left[\frac{1}{2} (\mathbf{C}_{KQ}^{-1} + (\mathbf{C}_{KQ}^{-1})^t) \right] \mathbf{y}.$$

Clearly, $(\mathbf{x}, \mathbf{x}) = q_Q(\mathbf{x})$ for all $\mathbf{x}$, and $(\mathbf{x}, \mathbf{y}) = \frac{1}{4}[q_Q(\mathbf{x} + \mathbf{y}) - q_Q(\mathbf{x} - \mathbf{y})]$ for all $\mathbf{x}, \mathbf{y}$.

For example, if Q is the quiver

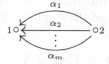

then $q_Q(\mathbf{x}) = x_1^2 + x_2^2 - m x_1 x_2 = (x_1 - \frac{m}{2} x_2)^2 + (1 - \frac{m^2}{4}) x_2^2$. Consequently, q_Q is positive definite if $m = 1$, semidefinite of corank 1 if $m = 2$, and indefinite if $m \geq 3$. Observe also that for $m \geq 2$ and $\mathbf{x} = (m, m)^t$ we have $q_Q(\mathbf{x}) \leq 0$, and hence q_Q is not weakly positive.

We saw in Section 3 that if q_Q is positive semidefinite then its radical $\operatorname{rad} q_Q = \{x \in F;\ q_Q(x) = 0\}$ is a subgroup of $F \cong \mathbb{Z}^n$. After tensoring by $_{\mathbb{Z}}\mathbb{Q}$, it yields a subspace of the $\mathbb{Q}$-vector space

$$E = K_0(KQ) \otimes_{\mathbb{Z}} \mathbb{Q} \cong \mathbb{Q}^n,$$

denoted by $(\operatorname{rad} q_Q)\mathbb{Q}$. The dimension of this subspace $(\operatorname{rad} q_Q)\mathbb{Q}$ equals the corank of q_Q. The following purely computational lemma provides many examples of quivers with positive semidefinite form.

4.2. Lemma. *Let Q be a quiver whose underlying graph $\overline{Q}$ is Euclidean. Then q_Q is positive semidefinite of corank one and $\operatorname{rad} q_Q = \mathbb{Z}\mathbf{h}_Q$, where $\mathbf{h}_Q$ is the vector*

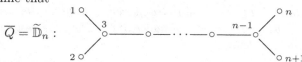

in case $\overline{Q}$ is the graph $\widetilde{\mathbb{A}}_m$, $\widetilde{\mathbb{D}}_m$, $\widetilde{\mathbb{E}}_6$, $\widetilde{\mathbb{E}}_7$, and $\widetilde{\mathbb{E}}_8$, respectively.

Proof. (i) Assume that

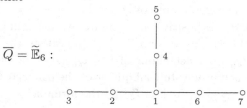

for some $m \geq 1$. Then $2q_Q(\mathbf{x}) = \sum_{i-j}(x_i - x_j)^2$, where the sum is taken over all edges $i{-}j$ in $\overline{Q}$. It follows that q_Q is positive semidefinite of corank 1 and a generator of $\operatorname{rad} q_Q$ is given by $\begin{smallmatrix} &1&\ldots&1& \\ 1&1&\ldots&1&1 \end{smallmatrix}$.

(ii) Assume that

$$\overline{Q} = \widetilde{\mathbb{D}}_n :$$

for some $n \geq 4$. Then $4q_Q(\mathbf{x}) = (2x_1 - x_3)^2 + (2x_2 - x_3)^2 + (x_{n-1} - 2x_n)^2 + (x_{n-1} - 2x_{n+1})^2 + 2\sum_{i=3}^{n-2}(x_i - x_{i+1})^2$. It follows that q_Q is positive semidefinite of corank 1 and a generator of $\operatorname{rad} q_Q$ is given by $\begin{smallmatrix} 1 \\ \ \\ 1 \end{smallmatrix}2\ldots2\begin{smallmatrix} 1 \\ \ \\ 1 \end{smallmatrix}$.

(iii) Assume that

$$\overline{Q} = \widetilde{\mathbb{E}}_6 :$$

Then $36q_Q(\mathbf{x}) = (6x_3 - 3x_2)^2 + (6x_7 - 3x_6)^2 + (6x_5 - 3x_4)^2 + 3[(3x_2 - 2x_1)^2 + (3x_6 - 2x_1)^2 + (3x_4 - 2x_1)^2]$. It follows that q_Q is positive semidefinite of corank 1 and a generator of $\operatorname{rad} q_Q$ is given by $1\,2\overset{1}{\underset{3}{}}\overset{2}{}2\,1$.

(iv) Assume that

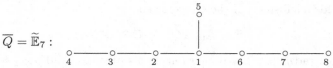

$$\overline{Q} = \widetilde{\mathbb{E}}_7 :$$

Then $24q_Q(\mathbf{x}) = 6[(2x_4 - x_3)^2 + (2x_8 - x_7)^2] + 2[(3x_3 - 2x_2)^2 + (3x_7 - 2x_6)^2] + (4x_2 - 3x_1)^2 + (4x_6 - 3x_1)^2 + 6(2x_5 - x_1)^2$. Here, q_Q is positive semidefinite of corank 1, a generator of $\operatorname{rad} q_Q$ is given by $1\,2\,3\,\overset{2}{4}\,3\,2\,1$.

(v) Assume that

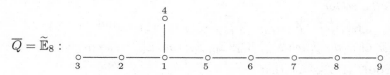

$$\overline{Q} = \widetilde{\mathbb{E}}_8 :$$

Then $120q_Q(\mathbf{x}) = 30(2x_9 - x_8)^2 + 10(3x_8 - 2x_7)^2 + 5(4x_7 - 3x_6)^2 + 3(5x_6 - 4x_5)^2 + 30(2x_3 - x_2)^2 + 2(6x_5 - 5x_1)^2 + 10(3x_2 - 2x_1)^2 + 30(2x_4 - x_1)^2$. It follows that q_Q is positive semidefinite of corank 1 and a generator of $\operatorname{rad} q_Q$ is given by $2\,4\,\overset{3}{6}\,5\,4\,3\,2\,1$. $\qquad\square$

We show later that the Dynkin and Euclidean graphs can in fact be characterised by the positivity of their quadratic forms. We need the following lemma.

4.3. Lemma. *Let Q be a connected quiver such that q_Q is positive semidefinite and Q' be a proper full subquiver of Q. Then the restriction $q_{Q'}$ of q_Q to Q' is positive definite.*

Proof. The form $q_{Q'}$ is certainly positive semidefinite, for every full subquiver Q' of Q. Let then Q' be a proper full subquiver of Q such that $q_{Q'}$ is not positive definite. We may, without loss of generality, assume Q' to be minimal with this property. Let $\mathbf{x}' = \sum x_i' \mathbf{e}_i$ be a nonzero vector such that $q_{Q'}(\mathbf{x}') = 0$. The minimality of Q' implies that $x_i' \neq 0$ for each $i \in Q_0'$. Actually, because $q_{Q'}$ is positive semidefinite, we may suppose that $x_i' > 0$ for each $i \in Q_0'$; indeed, the vector $\mathbf{x}'' = \sum |x_i'| \mathbf{e}_i$ satisfies $q_{Q'}(\mathbf{x}'') \le q_{Q'}(\mathbf{x}')$.

Let $j \in Q_0 \setminus Q_0'$ be a neighbour of $k \in Q_0'$ (such points j, k certainly exist, because Q' is a proper full subquiver of the connected quiver Q). We define a vector $\mathbf{x} = \sum x_i \mathbf{e}_i$ in $E = \mathbb{Q}^n$ by the formula

VII.4. THE QUADRATIC FORM OF A QUIVER

$$x_i = \begin{cases} x'_i & \text{if } i \in Q'_0, \\ \frac{1}{2}x'_k & \text{if } i = j, \\ 0 & \text{otherwise.} \end{cases}$$

Then $q_Q(\mathbf{x}) = q_Q(\mathbf{x}'+x_j\mathbf{e}_j) = q_{Q'}(\mathbf{x}')+x_j^2-\sum_{l-j} x'_l x_j = x_j^2-\sum_{l-j} x'_l x_j \le x_j^2 - x'_k x_j = \frac{1}{4}x'^2_k - \frac{1}{2}x'^2_k = -\frac{1}{4}x'^2_k < 0$, which is a contradiction. □

4.4. Corollary. *Let Q be a quiver whose underlying graph is Dynkin. Then q_Q is positive definite.*

Proof. This follows from (4.2), (4.3), and the observation that each quiver whose underlying graph is Dynkin is a proper full subquiver of a quiver whose underlying graph is Euclidean. □

We are now able to prove the characterisation of the Dynkin and Euclidean graphs by means of their quadratic forms.

4.5. Proposition. *Let Q be a finite, connected, and acyclic quiver and let $\overline{Q}$ be the underlying graph of Q.*

(a) *$\overline{Q}$ is a Dynkin graph if and only if q_Q is positive definite.*

(b) *$\overline{Q}$ is a Euclidean graph if and only if q_Q is positive semidefinite but not positive definite.*

(c) *$\overline{Q}$ is neither a Dynkin nor a Euclidean graph if and only if q_Q is indefinite.*

Proof. The necessity of (a) follows from (4.4) and the necessity of (b) follows from (4.2). Conversely, assume q_Q to be positive semidefinite. Then it follows from the example preceding (4.2) that $\overline{Q}$ does not contain a full subgraph consisting of two points connected by more than two edges. Hence, if $\overline{Q}$ is not Dynkin, then, by (2.1), $\overline{Q}$ contains a Euclidean graph as a full subgraph. By (4.3), this Euclidean subgraph cannot be proper. Hence $\overline{Q}$ is Euclidean. This shows (a) and (b).

Let Q be such that $\overline{Q}$ is neither a Dynkin nor a Euclidean graph. By (a) and (b), q_Q is not positive semidefinite. Consequently, it is indefinite. The converse follows clearly from the sufficiency parts of (a) and (b). □

We may clearly strengthen condition (b) as follows: $\overline{Q}$ is a Euclidean graph if and only if q_Q is positive semidefinite of corank one.

4.6. Corollary. *Let Q be a finite, connected, and acyclic quiver. The following conditions are equivalent:*

(a) *q_Q is weakly positive.*

(b) q_Q *is positive definite.*

(c) *The underlying graph* $\overline{Q}$ *of* Q *is a Dynkin graph.*

Proof. We have seen that (b) and (c) are equivalent, and (b) implies (a) trivially. Assume that q_Q is weakly positive. Then again $\overline{Q}$ does not contain a full subgraph consisting of two vertices connected by at least two edges. Hence, if q_Q is not positive definite, then $\overline{Q}$ is not Dynkin so that, by (2.1), Q contains a full subquiver Q' whose underlying graph is Euclidean. We computed in (4.2) generators for the (one-dimensional) radical subspaces of the forms arising from Euclidean graphs. Let $\mathbf{x}'$ be the generator of the radical subspace of $q_{Q'}$. As seen in (4.2), $\mathbf{x}'$ is positive. Consider the vector $\mathbf{x}$ defined by

$$\mathbf{x}_i = \begin{cases} x_i' & \text{if } i \in Q_0', \\ 0 & \text{if } i \notin Q_0'. \end{cases}$$

Clearly, $\mathbf{x}$ is positive and $q_Q(\mathbf{x}) = 0$. Thus q_Q is not weakly positive. $\square$

A consequence of this corollary and the results of Section 3 is that if $\overline{Q}$ is a Dynkin graph, then the positive roots of q_Q are Weyl roots and there are only finitely many such positive roots. We thus proceed to define reflections and the Weyl roots for the quadratic form q_Q of a finite, connected, and acyclic quiver Q . We recall that $E = \mathbb{Q}^n$ and $F = \mathbb{Z}^n$. For each point $i \in Q_0$, we define the **reflection** $s_i : E \to E$ at i to be the $\mathbb{Q}$ -linear map given by

$$s_i(\mathbf{x}) = \mathbf{x} - 2(\mathbf{x}, \mathbf{e}_i)\mathbf{e}_i$$

for $\mathbf{x} \in E$. In terms of the coordinates x_i of $\mathbf{x}$ in the canonical basis $\{\mathbf{e}_1, \ldots, \mathbf{e}_n\}$ of E , we see that $\mathbf{y} = s_i(\mathbf{x})$ has coordinates

$$y_j = \begin{cases} x_j & \text{if } j \neq i, \\ -x_i + \sum_{k \text{—} i} x_k & \text{if } j = i, \end{cases}$$

where the sum is taken over all edges k — i . Because $s_i(F) \subseteq F$, we see that s_i is indeed a reflection in the sense of Section 3.

For example, if Q is the quiver

$$\overset{1}{\circ} \longleftarrow \overset{3}{\circ} \longrightarrow \overset{2}{\circ}$$

whose underlying graph is the Dynkin graph $\mathbb{A}_3$, then $E \cong \mathbb{Q}^3$ and the reflections s_1 , s_2 , s_3 are expressed by their matrices in the canonical basis as

$$\mathbf{s}_1 = \begin{bmatrix} -1 & 0 & 1 \\ 0 & 1 & 0 \\ 0 & 0 & 1 \end{bmatrix}, \qquad \mathbf{s}_2 = \begin{bmatrix} 1 & 0 & 0 \\ 0 & -1 & 1 \\ 0 & 0 & 1 \end{bmatrix}, \qquad \mathbf{s}_3 = \begin{bmatrix} 1 & 0 & 0 \\ 0 & 1 & 0 \\ 1 & 1 & -1 \end{bmatrix}.$$

The **Weyl group** W_Q of Q is the Weyl group of the quadratic form q_Q, that is, the group of automorphisms of $E = \mathbb{Q}^n$ generated by the set of reflections $\{s_i\}_{i \in Q_0}$.

Because, by hypothesis, Q is acyclic, there exists a bijection between Q_0 and the set $\{1, \ldots, n\}$ such that if we have an arrow $j \to i$, then $j > i$; indeed, such a bijection is constructed as follows. Let 1 be any sink in Q, then consider the full subquiver $Q(1)$ of Q having as set of points $Q_0 \setminus \{1\}$; let 2 be a sink of $Q(1)$, and continue by induction. Such a numbering of the points of Q is called an **admissible numbering**. For instance, in the preceding example, the shown numbering of the points is admissible. Clearly, a given quiver Q usually admits many possible admissible numberings of the set of points.

Let $(a_1, \ldots, a_n)$ be an admissible numbering of the points of Q and let $E = \mathbb{Q}^n$. The element

$$c = s_{a_n} \ldots s_{a_2} s_{a_1} : E \longrightarrow E$$

of the Weyl group W_Q of Q is called the **Coxeter transformation** of Q (corresponding to the given admissible numbering). Because, for each i, we have $s_{a_i}^2 = 1$, clearly, $c^{-1} = s_{a_1} s_{a_2} \ldots s_{a_n}$. For instance, in the example, the matrices of c and c^{-1} in the canonical basis are

$$\mathbf{c} = \mathbf{s}_3 \mathbf{s}_2 \mathbf{s}_1 = \begin{bmatrix} 1 & 0 & 0 \\ 0 & 1 & 0 \\ 1 & 1 & -1 \end{bmatrix} \begin{bmatrix} 1 & 0 & 0 \\ 0 & -1 & 1 \\ 0 & 0 & 1 \end{bmatrix} \begin{bmatrix} -1 & 0 & 1 \\ 0 & 1 & 0 \\ 0 & 0 & 1 \end{bmatrix} = \begin{bmatrix} -1 & 0 & 1 \\ 0 & -1 & 1 \\ -1 & -1 & 1 \end{bmatrix}$$

and

$$\mathbf{c}^{-1} = \mathbf{s}_1 \mathbf{s}_2 \mathbf{s}_3 = \begin{bmatrix} -1 & 0 & 1 \\ 0 & 1 & 0 \\ 0 & 0 & 1 \end{bmatrix} \begin{bmatrix} 1 & 0 & 0 \\ 0 & -1 & 1 \\ 0 & 0 & 1 \end{bmatrix} \begin{bmatrix} 1 & 0 & 0 \\ 0 & 1 & 0 \\ 1 & 1 & -1 \end{bmatrix} = \begin{bmatrix} 0 & 1 & -1 \\ 1 & 0 & -1 \\ 1 & 1 & -1 \end{bmatrix}.$$

It turns out that the Coxeter transformation only depends on the quiver Q, not on the admissible numbering chosen. Indeed, if $(a_1, \ldots, a_n)$ and $(b_1, \ldots, b_n)$ are two admissible numberings of the points of Q, then there exists an i with $1 \leq i \leq n$ such that $b_1 = a_i$; because b_1 is a sink, there exists no edge $a_j - a_i$ with $j < i$ and, because it is easily seen that reflections corresponding to non-neighbours commute, we have $s_{a_j} s_{a_i} = s_{a_i} s_{a_j}$ for all $j < i$. The numbering $(a_i, a_1, \ldots, a_{i-1}, a_{i+1}, \ldots, a_n)$ is admissible and an obvious induction implies that $s_{a_n} \ldots s_{a_1} = s_{b_n} \ldots s_{b_1}$. We thus refer to c as being the Coxeter transformation of the quiver Q.

The matrix of the Coxeter transformation c, as defined earlier, is just the Coxeter matrix $\mathbf{\Phi}_{KQ}$ of KQ, as defined in (III.3.14).

4.7. Proposition. *The matrix of the Coxeter transformation* $c : E \to E$ *of a quiver Q in the canonical basis of E is equal to the Coxeter matrix* Φ_{KQ} *of KQ.*

Proof. We recall that $\Phi_{KQ} = -\mathbf{C}_{KQ}^t \mathbf{C}_{KQ}^{-1}$, where $\mathbf{C}_{KQ}$ denotes the Cartan matrix of KQ. Assume that $(1, \ldots, n)$ is an admissible numbering of Q_0. Identifying the reflections s_i and the Coxeter transformation c to their matrices in the canonical basis of the $\mathbb{Q}$-vector space $E = \mathbb{Q}^n$, we must show that $-\mathbf{C}_{KQ}^t \mathbf{C}_{KQ}^{-1} = s_n \ldots s_1$. For this purpose, it suffices to show that $-\mathbf{C}_{KQ}^t = s_n \ldots s_1 \mathbf{C}_{KQ}$, or, equivalently, that

$$\mathbf{C}_{KQ}^t s_1^t \ldots s_n^t = -\mathbf{C}_{KQ}.$$

We show by induction on k that

$$\mathbf{C}_{KQ}^t s_1^t \ldots s_k^t = [-\mathbf{C}_k \mid \mathbf{C}_{n-k}^t],$$

where $\mathbf{C}_k$ (or $\mathbf{C}_{n-k}^t$) is the matrix formed by the k first columns of $\mathbf{C}_{KQ}$ (or of the $(n - k)$ last columns of $\mathbf{C}_{KQ}^t$, respectively). Recall that $c_{ij} = \dim_K \varepsilon_j(KQ)\varepsilon_i$ is the (i,j)-coefficient of $\mathbf{C}_{KQ}$. Moreover, let a_{ij} be the number of arrows from j to i. It is easily seen that:

(1) $a_{ij} = 0$ for $i \geq j$ (because $(1, \ldots, n)$ is an admissible ordering of Q_0);
(2) $c_{i,i+1} = a_{i,i+1}$, for each i;
(3) $c_{ii} = 1$, for each i; and
(4) $c_{ij} = \sum_{i \leq k \leq j} a_{ik} c_{kj}$, for $i < j$.

For $k = 1$, we then have

$$
\mathbf{C}_{KQ}^t s_1^t =
\begin{bmatrix}
1 & 0 & 0 & \ldots & 0 \\
c_{12} & 1 & 0 & \ldots & 0 \\
c_{13} & c_{23} & 1 & \ldots & 0 \\
\vdots & \vdots & \vdots & \ddots & \vdots \\
c_{1n} & c_{2n} & c_{3n} & \ldots & 1
\end{bmatrix}
\cdot
\begin{bmatrix}
-1 & 0 & 0 & \ldots & 0 \\
a_{12} & 1 & 0 & \ldots & 0 \\
a_{13} & 0 & 1 & \ldots & 0 \\
\vdots & \vdots & \vdots & \ddots & \vdots \\
a_{1n} & 0 & 0 & \ldots & 1
\end{bmatrix}
$$

$$
=
\begin{bmatrix}
-1 & 0 & 0 & \ldots & 0 \\
-c_{12} + a_{12} & 1 & 0 & \ldots & 0 \\
-c_{13} + a_{12}c_{23} + a_{13} & c_{23} & 1 & \ldots & 0 \\
\vdots & \vdots & \vdots & \ddots & \vdots \\
-c_{1n} + \sum_{1 \leq i \leq n} a_{1i}c_{in} & c_{2n} & c_{3n} & \ldots & 1
\end{bmatrix}.
$$

Using (2), (3) and (4), we get $\mathbf{C}_{KQ}^t s_1^t = [-\mathbf{C}_1 \mid \mathbf{C}_{n-1}^t]$.

Assume the result to hold for $k - 1$. Then

$$\mathbf{C}_{KQ}^t s_1^t \dots s_k^t = [-\mathbf{C}_{k-1} \mid \mathbf{C}_{n-k+1}^t] s_k^t$$

$$= \begin{bmatrix} -1 & -c_{12} & \cdots & -c_{1,k-2} & -c_{1,k-1} & 0 & 0 & \cdots & 0 \\ 0 & -1 & \cdots & -c_{2,k-2} & -c_{2,k-1} & 0 & 0 & \cdots & 0 \\ \vdots & \vdots & \ddots & \vdots & \vdots & \vdots & \vdots & & \vdots \\ 0 & 0 & \cdots & -1 & -c_{k-2,k-1} & 0 & 0 & \cdots & 0 \\ 0 & 0 & \cdots & 0 & -1 & 0 & 0 & \cdots & 0 \\ 0 & 0 & \cdots & 0 & 0 & 1 & 0 & \cdots & 0 \\ 0 & 0 & \cdots & 0 & 0 & c_{k,k+1} & 1 & \cdots & 0 \\ \vdots & \vdots & & \vdots & \vdots & \vdots & \vdots & \ddots & \vdots \\ 0 & 0 & \cdots & 0 & 0 & c_{k,n} & c_{k+1,n} & \cdots & 1 \end{bmatrix} \cdot \begin{bmatrix} 1 & 0 & \cdots & 0 & a_{1k} & 0 & \cdots & 0 \\ 0 & 1 & \cdots & 0 & a_{2k} & 0 & \cdots & 0 \\ \vdots & \vdots & \ddots & \vdots & \vdots & \vdots & & \vdots \\ 0 & 0 & \cdots & 1 & a_{k-1,k} & 0 & \cdots & 0 \\ 0 & 0 & \cdots & 0 & -1 & 0 & \cdots & 0 \\ 0 & 0 & \cdots & 0 & 0 & 1 & \cdots & 0 \\ \vdots & \vdots & & \vdots & \vdots & \vdots & \ddots & \vdots \\ 0 & 0 & \cdots & 0 & 0 & 0 & \cdots & 1 \end{bmatrix}$$

$$= \begin{bmatrix} & & -\sum_{1 \le i \le k} a_{ik} c_{1i} & \\ & & -\sum_{2 \le i \le k} a_{ik} c_{2i} & \\ & & \vdots & \\ -\mathbf{C}_{k-1} & & -\sum_{k-2 \le i \le k} a_{ik} c_{k-2,i} & \mathbf{C}_{n-k}^t \\ & & -a_{k-1,k} & \\ & & -1 & \\ & & 0 & \\ & & \vdots & \\ & & 0 & \end{bmatrix}.$$

The conclusion follows from (2), (3), and (4). $\square$

For the rest of this section, we assume that Q is a quiver whose underlying graph $\overline{Q}$ is Dynkin. Then q_Q is positive definite and hence weakly positive. We denote by R, R^+, R^-, $R(W_Q)$, respectively, the sets of all roots, all positive roots, all negative roots, and all Weyl roots of q_Q. It follows from (3.4) that R^+ is a finite set and, from (3.9), that $R^+ \subseteq R(W_Q)$. We note that, if $\mathbf{x} \in F = \mathbb{Z}^n$ is a root, the vector $-\mathbf{x}$ is also a root, because $q_Q(-\mathbf{x}) = q_Q(\mathbf{x})$. In particular, the assignment $\mathbf{x} \mapsto -\mathbf{x}$ induces a bijection between R^+ and R^- (so that R^- is also finite).

4.8. Lemma. *Let Q be a quiver whose underlying graph is Dynkin. Then $R = R^+ \cup R^- = R(W_Q)$.*

Proof. To show that $R = R^+ \cup R^-$, it suffices to show that every root $\mathbf{x}$ of q_Q is either positive or negative. We may write $\mathbf{x} = \mathbf{x}^+ + \mathbf{x}^-$, where $\mathbf{x}^+$ is a vector all of whose nonzero coordinates are positive, while $\mathbf{x}^-$ is a vector all of whose nonzero coordinates are negative. Put $|\mathbf{x}| = \mathbf{x}^+ - \mathbf{x}^-$. Because $\mathbf{x}$ is a root, we have $\mathbf{x} \ne 0$. Hence $|\mathbf{x}| \ne 0$ and therefore, $|\mathbf{x}| > 0$. The inequalities $|\mathbf{x}|_j \ge \mathbf{x}_j$ and the equalities $|\mathbf{x}|_j^2 = \mathbf{x}_j^2$ for all $j \in Q_0$ yield

$$0 < q_Q(|\mathbf{x}|) = \sum_{i \in Q_0} |\mathbf{x}|_i^2 - \sum_{\alpha \in Q_1} |\mathbf{x}|_{s(\alpha)} |\mathbf{x}|_{t(\alpha)}$$

$$\le \sum_{i \in Q_0} \mathbf{x}_i^2 - \sum_{\alpha \in Q_1} \mathbf{x}_{s(\alpha)} \mathbf{x}_{t(\alpha)} = q_Q(\mathbf{x}) = 1,$$

and therefore $q_Q(|\mathbf{x}|) = 1$, that is, $|\mathbf{x}|$ is a root. Consequently the equalities

$$2 = q_Q(\mathbf{x}) + q_Q(|\mathbf{x}|) = q_Q(\mathbf{x}^+ + \mathbf{x}^-) + q_Q(\mathbf{x}^+ - \mathbf{x}^-) = 2[q_Q(\mathbf{x}^+) + q_Q(\mathbf{x}^-)]$$

yield $q_Q(\mathbf{x}^+) + q_Q(\mathbf{x}^-) = 1$. Because q_Q is positive definite, we have either $q_Q(\mathbf{x}^+) = 1$ and $q_Q(\mathbf{x}^-) = 0$ (hence $\mathbf{x} = \mathbf{x}^+ \in R^+$) or $q_Q(\mathbf{x}^-) = 1$ and $q_Q(\mathbf{x}^+) = 0$ (hence $\mathbf{x} = \mathbf{x}^- \in R^-$). This completes the proof that $R = R^+ \cup R^-$.

We have $R^+ \subseteq R(W_Q)$. Similarly, if $\mathbf{x} \in R^-$, then $\mathbf{x} \in R(W_Q)$; indeed, $-\mathbf{x} \in R^+$ gives $-\mathbf{x} = w\mathbf{e}_i$, for some $w \in W_Q$ and $i \in Q_0$, hence $\mathbf{x} = w(-\mathbf{e}_i) = ws_i(\mathbf{e}_i) \in R(W_Q)$. Thus $R^- \subseteq R(W_Q)$ and $R = R^+ \cup R^- \subseteq R(W_Q)$. Because, trivially, $R(W_Q) \subseteq R$, we have indeed $R = R(W_Q)$. $\square$

4.9. Proposition. *Let Q be a quiver whose underlying graph is Dynkin. Then the Weyl group W_Q of Q is finite.*

Proof. We show that W_Q is isomorphic to a subgroup of the group of permutations of R. Because, by (4.8), $R = R^+ \cup R^-$ is finite, this implies the statement.

We first observe that W_Q permutes the roots of q_Q because $q_Q(\mathbf{x}) = 1$ implies $q_Q(w\mathbf{x}) = 1$ for every $w \in W_Q$ (by (3.7)(b)). On the other hand, the action of W_Q on R is faithful, that is, the mapping $w \mapsto (\sigma_w : \mathbf{x} \mapsto w\mathbf{x})$, from W_Q into the group of permutations of R is injective; indeed, $\sigma_w = \sigma_v$ (for $w, v \in W_Q$) implies $w\mathbf{x} = v\mathbf{x}$ and hence $w^{-1}v\mathbf{x} = \mathbf{x}$ for every $\mathbf{x} \in R$. In particular, $w^{-1}v\mathbf{e}_i = \mathbf{e}_i$ for every $i \in Q_0$, which implies, by linearity, $w^{-1}v\mathbf{x} = \mathbf{x}$ for every $\mathbf{x} \in E$, that is, $w^{-1}v = 1$ and $w = v$. This proves our claim. $\square$

We need the following lemma.

4.10. Lemma. *Let Q be a quiver whose underlying graph is Dynkin, $\mathbf{x}$ be a positive root of q_Q, and i be a vertex of Q. Then either $s_i(\mathbf{x})$ is positive or $\mathbf{x} = \mathbf{e}_i$.*

Proof. From (3.7)(b), we know that $s_i(\mathbf{x})$ is a root of q_Q. Because q_Q is positive definite, we get the following:

$$0 \le q_Q(\mathbf{x} \pm \mathbf{e}_i) = (\mathbf{x} \pm e_i, \mathbf{x} \pm e_i) = q_Q(\mathbf{x}) + q_Q(\mathbf{e}_i) \pm 2(\mathbf{x}, \mathbf{e}_i) = 2(1 \pm (\mathbf{x}, \mathbf{e}_i)).$$

Hence $-1 \le (\mathbf{x}, \mathbf{e}_i) \le 1$. If $(\mathbf{x}, \mathbf{e}_i) = 1$, then $q_Q(\mathbf{x} - \mathbf{e}_i) = 0$ and consequently $\mathbf{x} = \mathbf{e}_i$. On the other hand, if $(\mathbf{x}, \mathbf{e}_i) \le 0$, then $s_i(\mathbf{x}) = \mathbf{x} - 2(\mathbf{x}, \mathbf{e}_i) > 0$, because $\mathbf{x} > 0$. This proves our claim. $\square$

4.11. Lemma. *Let Q be a finite, connected, and acyclic quiver; c be its Coxeter transformation; s_i be the reflection at i; and $\mathbf{x} \in E = \mathbb{Q}^n$. The following conditions are equivalent:*

(a) $c\mathbf{x} = \mathbf{x}$,

(b) $s_i\mathbf{x} = \mathbf{x}$ *for each point* $i \in Q_0$, *and*

(c) $(\mathbf{x}, \mathbf{y}) = 0$ *for each vector* $\mathbf{y} \in E$.

If, moreover, the underlying graph $\overline{Q}$ of Q is Dynkin or Euclidean, then the preceding conditions are equivalent to the following one:

(d) $q_Q(\mathbf{x}) = 0$.

Proof. Clearly, (b) implies (a). Conversely, if $(1, \ldots, n)$ is an admissible numbering of the points of Q, $c = s_n \ldots s_1$ and $c\mathbf{x} = \mathbf{x}$ holds, then, for any $i \in \{1, \ldots, n\}$, we have $x_i = (c\mathbf{x})_i = (s_n \ldots s_i\mathbf{x})_i$. Hence, by descending induction on i, we get $s_1\mathbf{x} = \ldots = s_n\mathbf{x} = \mathbf{x}$. The equivalence of (b) and (c) follows from the fact that $s_i\mathbf{x} = \mathbf{x}$ for each point $i \in Q_0$ is equivalent to $(\mathbf{x}, \mathbf{e}_i) = 0$ for each point $i \in Q_0$, where $\mathbf{e}_1, \ldots, \mathbf{e}_n$ is the standard basis of E.

If $\overline{Q}$ is Dynkin or Euclidean, then, by (4.5), the quadratic form q_Q is positive semidefinite. Therefore $|(\mathbf{x}, \mathbf{y})|^2 \leq q_Q(\mathbf{x})q_Q(\mathbf{y})$ for each vector $\mathbf{y} \in E$, so that (d) implies (c). The converse implication follows from the equality $q_Q(\mathbf{x}) = (\mathbf{x}, \mathbf{x})$. $\qquad\square$

4.12. Corollary. *Let Q be a quiver whose underlying graph is Dynkin and c be its Coxeter transformation.*

(a) *If $c\mathbf{x} = \mathbf{x}$ for a vector $\mathbf{x} \in E$, then $\mathbf{x} = 0$.*

(b) *For every positive vector $\mathbf{x}$, there exist $s \geq 0$ such that $c^s\mathbf{x} > 0$ but $c^{s+1}\mathbf{x} \not> 0$, and $t \geq 0$ such that $c^{-t}\mathbf{x} > 0$ but $c^{-t-1}\mathbf{x} \not> 0$.*

Proof. (a) If $c\mathbf{x} = \mathbf{x}$ then, by (4.11), we get $q_Q(\mathbf{x}) = 0$. Because, by (4.5), q_Q is positive definite, this implies $\mathbf{x} = 0$.

(b) Because W_Q is a finite group, c has finite order m (say). Consider the vector $\mathbf{y} = \mathbf{x} + c\mathbf{x} + \ldots + c^{m-1}\mathbf{x}$. Then $c\mathbf{y} = \mathbf{y}$. By (a), $\mathbf{y} = 0$. Therefore, there exists a least integer $s \geq 0$ such that $c^{s+1}\mathbf{x} \not> 0$ (and then $c^s\mathbf{x} > 0$). Similarly, one finds t as required. $\qquad\square$

The preceding corollary implies that one should look at those positive roots that become nonpositive after application of the Coxeter transformation.

4.13. Lemma. *Let Q be a quiver whose underlying graph is Dynkin and c be its Coxeter transformation. For a positive root $\mathbf{x}$, we have*

(a) $c\mathbf{x} \not> 0$ *if and only if $\mathbf{x} = \mathbf{p}_i$ for some i such that $1 \leq i \leq n$, where $\mathbf{p}_i = s_1 \ldots s_{i-1}\mathbf{e}_i$.*

(b) $c^{-1}\mathbf{x} \not> 0$ *if and only if* $\mathbf{x} = \mathbf{q}_i$ *for some* i *such that* $1 \leq i \leq n$, *where* $\mathbf{q}_i = s_n \ldots s_{i+1}\mathbf{e}_i$.

Proof. We only prove part (a); the proof of (b) is similar. If $c\mathbf{x} = s_n \ldots s_1\mathbf{x} \not> 0$, there exists a least integer $i \leq n$ such that $s_{i-1} \ldots s_1\mathbf{x} > 0$ and $s_i \ldots s_1\mathbf{x} \not> 0$. Then, invoking (4.10), we get $s_{i-1} \ldots s_1\mathbf{x} = \mathbf{e}_i$ and so $\mathbf{x} = (s_{i-1} \ldots s_1)^{-1}\mathbf{e}_i = s_1 \ldots s_{i-1}\mathbf{e}_i = \mathbf{p}_i$. Conversely, it is clear that $c\mathbf{p}_i \not> 0$. □

The last two results yield an algorithm allowing us to compute all the positive roots of the quadratic form of a quiver whose underlying graph is Dynkin.

4.14. Proposition. *Let Q be a quiver whose underlying graph is Dynkin and c be the Coxeter transformation of Q.*

(a) *If m_i is the least integer such that $c^{-m_i-1}\mathbf{p}_i \not> 0$, then the set*

$$\{c^{-s}\mathbf{p}_i \mid 1 \leq i \leq n,\ 0 \leq s \leq m_i\}$$

equals the set of all the positive roots of q_Q.

(b) *If n_i is the least integer such that $c^{n_i+1}\mathbf{q}_i \not> 0$, then the set*

$$\{c^t\mathbf{q}_i \mid 1 \leq i \leq n,\ 0 \leq t \leq n_i\}$$

equals the set of all the positive roots of q_Q.

Proof. We only prove (a). The proof of (b) is similar. Because it is clear that each $c^{-s}\mathbf{p}_i$, with $1 \leq i \leq n$, $0 \leq s \leq m_i$ is a positive root, it remains to show that each positive root is of this form. Let $\mathbf{x}$ be a positive root. By (4.12), there exists $s \geq 0$ such that $c^s\mathbf{x} > 0$ but $c^{s+1}\mathbf{x} \not> 0$. By (4.13), we have $c^s\mathbf{x} = \mathbf{p}_i$ for some $1 \leq i \leq n$. Therefore $\mathbf{x} = c^{-s}\mathbf{p}_i$ and clearly $s \leq m_i$. □

4.15. Examples. (a) Let Q be the quiver $\overset{1}{\circ} \longleftarrow \overset{3}{\circ} \longrightarrow \overset{2}{\circ}$ whose underlying graph is the Dynkin graph $\mathbb{A}_3$. Then $E \cong \mathbb{Q}^3$ and, as before,

$$s_1 = \begin{bmatrix} -1 & 0 & 1 \\ 0 & 1 & 0 \\ 0 & 0 & 1 \end{bmatrix}, \quad s_2 = \begin{bmatrix} 1 & 0 & 0 \\ 0 & -1 & 1 \\ 0 & 0 & 1 \end{bmatrix}, \quad s_3 = \begin{bmatrix} 1 & 0 & 0 \\ 0 & 1 & 0 \\ 1 & 1 & -1 \end{bmatrix},$$

$$c = \begin{bmatrix} -1 & 0 & 1 \\ 0 & -1 & 1 \\ -1 & -1 & 1 \end{bmatrix}, \quad c^{-1} = \begin{bmatrix} 0 & 1 & -1 \\ 1 & 0 & -1 \\ 1 & 1 & -1 \end{bmatrix}.$$

We have thus

$$\mathbf{p}_1 = \mathbf{e}_1 = \begin{bmatrix} 1 \\ 0 \\ 0 \end{bmatrix}, \quad \mathbf{p}_2 = s_1\mathbf{e}_2 = \begin{bmatrix} 0 \\ 1 \\ 0 \end{bmatrix}, \quad \mathbf{p}_3 = s_1 s_2\mathbf{e}_3 = \begin{bmatrix} 1 \\ 1 \\ 1 \end{bmatrix}.$$

Consequently,

$$\mathbf{c}^{-1}\mathbf{p}_1 = \begin{bmatrix} 0 \\ 1 \\ 1 \end{bmatrix}, \qquad \mathbf{c}^{-1}\mathbf{p}_2 = \begin{bmatrix} 1 \\ 0 \\ 1 \end{bmatrix}, \qquad \mathbf{c}^{-1}\mathbf{p}_3 = \begin{bmatrix} 0 \\ 0 \\ 1 \end{bmatrix},$$

$$\mathbf{c}^{-2}\mathbf{p}_1 = \begin{bmatrix} 0 \\ -1 \\ 0 \end{bmatrix} \not> 0, \quad \mathbf{c}^{-2}\mathbf{p}_2 = \begin{bmatrix} -1 \\ 0 \\ 0 \end{bmatrix} \not> 0, \quad \mathbf{c}^{-2}\mathbf{p}_3 = \begin{bmatrix} -1 \\ -1 \\ -1 \end{bmatrix} \not> 0.$$

Hence all the positive roots are $\begin{bmatrix} 1 \\ 0 \\ 0 \end{bmatrix}, \begin{bmatrix} 0 \\ 1 \\ 0 \end{bmatrix}, \begin{bmatrix} 1 \\ 1 \\ 1 \end{bmatrix}, \begin{bmatrix} 0 \\ 1 \\ 1 \end{bmatrix}, \begin{bmatrix} 1 \\ 0 \\ 1 \end{bmatrix}, \begin{bmatrix} 0 \\ 0 \\ 1 \end{bmatrix}.$

(b) Let Q be the quiver

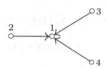

whose underlying graph is the Dynkin graph $\mathbb{D}_4$. Then $E \cong \mathbb{Q}^4$ and the reflections are expressed by the following matrices (in the canonical basis):

$$s_1 = \begin{bmatrix} -1 & 1 & 1 & 1 \\ 0 & 1 & 0 & 0 \\ 0 & 0 & 1 & 0 \\ 0 & 0 & 0 & 1 \end{bmatrix}, \qquad s_2 = \begin{bmatrix} 1 & 0 & 0 & 0 \\ 1 & -1 & 0 & 0 \\ 0 & 0 & 1 & 0 \\ 0 & 0 & 0 & 1 \end{bmatrix},$$

$$s_3 = \begin{bmatrix} 1 & 0 & 0 & 0 \\ 0 & 1 & 0 & 0 \\ 1 & 0 & -1 & 0 \\ 0 & 0 & 0 & 1 \end{bmatrix}, \qquad s_4 = \begin{bmatrix} 1 & 0 & 0 & 0 \\ 0 & 1 & 0 & 0 \\ 0 & 0 & 1 & 0 \\ 1 & 0 & 0 & -1 \end{bmatrix}.$$

Then

$$\mathbf{c}^{-1} = s_1 s_2 s_3 s_4 = \begin{bmatrix} 2 & -1 & -1 & -1 \\ 1 & -1 & 0 & 0 \\ 1 & 0 & -1 & 0 \\ 1 & 0 & 0 & -1 \end{bmatrix}.$$

We have

$$\mathbf{p}_1 = \mathbf{e}_1 = \begin{bmatrix} 1 \\ 0 \\ 0 \\ 0 \end{bmatrix}, \qquad \mathbf{p}_2 = s_1 \mathbf{e}_2 = \begin{bmatrix} 1 \\ 1 \\ 0 \\ 0 \end{bmatrix},$$

$$\mathbf{p}_3 = s_1 s_2 \mathbf{e}_3 = \begin{bmatrix} 1 \\ 0 \\ 1 \\ 0 \end{bmatrix}, \qquad \mathbf{p}_4 = s_1 s_2 s_3 \mathbf{e}_4 = \begin{bmatrix} 1 \\ 0 \\ 0 \\ 1 \end{bmatrix}.$$

Hence the complete list of the positive roots, given by the action of $\mathbf{c}^{-1}$ on the $\mathbf{p}_i$:

$$\begin{bmatrix} 1 \\ 0 \\ 0 \\ 0 \end{bmatrix} \xrightarrow{c^{-1}} \begin{bmatrix} 2 \\ 1 \\ 1 \\ 1 \end{bmatrix} \xrightarrow{c^{-1}} \begin{bmatrix} 1 \\ 1 \\ 1 \\ 1 \end{bmatrix} \xrightarrow{c^{-1}} \not> 0,$$

$$\begin{bmatrix} 1 \\ 1 \\ 0 \\ 0 \end{bmatrix} \xrightarrow{c^{-1}} \begin{bmatrix} 1 \\ 0 \\ 1 \\ 1 \end{bmatrix} \xrightarrow{c^{-1}} \begin{bmatrix} 0 \\ 1 \\ 0 \\ 0 \end{bmatrix} \xrightarrow{c^{-1}} \not\geq 0,$$

$$\begin{bmatrix} 1 \\ 0 \\ 1 \\ 0 \end{bmatrix} \xrightarrow{c^{-1}} \begin{bmatrix} 1 \\ 1 \\ 0 \\ 1 \end{bmatrix} \xrightarrow{c^{-1}} \begin{bmatrix} 0 \\ 0 \\ 1 \\ 0 \end{bmatrix} \xrightarrow{c^{-1}} \not\geq 0,$$

$$\begin{bmatrix} 1 \\ 0 \\ 0 \\ 1 \end{bmatrix} \xrightarrow{c^{-1}} \begin{bmatrix} 1 \\ 1 \\ 1 \\ 0 \end{bmatrix} \xrightarrow{c^{-1}} \begin{bmatrix} 0 \\ 0 \\ 0 \\ 1 \end{bmatrix} \xrightarrow{c^{-1}} \not\geq 0.$$

VII.5. Reflection functors and Gabriel's theorem

We now return to the proof of Gabriel's theorem. As said before, the latter states that the path algebra of a connected quiver is representation–finite if and only if the underlying graph of this quiver is a Dynkin diagram. In particular, the representation–finiteness of a path algebra is independent of the orientation of its quiver. This remark led to the definition of reflection functors [32], which are now understood as APR-tilts (see [18]). Before introducing these, we need some combinatorial considerations meant to make more precise the idea of a change of orientation.

Let $Q = (Q_0, Q_1, s, t)$ be a finite, connected, and acyclic quiver and let $n = |Q_0|$. For every point $a \in Q_0$, we define a new quiver

$$\sigma_a Q = (Q_0', Q_1', s', t')$$

as follows: All the arrows of Q having a as source or as target are reversed, all other arrows remain unchanged. More precisely, $Q_0' = Q_0$ and there exists a bijection $Q_1 \to Q_1'$ such that if $\alpha' \in Q_1'$ denotes the arrow corresponding to $\alpha \in Q_1$ under this bijection, then:

(i) if $s(\alpha) \neq a$ and $t(\alpha) \neq a$, then $s'(\alpha') = s(\alpha)$ and $t'(\alpha') = t(\alpha)$; whereas

(ii) if $s(\alpha) = a$ or $t(\alpha) = a$, then $s'(\alpha') = t(\alpha)$ and $t'(\alpha') = s(\alpha)$.

For instance, if Q is the quiver

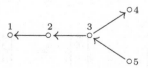

then $\sigma_3 Q$ is the quiver

We defined, in the previous section, the notion of an admissible number-ing of the points of a quiver. We now need a reformulation of this concept.

An **admissible sequence of sinks** in a quiver Q is defined to be a total ordering $(a_1, \ldots, a_n)$ of all the points in Q such that:

(i) a_1 is a sink in Q; and

(ii) a_i is a sink in $\sigma_{a_{i-1}} \ldots \sigma_{a_1} Q$, for every $2 \leq i \leq n$.

Dually, an **admissible sequence of sources** in Q is a total ordering $(b_1, \ldots, b_n)$ of all the points in Q such that:

(i) b_1 is a source in Q; and

(ii) b_i is a source in $\sigma_{b_{i-1}} \ldots \sigma_{b_1} Q$, for every $2 \leq i \leq n$.

It is clear that if $(a_1, \ldots, a_n)$ is an admissible sequence of sinks, then $(a_n, \ldots, a_1)$ is an admissible sequence of sources, and conversely. Because, by hypothesis, Q is acyclic, there exists an admissible numbering $(1, \ldots, n)$ of its points. Such an admissible numbering is always an admissible sequence of sinks and, conversely, if $(a_1, \ldots, a_n)$ is an admissible sequence of sinks, then an admissible numbering of the points in Q is given by the mapping $a_i \mapsto i$. In general, a given quiver admits many admissible sequences of sinks.

5.1. Lemma. *Let Q be a finite, connected, and acyclic quiver whose n points are admissibly numbered as $(a_1, \ldots, a_n)$.*

(a) *If $1 \leq i \leq n$, then a_i is a source and a_{i+1} is a sink in $\sigma_{a_i} \ldots \sigma_{a_1} Q$.*

(b) *If $1 \leq i \leq n$, then a_i is a sink and a_{i-1} is a source in $\sigma_{a_i} \ldots \sigma_{a_n} Q$.*

(c) *$\sigma_{a_n} \ldots \sigma_{a_1} Q = Q = \sigma_{a_1} \ldots \sigma_{a_n} Q$.*

Proof. For (a) and (b), an obvious induction on i yields the result. For (c), we need only observe that each arrow in Q is reversed exactly twice. □

5.2. Lemma. *Let Q and Q' be two trees having the same underlying graph. There exists a sequence $i_1, \ldots, i_t$ of points of Q such that*

(a) *for each s such that $1 \leq s \leq t$, i_s is a sink in $\sigma_{i_{s-1}} \ldots \sigma_{i_1} Q$; and*

(b) *$\sigma_{i_t} \ldots \sigma_{i_1} Q = Q'$.*

Proof. It suffices to prove the result if Q and Q' differ in the orientation of exactly one arrow. Let thus $\alpha : i \to j$ be an arrow in Q_1 such that the corresponding arrow in Q_1' is $\alpha' : j \to i$ whereas if $\beta \in Q_1$, $\beta \neq \alpha$, then the corresponding arrow $\beta' \in Q_1'$ has the same source and target, respectively, as β. Let $Q'' = (Q_0, Q_1 \setminus \{\alpha\})$; then Q'' is a (common) subquiver of (both of the trees) Q and Q' and it is not connected. Indeed, i and j belong to distinct connected components of Q''. We may thus write $Q'' = Q^i \cup Q^j$, where Q^i and Q^j are connected subquivers of Q'' containing i and j, respectively. Because Q^i and Q^j are trees, we may assume both to be admissibly numbered with $Q_0^i = \{1, \dots, m\}$ and $Q_0^j = \{m+1, \dots, n\}$. Because, by (5.1), for each k such that $1 \leq k \leq m$, k is a sink in $\sigma_{k-1} \dots \sigma_1 Q^i$, hence a sink in $\sigma_{k-1} \dots \sigma_1 Q$, and moreover we have $\sigma_m \dots \sigma_1 Q = Q'$, the statement follows. $\square$

We now come to the definition of reflection functors. Let A be a hereditary algebra, which we can assume to be nonsimple. By (1.7), there exists an algebra isomorphism $A \cong KQ_A$, where Q_A is a finite, connected, and acyclic quiver, with $n = |(Q_A)_0| > 1$. Then there exists a sink $a \in (Q_A)_0$ that is not a source, so that the simple A-module $S(a)_A$ is projective and noninjective. Let

$$T[a]_A = \tau^{-1} S(a) \oplus \left(\bigoplus_{b \neq a} P(b) \right)$$

denote the APR-tilting module at a (see (VI.2.8)(c)) and $B = \operatorname{End} T[a]_A$.

It also follows from the tilting theorem (VI.3.8) that the left B-module ${}_B T[a]$ is a tilting module and that $A \cong \operatorname{End}_B(T[a])^{\mathrm{op}}$. We will show that $Q_B = \sigma_a Q_A$, and therefore a is a source in Q_B. The functors

$$\operatorname{mod} A \underset{S_a^-}{\overset{S_a^+}{\rightleftarrows}} \operatorname{mod} B$$

defined by the formulas $S_a^+ = \operatorname{Hom}_A(T[a], -)$ and $S_a^- = (-) \otimes_B T[a]$ are called, respectively, the **reflection functor** at the sink $a \in (Q_A)_0$ and the **reflection functor** at the source $a \in (Q_B)_0$. The following theorem shows that passing from A to B amounts to passing from Q_A to $\sigma_a Q_A$; hence the reflection functors correspond to changes of orientation in the quiver Q_A.

5.3. Theorem. *Let A be a basic hereditary and nonsimple algebra, a be a sink in its quiver Q_A, and $T[a]$ be the APR-tilting A-module at a.*

(a) *The algebra $B = \operatorname{End} T[a]_A$ is isomorphic to $K(\sigma_a Q_A)$, a is a source in Q_B, the simple B-module $S(a)_B$ is injective and isomorphic to*

$\operatorname{Ext}^1_A(T[a], S(a))$, *the left B-module $_BT[a]$ is a tilting module, and*
$A \cong \operatorname{End}_B(T[a])^{\mathrm{op}}$.

(b) *The reflection functor $S^+_a : \operatorname{mod} A \to \operatorname{mod} B$ induces an equivalence
between the K-linear full subcategory of $\operatorname{mod} A$ of all A-modules
without direct summand isomorphic to the simple projective module
$S(a)_A$ and the K-linear full subcategory of $\operatorname{mod} B$ of all B-modules
without direct summand isomorphic to the simple injective B-module
$S(a)_B$. The quasi-inverse equivalence is induced by the reflection
functor $S^-_a : \operatorname{mod} B \to \operatorname{mod} A$.*

Proof. Throughout this proof, we denote the APR-tilting A-module
$T[a]$ briefly by T, and we use the notation introduced in (VI.3.10).

By our assumption and (1.7), the quiver Q_A of A is finite, connected,
and acyclic; $|(Q_A)_0| \geq 2$; and we may suppose, without loss of generality,
that $A = KQ_A$. Note that $S(a) = P(a) = \varepsilon_a A$, where ε_c is the stationary
path at c in Q_A.

By (VI.2.8)(c), we have $T = \bigoplus\limits_{c \in (Q_A)_0} T_c$, where $T_a = \tau^{-1}\varepsilon_a A = \tau^{-1}P(a)$
and $T_c = \varepsilon_c A$ for $c \neq a$. By (VI.3.1)(b), the right B-modules $\operatorname{Hom}_A(T, T_a)$
and $\operatorname{Hom}_A(T, T_b)$, for $b \neq a$, form a complete set of pairwise nonisomorphic
indecomposable projective modules. For each $c \in (Q_A)_0$, denote by $e_c \in
\operatorname{End} T_A$ the composition of the canonical projection $p_c : T \to T_c$ with the
canonical injection $u_c : T_c \to T$. According to (3.10), we have $e_c B \cong
\operatorname{Hom}_A(T, T_c)$ for all $c \in (Q_A)_0$ and the elements e_c are primitive orthogonal
idempotents of $B = \operatorname{End} T_A$ such that

$$B = \bigoplus_{c \in (Q_A)_0} e_c B.$$

It follows directly from the tilting theorem (VI.3.8) that the left B-module
$_BT$ is a tilting module and that $A \cong \operatorname{End}_B(T)^{\mathrm{op}}$.

We claim that the simple B-module $S(a)_B = e_a B/\operatorname{rad} e_a B$ is isomorphic
to $\operatorname{Ext}^1_A(T, S(a))$. For this, we notice first that

$$\operatorname{Ext}^1_A(T, S(a)) \cong D\operatorname{Hom}_A(S(a), \tau T) \cong D\operatorname{Hom}_A(S(a), S(a)) \cong K.$$

Hence $\operatorname{Ext}^1_A(T, S(a))$ is a one-dimensional K-vector space and is therefore
simple as a B-module. On the other hand, (VI.3.10)(a) yields

$$\begin{aligned}
\operatorname{Ext}^1_A(T, S(a))e_a &\cong \operatorname{Ext}^1_A(e_a T, S(a)) \\
&\cong \operatorname{Ext}^1_A(\tau^{-1}S(a), S(a)) \cong D\operatorname{Hom}_A(S(a), S(a)) \cong K.
\end{aligned}$$

This establishes our claim.

By (VI.2.8)(c), the tilting module T_A is separating, and

$$\mathcal{F}(T) = \operatorname{add} S(a)_A,$$

whereas $\mathcal{T}(T)$ is the full subcategory of $\operatorname{mod} A$ generated by the remaining indecomposable modules. On the other hand, by (VI.5.6)(b), T_A is also splitting, so that $\mathcal{X}(T_A) = \operatorname{add} S(a)_B$, whereas $\mathcal{Y}(T)$ is the full subcategory $\operatorname{mod} B$ generated by the remaining indecomposable modules. Then (b) follows at once from the tilting theorem (VI.3.8).

To prove that B is hereditary it suffices, by (1.4), to show that, for each simple B-module S_B, we have $\operatorname{pd} S_B \le 1$. If $S_B \not\cong S(a)_B$, then $S_B \in \mathcal{Y}(T)$; hence there exists $M \in \mathcal{T}(T)$ such that $S_B \cong \operatorname{Hom}_A(T, M)$. By (VI.4.1), we have $\operatorname{pd} S_B \le \operatorname{pd} M_A \le 1$, because A is hereditary. On the other hand, we know from (IV.3.9) and (IV.4.4) that the almost split sequence in $\operatorname{mod} A$ starting with $S(a)_A = P(a)$ is of the form

$$0 \longrightarrow S(a) \longrightarrow \bigoplus_{c \ne a} P(c)^{m_c} \longrightarrow \tau^{-1} S(a) \longrightarrow 0,$$

where $P(c) = \varepsilon_c A$ and $m_c = \dim_K \operatorname{Irr}(S(a), P(c)) = \dim_K \varepsilon_c(\operatorname{rad} A/\operatorname{rad}^2 A)\varepsilon_a$, by (1.6). In particular, m_c equals the number of arrows from c to a in Q_A. Thus the direct sum in the almost split sequence is taken over all $c \in (Q_A)_0$ that are neighbours of the sink a. Applying the functor $S_a^+ = \operatorname{Hom}_A(T, -)$ to this almost split sequence yields a short exact sequence

$$0 \to \operatorname{Hom}_A(T, \bigoplus_{c \to a} P(c)^{m_c}) \longrightarrow \operatorname{Hom}_A(T, \tau^{-1}S(a)) \longrightarrow S(a)_B \to 0$$

in $\operatorname{mod} B$, because $\operatorname{Hom}_A(T, S(a)) = 0$, $\operatorname{Ext}^1_A(T, S(a)) \cong S(a)_B$ and $\operatorname{Ext}^1_A(T, P(c)) \cong D\operatorname{Hom}_A(P(c), S(a)) = 0$ for any $c \ne a$. Because the B-modules $\operatorname{Hom}_A(T, \tau^{-1}S(a))$ and $\operatorname{Hom}_A(T, P(c)) \cong e_c B$ for $c \ne a$ are projective, we infer that $\operatorname{pd} S(a)_B \le 1$.

It remains to show that $Q_B = \sigma_a Q$. Clearly, $(Q_B)_0 = (Q_A)_0 = (\sigma_a Q_A)_0$. On the other hand, it follows from the tilting theorem (VI.3.8) that the functor $S_a^+ = \operatorname{Hom}_A(T, -) : \operatorname{mod} A \longrightarrow \operatorname{mod} B$ induces isomorphisms of K-vector spaces

$$\operatorname{Hom}_A(\varepsilon_c A, \tau^{-1}S(a)) \cong \operatorname{Hom}_B(e_c B, e_a B), \text{ and}$$

$$\operatorname{Hom}_A(\tau^{-1}S(a), \varepsilon_c A) \cong \operatorname{Hom}_B(e_a B, e_c B).$$

Also, $\operatorname{Hom}_B(e_a B, e_b B) = 0$ for all $b \ne a$. Indeed, there are isomorphisms

$$\operatorname{Hom}_B(e_a B, e_b B) \cong \operatorname{Hom}_B(\operatorname{Hom}_A(T, \tau^{-1}S(a)), \operatorname{Hom}_A(T, P(b)))$$

$$\cong \operatorname{Hom}_A(\tau^{-1}S(a), P(b)),$$

and there is no nonzero homomorphism $h : \tau^{-1}S(a) \to P(b)$, because otherwise, by (1.4), the A-module Im h is projective; hence $\tau^{-1}S(a)$ is projective, and we get a contradiction. This shows our claim, which implies that $\mathrm{Irr}(e_aB, e_bB) = 0$ for all $b \neq a$. Then, by (1.6), a is a source in Q_B.

We now show that $S_a^+ = \mathrm{Hom}_A(T, -)$ induces, for all $b \neq a$ and $c \neq a$, an isomorphism of K-vector spaces $\mathrm{Irr}(\varepsilon_bA, \varepsilon_cA) \cong \mathrm{Irr}(e_bB, e_cB)$. Because, by (1.7), the quivers Q_A and Q_B are acyclic, we may suppose that $b \neq c$. Then $\varepsilon_bA \not\cong \varepsilon_cA$ and (consequently) $e_bB \not\cong e_cB$. Therefore, $\mathrm{rad}_A(\varepsilon_bA, \varepsilon_cA) = \mathrm{Hom}_A(\varepsilon_bA, \varepsilon_cA)$ and $\mathrm{rad}_B(e_bB, e_cB) = \mathrm{Hom}_B(e_bB, e_cB)$, so that the functor $\mathrm{Hom}_A(T, -)$ induces an isomorphism $\mathrm{rad}_A(\varepsilon_bA, \varepsilon_cA) \cong \mathrm{rad}_B(e_bB, e_cB)$.

We claim that it also induces an isomorphism between the subspaces $\mathrm{rad}_A^2(\varepsilon_bA, \varepsilon_cA)$ and $\mathrm{rad}_B^2(e_bB, e_cB)$. Indeed, assume that f belongs to $\mathrm{rad}_A^2(\varepsilon_bA, \varepsilon_cA)$. Then there exist indecomposable A-modules $M_1, \ldots, M_t$ and homomorphisms $f_j' \in \mathrm{rad}_A(\varepsilon_bA, M_j)$, $f_j \in \mathrm{rad}_A(M_j, \varepsilon_cA)$ such that $f = f_1f_1' + \ldots + f_rf_t'$. For any $j \in \{1, \ldots, t\}$, Im f_j is a submodule of the projective module ε_cA and hence is projective by (1.4). Then Im f_j is isomorphic to a direct summand of the indecomposable module M_j and therefore $M_j \cong$ Im f_j. Consequently, M_j is projective and, by (I.5.17), there exists $a_j \in (Q_A)_0$ such that $M_j \cong \varepsilon_{a_j}A$. Note that $a_j \neq c$, because f_j' is a nonisomorphism.

The additivity of $\mathrm{Hom}_A(T, -)$ yields

$$\mathrm{Hom}_A(T, f) = \mathrm{Hom}_A(T, \sum_{j=1}^{t} f_j f_j') = \sum_{j=1}^{t} \mathrm{Hom}_A(T, f_j)\mathrm{Hom}_A(T, f_j').$$

Now $f_j \in \mathrm{rad}_A(M_j, \varepsilon_cA)$ implies that $\mathrm{Hom}_A(T, f_j) \in \mathrm{rad}_B(e_{a_j}B, e_cB)$, by the observation. Similarly, $\mathrm{Hom}_A(T, f_j') \in \mathrm{rad}_B(e_bB, e_{a_j}B)$, and consequently, $\mathrm{Hom}_A(T, f) \in \mathrm{rad}_B^2(e_bB, e_cB)$. Similarly, one shows that the reflection functor $S_a^- = - \otimes_B T : \mathrm{mod}\ B \longrightarrow \mathrm{mod}\ A$ applies $\mathrm{rad}_B^2(e_bB, e_cB)$ into $\mathrm{rad}_A^2(\varepsilon_bA, \varepsilon_cA)$. This shows our claim.

Applying (1.6) yields

$$\varepsilon_c(\mathrm{rad}\ A/\mathrm{rad}^2 A)\varepsilon_b \cong \mathrm{Irr}(\varepsilon_bA, \varepsilon_cA) \cong \mathrm{Irr}(e_bB, e_cB) \cong e_c(\mathrm{rad}\ B/\mathrm{rad}^2 B)e_b.$$

Therefore, if $b, c \neq a$, then there is a bijection between the set of arrows from c to b in Q_A and in Q_B.

The same arguments as earlier show the existence of an isomorphism of K-vector spaces $\mathrm{Irr}(\varepsilon_bA, \tau^{-1}S(a)) \cong \mathrm{Irr}(e_bB, e_aB)$ for all $b \neq a$. Applying (1.6) and (IV. 4.4), we get

$$\varepsilon_b(\mathrm{rad}\ A/\mathrm{rad}^2 A)\varepsilon_a \cong \mathrm{Irr}(\varepsilon_aA, \varepsilon_bA) \cong \mathrm{Irr}(S(a), \varepsilon_bA) \cong \mathrm{Irr}(\varepsilon_bA, \tau^{-1}S(a))$$
$$\cong \mathrm{Irr}(e_bB, e_aB) \cong e_a(\mathrm{rad}\ B/\mathrm{rad}^2 B)e_b.$$

This defines a bijection between the set of arrows from a to b in Q_A and the set of arrows from b to a in Q_B, and it finishes the proof of the equality $\sigma_a Q = Q_B$.

In particular, while $S(a)_A$ is a simple projective noninjective module, we have that $S(a)_B$ is a simple injective nonprojective module (because a becomes a source in Q_B). □

Now we show that the reflection functors S_a^+ and S_a^-, when applied to indecomposable modules M, correspond to the reflection homomorphism $s_a : \mathbb{Z}^n \longrightarrow \mathbb{Z}^n$ (as defined in Section 4) applied to their dimension vectors $\dim M$, where $n = |Q_0|$.

5.4. Proposition. *Let A be a basic hereditary and nonsimple algebra, a be a sink in its quiver Q_A, and $n = |Q_0|$. Let $T[a]$ be the APR-tilting A-module at a, $B = \operatorname{End} T[a]$, S_a^+, S_a^- the reflection functors at a, and $s_a : \mathbb{Z}^n \longrightarrow \mathbb{Z}^n$ the reflection at a.*

(a) *Let M be an indecomposable A-module. Then M is isomorphic to $S(a)_A$ if and only if $S_a^+ M = 0$ (or equivalently, $s_a(\dim M) \not> 0$). If $M \not\cong S(a)_A$, then $S_a^+ M$ is an indecomposable B-module and $\dim(S_a^+ M) = s_a(\dim M)$.*

(b) *Let N be an indecomposable B-module. Then N is isomorphic to $S(a)_B$ if and only if $S_a^- N = 0$ (or equivalently, $s_a(\dim N) \not> 0$). If $N \not\cong S(a)_B$, then $S_a^- N$ is an indecomposable A-module and $\dim(S_a^- N) = s_a(\dim N)$.*

Proof. We only prove (a); the proof of (b) is similar. We denote the APR-tilting A-module $T[a]$ by T. Because T_A is an APR-tilting module, $\mathcal{F}(T) = \operatorname{add} S(a)_A$, by (VI.2.8)(c). It follows from (VI.2.3) that if M is an indecomposable A-module, then $S_a^+ M = \operatorname{Hom}_A(T, M) = 0$ if and only if M is isomorphic to $S(a)_A$.

Assume that M is an indecomposable module nonisomorphic to $S(a)_A$. By (5.3), the B-module $S_a^+ M = \operatorname{Hom}_A(T, M)$ is indecomposable. Let $b \neq a$ be a point in $Q = Q_A$. By (VI.3.10), the fact that $M \in \mathcal{T}(T)$ implies that

$$
\begin{aligned}
(\dim S_a^+ M)_b &= \dim_K \operatorname{Hom}_A(\operatorname{Hom}_A(T, \varepsilon_b A), \operatorname{Hom}_A(T, M)) \\
&= \dim_K \operatorname{Hom}_A(\varepsilon_b A, M) \\
&= \dim_K M \varepsilon_b = (\dim M)_b = (s_a(\dim M))_b.
\end{aligned}
$$

On the other hand, if $b = a$, we have isomorphisms

$$(S_a^+ M)e_a \cong \mathrm{Hom}_B(e_a B, S_a^+ M)$$
$$\cong \mathrm{Hom}_B(\mathrm{Hom}_A(T, \tau^{-1} S(a)), \mathrm{Hom}_A(T, S_a^+ M))$$
$$\cong \mathrm{Hom}_A(\tau^{-1} S(a), M).$$

Consider the almost split sequence

$$0 \longrightarrow S(a) \longrightarrow \bigoplus_{c \to a} P(c)^{m_c} \longrightarrow \tau^{-1} S(a) \longrightarrow 0$$

constructed in the proof of (5.3), where m_c equals the number of arrows from c to a in Q_A. Because M is indecomposable, $S(a)$ is projective, and $M \not\cong S(a)$, there is no nonzero homomorphism $M \to S(a)_A$ and therefore $\mathrm{Ext}_A^1(\tau^{-1} S(a), M) \cong D\mathrm{Hom}_A(M, S(a)) = 0$. It follows that applying $\mathrm{Hom}_A(-, M)$ to the almost split sequence yields the exact sequence

$$0 \to \mathrm{Hom}_A(\tau^{-1} S(a), M) \to \mathrm{Hom}_A(\bigoplus_{c \to a} P(c)^{m_c}; M) \to \mathrm{Hom}_A(S(a), M) \to 0.$$

Therefore

$$(\mathbf{dim}\, S_a^+ M)_a = \dim_K (S_a^+ M)e_a = \dim_K \mathrm{Hom}_A(\tau^{-1} S(a), M)$$
$$= -\dim_K \mathrm{Hom}_A(S(a), M) + \sum_{c \to a} m_c \dim_K \mathrm{Hom}_A(P(c), M)$$
$$= -\dim_K M\varepsilon_a + \sum_{c \to a} m_c(\dim_K M\varepsilon_c) = (s_a(\mathbf{dim}\, M))_a.$$

We have thus shown that $\mathbf{dim}\, S_a^+ M = s_a(\mathbf{dim}\, M)$.

It remains to show that there is an isomorphism $M \cong S(a)_A$ if and only if the vector $s_a(\mathbf{dim}\, M)$ is not positive. If $M \cong S(a)_A$, then the ath coordinate of $s_a(\mathbf{dim}\, M) = s_a(\mathbf{e}_a)$ equals -1. Conversely, if $M \not\cong S(a)_A$, then $s_a(\mathbf{dim}\, M) = \mathbf{dim}\, S_a^+ M > 0$, and we are done. $\qquad \square$

As shown in (III.1.7), a module over a path K-algebra KQ can be thought of as a K-linear representation of the quiver Q. We now present the original construction of reflection functors given by Bernstein, Gelfand, and Ponomarev [32] for linear representations of quivers. Here we get it by translating, in terms of representations of the quivers Q_A and $Q_B = \sigma_a Q_A$, the effect of the tilting functors S_a^+, S_a^- between the categories of A-modules and B-modules.

5.5. Definition. Let Q be a finite connected quiver, a a sink in Q, and $Q' = \sigma_a Q$. We define the **reflection functor**

$$\mathcal{S}_a^+ : \mathrm{rep}_K(Q) \longrightarrow \mathrm{rep}_K(Q')$$

between the categories of finite dimensional K-linear representations of the quivers Q and Q' as follows. Let $M = (M_i, \varphi_\alpha)_{i \in Q_0, \alpha \in Q_1}$ be an object in

$\text{rep}_K(Q)$. We define the object $\mathcal{S}_a^+ M = (M_i', \varphi_\alpha')_{i \in Q_0', \alpha \in Q_1'}$ in $\text{rep}_K(Q')$ as follows:

(a) $M_i' = M_i$ for $i \neq a$, whereas M_a' is the kernel of the K-linear map $(\varphi_\alpha)_\alpha : \bigoplus_{\alpha:s(\alpha) \to a} M_{s(\alpha)} \longrightarrow M_a$ (the direct sum is being taken over all arrows α in Q with target a);

(b) $\varphi_\alpha' = \varphi_\alpha$ for all arrows $\alpha : i \to j$ in Q with $j \neq a$, whereas, if $\alpha : i \to a$ is an arrow in Q, then $\varphi_\alpha' : M_a' \to M_i' = M_i$ is the composition of the inclusion of M_a' into $\bigoplus_{\alpha:s(\beta) \to a} M_{s(\beta)}$ with the projection onto the direct summand M_i.

Let $f = (f_i)_{i \in Q_0} : M \longrightarrow N$ be a morphism in $\text{rep}_K(Q)$, where $M = (M_i, \varphi_\alpha)$ and $N = (N_i, \psi_\alpha)$. We define the morphism

$$\mathcal{S}_a^+ f = f' = (f_i')_{i \in Q_0'} : \mathcal{S}_a^+ M \to \mathcal{S}_a^+ N$$

in $\text{rep}_K(Q')$ as follows. For all $i \neq a$, we let $f_i' = f_i$, whereas f_a' is the unique K-linear map, making the following diagram commutative

$$
\begin{array}{ccccccc}
0 & \longrightarrow & (\mathcal{S}_a^+ M)_a & \longrightarrow & \bigoplus_{\alpha:s(\alpha) \to a} M_{s(\alpha)} & \xrightarrow{(\varphi_\alpha)_\alpha} & M_a \\
 & & \Big\downarrow f_a' & & \Big\downarrow \oplus_\alpha f_{s(\alpha)} & & \Big\downarrow f_a \\
0 & \longrightarrow & (\mathcal{S}_a^+ N)_a & \longrightarrow & \bigoplus_{\alpha:s(\alpha) \to a} N_{s(\alpha)} & \xrightarrow{(\psi_\alpha)_\alpha} & N_a
\end{array}
$$

Now we define the reflection functor attached to a source.

Let Q' be a finite connected quiver, a be a source in Q', and $Q = \sigma_a Q'$. We define a **reflection functor**

$$\mathcal{S}_a^- : \text{rep}_K(Q') \longrightarrow \text{rep}_K(Q)$$

between the categories of finite dimensional K-linear representations of the quivers Q' and Q as follows. Let $M' = (M_i', \varphi_\alpha')_{i \in Q_0', \alpha \in Q_1'}$ be an object in $\text{rep}_K(Q')$. We define the object $\mathcal{S}_a^- M' = (M_i, \varphi_\alpha)_{i \in Q_0, \alpha \in Q_1}$ in $\text{rep}_K(Q')$ as follows:

(a') $M_i = M_i'$ for all $i \neq a$, whereas M_a is the cokernel of the K-linear map $(\varphi_\alpha')_\alpha : M_a' \longrightarrow \bigoplus_{\alpha:a \to t(\alpha)} M_{t(\alpha)}'$ (the direct sum is being taken over all arrows α in Q' with source a);

(b') $\varphi_\alpha = \varphi_\alpha'$ for all arrows $\alpha : i \to j$ in Q' with $i \neq a$, whereas, if $\alpha : a \to j$ is an arrow in Q', then $\varphi_\alpha : M_j = M_j' \to M_a$ is the composition of the inclusion of M_j' into $\bigoplus_{\alpha:a \to t(\beta)} M_{t(\beta)}'$ with the cokernel projection onto M_a.

Let $f' = (f_i')_{i \in Q_0'} : M' \longrightarrow N'$ be a morphism in $\text{rep}_K(Q')$, where $M' = (M_i', \varphi_\alpha')$ and $N' = (N_i', \psi_\alpha')$. We define the morphism $\mathcal{S}_a^- f' =$

$f = (f_i)_{i \in Q_0} : \mathcal{S}_a^- M' \to \mathcal{S}_a^- N'$ in $\text{rep}_K(Q)$ as follows. For all $i \neq a$, we let $f_i = f_i'$, whereas f_a is the unique K-linear map, making the following diagram commutative

$$
\begin{array}{ccccccc}
M_a' & \longrightarrow & \displaystyle\bigoplus_{\alpha:a \to t(\alpha)} M_{t(\alpha)}' & \xrightarrow{(\varphi_\alpha)_\alpha} & (\mathcal{S}_a^- M')_a & \longrightarrow & 0 \\
\downarrow{f_a'} & & \downarrow{\oplus_\alpha f_{t(\alpha)}'} & & \vdots\downarrow f_a & & \\
N_a' & \longrightarrow & \displaystyle\bigoplus_{\alpha:a \to t(\alpha)} N_{t(\alpha)}' & \xrightarrow{(\psi_\alpha)_\alpha} & (\mathcal{S}_a^- N')_a & \longrightarrow & 0
\end{array}
$$

The following proposition shows that, up to the equivalences of categories (constructed in (III.1.6)) between modules over a path algebra and representations of its quiver, the reflection functors $\mathcal{S}_a^+$ and $\mathcal{S}_a^-$ coincide respectively with the reflection functors S_a^+ and S_a^- defined earlier.

5.6. Proposition. *Let Q be a finite, connected, and acyclic quiver; a be a sink in Q; and $Q' = \sigma_a Q$. Then the following diagram is commutative*

$$
\begin{array}{ccc}
\text{mod}\, KQ & \underset{S_a^-}{\overset{S_a^+}{\rightleftarrows}} & \text{mod}\, KQ' \\
F \downarrow \cong & & F' \downarrow \cong \\
\text{rep}_K(Q) & \underset{\mathcal{S}_a^-}{\overset{\mathcal{S}_a^+}{\rightleftarrows}} & \text{rep}_K(Q')
\end{array}
$$

that is, $\mathcal{S}_a^+ F \cong F' S_a^-$ and $\mathcal{S}_a^- F' \cong F S_a^+$, where F and F' are the category equivalences defined in (III.1.6) for KQ and KQ', respectively.

Proof. We only prove that $\mathcal{S}_a^+ F \cong F' S_a^+$; the proof of the second statement is similar. We let $A = KQ$ and $B = KQ'$, and we use freely the notation of (5.1)–(5.5). We recall from (III.1.6) that the functor F associates with any module M in $\text{mod}\, A$ the representation $FM = ((FM)_i, \varphi_\alpha)$ in $\text{rep}_K(Q)$, where $(FM)_i = M\varepsilon_i$ and, for an arrow $\alpha : i \to j$ in Q, the K-linear map $\varphi_\alpha : M\varepsilon_i \to M\varepsilon_j$ is defined by $x \mapsto x\alpha = x\alpha\varepsilon_j$. The functor F' is defined analogously, with ε_i and e_i interchanged.

Let $b \neq a$ be a point in Q. It follows from (5.3) and (I.4.2), that

$$
(F' S_a^+ M)_b = (S_a^+ M)e_b \cong \text{Hom}_B(e_b B, S_a^+ M) \cong \text{Hom}_B(S_a^+(e_b A), S_a^+ M)
$$
$$
\cong \text{Hom}_A(\varepsilon_b A, M) \cong M\varepsilon_b = (\mathcal{S}_a^+ FM)_b,
$$

and the composed isomorphism $(F'S_a^+ M)_b \cong (S_a^+ FM)_b$ is obviously functorial. On the other hand, if $b = a$, we have vector space isomorphisms

$$(F'S_a^+ M)_a = (S_a^+ M)e_a \cong \mathrm{Hom}_B(e_a B, S_a^+ M)$$
$$\cong \mathrm{Hom}_B(S_a^+(\tau^{-1}S(a)), S_a^+ M)) \cong \mathrm{Hom}_A(\tau^{-1}S(a), M).$$

We recall that the almost split sequence in $\mathrm{mod}\, A$ starting from the simple projective module $S(a) = P(a)$ is of the form

$$0 \longrightarrow S(a) \overset{u}{\longrightarrow} \bigoplus_{c \neq a} P(c)^{m_c} \longrightarrow \tau^{-1}S(a) \longrightarrow 0,$$

where $P(c) = \varepsilon_c A$, $m_c = \dim_K \mathrm{Irr}(S(a), P(c)) = \dim_K \varepsilon_c(\mathrm{rad}\, A/\mathrm{rad}^2 A)\varepsilon_a$ is the number of arrows $\alpha : c \to a$ in Q. Hence, there are K-linear isomorphisms $\mathrm{Irr}(S(a), P(c)) \cong \varepsilon_c(\mathrm{rad}\, A/\mathrm{rad}^2 A)\varepsilon_a \cong \bigoplus_{\alpha:c\to a} K\alpha$, because the set of all arrows $\alpha : c \to a$ in $Q_A = Q$ gives (by definition) a basis of the K-vector space $\varepsilon_c(\mathrm{rad}\, A/\mathrm{rad}^2 A)\varepsilon_a$. The left minimal almost split morphism $u = (u_c)_c : S(a) \longrightarrow \bigoplus_{c \neq a} P(c)^{m_c}$ is such that, for each c, the homomorphism $u_c = [u_{c_1} \ldots u_{c_{m_c}}]^t : S(a) \longrightarrow P(c)^{m_c}$ is given by a basis $\{u_{c_1} \ldots u_{c_{m_c}}\}$ of the K-vector space $\mathrm{Irr}(S(a), P(c))$. We may therefore rewrite u_c as (u_α), where α runs over all arrows $c \to a$, so that the almost split sequence becomes

$$0 \longrightarrow S(a) \xrightarrow{u=(u_\alpha)_\alpha} \bigoplus_{\alpha:s(\alpha)\to a} P(s(\alpha)) \overset{v}{\longrightarrow} \tau^{-1}S(a) \longrightarrow 0$$

where the direct sum is being taken over all arrows α in $Q_A = Q$ having a as a target. Applying $\mathrm{Hom}_A(-, M)$ yields the top left exact sequence in the commutative diagram

$$0 \to \mathrm{Hom}_A(\tau^{-1}S(a), M) \to \mathrm{Hom}_A \bigoplus_{\alpha:s(\alpha)\to a} P(s(\alpha)), M) \xrightarrow{\mathrm{Hom}_A(u,M)} \mathrm{Hom}_A(S(a), M)$$

$$\cong \downarrow \qquad\qquad\qquad\qquad \cong \downarrow$$

$$0 \longrightarrow (S_a^+ FM)_a \longrightarrow \bigoplus_{\alpha:s(\alpha)\to a} (FM)_{s(\alpha)} \xrightarrow{(\varphi_\alpha)_\alpha} (FM)_a$$

where $(FM)_j = M\varepsilon_j$, $\mathrm{Hom}_A(u, M) = (\mathrm{Hom}_A(u, M)_\alpha)_{\alpha:s(\alpha)\to a}$, and the vertical isomorphisms are induced by the isomorphism $\mathrm{Hom}_A(eA, L) \cong Le$ of (I.4.2), where L is an A-module and e is an idempotent of A. The lower row is (left) exact by definition of S_a^+. Therefore there exists a K-vector space isomorphism $\mathrm{Hom}_A(\tau^{-1}S(a), M) \cong (S_a^+ FM)_a$ making the left-hand square commutative. Hence $(\mathcal{S}_a^+ FM)_a \cong (F'S_a^+ M)_a$. A simple calculation (left as an exercise) shows that the vector space isomorphisms $(\mathcal{S}_a^+ FM)_c \cong (F'S_a^+ M)_c$ for $c \in Q_0$ induce an isomorphism of representations $\mathcal{S}_a^+ FM \cong F'S_a^+ M$ in $\mathrm{rep}_K(Q')$. It is also easy to verify that this isomorphism is functorial, so that we have $F'S_a^+ \cong \mathcal{S}_a^+ F$.   □

The following corollary summarises the properties of the functors $\mathcal{S}_a^+$, $\mathcal{S}_a^-$ that translate those of the functors S_a^+, S_a^- into the language of representations of a quiver. The proof is easy and left as an exercise to the reader.

5.7. Corollary. *Let Q be a finite, connected, and acyclic quiver with at least two points; a a sink in Q; and $Q' = \sigma_a Q$. The reflection functors $\mathcal{S}_a^+ : \mathrm{rep}_K(Q) \to \mathrm{rep}_K(Q')$ and $\mathcal{S}_a^- : \mathrm{rep}_K(Q') \to \mathrm{rep}_K(Q)$ satisfy the following properties:*

(a) *The functor $\mathcal{S}_a^-$ is left adjoint to $\mathcal{S}_a^+$.*

(b) *If M is indecomposable in $\mathrm{rep}_K(Q)$, then the following three conditions are equivalent:*

   (i) $\mathcal{S}_a^+ M \neq 0$,

   (ii) $M \not\cong S(a)$,

   (iii) $s_a(\mathbf{dim}\ M) > 0$.

Moreover, if this is the case, then $\mathbf{dim}\, \mathcal{S}_a^+ M = s_a(\mathbf{dim}\ M)$, $\mathcal{S}_a^- \mathcal{S}_a^+ M \cong M$ and $\mathcal{S}_a^+$ induces an algebra isomorphism $\mathrm{End}\, M \cong \mathrm{End}(\mathcal{S}_a^+ M)$.

(c) *If M' is indecomposable in $\mathrm{rep}_K(Q')$, then the following three conditions are equivalent:*

   (i) $\mathcal{S}_a^- M' \neq 0$,

   (ii) $M' \not\cong S(a)$,

   (iii) $s_a(\mathbf{dim}\ M') > 0$.

Moreover, if this is the case, then $\mathbf{dim}\, \mathcal{S}_a^- M' = s_a(\mathbf{dim}\ M')$, $\mathcal{S}_a^+ \mathcal{S}_a^- M' \cong M'$, and $\mathcal{S}_a^-$ induces an algebra isomorphism $\mathrm{End}\, M' \cong \mathrm{End}(\mathcal{S}_a^- M')$.

(d) *The functors $\mathcal{S}_a^+$ and $\mathcal{S}_a^-$ induce quasi-inverse equivalences between the K-linear full subcategory of $\mathrm{rep}_K(Q)$ of the representations having no direct summand isomorphic to the simple projective representation $S(a)$, and the K-linear full subcategory of $\mathrm{rep}_K(Q')$ of the representations having no direct summand isomorphic to the simple injective representation $S(a)$.*   □

Let A be a hereditary nonsimple algebra and $(j_1, \ldots, j_n)$ be an admissible numbering of the points of Q_A. It follows from (5.1)–(5.4) that the functors

$$C^+ = S_{j_n}^+ \ldots S_{j_1}^+ \quad \text{and} \quad C^- = S_{j_1}^- \ldots S_{j_n}^-$$

are endofunctors of mod A. They are called the **Coxeter functors**. The definition of C^+ and C^- does not depend on the choice of the admissible numbering $(j_1, \ldots, j_n)$ of the points of Q_A, because of the following interpretation of the Coxeter functors in terms of the Auslander–Reiten translation.

5.8. Lemma. *Let A be a hereditary and nonsimple K-algebra, and let $(j_1, \ldots, j_n)$ be an admissible numbering of the points of Q_A.*

(a) *If M is an indecomposable nonprojective A-module, then there are A-module isomorphisms $C^+M \cong \tau M$ and $C^-C^+M \cong M$.*

(b) *If N is an indecomposable noninjective A-module, then there are A-module isomorphisms $C^-N \cong \tau^{-1}N$ and $C^+C^-N \cong N$.*

Proof. In view of (IV.2.10), it suffices to prove the first statements in (a) and (b). We only prove (a); the proof of (b) is similar. We may assume the points of Q_A to be admissibly numbered as $(1, \ldots, n)$. Applying repeatedly (5.3) to the admissible sequence of sinks $(1, \ldots, n)$, we see that for each i such that $1 \leq i \leq n$, the module $P(i)$ is simple projective over $K(\sigma_{i-1} \ldots \sigma_1 Q_A)$ and that, for every indecomposable nonprojective A-module M, we have

$$\mathrm{Hom}_A \left(\tau^{-1}(\bigoplus_{k=1}^{i} P(k)) \oplus (\bigoplus_{l=i+1}^{n} P(l)), \ M \right) \cong S_i^+ \ldots S_1^+ M.$$

Therefore $C^+M = S_n^+ \ldots S_1^+ M \cong \mathrm{Hom}_A(\tau^{-1}A, M)$. Because the algebra A is hereditary, (IV.2.14) applies to A and M, and we get A-module isomorphisms $C^+M \cong \mathrm{Hom}_A(\tau^{-1}A, M) \cong \mathrm{Hom}_A(A, \tau M) \cong \tau M$. $\square$

We also need the following technical result.

5.9. Lemma. *Let A be a hereditary and nonsimple algebra, $(j_1, \ldots, j_n)$ be an admissible numbering of the points of Q_A, and M be an indecomposable module in mod A.*

(a) *If $b \leq a \leq n$ and $s_{j_a} \ldots s_{j_1}(\dim M) > 0$, then $s_{j_b} \ldots s_{j_1}(\dim M) > 0$, the module $S_{j_b}^+ \ldots S_{j_1}^+ M$ over the algebra $K(\sigma_{j_b} \ldots \sigma_{j_1} Q_A)$ is indecomposable, and $\dim S_{j_b}^+ \ldots S_{j_1}^+ M = s_{j_b} \ldots s_{j_1}(\dim M)$.*

(b) *If $c(\dim M) > 0$, then the module C^+M is indecomposable and $\dim C^+M = c(\dim M)$.*

Proof. We assume for simplicity that the points of Q_A are admissibly numbered as $(1, \ldots, n)$. Assume to the contrary that there exists $b \leq a$ such that $s_b \ldots s_1(\dim M) \not> 0$. We clearly may suppose that b is minimal

with this property, that is, that $s_c \ldots s_1(\dim M) > 0$ for all $c \leq b - 1$.
It follows from (5.4)(a) and an obvious induction, that for any $c \leq b - 1$,
the module $S_c^+ \ldots S_1^+ M$ over the algebra $K(\sigma_c \ldots \sigma_1 Q_A)$ is indecomposable and $\dim(S_c^+ \ldots S_1^+ M) = s_c \ldots s_1(\dim M)$. Furthermore, the module
$S_{b-1}^+ \ldots S_1^+ M \cong S(b)$ is simple projective over the algebra $K(\sigma_b \ldots \sigma_1 Q_A)$.
Therefore $\dim(S_{b-1}^+ \ldots S_1^+ M)$ is the canonical basis vector $\mathbf{e}_b$ of $\mathbb{Z}^n$ so that
$s_a \ldots s_1(\dim M) = s_a \ldots s_b(\mathbf{e}_b) = s_a \ldots s_{b+1}(-\mathbf{e}_b) = -\mathbf{e}_b \not> 0$, which is a
contradiction.

This shows indeed that $s_b \ldots s_1(\dim M) > 0$ for all $b \leq a$, but also that,
for any $b \leq a$, the module $S_b^+ \ldots S_1^+ M$ over the algebra $K(\sigma_b \ldots \sigma_1 Q_A)$ is indecomposable and $\dim(S_b^+ \ldots S_1^+ M) = s_b \ldots s_1(\dim M)$. This completes
the proof of (a). To prove (b), we apply (a) to the case where $a = n$. □

We are now able to prove Gabriel's theorem.

5.10. Theorem. *Let Q be a finite, connected, and acyclic quiver; K be
an algebraically closed field; and $A = KQ$ be the path K-algebra of Q.*

(a) *The algebra A is representation-finite if and only if the underlying
graph $\overline{Q}$ of Q is one of the Dynkin diagrams $\mathbb{A}_n$, $\mathbb{D}_n$, with $n \geq 4$,
$\mathbb{E}_6$, $\mathbb{E}_7$, and $\mathbb{E}_8$.*

(b) *If $\overline{Q}$ is a Dynkin graph, then the mapping $\dim : M \mapsto \dim M$
induces a bijection between the set of isomorphism classes of indecomposable A-modules and the set $\{\mathbf{x} \in \mathbb{N}^n; q_Q(\mathbf{x}) = 1\}$ of positive
roots of the quadratic form q_Q of Q.*

(c) *The number of the isomorphism classes of indecomposable A-modules equals $\frac{1}{2}n(n+1)$, $n^2 - n$, 36, 63, and 120, if $\overline{Q}$ is the Dynkin
graph $\mathbb{A}_n$, $\mathbb{D}_n$, with $n \geq 4$, $\mathbb{E}_6$, $\mathbb{E}_7$, and $\mathbb{E}_8$, respectively.*

Proof. Necessity of (a). Assume that $\overline{Q}$ is not a Dynkin diagram. By
(2.1), $\overline{Q}$ contains a Euclidean graph as a subgraph. By (2.2), we may assume
that $\overline{Q}$ is itself Euclidean. If $\overline{Q} = \widetilde{\mathbb{A}}_m$ for some $m \geq 1$, then (2.3) gives
that KQ is representation-infinite. Otherwise, we observe that, according
to (5.3), the algebra KQ is representation–infinite if and only if $K(\sigma_a Q)$
is representation–infinite for each sink (or source) a of Q. Thus, if $\overline{Q}$ is
Euclidean of type $\widetilde{\mathbb{D}}_n$ (for some $n \geq 4$) or $\widetilde{\mathbb{E}}_p$ (for $p = 6$, 7, or 8), it follows
from (2.7) and (5.2) that KQ is representation–infinite. We have thus shown
that if KQ is representation–finite, then $\overline{Q}$ is a Dynkin graph.

Sufficiency of (a). Assume that Q is a quiver whose underlying graph is
a Dynkin graph. We must show that $A = KQ$ is representation–finite. We
may assume the points of Q to be admissibly numbered as $(1, \ldots, n)$. Let

M be an indecomposable A-module. We claim that the vector $\mathbf{x} = \mathbf{dim}\, M$ is a positive root of the quadratic form q_Q of the quiver Q.

Let $c = s_n \ldots s_1$ denote the Coxeter transformation of Q and $C^+ = S_n^+ \ldots S_1^+$, $C^- = S_1^- \ldots S_n^-$ be the Coxeter functors defined with respect to the admissible numbering $(1, \ldots, n)$ of points of Q. By (4.12), there exists a least $t \geq 0$ such that $c^t \mathbf{x} > 0$ but $c^{t+1} \mathbf{x} \not> 0$. Because $c = s_n \ldots s_1$, there also exists a least i such that $0 \leq i \leq n-1$, $s_i \ldots s_1 c^t \mathbf{x} > 0$, but $s_{i+1} s_i \ldots s_1 c^t \mathbf{x} \not> 0$.

By applying (5.9)(b) repeatedly, we prove that the right A-modules $C^+ M, C^{+2} M, \ldots, C^{+t} M$ are indecomposable and that

$$\mathbf{dim}\, C^{+j} M = c^j (\mathbf{dim}\, M)$$

for all $j \leq t$. Then applying (5.9)(a) to $C^{+t} M$ we conclude that $M' = S_i^+ \ldots S_1^+ C^{+t} M$ is an indecomposable module over $K(\sigma_i \ldots \sigma_1 Q)$ and

$$\mathbf{dim}\, (S_i^+ \ldots S_1^+ C^{+t} M) = s_i \ldots s_1 c^t (\mathbf{dim}\, M) = s_i \ldots s_1 c^t \mathbf{x}.$$

Because $s_{i+1}(\mathbf{dim}\, M') \not> 0$, there is an isomorphism $M' \cong S(i+1)$, by (5.4)(a). But then $s_i \ldots s_1 c^t \mathbf{x} = \mathbf{e}_{i+1}$, and according to (4.14) the vector $\mathbf{x} = c^{-t} s_1 \ldots s_i \mathbf{e}_{i+1} = c^{-t} \mathbf{p}_{i+1}$ (in the notation of (4.13)) is a positive root of q_Q. Furthermore, in view of (5.8) and (5.3)(b), the isomorphism $S_i^+ \ldots S_1^+ C^{+t} M \cong S(i+1)$ yields $M \cong C^{-t} S_1^- \ldots S_i^- S(i+1)$.

We have shown that the mapping $\mathbf{dim}\, : M \mapsto \mathbf{dim}\, M$ takes an indecomposable A-module to a positive root of q_Q. Moreover, the integers i and t as defined earlier, only depend on the vector $\mathbf{x} = \mathbf{dim}\, M$. Thus, if M, N are two indecomposable A-modules such that $\mathbf{dim}\, M = \mathbf{x} = \mathbf{dim}\, N$, we have, as earlier $S_i^+ \ldots S_1^+ C^{+t} M \cong S(i+1) \cong S_i^+ \ldots S_1^+ C^{+t} N$ so that $M \cong C^{-t} S_1^- \ldots S_i^- S(i+1) \cong N$. Thus $\mathbf{dim}$ is an injective mapping from the set of isomorphism classes of indecomposable A-modules to the set of positive roots of q_Q.

Finally, the mapping is surjective because, by (4.14), every positive root $\mathbf{x}$ of q_Q is of the form $\mathbf{x} = c^{-t} \mathbf{p}_{i+1} = c^{-t} s_1 \ldots s_i \mathbf{e}_{i+1}$, for some i and t. But then the indecomposable module $M = C^{-t} S_1^- \ldots S_i^- S(i+1)$ satisfies $\mathbf{x} = \mathbf{dim}\, M$. Because q_Q has only finitely many positive roots, by (3.4) and (4.6), A has only finitely many nonisomorphic indecomposable modules. This finishes the proof of (a) and (b).

The statement (c) follows from (b) and the fact that the number of positive roots of q_Q equals $\frac{1}{2} n(n+1)$, $n^2 - n$, 36, 63, and 120 if $\overline{Q}$ is the Dynkin graph $\mathbb{A}_n$, $\mathbb{D}_n$, with $n \geq 4$, $\mathbb{E}_6$, $\mathbb{E}_7$, and $\mathbb{E}_8$, respectively (see [41], [95], and Exercises 10, 11, and 12). $\qquad\square$

The reader may have observed that we have shown in the course of the proof (with the preceding notation) the following useful fact.

5.11. Corollary. *For any indecomposable module M over a representation–finite hereditary algebra A, there exist integers $t \geq 0$ and i with $0 \leq i \leq n - 1$ (depending only on the vector $\dim M$) such that*

$$M \cong C^{-t} S_1^- \ldots S_i^- S(i+1).$$ □

We can deduce from Gabriel's theorem the shape of the Auslander–Reiten quiver of a representation–finite hereditary algebra. We first obtain an expression of the indecomposable projective and injective modules by means of the reflection functors.

5.12. Corollary. *Let Q be a Dynkin quiver with n points admissibly numbered as $(1, \ldots, n)$ and let i be such that $1 \leq i \leq n$. Denote by $P(i)$ and $I(i)$, respectively, the corresponding indecomposable projective and injective KQ-modules corresponding to the point $i \in Q_0$.*

(a) *If $S(i)$ denotes the simple $K(\sigma_i \ldots \sigma_n Q)$-module corresponding to i in $\sigma_i \ldots \sigma_n Q$, then $P(i) \cong S_1^- \ldots S_{i-1}^- S(i)$ and $\mathbf{p}_i = \dim P(i)$.*

(b) *If $S(i)$ denotes the simple $K(\sigma_i \ldots \sigma_1 Q)$-module corresponding to i in $\sigma_i \ldots \sigma_1 Q$, then $I(i) \cong S_n^+ \ldots S_{i+1}^+ S(i)$ and $\mathbf{q}_i = \dim I(i)$.*

Proof. We only prove (a); the proof of (b) is similar. By Gabriel's theorem (5.10), the indecomposable KQ-modules are uniquely determined up to isomorphism by their dimension vectors; hence it suffices to show that

$$\mathbf{p}_i = s_1 \ldots s_{i-1}(\mathbf{e}_i) = \dim P(i).$$

We show by descending induction on k with $1 \leq k \leq i$ that $s_k \ldots s_{i-1}(\mathbf{e}_i)_j$ equals 1 if $k \leq j \leq i$ and there exists a path from i to k through j, and equals 0 otherwise. There is nothing to show if $k = i$. Assume $k < i$ and that the statement holds for all $k < j \leq i$. There is at most one point j in Q such that $k < j \leq i$ and there is an arrow $j \to k$ and a path from i to j. Indeed, the existence of two such points j would contradict the fact that Q is a tree. Hence it follows from the definition of s_k that $s_k \ldots s_{i-1}(\mathbf{e}_i)_k = 1$ if there exists $k < j \leq i$ such that there is an arrow $j \to k$ and a path from i to j (that is, if there exists a path from i to k), and $s_k \ldots s_{i-1}(\mathbf{e}_i)_k = 0$ otherwise. Because, by our inductive assumption, $[s_k s_{k+1} \ldots s_{i-1}(\mathbf{e}_i)]_j = [s_{k+1} \ldots s_{i-1}(\mathbf{e}_i)]_j$ for all $j \neq k$, this shows our claim. The result follows after setting $k = 1$. □

5.13. Proposition. *Let A be a representation–finite hereditary algebra.*

(a) *For every indecomposable A-module M, there exist $t \geq 0$ and an indecomposable projective A-module P such that $M \cong \tau^{-t}P$.*

(b) *The Auslander–Reiten quiver $\Gamma(\operatorname{mod} A)$ of A is acyclic.*

Proof. We assume for simplicity that the points of Q_A are admissibly numbered as $(1, \dots, n)$. Let $C^- = S_1^- \dots S_n^-$ be the Coxeter functor.

(a) By (5.11), there exists a pair of integers $t \geq 0$ and $0 \leq i \leq n-1$ such that $M \cong C^{-t} S_1^- \dots S_i^- S(i+1)$. The result follows from (5.8) and (5.12).

(b) Assume that

$$M_0 \to M_1 \to \dots \to M_s = M_0$$

is a cycle in $\Gamma(\operatorname{mod} A)$. By (a), for each i with $0 \leq i < s$, there exist $t_i \geq 0$ and $a_i \in (Q_A)_0$ such that $M_i \cong \tau^{-t_i}P(a_i)$. Let $t = \min\{t_i \mid 0 \leq i < s\}$. Then the previous cycle induces a cycle

$$\tau^t M_0 \to \tau^t M_1 \to \dots \to \tau^t M_s = \tau^t M_0$$

in $\Gamma(\operatorname{mod} A)$, because it follows from (IV.2.15) that $\operatorname{Irr}(X, Y) \cong \operatorname{Irr}(\tau X, \tau Y)$ for any pair of indecomposable nonprojective modules X and Y. Moreover, by definition of t, this cycle passes through a projective A-module. Because A is hereditary, by (1.10), the cycle consists of indecomposable projective modules connected by irreducible monomorphisms, which is a contradiction. $\square$

5.14. Corollary. *Let M be an indecomposable module over a representa-tion–finite hereditary algebra A. Then $\operatorname{End}_A M \cong K$ and $\operatorname{Ext}_A^1(M, M) = 0$.*

Proof. By (5.13)(a), there exist $t \geq 0$ and an indecomposable projec-tive A-module P such that $M \cong \tau^t P$. Applying (IV.2.14) and (IV.2.15) we get a sequence of isomorphisms $\operatorname{Hom}_A(M, M) \cong \operatorname{Hom}_A(\tau^t P, \tau^t P) \cong \operatorname{Hom}_A(P, P) \cong K$ (by (1.5)) and $\operatorname{Ext}_A^1(M, M) \cong D\operatorname{Hom}_A(M, \tau M) \cong D\operatorname{Hom}_A(\tau^t P, \tau^{t+1}P) \cong D\operatorname{Hom}_A(P, \tau P) \cong \operatorname{Ext}_A^1(P, P) = 0$. $\square$

By (IV.2.14), the fact that each indecomposable module over a represen-tation–finite hereditary algebra A is a brick implies that $\operatorname{Ext}_A^1(M, \tau M)$ is one-dimensional for each indecomposable nonprojective module M_A and, hence, any nonsplit short exact sequence $0 \to \tau M \to L \to M \to 0$ is almost split.

We also note that it follows from (1.10) and (5.14) that the combinatorial method of constructing the Auslander–Reiten quiver explained in Examples (IV.4.10)–(IV.4.14) works perfectly well for representation–finite hereditary algebras.

5.15. Examples. (a) Let Q be the quiver $\overset{1}{\circ}\longleftarrow\overset{3}{\circ}\longrightarrow\overset{2}{\circ}$ whose underlying graph is the Dynkin graph $\mathbb{A}_3$. We wish to construct a complete list of the nonisomorphic indecomposable KQ-modules.

The simple representations are:

$$S(1) = (K\longleftarrow 0\longrightarrow 0), \ S(2) = (0\longleftarrow 0\longrightarrow K), \text{ and } S(3) = (0\longleftarrow K\longrightarrow 0).$$

The indecomposable projective representations are:

$$P(1) = S(1), \ P(2) = S(2), \text{ and } P(3) = (K\overset{1}{\longleftarrow}K\overset{1}{\longrightarrow}K).$$

The indecomposable injective representations are: $I(3) = S(3)$,

$$I(1) = (K\overset{1}{\longleftarrow}K\longrightarrow 0), \text{ and } I(2) = (0\longleftarrow K\overset{1}{\longrightarrow}K).$$

The positive roots of q_Q have been computed in (4.15)(a). We see in particular that every indecomposable KQ-module is either projective or injective. To construct $\Gamma(\operatorname{mod} KQ)$ as in (IV.4.10), it suffices to observe that $\operatorname{rad} P(3) \cong P(1) \oplus P(2)$. The construction proceeds easily:

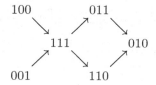

(b) Let Q be the quiver

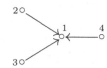

whose underlying graph is the Dynkin graph $\mathbb{D}_4$. We wish to construct a complete list of the nonisomorphic indecomposable KQ-modules.

The simple representations are:

$$S(1) = \begin{pmatrix} 0 \searrow \\ \quad \nearrow K\longleftarrow 0 \\ 0 \end{pmatrix} \qquad S(3) = \begin{pmatrix} 0 \searrow \\ \quad \nearrow 0\longleftarrow 0 \\ K \end{pmatrix}$$

$$S(2) = \begin{pmatrix} K \searrow \\ \quad \nearrow 0\longleftarrow 0 \\ 0 \end{pmatrix} \qquad S(4) = \begin{pmatrix} 0 \searrow \\ \quad \nearrow 0\longleftarrow K \\ 0 \end{pmatrix}$$

The indecomposable projective representations are: $P(1) = S(1)$ and

$$P(2) = \begin{pmatrix} K \searrow 1 \\ \quad \searrow K\longleftarrow 0 \\ 0 \nearrow \end{pmatrix} \quad P(3) = \begin{pmatrix} 0 \searrow \\ \quad \nearrow K\longleftarrow 0 \\ K \quad 1 \end{pmatrix} \quad P(4) = \begin{pmatrix} 0 \searrow \\ \quad \nearrow K\overset{1}{\longleftarrow}K \\ 0 \end{pmatrix}$$

The indecomposable injective representations are:

$$I(1) = \begin{pmatrix} K & & \\ & \searrow 1 & \\ & & K \xleftarrow{\;1\;} K \\ & \nearrow 1 & \\ K & & \end{pmatrix},$$

$I(2) = S(2)$, $I(3) = S(3)$, and $I(4) = S(4)$.

The positive roots of q_Q have been computed in (4.15)(b). To obtain the remaining indecomposable representations, it suffices, by Gabriel's theorem (5.10), to exhibit, for each positive root $\mathbf{x}$, an indecomposable representation having $\mathbf{x}$ as dimension vector. We thus have four other indecomposable representations, given respectively by:

(1) $\dim M_1 = \begin{pmatrix} 1 \\ 1 \end{pmatrix}\! 2\, 1$, then $M_1 = \begin{pmatrix} K & & \\ & \searrow 1 & \\ & & K^2 \xleftarrow{\left[\begin{smallmatrix}1\\1\end{smallmatrix}\right]} K \\ & \nearrow 1 & \\ K & & \end{pmatrix}$ (this is indeed an

indecomposable representation, by the proof of (2.6));

(2) $\dim M_2 = \begin{pmatrix} 0 \\ 1 \end{pmatrix}\! 1\, 1$, then $M_2 = \begin{pmatrix} 0 & & \\ & \searrow & \\ & & K \xleftarrow{\;1\;} K \\ & \nearrow 1 & \\ K & & \end{pmatrix}$

(3) $\dim M_3 = \begin{pmatrix} 1 \\ 0 \end{pmatrix}\! 1\, 1$, then $M_3 = \begin{pmatrix} K & & \\ & \searrow 1 & \\ & & K \xleftarrow{\;1\;} K \\ & \nearrow 0 & \\ 0 & & \end{pmatrix}$

(4) $\dim M_4 = \begin{pmatrix} 1 \\ 1 \end{pmatrix}\! 1\, 0$, then $M_4 = \begin{pmatrix} K & & \\ & \searrow 1 & \\ & & K \xleftarrow{\quad} 0 \\ & \nearrow 1 & \\ K & & \end{pmatrix}$

(indeed, M_2, M_3, and M_4 are indecomposable, because each has a simple socle isomorphic to $S(1)$).

To construct $\Gamma(\operatorname{mod} KQ)$, we note that there are isomorphisms

$$\operatorname{rad} P(2) \cong \operatorname{rad} P(3) \cong \operatorname{rad} P(4) \cong P(1).$$

The construction then proceeds easily, as in (IV.4.10)–(IV.4.14):

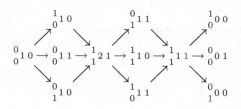

(c) Let Q be the quiver

with underlying graph $\mathbb{E}_6$. Then $\Gamma(\operatorname{mod} KQ)$ is the quiver

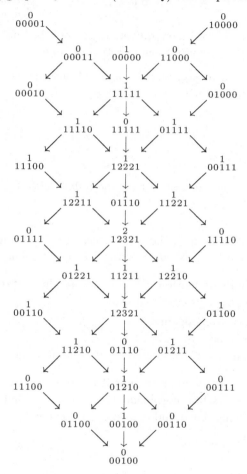

We leave to the reader as an exercise to describe explicitly each of the indecomposable KQ-modules as a representation. Notice that the largest root $_{1}2\overset{2}{3}_{2}1$ has already been described in the proof of (2.6).

VII.6. Exercises

1. Show that each of the following matrix algebras is hereditary:

(a) $\begin{bmatrix} K & 0 & 0 & 0 \\ K & K & 0 & 0 \\ K & K & K & 0 \\ K & K & 0 & K \end{bmatrix}$ (b) $\begin{bmatrix} K & 0 & 0 & 0 \\ K & K & 0 & 0 \\ K & 0 & K & 0 \\ K & 0 & 0 & K \end{bmatrix}$ (c) $\begin{bmatrix} K & 0 & 0 & 0 & 0 & 0 \\ K & K & 0 & 0 & 0 & 0 \\ K & 0 & K & 0 & 0 & 0 \\ K & 0 & K & K & 0 & 0 \\ K & 0 & K & 0 & K & 0 \\ K & 0 & K & 0 & K & K \end{bmatrix}.$

In each case, give the ordinary quiver, then describe the indecomposable projective and the indecomposable injective modules.

2. Construct, as a matrix algebra, a hereditary algebra whose ordinary quiver is one of the following:

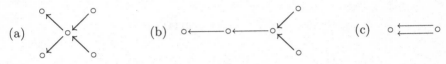

(a)   (b)   (c)

3. Let A be an algebra. Show that the following conditions are equivalent:

(a) A is hereditary.
(b) For each module M_A, the functor $\mathrm{Ext}^1_A(M, -)$ is right exact.
(b) For each module ${}_A N$, the functor $\mathrm{Tor}^A_1(-, N)$ is left exact.

4. Let A be a finite dimensional basic connected hereditary algebra. Show that the following conditions are equivalent:

(a) A is a Nakayama algebra.
(b) $A \cong \mathbb{T}_n(K)$ for some $n \geq 1$.
(c) A admits a projective-injective indecomposable module.

5. An algebra A is called triangular if there exists a hereditary algebra H and a surjective algebra morphism $\varphi : H \to A$ such that $\mathrm{Ker}\, \varphi \subseteq \mathrm{rad}^2 H$. Show that A is triangular if and only if Q_A is acyclic.

6. Let Q be the quiver

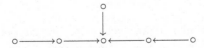

Construct bricks having as dimension vectors ${}_{11}{}^{1}_{2}{}^{1}_{10}$, $12\overset{1}{3}21$, and ${}_{01}\overset{1}{2}21$, respectively.

7. Show that each of the following integral quadratic forms is positive definite:

(a) $x_1^2 + x_2^2 + x_3^2 + x_4^2 - x_1x_2 - x_1x_3 - x_2x_4 - x_3x_4 + x_1x_4$.

(b) $x_1^2 + x_2^2 + x_3^2 + x_4^2 - x_1x_2 + x_1x_3 - x_1x_4 - x_2x_3 + x_2x_4 - x_3x_4$.

8. Show that each of the following integral quadratic forms is weakly positive but not positive definite.

(a) $x_1^2 + x_2^2 + x_3^2 - x_1x_2 + x_1x_3 + x_2x_3$.

(b) $x_1^2 + x_2^2 + x_3^2 + x_4^2 - x_1x_2 - x_1x_3 - x_2x_4 - x_3x_4 + 2x_1x_4$.

(c) $x_1^2 + x_2^2 + x_3^2 + x_4^2 + x_5^2 + x_6^2 - x_1x_4 - x_2x_4 - x_3x_4 - x_4x_5 - x_4x_6$
$+ x_1x_5 + x_1x_6 + x_2x_5 + x_2x_6 + x_3x_5 + x_3x_6$.

Show that the quadratic form (c) is not positive semidefinite.

9. A vector $\mathbf{x} \in \mathbb{Z}^n$ is called **sincere** if all its coordinates are nonzero . Let $\mathbf{x}$ be a sincere positive root of a weakly positive integral quadratic form q. Show that the following conditions are equivalent:

(a) $\mathbf{x}$ is a maximal root.

(b) $s_i(\mathbf{x}) \leq \mathbf{x}$ for each i.

(c) $D_i q(\mathbf{x}) \geq 0$ for each i.

10. Let Q be a quiver with underlying graph

$$\mathbb{A}_m : \quad \overset{1}{\circ}\!\!-\!\!-\!\!\overset{2}{\circ}\!\!-\!\!-\!\!-\!\!-\!\!\cdots\!\!-\!\!-\!\!\overset{m-1}{\circ}\!\!-\!\!\overset{m}{\circ} \quad (m \geq 1).$$

Show that the positive roots of q_Q in $F = \bigoplus_{i=1}^m \mathbf{e}_i\mathbb{Z}$ are just the vectors $\mathbf{e}_1, \ldots, \mathbf{e}_n$ and $\mathbf{e}_i + \mathbf{e}_{i+1} + \ldots + \mathbf{e}_j$, where $1 \leq i < j \leq m$. Thus Q affords $\frac{m(m+1)}{2}$ positive roots.

11. Let Q be the quiver with underlying graph

$$\widetilde{\mathbb{D}}_n : \quad (n \geq 4).$$

Show that the positive roots of q_Q in $F = \bigoplus_{i=1}^n \mathbf{e}_i\mathbb{Z}$ are just the vectors $\mathbf{e}_1, \ldots, \mathbf{e}_n$, $\mathbf{e}_i + \mathbf{e}_{i+1} + \ldots + \mathbf{e}_j$, where $1 \leq i < j \leq n$, and $j \geq 3$, $\mathbf{e}_1 + \mathbf{e}_3 + \ldots + \mathbf{e}_j$, where $j \geq 3$, $\mathbf{e}_1 + \mathbf{e}_2 + 2(\mathbf{e}_3 + \ldots + \mathbf{e}_i) + \mathbf{e}_{i+1} + \ldots + \mathbf{e}_j$, where $3 \leq i < j \leq n$. Thus Q affords $n(n-1)$ positive roots.

12. Compute all the positive roots for $\mathbb{E}_6$, $\mathbb{E}_7$, and $\mathbb{E}_8$ (one finds, respectively, 36, 63, and 120 positive roots).

13. Show that for each $m \geq 1$, the Weyl group W_Q of a quiver Q with underlying graph $\mathbb{A}_m$ is the symmetric group S_{m+1}. **Hint:** Define the isomorphism $W_Q \to S_{m+1}$ by the formula $s_j \mapsto (j, j+1)$.

14. Show that the Weyl group W_Q of a quiver Q with underlying graph $\mathbb{D}_n$ (for some $n \geq 4$) is isomorphic to the semidirect product $S_n \ltimes (\mathbb{Z}/2\mathbb{Z})^{n-1}$ of the symmetric group S_n and the group $(\mathbb{Z}/2\mathbb{Z})^{n-1}$ (the direct sum of $n-1$ copies of $\mathbb{Z}/2\mathbb{Z}$), see [95, sec.12.1].

15. For each of the following quivers, compute all the indecomposable representations and the Auslander–Reiten quiver of the corresponding path algebra:

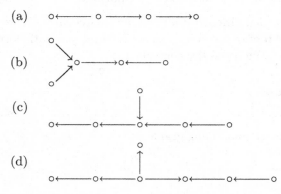

(a)

(b)

(c)

(d)

16. Let A be given by the quiver

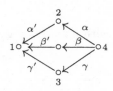

bound by $\alpha\alpha' + \beta\beta' + \gamma\gamma' = 0$. Show that A is not hereditary but, for every idempotent $e \in A$ such that $e \neq 1$ and $e \neq 0$, the algebra eAe is hereditary.

17. Complete the proof of (5.6) by proving the functoriality of the isomorphisms involved.

18. Prove (5.7).

Chapter VIII

Tilted algebras

As seen in the preceding chapters, the Auslander–Reiten quiver of an algebra is a very useful combinatorial invariant allowing us to store algebraic information about the module category. We were, for instance, able to use it to compute homomorphisms and extensions between modules, as well as to construct an algebra obtained by tilting from one that was known before. However, its usefulness is not restricted to being a device for storing information. As we shall see in this chapter, its combinatorial properties can be used to characterise classes of algebras.

We start from the results of Chapter VII on the Auslander–Reiten quiver of a representation–finite hereditary algebra A; it follows from these results that the full subquiver of $\Gamma(\operatorname{mod} A)$ consisting of the projective points is connected, acyclic, and meets each τ-orbit of $\Gamma(\operatorname{mod} A)$ exactly once and every path in $\Gamma(\operatorname{mod} A)$ having its source and target in it must entirely lie in it. These three properties characterise what is called a section in a (generally infinite) component of the Auslander–Reiten quiver.

We first generalise this remark by showing that any representation–infinite hereditary algebra has sections in two infinite components, which we call postprojective and preinjective. We then define a new class of algebras, the so-called tilted algebras, which now play a prominent rôle in the representation theory of algebras and which are obtained from hereditary algebras by tilting. The main result of this chapter is a handy criterion, independently obtained by Liu [111] and Skowroński [156], which characterises the tilted algebras as being those algebras B having a faithful section Σ in a component $\mathcal{C}$ of $\Gamma(\operatorname{mod} B)$ such that $\operatorname{Hom}_B(U, \tau V) = 0$ for all modules U, V from Σ. Throughout this chapter, and contrary to the previous ones, our emphasis is on studying representation–infinite algebras rather than representation–finite ones.

VIII.1. Sections in translation quivers

Because our objective in this chapter is to describe combinatorial properties of connected components of the Auslander–Reiten quiver of a (not necessarily representation–finite) hereditary algebra or of an algebra "close" to being hereditary, we recall that such a component has the combinatorial structure of a translation quiver, as defined in (IV.4.7). We need a special type of translation quiver.

1.1. Definition. Let $\Sigma = (\Sigma_0, \Sigma_1)$ be a connected and acyclic quiver. We define an **infinite translation quiver** $(\mathbb{Z}\Sigma, \tau)$ as follows. The set of points of $\mathbb{Z}\Sigma$ is $(\mathbb{Z}\Sigma)_0 = \mathbb{Z} \times \Sigma_0 = \{(n, x) \mid n \in \mathbb{Z},\ x \in \Sigma_0\}$ and, for each arrow $\alpha : x \to y$ in Σ_1, there exist two arrows

$$(n, \alpha) : (n, x) \to (n, y) \quad \text{and} \quad (n, \alpha') : (n+1, y) \to (n, x)$$

in $(\mathbb{Z}\Sigma)_1$, and these are all the arrows in $(\mathbb{Z}\Sigma)_1$. We define the translation τ on $\mathbb{Z}\Sigma$ by $\tau(n, x) = (n+1, x)$ for all $(n, x) \in (\mathbb{Z}\Sigma)_0$.

For every $(n, x) \in (\mathbb{Z}\Sigma)_0$, we define a bijection between the set of arrows of target (n, x) and the set of arrows of source $(n+1, x)$ by the formulas

$$\sigma(n, \alpha) = (n, \alpha') \quad \text{and} \quad \sigma(n, \alpha') = (n+1, \alpha).$$

For example, let Σ be the quiver

Then $\mathbb{Z}\Sigma$ is the translation quiver

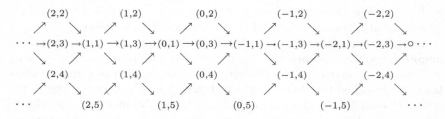

We denote by $\mathbb{N}\Sigma$ the full translation subquiver of $\mathbb{Z}\Sigma$ consisting of all points $(n, x) \in (\mathbb{Z}\Sigma)_0$ with $n \geq 0$ and, similarly, by $(-\mathbb{N})\Sigma$ the full translation subquiver of $\mathbb{Z}\Sigma$ consisting of all points $(n, x) \in (\mathbb{Z}\Sigma)_0$ with $n \leq 0$.

Clearly, the quiver $\mathbb{Z}\Sigma$ thus defined is a translation quiver with neither projectives nor injectives, and the maps $\tau : (\mathbb{Z}\Sigma)_0 \to (\mathbb{Z}\Sigma)_0$ and $\sigma : (\mathbb{Z}\Sigma)_1 \to (\mathbb{Z}\Sigma)_1$ are bijective. Moreover, it is easily verified that the quiver Σ, identified with the full translation subquiver of $\mathbb{Z}\Sigma$ consisting of the points $(0, x)$, with $x \in \Sigma_0$, and of the arrows $(0, \alpha)$, with $\alpha \in \Sigma_1$, is a section of $\mathbb{Z}\Sigma$ in the sense of the following definition.

1.2. Definition. Let (Γ, τ) be a connected translation quiver. A connected full subquiver Σ of Γ is a **section** of Γ if the following conditions are satisfied:

(S1) Σ is acyclic.

(S2) For each $x \in \Gamma_0$, there exists a unique $n \in \mathbb{Z}$ such that $\tau^n x \in \Sigma_0$.

(S3) If $x_0 \to x_1 \to \cdots \to x_t$ is a path in Γ with $x_0, x_t \in \Sigma_0$, then $x_i \in \Sigma_0$ for all i such that $0 \le i \le t$.

For a translation quiver (Γ, τ), the τ-**orbit** of a point $x \in \Gamma_0$ is defined to be the set of all points of the form $\tau^n x$, with $n \in \mathbb{Z}$. With this terminology, (S2) can be restated to say that Σ meets each τ-orbit exactly once.

A full subquiver Σ of a quiver Γ is defined to be **convex** in Γ if, for any path $x_0 \to x_1 \to \cdots \to x_t$ in Γ with $x_0, x_t \in \Sigma_0$, we have $x_i \in \Sigma_0$ for all i such that $0 \le i \le t$. Thus, (S3) says that a section of Γ is convex in Γ.

1.3. Examples. (a) Let A be a connected hereditary algebra and Σ_A be the full subquiver of the Auslander–Reiten quiver $\Gamma(\operatorname{mod} A)$ consisting of the points corresponding to the isomorphism classes of all the indecomposable projective A-modules. We know, by (VII.1.4)(g), that any indecomposable projective A-module has only projective predecessors. Because, according to (VII.1.6), for any two indecomposable projective A-modules $P(a) = e_a A$ and $P(b) = e_b A$, there exists a K-linear isomorphism

$$e_a(\operatorname{rad} A / \operatorname{rad}^2 A) e_b \cong \operatorname{Irr}(P(b), P(a)),$$

then $\Sigma_A \cong Q_A^{\mathrm{op}}$. In particular, Σ is connected.

Similarly, $\Gamma(\operatorname{mod} A)$ contains a section induced by the indecomposable injective A-modules. Indeed, let Σ'_A be the full subquiver of the Auslander–Reiten quiver $\Gamma(\operatorname{mod} A)$ consisting of the points corresponding to the isomorphism classes of indecomposable injective A-modules. Then the duality $D : \operatorname{mod} A \to \operatorname{mod} A^{\mathrm{op}}$ carries Σ'_A to $D(\Sigma'_A) = \Sigma_{A^{\mathrm{op}}}$. By applying these arguments to A^{op}, we get $\Sigma_{A^{\mathrm{op}}} \cong Q_A$, and consequently

$$\Sigma'_A \cong Q_A^{\mathrm{op}} \cong \Sigma_A.$$

Assume now that A is representation–finite. We claim that Σ_A is a section of $\Gamma(\operatorname{mod} A)$. Indeed, because Q_A is acyclic, so is Σ_A. The convexity of Σ_A follows from (VII.1.9), because A is hereditary, and therefore the indecomposable projectives have only projective predecessors. Finally, it follows from (VII.5.12)(a) that Σ_A meets each τ-orbit exactly once, proving our claim.

As we shall see in Section 2, the same statement holds for representation–infinite hereditary algebras.

(b) We now give an example of a nonhereditary representation–finite algebra having a section in its Auslander–Reiten quiver. Let A be given by the quiver

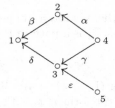

bound by $\alpha\beta = \gamma\delta$, $\varepsilon\delta = 0$. Then $\Gamma(\mathrm{mod}\,A)$ is given by

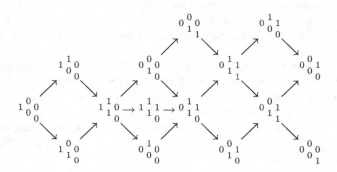

where indecomposable modules are represented by their dimension vectors. We notice that each of the following two sets of indecomposable modules

$$\left\{{}_1{}^1_1{}^0_0{}0,\ 0{}^0_1{}^0_0{}0,\ 0{}^1_0{}^0_0{}0,\ 1{}^1_1{}^1_0{}1,\ 0{}^0_1{}^0_0{}0\right\} \quad \text{and} \quad \left\{0{}^0_1{}^0_0{}0,\ 1{}^1_1{}^1_0{}1,\ 0{}^1_0{}^0_0{}0,\ 0{}^1_1{}^1_0{}0,\ 0{}^0_1{}^0_0{}1\right\}$$

defines a section of $\Gamma(\mathrm{mod}\,A)$.

It turns out that the mere existence of a section Σ in a translation quiver (Γ, τ) implies that (Γ, τ) can be fully embedded in $\mathbb{Z}\Sigma$. Before proving this statement, we need an easy lemma.

1.4. Lemma. *Let (Γ, τ) be a connected translation quiver and Σ be a section of (Γ, τ). Then the following hold:*

(a) *If $x \to y$ is an arrow in Γ and $x \in \Sigma_0$, then $y \in \Sigma_0$ or $\tau y \in \Sigma_0$.*

(b) *If $x \to y$ is an arrow in Γ and $y \in \Sigma_0$, then $x \in \Sigma_0$ or $\tau^{-1}x \in \Sigma_0$.*

Proof. We only prove (a); the proof of (b) is similar. By (S2), there exists $m \in \mathbb{Z}$ such that $\tau^m y \in \Sigma_0$. Assume that $m \leq 0$; then there exists a path in (Γ, τ) of the form $x \to y \to \cdots \to \tau^m y$ with both ends in Σ. By (S3), we have $y \in \Sigma_0$. Hence, by (S2), $m = 0$. Similarly, $m > 0$ yields $\tau y \in \Sigma_0$. $\qquad\square$

1.5. Proposition. *Let (Γ, τ) be a connected translation quiver and Σ be a section of Γ. Then Γ is isomorphic to the full translation subquiver of $\mathbb{Z}\Sigma$ consisting of the points (n, x) with $n \in \mathbb{Z}$, $x \in \Sigma_0$ such that $\tau^n x$ is defined in Γ. In particular, Γ is acyclic.*

Proof. Let Ω be the full translation subquiver of $\mathbb{Z}\Sigma$ consisting of all pairs $(n, x) \in (\mathbb{Z}\Sigma)_0$ such that $\tau^n x$ is defined in Γ. Considering Ω as a subquiver of Γ, we see that Ω is the translation subquiver of Γ such that $\Omega_0 = \Gamma_0$ and Ω_1 consists of all possible arrows of Γ_1 of the forms

$$\tau^n \alpha = \sigma^{2n} \alpha : \tau^n x \to \tau^n y \quad \text{and} \quad \sigma \tau^n \alpha = \sigma^{2n+1} \alpha : \tau^{n+1} x \to \tau^n y,$$

where $n \in \mathbb{Z}$ and $\alpha : x \to y$ is an arrow in Σ_1. We need to show that in fact $\Omega_1 = \Gamma_1$, that is, each arrow in Γ_1 lies in Ω_1.

Let $\alpha : a \to b$ be an arrow in Γ. By (S2), there exist $x, y \in \Sigma_0$ and $m, n \in \mathbb{Z}$ such that $a = \tau^m x$ and $b = \tau^n y$. Assume $m = 0$. Then $a = x \in \Sigma_0$. By (1.4), b or τb belongs to Σ_0. In either case, $\alpha \in \Omega_1$. Because the case $n = 0$ is similar, assume that $m \neq 0$ and $n \neq 0$. Suppose first $m > 0$ and $n > 0$. Because all $\tau^i x$, $\tau^j y$, with $0 \leq i \leq m$, $0 \leq j \leq n$ are defined, this implies that there exists in Γ_1 an arrow of the form

$$\sigma^{-2m+1} \alpha : \tau^{n-m+1} y \to x \quad \text{or} \quad \sigma^{-2n} \alpha : \tau^{m-n} x \to y.$$

In the first case, (1.4) yields that $\tau^{n-m+1} y \in \Sigma_0$ or $\tau^{n-m} y \in \Sigma_0$. By (S3), this implies $\tau^{n-m+1} y = y$ or $\tau^{n-m} y = y$; thus, by (S1), $m = n+1$ or $m = n$. Hence $\alpha \in \Omega_1$. We proceed analogously in the second case.

The case where $m < 0$ and $n < 0$ being similar, we may suppose that $m > 0$ and $n < 0$. Then Γ contains a path of the form

$$y \longrightarrow \cdots \longrightarrow \tau^n y = b \longrightarrow \tau^{-1} a = \tau^{m-1} x \longrightarrow \cdots \longrightarrow x.$$

By (S3), all points on this path belong to Σ and, in particular, Σ contains two points of the τ-orbit of x (or y), a contradiction to (S2). Finally, the case where $m > 0$ and $n < 0$ is treated in the same way. $\square$

For example, if A is the algebra of Example 1.3 (b) and Σ is one of the two sections of $\Gamma(\text{mod } A)$, then it is readily seen that $\Gamma(\text{mod } A)$ is isomorphic to the connected full subquiver of $\mathbb{Z}\Sigma$ consisting of all $(n, x) \in (\mathbb{Z}\Sigma)_0$ such that $\tau^n x$ corresponds to an indecomposable A-module (thus, for instance, if $x = 0\,{}_1^0\,0$, then only τx, $\tau^{-1} x$ and $\tau^{-2} x$ are defined). We have the following obvious corollary.

1.6. Corollary. *Let Δ be a section in $\mathbb{Z}\Sigma$. Then $\mathbb{Z}\Delta \cong \mathbb{Z}\Sigma$.* $\square$

In particular, if a is a sink in a finite, connected, and acyclic quiver Σ, then $\sigma_a \Sigma$ (see (VII.5)) is isomorphic to the full translation subquiver of $\mathbb{Z}\Sigma$ consisting of the points $(1, a)$ and $\{(0, b) \mid b \in \Sigma_0,\ b \neq a\}$. Clearly, this is a section in $\mathbb{Z}\Sigma$, so that $\mathbb{Z}(\sigma_a \Sigma) \cong \mathbb{Z}\Sigma$. Inductively, if $(a_1, \ldots, a_n)$ is an admissible sequence of sinks in Σ, then $\mathbb{Z}(\sigma_{a_n} \ldots \sigma_{a_1} \Sigma) \cong \mathbb{Z}\Sigma$. These remarks, together with (VII.5.2), imply the following lemma.

1.7. Lemma. *Let Σ and Δ be two trees having the same underlying graph. Then $\mathbb{Z}\Sigma \cong \mathbb{Z}\Delta$.* $\qquad\square$

The statement of (1.7) characterises trees. Indeed, we have the following result.

1.8. Lemma. *Let Σ and Σ' be quivers having the same underlying graph of type $\widetilde{\mathbb{A}}_m$. Then $\mathbb{Z}\Sigma \cong \mathbb{Z}\Sigma'$ if and only if the quivers Σ and Σ' have the same number of clockwise-oriented arrows and the same number of counterclockwise-oriented arrows.*

Proof. Let a be a sink in Σ. Then $\mathbb{Z}\Sigma$ contains a unique section Δ such that a is the unique sink in Δ and Δ has a unique source. By (1.6), $\mathbb{Z}\Delta \cong \mathbb{Z}\Sigma$. We may thus assume from the start that each of Σ and Σ' has a unique source and a unique sink. But then the statement is clear. $\qquad\square$

VIII.2. Representation–infinite hereditary algebras

We know from Chapter VII that the representation–finite hereditary algebras coincide with the path algebras of Dynkin quivers and that their Auslander–Reiten quivers are finite and acyclic and (by (1.3)(a)) have at least two sections consisting of, respectively, the indecomposable projective modules and the indecomposable injective modules. Furthermore, the sections are Dynkin quivers. We now generalise these statements to hereditary algebras, which are not necessarily representation–finite.

2.1. Proposition. *Let $A = KQ$, where Q is a finite, connected, and acyclic quiver, and let $\Gamma(\mathrm{mod}\,A)$ be the Auslander–Reiten quiver of A.*

(a) *$\Gamma(\mathrm{mod}\,A)$ contains a connected component $\mathcal{P}(A)$ such that*
    (i) *for every indecomposable A-module M in $\mathcal{P}(A)$, there exist a unique $t \geq 0$ and a unique $a \in Q_0$ such that $M \cong \tau^{-t} P(a)$;*
    (ii) *$\mathcal{P}(A)$ contains a section consisting of all the indecomposable projective A-modules; and*
    (iii) *$\mathcal{P}(A)$ is acyclic.*

(b) *$\Gamma(\mathrm{mod}\,A)$ contains a connected component $\mathcal{Q}(A)$ such that*
    (i) *for every indecomposable A-module N in $\mathcal{Q}(A)$, there exist a unique $s \geq 0$ and a unique $b \in Q_0$ such that $N \cong \tau^s I(b)$;*
    (ii) *$\mathcal{Q}(A)$ contains a section consisting of the indecomposable injective A-modules; and*
    (iii) *$\mathcal{Q}(A)$ is acyclic.*

(c) $\mathcal{P}(A) = \mathcal{Q}(A)$ *if and only if A is representation–finite.*

Proof. (a) Let Σ be the full subquiver of the Auslander–Reiten quiver $\Gamma(\text{mod } A)$ of A consisting of the points corresponding to the indecomposable projective A-modules. As pointed out in (1.3)(a), there is a quiver isomorphism $\Sigma \cong Q^{\text{op}}$. We let $\mathcal{P}(A)$ be the connected component of $\Gamma(\text{mod } A)$ containing Σ.

(i) We claim that any indecomposable module in $\mathcal{P}(A)$ is isomorphic to a module of the form $\tau^{-t}P(a)$, with $t \geq 0$ and $a \in Q_0$. Indeed, we first show, by induction on t, that if $f : M \to \tau^{-t}P(a)$ is an irreducible morphism, with M indecomposable, then M is of the wanted form. If M is projective, there is nothing to prove. So assume it is not. Because predecessors of projectives are projective, we have $t \geq 1$ and there exists an irreducible morphism $\sigma^2 f = \tau f : \tau M \to \tau^{-t+1}P(a)$. By the induction hypothesis, there exist $r \geq 0$ and $b \in Q_0$ such that $\tau M \cong \tau^{-r}P(b)$; hence $M \cong \tau^{-r-1}P(b)$. Next, assume that there exists an irreducible morphism $g : \tau^{-t}P(a) \to M$. If M is projective, there is nothing to prove. So assume it is not. There exists an irreducible morphism $\sigma g : \tau M \to \tau^{-t}P(a)$. By the preceding argument, τM is of the required form, hence so is M. These two statements and induction imply our claim.

We now prove that t and a are uniquely determined. If $\tau^{-t}P(a) \cong \tau^{-r}P(b)$, then assuming, without loss of generality, that $t \geq r$, we have $P(a) \cong \tau^{t-r}P(b)$, hence $t = r$ and $a = b$.

(ii) We must show that Σ is a section in $\mathcal{P}(A)$. Because $\Sigma \cong Q^{\text{op}}$, we have that Σ is acyclic. Because predecessors of projectives are projective, we also have that Σ is convex in $\mathcal{P}(A)$. Finally, it follows from (i) that Σ meets each τ-orbit of $\mathcal{P}(A)$ exactly once.

(iii) This follows from (ii) and (1.5).

(b) The proof is entirely similar to that of (a) and is omitted.

(c) Clearly, if A is representation–finite, then $\Gamma(\text{mod } A)$ is connected, and so $\mathcal{P}(A) = \mathcal{Q}(A)$. Assume conversely that $\mathcal{P}(A) = \mathcal{Q}(A)$. Then, in particular, $\mathcal{P}(A)$ contains all the indecomposable injective A-modules. Let $m = \max\{t \geq 0 \mid \tau^{-t}P(a) \text{ be injective for some } a \in Q_0\}$ and n denote the number of points in Q. Then $\mathcal{P}(A)$ contains at most mn indecomposable modules so that it is a finite component of $\Gamma(\text{mod } A)$. By (IV.5.4), A is representation–finite. $\square$

2.2. Definition. Let A be an arbitrary (not necessarily hereditary) K-algebra, and $\Gamma(\text{mod } A)$ the Auslander–Reiten quiver of A.

(a) A connected component $\mathcal{P}$ of $\Gamma(\text{mod } A)$ is called **postprojective** if $\mathcal{P}$ is acyclic and, for any indecomposable module M in $\mathcal{P}$, there exist $t \geq 0$ and $a \in (Q_A)_0$ such that $M \cong \tau^{-t}P(a)$. An indecomposable A-module is called **postprojective** if it belongs to a postprojective component of $\Gamma(\text{mod } A)$, and an arbitrary A-module is called **postprojective** if it is a direct sum of indecomposable postprojective A-modules.

(b) A connected component $\mathcal{Q}$ of $\Gamma(\operatorname{mod} A)$ is called **preinjective** if $\mathcal{Q}$ is acyclic and, for any indecomposable module N in $\mathcal{Q}$, there exist $s \geq 0$ and $b \in (Q_A)_0$ such that $N \cong \tau^s I(b)$. An indecomposable A-module is called **preinjective** if it belongs to a preinjective component of $\Gamma(\operatorname{mod} A)$, and an arbitrary A-module is called **preinjective** if it is a direct sum of indecomposable preinjective A-modules.

The postprojective components and the postprojective modules are also sometimes called the preprojective components and the preprojective modules, respectively (see [21]). Here we use the term "postprojective" introduced by Gabriel and Roiter in [77].

With this terminology, we have the following obvious corollary of (2.1) and its proof.

2.3. Corollary. *Let Q be a finite, connected, and acyclic quiver that is not a Dynkin quiver, and let $A = KQ$.*

  (a) *The quiver $\Gamma(\operatorname{mod} A)$ contains a postprojective component $\mathcal{P}(A)$ that is isomorphic to $(-\mathbb{N})Q^{\mathrm{op}}$ and contains all the indecomposable projective A-modules.*

  (b) *The quiver $\Gamma(\operatorname{mod} A)$ contains a preinjective component $\mathcal{Q}(A)$ that is isomorphic to $\mathbb{N}Q^{\mathrm{op}}$ and contains all the indecomposable injective A-modules.* □

Clearly, the assumption that Q is not a Dynkin quiver is equivalent to saying that A is a representation–infinite hereditary algebra. We notice that, because $\mathcal{P}(A)$ contains all the indecomposable projective A-modules, it is necessarily the unique postprojective component of $\Gamma(\operatorname{mod} A)$. Similarly, $\mathcal{Q}(A)$ is the unique preinjective component of $\Gamma(\operatorname{mod} A)$.

2.4. Examples. (a) Let A be the path algebra of the Kronecker quiver

$$\circ \rightleftarrows \circ$$

Then the postprojective component $\mathcal{P}(A)$ is given by

and the preinjective component $\mathcal{Q}(A)$ is given by

where indecomposable modules are represented by their dimension vectors. These components are easily computed starting, respectively, from the indecomposable projectives and injectives and using the procedure used in (IV.4.10)–(IV.4.14).

(b) Let A be the path algebra of the quiver

Then $\mathcal{P}(A)$ is given by

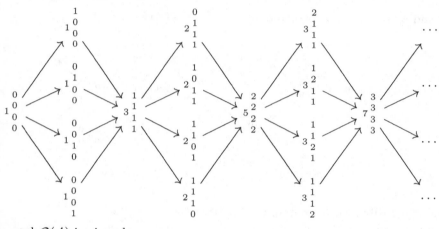

and $\mathcal{Q}(A)$ is given by

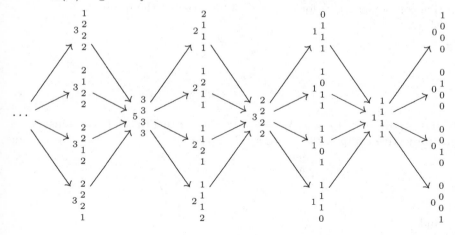

(c) If A is not hereditary, its postprojective component may contain injectives. This is clear if A is representation–finite (see, for instance, Example 1.3 (b)). The following is an example of a representation–infinite algebra having a postprojective component containing all projectives and one injective. Let A be given by the quiver

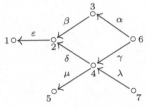

bound by $\alpha\beta = \gamma\delta$, $\beta\varepsilon = 0$, and $\delta\varepsilon = 0$. Then $\mathcal{P}(A)$ is given by

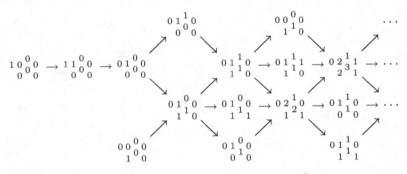

and it is easily seen to contain the injective $I(1) = 1\,1\,{}^{0}_{0}\,{}^{0}_{0}$.

(d) If A is not hereditary, then it may contain more than one postprojective component. Let, for instance, A be given by the quiver

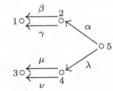

bound by $\alpha\beta = 0$, $\alpha\gamma = 0$, $\lambda\mu = 0$, $\lambda\nu = 0$. Then $\Gamma(\mathrm{mod}\,A)$ contains two postprojective components, respectively given by

and

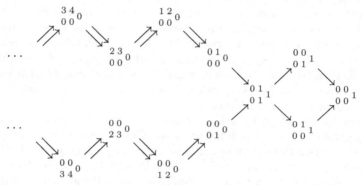

but it contains only one preinjective component, given by

We notice that the preinjective component contains all injectives and one projective $P(5) = \begin{smallmatrix} 0 & 1 \\ 0 & 1 \end{smallmatrix} 1$. Because all indecomposable projective and injective A-modules appear in these three components, these are all the postprojective and preinjective components of $\Gamma(\operatorname{mod} A)$.

We now let A be an arbitrary (not necessarily hereditary) algebra and record some of the properties of the postprojective and preinjective modules in the following lemmas.

2.5. Lemma. *Let A be an arbitrary (not necessarily hereditary) algebra.*

(a) *Let $\mathcal{P}$ be a postprojective component of the quiver $\Gamma(\operatorname{mod} A)$ and M be an indecomposable module in $\mathcal{P}$. Then the number of predecessors of M in $\mathcal{P}$ is finite and any indecomposable A-module L such that $\operatorname{Hom}_A(L, M) \neq 0$ is a predecessor of M in $\mathcal{P}$. In particular, $\operatorname{Hom}_A(L, M) = 0$ for all but finitely many nonisomorphic indecomposable A-modules L.*

(b) *Let $\mathcal{Q}$ be a preinjective component of the quiver $\Gamma(\operatorname{mod} A)$ and N be an indecomposable module in $\mathcal{Q}$. Then the number of successors of N in $\mathcal{Q}$ is finite and any indecomposable A-module L such that $\operatorname{Hom}_A(N, L) \neq 0$ is a successor of N in $\mathcal{Q}$. In particular, $\operatorname{Hom}_A(N, L) = 0$ for all but finitely many nonisomorphic indecomposable A-modules L.*

Proof. We only prove (a); the duality reduces (b) to (a).

First, we show that there is a simple projective predecessor of M in $\mathcal{P}$. Because $M \cong \tau^{-t_0} P(a_0)$ for some $t_0 \geq 0$ and an indecomposable projective A-module $P(a_0)$, according to (IV.4.3) and (IV.4.4), the modules $\tau M, \tau^2 M, \ldots, \tau^{t_0} M \cong P(a_0)$ are predecessors of M in $\mathcal{P}$. If $P(a_0)$ is simple, we are done. If $P(a_0)$ is not simple, the radical $\operatorname{rad} P(a_0)$ of $P(a_0)$ is nonzero and, by (IV.4.3), every indecomposable summand M_1 of $\operatorname{rad} P(a_0)$ is a predecessor of $P(a_0)$ and of M in $\mathcal{P}$. By our assumption, $M_1 \cong \tau^{-t_1} P(a_1)$ for some $t_1 \geq 0$ and an indecomposable projective A-module $P(a_1)$, and we conclude, as earlier, that $P(a_1)$ is a predecessor of M in $\mathcal{P}$. Continuing in this way, we find a simple predecessor $P(a_r)$ of M in $\mathcal{P}$, because $\mathcal{P}$ is acyclic and contains only finitely many indecomposable projective A-modules.

Denote by $\mathbf{h}(M)$ the length of a longest path connecting M with a simple projective module in $\mathcal{P}$. We prove the remaining statements in (a) for all modules M in $\mathcal{P}$ by induction on $\mathbf{h}(M)$.

Assume that $\mathbf{h}(M) \geq 1$, because if $\mathbf{h}(M) = 0$, then the module M is simple projective and there is nothing to show. Then M is not simple projective and there exists a right mimimal almost split morphism $M' \to M$. If N_1 is any indecomposable summand of M', then (IV.4.2)(b) yields $\mathbf{h}(N_1) < \mathbf{h}(M)$ and N_1 belongs to $\mathcal{P}$. By the induction hypothesis, the statement (a) holds for N_1. Because, by (IV.4.2)(b), all immediate predecessors N of M in $\mathcal{P}$ are isomorphic to direct summands of M', $\mathbf{h}(N) < \mathbf{h}(M)$. Moreover, if $L \not\cong M$ is an indecomposable module such that $\operatorname{Hom}_A(L, M) \neq 0$, then any nonzero homomorphism $f : L \to M$ factors through $M' \to M$ and therefore there exists an indecomposable summand N of M' such that $\operatorname{Hom}_A(L, N) \neq 0$. In view of $\mathbf{h}(N) < \mathbf{h}(M)$, it follows from the induction hypothesis that (a) holds for all indecomposable summands N of M', and therefore (a) holds for M. This completes the proof. □

In the course of the proof, we showed that any indecomposable A-module M in $\mathcal{P}$ has a simple projective predecessor, and any indecomposable A-module N in $\mathcal{Q}$ has a simple injective successor.

We restate the results of (2.5) in slightly different terms. Let M, N be two indecomposable A-modules. A **path** in $\operatorname{mod} A$ from M to N is a sequence

$$M = M_0 \xrightarrow{f_1} M_1 \xrightarrow{f_2} M_2 \longrightarrow \cdots \xrightarrow{f_t} M_t = N$$

where all the M_i are indecomposable, and all the f_i are nonzero nonisomorphisms. In this case, M is called a **predecessor** of N in $\operatorname{mod} A$ and N is called a **successor** of M in $\operatorname{mod} A$. A path from an indecomposable A-module M to itself, that is, a sequence of nonzero nonisomorphisms between indecomposables of the form

$$M = M_0 \xrightarrow{f_1} M_1 \xrightarrow{f_2} M_2 \longrightarrow \cdots \xrightarrow{f_t} M_t = M,$$

is called a **cycle** in $\operatorname{mod} A$. Then (2.5) says that, in the case of modules lying in postprojective or preinjective components, these module-theoretical notions can be expressed graphically.

2.6. Corollary. *Let A be an arbitrary (not necessarily hereditary) K-algebra.*

(a) *Let $\mathcal{P}$ be a postprojective component of $\Gamma(\operatorname{mod} A)$ and M be an indecomposable module in $\mathcal{P}$. Then*

  (i) *any predecessor L of M in $\operatorname{mod} A$ is postprojective and there is a path in $\mathcal{P}$ from L to M, and*

  (ii) *M lies on no cycle in $\operatorname{mod} A$.*

(b) *Let $\mathcal{Q}$ be a preinjective component of $\Gamma(\operatorname{mod} A)$ and N be an indecomposable module in $\mathcal{Q}$. Then*

  (i) *any successor N of L in $\operatorname{mod} A$ is preinjective and there is a path in $\mathcal{Q}$ from N to L, and*

  (ii) *N lies on no cycle in $\operatorname{mod} A$.*

Proof. We only prove (a); the duality reduces (b) to (a).

(i) Let L be a predecessor of M in $\operatorname{mod} A$ and

$$L = M_0 \to M_1 \to \cdots \to M_{t-1} \to M_t = M$$

be a path in $\operatorname{mod} A$. By (2.5), M_{t-1} lies in $\mathcal{P}$ and is a predecessor of M in $\mathcal{P}$. The statement now follows by induction.

(ii) follows from (i) and the acyclicity of $\mathcal{P}$. $\square$

2.7. Lemma. *Let A be an arbitrary (not necessarily hereditary) algebra and M be an indecomposable postprojective, or preinjective, A-module. Then $\operatorname{End} M \cong K$ and $\operatorname{Ext}_A^1(M, M) = 0$.*

Proof. Let M be an indecomposable postprojective or preinjective A-module. Assume to the contrary that $\dim_K \operatorname{End} M > 1$. Because $\operatorname{End} M$ is local, this implies $\operatorname{rad} \operatorname{End} M \neq 0$, thus there exists a nonzero nonisomorphism $f : M \to M$. It follows from (i) of (2.6)(a) and (2.6)(b) that M lies on a cycle in $\operatorname{mod} A$, a contradiction with the statements (a)(ii) and (b)(ii) of (2.6).

Next suppose that $\operatorname{Ext}_A^1(M, M) \neq 0$. By the Auslander–Reiten formula (IV.2.13), we have

$$\operatorname{Ext}_A^1(M, M) \cong D\overline{\operatorname{Hom}}_A(M, \tau M) \subseteq D\operatorname{Hom}_A(M, \tau M).$$

Hence there exists a nonzero homomorphism $M \to \tau M$ and thus a cycle

$$M \to \tau M \to * \to M$$

in $\operatorname{mod} A$. Hence we again get a contradiction with the statements (a)(ii) and (b)(ii) of (2.6). $\square$

Our next aim is to show that any representation–infinite hereditary algebra has indecomposable modules that are neither postprojective nor preinjective. For this purpose, we need the following lemma, valid over an arbitrary (not necessarily hereditary) algebra.

2.8. Lemma. *Let A be an arbitrary (not necessarily hereditary) algebra and*

$$0 \to L \to M \to N \to 0$$

be a nonsplit short exact sequence of A-modules. Then

$$\dim_K \operatorname{End} M < \dim_K \operatorname{End}(L \oplus N).$$

Proof. We have the following commutative diagram with exact columns and rows

$$
\begin{array}{ccccc}
& 0 & & 0 & & 0 \\
& \downarrow & & \downarrow & & \downarrow \\
0 \longrightarrow & \operatorname{Hom}_A(N,L) & \longrightarrow & \operatorname{Hom}_A(N,M) & \longrightarrow & \operatorname{Hom}_A(N,N) \\
& \downarrow & & \downarrow & & \downarrow \\
0 \longrightarrow & \operatorname{Hom}_A(M,L) & \longrightarrow & \operatorname{Hom}_A(M,M) & \longrightarrow & \operatorname{Hom}_A(M,N) \\
& \downarrow & & \downarrow & & \downarrow \\
0 \longrightarrow & \operatorname{Hom}_A(L,L) & \longrightarrow & \operatorname{Hom}_A(L,M) & \longrightarrow & \operatorname{Hom}_A(L,N) \\
& {\scriptstyle \delta} \downarrow & & & & \\
& \operatorname{Ext}^1_A(N,L) & & & &
\end{array}
$$

such that the connecting homomorphism $\delta : \operatorname{Hom}_A(L,L) \longrightarrow \operatorname{Ext}^1_A(N,L)$ maps the identity homomorphism on L to the class of the given nonsplit short exact sequence $0 \to L \to M \to N \to 0$. In particular, $\delta \neq 0$ and hence

$$\dim_K \operatorname{Hom}_A(M,L) < \dim_K \operatorname{Hom}_A(N,L) + \dim_K \operatorname{Hom}_A(L,L).$$

Consequently,

$$
\begin{aligned}
\dim_K \operatorname{End} M &\leq \dim_K \operatorname{Hom}_A(M,L) + \dim_K \operatorname{Hom}_A(M,N) \\
&< \dim_K \operatorname{Hom}_A(N,L) + \dim_K \operatorname{End} L + \dim_K \operatorname{End} N \\
&\quad + \dim_K \operatorname{Hom}_A(L,N) \\
&= \dim_K \operatorname{End}(L \oplus N). \qquad \square
\end{aligned}
$$

2.9. Proposition. *Let A be a representation–infinite hereditary algebra. Then there exists an indecomposable A-module M such that $\mathrm{Ext}^1_A(M, M) \neq 0$. In particular, M is neither postprojective nor preinjective.*

Proof. Because A is representation–infinite, it follows from (VII.4.6) that q_A is not weakly positive. Hence there exists a positive vector $\mathbf{x} \in K_0(A)$ such that $q_A(\mathbf{x}) \leq 0$. Clearly, there exists a nonzero (not necessarily indecomposable) A-module N such that $\mathbf{x} = \dim N$. Let thus N be an A-module such that $\mathbf{x} = \dim N$ and $\dim_K \mathrm{End}\, N$ is the smallest possible. We notice that, in view of (III.3.13), we have

$$\dim_K \mathrm{End}\, N - \dim_K \mathrm{Ext}^1_A(N, N) = q_A(\dim N) \leq 0,$$

and consequently

$$\dim_K \mathrm{Ext}^1_A(N, N) \geq \dim_K \mathrm{End}\, N \geq 1,$$

so that $\mathrm{Ext}^1_A(N, N) \neq 0$. Hence there exists an indecomposable summand M of N such that $\mathrm{Ext}^1_A(M, N) \neq 0$. We claim that $\mathrm{Ext}^1_A(M, M) \neq 0$. Indeed, if this is not the case, then, writing $N = M \oplus L$, we have $\mathrm{Ext}^1_A(M, L) \cong \mathrm{Ext}^1_A(M, N) \neq 0$ so that there exists a nonsplit short exact sequence $0 \longrightarrow L \longrightarrow E \longrightarrow M \longrightarrow 0$. By (2.8), we get

$$\dim_K \mathrm{End}\, E < \dim_K \mathrm{End}\, (L \oplus M) = \dim_K \mathrm{End}\, N.$$

Because

$$\dim E = \dim L + \dim M = \dim (L \oplus M) = \dim N = \mathbf{x},$$

this contradicts the minimality of N, thus showing our claim. Finally, the last statement follows from (2.7). □

2.10. Corollary. *Let A be a representation–infinite hereditary algebra. Then there exists an infinite family of pairwise nonisomorphic indecomposable A-modules that are neither postprojective nor projective.*

Proof. It follows from (2.9) that $\Gamma(\mathrm{mod}\, A)$ has a component $\mathcal{C}$ that is different from the unique postprojective component and the unique preinjective component. Because A is representation–infinite, $\mathcal{C}$ is infinite by (IV.5.4) and, clearly, no module in $\mathcal{C}$ is postprojective or preinjective. □

2.11. Example. Let A be given by the Kronecker quiver $\circ \mathrel{\overset{\longleftarrow}{\longleftarrow}} \circ$. Then A is a representation–infinite hereditary algebra. Let $m \geq 1$ and $\lambda \in K$ be arbitrary; then consider the module $H_m(\lambda)$ given by

$$K^m \xleftarrow[J_{m,\lambda}]{1} K^m,$$

where $J_{m,\lambda}$ denotes the Jordan block corresponding to the eigenvalue λ. Then it is easily seen that $H_m(\lambda)$ is indecomposable (this was done in

(III.1.8) for $\lambda = 0$ and is done in exactly the same way for any value of λ). On the other hand, comparing

$$\operatorname{\mathbf{dim}} H_m(\lambda) = (m, m)$$

with the dimension vectors of the postprojective and preinjective A-modules as computed in (2.4)(a), we see that $H_m(\lambda)$ is neither postprojective nor preinjective. Because it is easily seen that $H_m(\lambda) \cong H_n(\mu)$ if and only if $m = n$ and $\lambda = \mu$, we obtain an infinite family of indecomposable modules that are neither postprojective nor preinjective.

It follows from (2.10) that the Auslander–Reiten quiver of a representation–infinite hereditary algebra has components containing neither projective nor injective modules.

2.12. Definition. Let A be an arbitrary (not necessarily hereditary) algebra. A connected component $\mathcal{C}$ of $\Gamma(\operatorname{mod} A)$ is called **regular** if $\mathcal{C}$ contains neither projective nor injective modules. An indecomposable A-module is called **regular** if it belongs to a regular component of $\Gamma(\operatorname{mod} A)$ and an arbitrary A-module is called **regular** if it is a direct sum of indecomposable regular A-modules.

Let A be a representation–infinite hereditary algebra. We denote by $\mathcal{R}(A)$ the family of all the regular components of $\Gamma(\operatorname{mod} A)$ and by $\operatorname{add} \mathcal{R}(A)$ the full subcategory of $\operatorname{mod} A$ whose objects are all the regular A-modules. We may visualise the shape of $\Gamma(\operatorname{mod} A)$ as follows:

We now show that in this picture, the homomorphisms can only go from left to right.

2.13. Corollary. *Let A be a representation–infinite hereditary algebra and L, M, and N be three indecomposable A-modules.*

(a) *If L is postprojective and M is regular, then $\operatorname{Hom}_A(M, L) = 0$.*
(b) *If L is postprojective and N is preinjective, then $\operatorname{Hom}_A(N, L) = 0$.*
(c) *If M is regular and N is preinjective, then $\operatorname{Hom}_A(N, M) = 0$.*

Proof. This easily follows from (2.5).   □

The statement of (2.13) is more briefly expressed by writing

$$\operatorname{Hom}_A(\mathcal{R}(A), \mathcal{P}(A)) = 0, \ \operatorname{Hom}_A(\mathcal{Q}(A), \mathcal{P}(A)) = 0, \ \operatorname{Hom}_A(\mathcal{Q}(A), \mathcal{R}(A)) = 0.$$

2.14. Corollary. *Let A be a representation–infinite hereditary algebra. Then the mutually inverse equivalences* $\underline{\mathrm{mod}}\, A \underset{\tau^{-1}}{\overset{\tau}{\rightleftarrows}} \overline{\mathrm{mod}}\, A$ *(IV.2.11), induced by the Auslander–Reiten translations τ and τ^{-1}, induce mutually inverse equivalences of categories*

$$\mathrm{add}\,\mathcal{R}(A) \underset{\tau^{-1}}{\overset{\tau}{\rightleftarrows}} \mathrm{add}\,\mathcal{R}(A).$$

Proof. Let M, N be two regular nonzero A-modules. It follows from the definition that the modules τM, τN, $\tau^{-1}M$, $\tau^{-1}N$ are nonzero and regular. By (2.13), no homomorphism from M to N factors through a projective or injective module. Hence $\underline{\mathrm{Hom}}_A(M, N) = \mathrm{Hom}_A(M, N) = \overline{\mathrm{Hom}}_A(M, N)$. The result follows from (IV.2.10) and (IV. 2.11). $\square$

The structure of the category $\mathrm{add}\,\mathcal{R}(A)$ will be discussed in detail in the second volume of this book.

VIII.3. Tilted algebras

As we have seen, each of the postprojective and preinjective components of the Auslander–Reiten quiver of a hereditary algebra is completely determined by a section. It follows from (2.3) that such a component is obtained by repeated applications of the Auslander–Reiten translations to the section consisting of the indecomposable projective modules or the indecomposable injective modules, respectively. The use of sections in acyclic components of Auslander–Reiten quivers is not limited to hereditary algebras. We now introduce a class of algebras (containing the class of hereditary algebras) that, as we shall see, are characterised by the property that their Auslander–Reiten quiver has an acyclic component containing a section satisfying reasonable properties.

3.1. Definition. Let Q be a finite, connected, and acyclic quiver. An algebra B is said to be **tilted of type** Q if there exists a tilting module T over the path algebra $A = KQ$ of Q such that $B = \mathrm{End}\,T_A$.

Because we are only interested in basic algebras, we may (and shall) always assume that T_A is multiplicity-free. We notice that, by (VI.3.5), a tilted algebra is always connected.

For instance, any hereditary algebra is tilted. Indeed, let Q be a finite, connected, and acyclic quiver and let $A = KQ$; then A_A is a tilting module so that $A = \mathrm{End}\,A_A$ is tilted of type Q. In Chapter VI, the examples (VI.3.11)(a) and (VI.3.11)(c) show endomorphism algebras of tilting modules over hereditary algebras, thus tilted algebras, which are not hereditary.

We now wish to list some elementary properties of tilted algebras that follow directly from the results of Chapter VI. One terminology is useful here. Let $\mathcal{A}$ be an additive full subcategory of $\mathrm{mod}\,A$, closed under isomorphic images and direct summands. We say that $\mathcal{A}$ is **closed under predecessors** if, for any path $L \to \cdots \to M$ in $\mathrm{mod}\,A$, with M in $\mathcal{A}$, the module L belongs to $\mathcal{A}$ as well; similarly, $\mathcal{A}$ is **closed under successors** if, for any path $L \to \cdots \to M$ in $\mathrm{mod}\,A$, with L in $\mathcal{A}$, the module M belongs to $\mathcal{A}$ as well.

3.2. Lemma. *Let A be a hereditary algebra, T_A be a tilting module, and $B = \mathrm{End}\,T_A$.*

(a) *The torsion pair $(\mathcal{X}(T), \mathcal{Y}(T))$ in $\mathrm{mod}\,B$ is splitting.*

(b) *$\mathcal{Y}(T)$ is closed under predecessors and $\mathcal{X}(T)$ is closed under successors.*

(c) *If A is representation–finite, then so is B.*

(d) *Any almost split sequence in $\mathrm{mod}\,B$ lies either entirely in $\mathcal{X}(T)$ or entirely in $\mathcal{Y}(T)$, or else it is a connecting sequence.*

(e) *gl.dim $B \leq 2$ and, for any indecomposable B-module Z, we have $\mathrm{pd}\,Z_B \leq 1$ or $\mathrm{id}\,Z_B \leq 1$.*

Proof. (a) Because A is hereditary, it follows from (VI.5.7) that T_A is a splitting tilting module.

(b) This follows from (a); indeed, let $Z = Z_0 \to Z_1 \to \cdots \to Z_{t-1} \to Z_t = Y$ be a path in $\mathrm{mod}\,A$, with $Y \in \mathcal{Y}(T)$. Then $\mathrm{Hom}_A(Z_{t-1}, Y) \neq 0$ implies that $Z_{t-1} \notin \mathcal{X}(T)$; hence, by (a), $Z_{t-1} \in \mathcal{Y}(T)$. An obvious induction completes the proof that $\mathcal{Y}(T)$ is closed under predecessors. The other statement is proved similarly.

(c) This also follows directly from (a).

(d) This follows from (a) and (VI.5.2).

(e) The first statement follows from (VI.4.2). Let Z be an indecomposable B-module. By (a), Z belongs to either $\mathcal{X}(T)$ or $\mathcal{Y}(T)$. If $Z \in \mathcal{Y}(T)$, there exists an indecomposable A-module $M \in \mathcal{T}(T)$ such that $Z \cong \mathrm{Hom}_A(T, M)$. But then by (VI.4.1), we get $\mathrm{pd}\,Z_B \leq \mathrm{pd}\,M_A \leq 1$. Assume $Z \in \mathcal{X}(T)$. Because $(\mathcal{X}(T), \mathcal{Y}(T))$ is splitting, it follows from (VI.1.7) that $\tau^{-1}Z \in \mathcal{X}(T)$. On the other hand, $B_B \in \mathcal{Y}(T)$. Hence $\mathrm{Hom}_B(\tau^{-1}Z, B) = 0$ and, by (IV.2.7)(b), we have $\mathrm{id}\,Z_B \leq 1$. $\qquad\square$

We notice that (c) can be reformulated by saying that any tilted algebra of Dynkin type (that is, whose type is a Dynkin quiver) is representation–finite.

We now wish to prove that the ordinary quiver of a tilted algebra is acyclic. This follows from the next lemma.

3.3. Lemma. *Let A be a hereditary algebra. If T_1 and T_2 are indecomposable A-modules such that $\mathrm{Ext}_A^1(T_2, T_1) = 0$, then any nonzero homomor-*

phism from T_1 to T_2 is a monomorphism or an epimorphism. In particular, if T_1 is indecomposable and $\mathrm{Ext}^1_A(T_1, T_1) = 0$, then $\mathrm{End}\, T_1 \cong K$.

Proof. Let $f : T_1 \to T_2$ be a nonzero homomorphism, and assume that f is neither a monomorphism nor an epimorphism. Letting $M = \mathrm{Im}\, f$, we can factor f as $f = gh$, where $h : T_1 \to M$ is the canonical epimorphism. Because f is neither a monomorphism nor an epimorphism, we have $\dim_K M < \dim_K T_1$ and $\dim_K M < \dim_K T_2$. In particular, M is isomorphic to neither T_1 nor T_2. Applying the functor $\mathrm{Ext}^1_A(T_2/M, -)$ to the short exact sequence $0 \longrightarrow \mathrm{Ker}\, h \longrightarrow T_1 \overset{h}{\longrightarrow} M \longrightarrow 0$, we obtain an exact sequence

$$\mathrm{Ext}^1_A(T_2/M, T_1) \xrightarrow{\mathrm{Ext}^1_A(T_2/M, h)} \mathrm{Ext}^1_A(T_2/M, M) \longrightarrow \mathrm{Ext}^2_A(T_2/M, \mathrm{Ker}\, h),$$

where the last term vanishes, because A is hereditary. Then $\mathrm{Ext}^1_A(T_2/M, h)$ is surjective. It follows that there exists an A-module N and a commutative diagram with exact rows

$$
\begin{array}{ccccccccc}
0 & \longrightarrow & T_1 & \overset{g'}{\longrightarrow} & N & \longrightarrow & T_2/M & \longrightarrow & 0 \\
& & {\scriptstyle h}\downarrow & & {\scriptstyle h'}\downarrow & & {\scriptstyle 1}\downarrow & & \\
0 & \longrightarrow & M & \overset{g}{\longrightarrow} & T_2 & \longrightarrow & T_2/M & \longrightarrow & 0.
\end{array}
$$

This implies that we have a short exact sequence

$$0 \longrightarrow T_1 \xrightarrow{\left[\begin{smallmatrix} h \\ -g' \end{smallmatrix}\right]} M \oplus N \xrightarrow{[g\ h']} T_2 \longrightarrow 0.$$

Because $\mathrm{Ext}^1_A(T_2, T_1) = 0$, by hypothesis, this sequence splits. Therefore $M \oplus N \cong T_1 \oplus T_2$. By the unique decomposition theorem (I.4.10), M is isomorphic to one of the indecomposable modules T_1 or T_2, and this is a contradiction.

The last statement follows from the fact that, because any nonzero homomorphism $T_1 \to T_1$ is a monomorphism or an epimorphism, it is an isomorphism. $\square$

3.4. Corollary. *If B is a tilted algebra, then the quiver Q_B of B is acyclic.*

Proof. Assume that $B = \mathrm{End}\, T_A$, where A is hereditary and T_A is a tilting module. Let T'_1, T'_2, T'_3 be three indecomposable direct summands of T, and $f : T'_1 \to T'_2$, $g : T'_2 \to T'_3$ be nonzero A-module homomorphisms. We claim that we cannot have that f is a proper epimorphism and g is a proper monomorphism. Indeed, if this is the case, then $gf : T'_1 \to T'_3$ is nonzero and is neither a monomorphism nor an epimorphism, and this contradicts (3.3). By (II.3.3) and (VI.3.10)(a), any cycle in Q_B induces a cycle

$$T_1 \overset{f_1}{\longrightarrow} T_2 \overset{f_2}{\longrightarrow} \cdots \longrightarrow T_r \overset{f_r}{\longrightarrow} T_1$$

in the category $\bmod A$, where $f_1, \ldots, f_r$ are nonzero nonisomorphisms and $T_1, T_2, \ldots, T_r$ are indecomposable direct summands of T. By our preceding claim, this cycle cannot involve an epimorphism followed by a monomorphism. Hence all the f_i are either epimorphisms or monomorphisms. It follows that $f_r \ldots f_1 \in \operatorname{End} T_{a_1}$ is an epimorphism or a monomorphism, hence an isomorphism. Consequently, f_1 is an isomorphism and we get a contradiction. $\square$

We now show, by applying (VI.5.4), that the Auslander–Reiten quiver of a tilted algebra has an acyclic component containing a finite section. To apply (VI.5.4) to the case where A is hereditary, we need only observe that if $I(a)$ is an indecomposable injective A-module, then any direct summand of $I(a)/\operatorname{soc} I(a)$ is injective; consequently, there exists an irreducible morphism $I(a) \to J$ in $\bmod A$, with J indecomposable, if and only if $J \cong I(b)$, and there exists an arrow $b \to a$ in Q_A.

3.5. Theorem. *Let A be a hereditary algebra, T_A be a tilting module, and $B = \operatorname{End} T_A$. Then the class Σ of all B-modules of the form $\operatorname{Hom}_A(T, I)$, where I is an indecomposable injective A-module, forms a section lying in an acyclic component $\mathcal{C}_T$ of $\Gamma(\bmod B)$. Moreover, Σ is isomorphic to Q_A^{op}, any predecessor of Σ in $\mathcal{C}_T$ lies in $\mathcal{Y}(T)$, and any proper successor of Σ in $\mathcal{C}_T$ lies in $\mathcal{X}(T)$.*

Proof. We first show that there exists a quiver isomorphism between Σ and the section in $\Gamma(\bmod A)$ consisting of the indecomposable injective A-modules (which, by (1.3), is isomorphic to Q_A^{op}). Indeed, let $\operatorname{Hom}_A(T, I) \to \operatorname{Hom}_A(T, I')$ be an irreducible morphism in $\bmod B$, where I and I' are indecomposable injective A-modules. By (VI.5.4)(a), I' is isomorphic to a direct summand of $I/\operatorname{soc} I$, so that there exists an irreducible morphism $I \to I'$ in $\bmod A$. Conversely, if there exists an irreducible morphism $I \to I'$ in $\bmod A$, then I' is isomorphic to a direct summand of $I/\operatorname{soc} I$ so, again by (VI.5.4)(a), there exists an irreducible morphism $\operatorname{Hom}_A(T, I) \to \operatorname{Hom}_A(T, I')$ in $\bmod B$. Because, in this case, the equivalence $\mathcal{Y}(T) \cong \mathcal{T}(T)$ yields an isomorphism $\operatorname{Irr}(I, I') \cong \operatorname{Irr}(\operatorname{Hom}_A(T, I), \operatorname{Hom}_A(T, I'))$, we are done.

This quiver isomorphism shows that Σ is a full connected subquiver of $\Gamma(\bmod B)$ and that Σ is acyclic. Let $\mathcal{C}_T$ denote the connected component of $\Gamma(\bmod B)$ containing Σ.

Because Σ consists of modules from $\mathcal{Y}(T)$, which is closed under predecessors, then any predecessor of Σ in $\mathcal{C}_T$ lies in $\mathcal{Y}(T)$. On the other hand, if there exists an irreducible morphism $Y \to X$ with Y in Σ, but X not in Σ, then, by (VI.5.4)(a), $X \notin \mathcal{Y}(T)$. Therefore $X \in \mathcal{X}(T)$. Because $\mathcal{X}(T)$ is closed under successors, this shows that any proper successor of Σ lies in $\mathcal{X}(T)$. This implies that Σ is convex in $\mathcal{C}_T$; let $Y_0 \to \cdots \to Y_t$ be a chain of irreducible morphisms, where Y_0, Y_t lie in Σ, then $Y_1 \in \mathcal{Y}(T)$ (because it precedes $Y_t \in \mathcal{Y}(T)$) hence (VI.5.4)(a) gives that Y_1 lies in Σ thus, inductively, all the Y_i lie in Σ.

We next observe that any indecomposable projective B-module lies in $\mathcal{Y}(T)$ and so cannot be a proper successor of Σ in $\mathcal{C}_T$. On the other hand, any indecomposable injective B-module that belongs to $\mathcal{Y}(T)$ must lie on Σ: indeed, if $\mathrm{Hom}_A(T, M)$ is indecomposable injective, let $j : M \to I$ be an injective envelope in $\mathrm{mod}\, A$, then $\mathrm{Hom}_A(T, j) : \mathrm{Hom}_A(T, M) \to \mathrm{Hom}_A(T, I)$ is a monomorphism (because j is), hence a section, so that $\mathrm{Hom}_A(T, M)$ is isomorphic to a direct summand of $\mathrm{Hom}_A(T, I)$, thus lies on Σ. This shows that no proper predecessor of Σ in $\mathcal{C}_T$ is injective.

We now prove that Σ intersects each τ-orbit in $\mathcal{C}_T$. We claim that if Y belongs to Σ and Z (in $\mathcal{C}_T$) belongs to a τ-orbit that is neighbouring to the τ-orbit of Y, then Σ intersects the τ-orbit of Z. This claim and induction clearly yield our statement. Thus, assume that there exist $n \in \mathbb{Z}$ and an irreducible morphism $\tau^n Y \to Z$ or $Z \to \tau^n Y$. We now show that we may suppose $n = 0$. If this is not the case, and $|n|$ is minimal, we have two cases:

(a) If $n < 0$, then $Z \in \mathcal{Y}(T)$. If not, $Z \in \mathcal{X}(T)$ implies that Z is not projective hence there exists an irreducible morphism $\tau^{n+1} Y \to \tau Z$ or $\tau Z \to \tau^{n+1} Y$, respectively, and this contradicts minimality. Now there exists a chain of irreducible morphisms $Y \to * \to \tau^{-1} Y$. Because $\mathcal{Y}(T)$ is closed under predecessors and, by (VI.5.2), $\tau^{-1} Y \in \mathcal{X}(T)$, then Z cannot be a successor of $\tau^{-1} Y$. Hence there exists an irreducible morphism $Z \to \tau^{-1} Y$, and so an irreducible morphism $Y \to Z$.

(b) If $n > 0$, then either Z belongs to Σ and we are done, or $Z \in \mathcal{X}(T)$. Indeed, if Z is in neither Σ nor $\mathcal{X}(T)$, then Z is not injective; hence there exists an irreducible morphism $\tau^{n-1} Y \to \tau^{-1} Z$ or $\tau^{-1} Z \to \tau^{n-1} Y$, respectively, and this contradicts minimality. If $Z \in \mathcal{X}(T)$, then Z is a neighbour of $\tau^n Y$, for some $n > 0$, so is a predecessor of $Y \in \mathcal{Y}(T)$, and we get a contradiction.

Consider thus the case $n = 0$, that is, there exists an irreducible morphism $Y \to Z$ or $Z \to Y$. In the first case, it follows from (VI.5.4) that either Z or τZ lies in Σ. In the second case, we have necessarily that $Z \in \mathcal{Y}(T)$. Thus, either Z belongs to Σ and we are done, or Z is not injective; hence there exists an irreducible morphism $Y \to \tau^{-1} Z$ and the first case shows that either Z or $\tau^{-1} Z$ lies on Σ.

Finally, Σ intersects each τ-orbit exactly once; indeed, if both Y and $\tau^{-t} Y$, with $t \geq 1$, belong to Σ, then $\tau^{-t} Y \in \mathcal{Y}(T)$ implies $\tau^{-1} Y \in \mathcal{Y}(T)$, and this contradicts (VI.5.2). This completes the proof that Σ is a section in $\mathcal{C}_T$. The acyclicity of $\mathcal{C}_T$ follows from (1.5). □

One may think of the component $\mathcal{C}_T$ of $\Gamma(\mathrm{mod}\, B)$ as connecting the torsion-free part $\mathcal{Y}(T)$ with the torsion part $\mathcal{X}(T)$ along the section Σ. For this reason, the component $\mathcal{C}_T$ is called the **connecting component** of $\Gamma(\operatorname{mod} B)$ **determined by** T.

We may visualise the situation as in the following picture:

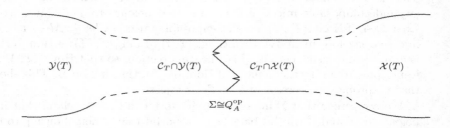

If B is representation–finite, then $\mathcal{C}_T = \Gamma(\operatorname{mod} B)$, so that we have the following easy corollary.

3.6. Corollary. *Let B be a representation–finite tilted algebra. Then the Auslander–Reiten quiver $\Gamma(\operatorname{mod} B)$ is acyclic and contains a section.* $\square$

3.7. Examples. (a) Let A be the path algebra of the quiver Q

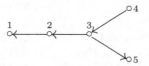

of type $\mathbb{D}_5$. Because A is a representation–finite hereditary algebra, its Auslander–Reiten quiver is easily computed to be

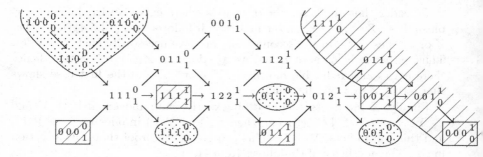

where the indecomposable modules are represented by their dimension vectors. Consider the module $T_A = \bigoplus_{i=1}^{5} T_i$, where

$$T_1 = P(5) = 0\ 0\ 0\ \begin{smallmatrix}0\\1\end{smallmatrix}, \qquad T_2 = P(4) = 1\ 1\ 1\ \begin{smallmatrix}1\\1\end{smallmatrix}, \qquad T_3 = 0\ 1\ 1\ \begin{smallmatrix}1\\1\end{smallmatrix},$$

$$T_4 = I(5) = 0\ 0\ 1\ \begin{smallmatrix}1\\1\end{smallmatrix}, \qquad T_5 = I(4) = 0\ 0\ 0\ \begin{smallmatrix}1\\0\end{smallmatrix}.$$

It is easily checked that T is a tilting A-module and that $B = \operatorname{End} T$ is given by the quiver

$$\underset{1}{\circ} \xleftarrow{\ \delta\ } \underset{2}{\circ} \xleftarrow{\ \gamma\ } \underset{3}{\circ} \xleftarrow{\ \beta\ } \underset{4}{\circ} \xleftarrow{\ \alpha\ } \underset{5}{\circ}$$

bound by $\alpha\beta\gamma\delta = 0$. Computing the Auslander–Reiten quiver of B yields

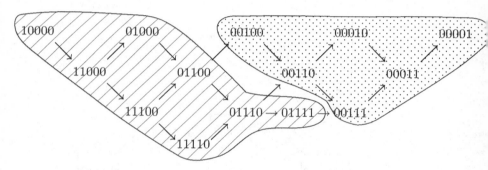

Here, as in Chapter VI, we denote by ⬭ the classes $\mathcal{T}(T)$ and $\mathcal{Y}(T)$ and by ⬭ the classes $\mathcal{F}(T)$ and $\mathcal{X}(T)$.

The section Σ consists of the indecomposable B-modules

$\operatorname{Hom}_A(T, I(1)) = 01000, \quad \operatorname{Hom}_A(T, I(2)) = 01100, \quad \operatorname{Hom}_A(T, I(3)) = 01110,$
$\operatorname{Hom}_A(T, I(4)) = 01111, \quad \operatorname{Hom}_A(T, I(5)) = 11110.$

We see that $\Sigma \cong Q_A^{\mathrm{op}}$.

(b) Let A be the path algebra of the quiver Q

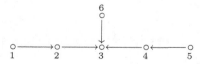

of type $\mathbb{E}_6$. Because A is a representation–finite hereditary algebra, its Auslander–Reiten quiver is easily computed to be

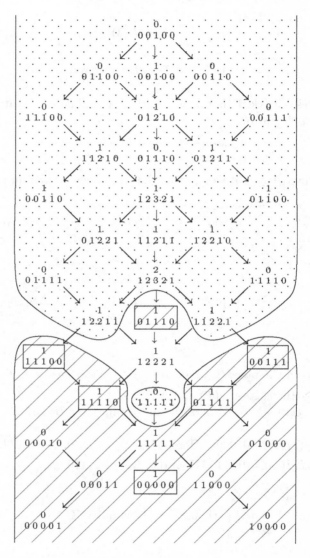

Consider the module $T_A = \bigoplus_{i=1}^{6} T_i$, where $T_1 = {1 \atop 0\ 1\ 1\ 1\ 0}$, $T_2 = {1 \atop 0\ 0\ 1\ 1\ 1}$,

$T_3 = {1 \atop 1\ 1\ 1\ 0\ 0}$, $T_4 = {1 \atop 0\ 1\ 1\ 1\ 1}$, $T_5 = {1 \atop 1\ 1\ 1\ 1\ 0}$, $T_6 = {1 \atop 0\ 0\ 0\ 0\ 0}$.

It is easily checked that T is a tilting A-module and that $B = \operatorname{End} T$ is given by the quiver

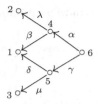

bound by $\alpha\beta = \gamma\delta$. Computing the Auslander–Reiten quiver of B yields

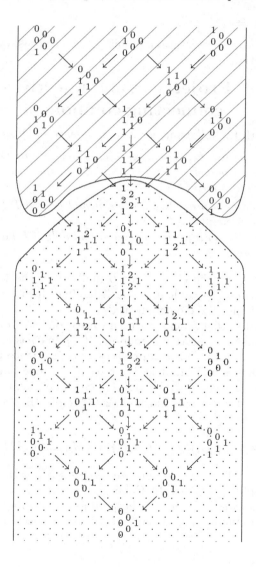

The section Σ consists of the indecomposable B-modules

$$\operatorname{Hom}_A(T, I(1)) = \begin{smallmatrix} & 0 & \\ 0 & {}^{0} & 0 \\ & 1 & \\ & 1 & \end{smallmatrix}\ ,\quad \operatorname{Hom}_A(T, I(2)) = \begin{smallmatrix} & 0 & \\ 1 & {}^{1} & 0 \\ & 1 & \\ & 1 & \end{smallmatrix}\ ,\quad \operatorname{Hom}_A(T, I(3)) = \begin{smallmatrix} & 1 & \\ 1 & {}^{1} & 0 \\ & 1 & \\ & 1 & \end{smallmatrix}\ ,$$

$$\operatorname{Hom}_A(T, I(4)) = \begin{smallmatrix} & 1 & \\ 1 & {}^{1} & 0 \\ & 0 & \\ & & \end{smallmatrix}\ ,\quad \operatorname{Hom}_A(T, I(5)) = \begin{smallmatrix} & 1 & \\ 0 & {}^{0} & 0 \\ & 0 & \\ & & \end{smallmatrix}\ ,\quad \operatorname{Hom}_A(T, I(6)) = \begin{smallmatrix} & 1 & \\ 1 & {}^{1} & 1 \\ & 1 & \\ & 1 & \end{smallmatrix}\ .$$

We note that $\Sigma \cong Q_A^{\mathrm{op}}$.

For examples of representation–infinite tilted algebras, we refer the reader to the next section.

VIII.4. Projectives and injectives in the connecting component

We start with the following useful consequence of (VI.5.3).

4.1. Proposition. *Let A be a representation–infinite hereditary algebra, T_A be a tilting module, $B = \operatorname{End} T_A$, and $\mathcal{C}_T$ be the connecting component of $\Gamma(\operatorname{mod} B)$ determined by T.*

(a) *$\mathcal{C}_T$ contains a projective module if and only if T has a preinjective direct summand.*

(b) *$\mathcal{C}_T$ contains an injective module if and only if T has a postprojective direct summand.*

Proof. Let Σ be the class of all B-modules of the form $\operatorname{Hom}_A(T, I)$, where I is an indecomposable injective A-module. It follows from (3.5) that Σ is a section lying in the component $\mathcal{C}_T$.

(a) We assume that T has no preinjective direct summand and claim that $\mathcal{C}_T$ contains no projective B-module. If Z_B in $\mathcal{C}_T$ is an indecomposable projective, then, by (3.5), it is a predecessor of Σ. Hence there exists $t \geq 0$ such that $\tau^{-t}Z$ lies in Σ, that is, there exists an indecomposable injective A-module I such that $Z \cong \tau^t\operatorname{Hom}_A(T, I)$. The assumption that T has no preinjective direct summand and (2.13) imply that all preinjective A-modules lie in

$$\mathcal{T}(T) = \big\{ M_A \mid \operatorname{Ext}_A^1(T, M) = 0 \big\} = \big\{ M_A \mid \operatorname{Hom}_A(M, \tau T) = 0 \big\}$$

and hence so do all the almost split sequences with preinjective end terms. Therefore, applying repeatedly (VI.5.3)(a) yields $Z \cong \tau^t\operatorname{Hom}_A(T, I) \cong \operatorname{Hom}_A(T, \tau^t I)$. Now, $\tau^t I$ lies in the preinjective component and hence is not a direct summand of T. Therefore Z is not projective.

Conversely, assume that T has a preinjective direct summand. Because the preinjective component is acyclic, there exists a "last" preinjective direct summand of T, that is, a preinjective indecomposable direct summand T_0 such that no proper successor of T_0 is a direct summand of T. This implies that all successors of T_0 lie in $\mathcal{T}(T)$ (for, if M is a successor of T_0, then M is a predecessor of no other indecomposable summand of T; however, $0 \neq \operatorname{Ext}_A^1(T, M) \cong D\operatorname{Hom}_A(M, \tau T)$ gives an indecomposable summand T_1 of T such that there exists a path $M \to \tau T_1 \to * \to T_1$). Because T_0 is preinjective, there exists $t \geq 0$ such that $\tau^{-t}T_0 = I$ is injective. Hence, applying (VI.5.3)(a) repeatedly, $\tau^{-t}\operatorname{Hom}_A(T, T_0) \cong \operatorname{Hom}_A(T, \tau^{-t}T_0) \cong \operatorname{Hom}_A(T, I)$ lies in Σ. But then the projective B-module $\operatorname{Hom}_A(T, T_0) \cong \tau^t\operatorname{Hom}_A(T, I)$ belongs to $\mathcal{C}_T$.

(b) We assume that T has no postprojective direct summand and claim that $\mathcal{C}_T$ contains no injective B-module. If Z_B in $\mathcal{C}_T$ is an indecomposable injective, then, by (3.5), it is a successor of Σ. Hence there exist $t \geq 0$ and an indecomposable injective A-module I such that $Z \cong \tau^{-t}\operatorname{Hom}_A(T, I)$. The assumption implies that no projective A-module is a direct summand of T. By (VI.4.9), $\operatorname{Hom}_A(T, I)$ is not injective. Hence $t \geq 1$ and, if P denotes the projective cover of soc I, we have $Z \cong \tau^{-t+1}\operatorname{Ext}_A^1(T, P)$. On the other hand, it follows from the assumption that T has no postprojective direct summand and (2.13) that all postprojective A-modules lie in $\mathcal{F}(T)$ and hence so do all the almost split sequences with postprojective end terms. Therefore, applying repeatedly (VI.5.3)(b) yields $Z \cong \tau^{-t+1}\operatorname{Ext}_A^1(T, P) \cong \operatorname{Ext}_A^1(T, \tau^{-t+1}P)$. Now $\tau^{-t+1}P$ is a postprojective A-module and hence cannot be injective. Therefore, Z is not injective either; see (VI.5.8).

Conversely, assume that T has a postprojective direct summand. If T has actually a projective summand P, then if I denotes the injective envelope of the top of P, we have, by (VI.4.9), that $\operatorname{Hom}_A(T, I)$ is injective and lies on Σ and hence in $\mathcal{C}_T$. We may thus assume that T has no projective direct summand. Because the postprojective component is acyclic, there exists a "first" postprojective direct summand of T, that is, a postprojective indecomposable direct summand T_0 of T such that no proper predecessor of T_0 is a direct summand of T. This implies that all proper predecessors of T_0 lie in $\mathcal{F}(T)$. Because T_0 is postprojective, there exists $t > 0$ such that $\tau^t T_0 = P$ is indecomposable projective. Let I denote the injective envelope of the top of P. Applying repeatedly (VI.5.3)(b) yields

$$\tau^{-t}\operatorname{Hom}_A(T, I) \cong \tau^{-t+1}\operatorname{Ext}_A^1(T, P) \cong \operatorname{Ext}_A^1(T, \tau^{-t+1}P) \cong \operatorname{Ext}_A^1(T, \tau T_0).$$

Thus, $\operatorname{Ext}_A^1(T, \tau T_0)$ belongs to $\mathcal{C}_T$. By (VI. 5.8), the right B-module $\operatorname{Ext}_A^1(T, \tau T_0)$ is injective. This completes the proof. $\qquad \square$

As a first corollary, we have the following.

4.2. Corollary. *Let A be a hereditary algebra, T_A be a tilting module, $B = \operatorname{End} T_A$, and $\mathcal{C}_T$ be the connecting component of $\Gamma(\operatorname{mod} B)$ determined by T. Then $\mathcal{C}_T$ is a regular component if and only if T is a regular module.*

Proof. Indeed, $\mathcal{C}_T$ is regular if and only if T has neither postprojective nor preinjective direct summands. Because over a hereditary algebra A all projective modules lie in the postprojective component and all injective modules lie in the preinjective component, the statement follows.   □

The existence of a regular tilting module over a (necessarily representation–infinite) hereditary algebra is far from obvious. In fact, as we shall see, there exists no regular tilting module if Q_A is a Euclidean quiver, while there exist regular tilting modules over the path algebras of quivers with at least three points that are neither Dynkin nor Euclidean.

4.3. Corollary. *Let A be a hereditary algebra, T be a tilting A-module, $B = \operatorname{End} T_A$, and $\mathcal{C}_T$ be the connecting component of $\Gamma(\operatorname{mod} B)$ determined by T.*

(a) *B is representation–finite if and only if $\mathcal{C}_T$ is both postprojective and preinjective.*

(b) *If B is representation–finite, then T_A has both a postprojective and a preinjective direct summand.*

Proof. (a) Assume that $\mathcal{C}_T$ is postprojective, then Σ has finitely many predecessors by (2.5). Similarly, if $\mathcal{C}_T$ is preinjective, then Σ has finitely many successors. Thus, $\mathcal{C}_T$ is finite. By (IV.5.4), B is representation–finite.

Conversely, if B is representation–finite, then $\Gamma(\operatorname{mod} B) = \mathcal{C}_T$ is acyclic. Every module Z in $\mathcal{C}_T$ can be written in the form $Z \cong \tau^t Y$, for some Y in Σ and some $t \in \mathbb{Z}$. Because B is representation–finite, there exist an indecomposable projective module P and $s \geq 0$ such that $\tau^s Y = P$. Therefore $Z \cong \tau^{t-s} P$. This shows that $\mathcal{C}_T$ is a postprojective component. Similarly, it is a preinjective component.

(b) This follows from (a) and (4.1).   □

We have already pointed out that any tilted algebra of Dynkin type is representation–finite. We present in (4.8) and (5.8) examples showing that we may obtain representation–finite tilted algebras by tilting representation–infinite hereditary algebras. In fact, for tilted algebras of Euclidean type, the converse of (4.3)(b) is also true.

4.4. Proposition. *Let Q be a Euclidean quiver, $A = KQ$, and T_A be a tilting module having both a postprojective and a preinjective direct summand. Then $B = \operatorname{End} T_A$ is representation–finite.*

Proof. Because T is a splitting tilting module, it suffices to show that each of $\mathcal{T}(T)$ and $\mathcal{F}(T)$ contains only finitely many nonisomorphic indecomposable modules.

Let T_0 be a postprojective indecomposable direct summand of T. We claim that there exist only finitely many nonisomorphic indecomposable modules M such that $\operatorname{Hom}_A(T_0, M) = 0$. This clearly would imply that $\mathcal{F}(T)$ has only finitely many nonisomorphic indecomposable modules. Because T_0 is postprojective, there exist $t \geq 0$ and $a \in Q_0$ such that $T_0 = \tau^{-t}P(a)$. Let M be an indecomposable A-module such that $\operatorname{Hom}_A(T_0, M) = 0$. Because A is hereditary, then (IV.2.15) yields

$$\operatorname{Hom}_A(P(a), \tau^t M) \cong \operatorname{Hom}_A(\tau^t T_0, \tau^t M) \cong \operatorname{Hom}_A(T_0, M) = 0.$$

This implies that $(\mathbf{dim}\,\tau^t M)_a = 0$, that is, $\tau^t M$ is annihilated by the idempotent e_a corresponding to $a \in Q_0$. Then, $\tau^t M$ is zero or an indecomposable module over the path algebra of the quiver $Q^{(a)}$ obtained from Q by deleting the point a and all the arrows having a as source or target. Because Q is a Euclidean quiver, $Q^{(a)}$ is a disjoint union of Dynkin quivers; hence its path algebra is representation–finite. This shows that there exist only finitely many nonisomorphic indecomposable A-modules M such that $\operatorname{Hom}_A(T_0, M) = 0$. Our claim follows.

Dually, let T_1 be a preinjective indecomposable direct summand of T. There exist $s \geq 0$ and $b \in Q_0$ such that $T_1 = \tau^s I(b)$. Thus, if N is an indecomposable A-module such that $\operatorname{Ext}_A^1(T_1, N) = 0$, then (IV.2.15) yields

$$\operatorname{Hom}_A(\tau^{-s-1}N, I(b)) \cong \operatorname{Hom}_A(\tau^{-s-1}N, \tau^{-s}T_1)$$
$$\cong \operatorname{Hom}_A(\tau^{-1}N, T_1) \cong D\operatorname{Ext}_A^1(T_1, N) = 0,$$

and $\tau^{-s-1}N$ is zero or an indecomposable module over the path algebra of the quiver $Q^{(b)}$ obtained from Q by deleting the point b and all the arrows having b as source or target. Because, as earlier, $Q^{(b)}$ is a disjoint union of Dynkin quivers, there exist only finitely many isomorphism classes of indecomposable A-modules N such that $\operatorname{Ext}_A^1(T_1, N) = 0$ and consequently of indecomposable modules in $\mathcal{T}(T)$. $\qquad\square$

We note that if we tilt a representation–infinite hereditary algebra A to a representation–finite algebra B by a tilting module T_A, then each of $\mathcal{T}(T)$ and $\mathcal{F}(T)$ contains only finitely many nonisomorphic indecomposable A-modules, and consequently there is usually a big difference between the categories $\operatorname{mod} A$ and $\operatorname{mod} B$. At the other extreme, we now exhibit a class of representation–infinite tilted algebras whose module categories are as close as possible to that of the hereditary algebra from which we tilt.

4.5. Theorem. *Let A be a representation–infinite hereditary algebra, T be a postprojective tilting A-module, and $B = \operatorname{End} T_A$.*

(a) *$\mathcal{T}(T)$ contains all but finitely many nonisomorphic indecomposable A-modules, and any indecomposable A-module not in $\mathcal{T}(T)$ is post-projective.*

(b) *$\mathcal{F}(T)$ contains only finitely many nonisomorphic indecomposable A-modules, and all of them are postprojective.*

(c) *The connecting component $\mathcal{C}_T$ of $\Gamma(\operatorname{mod} B)$ determined by T is a preinjective component $\mathcal{Q}(B)$ containing all indecomposable injective modules and all indecomposable modules from $\mathcal{X}(T)$ but no projective module.*

(d) *The images under the functor $\operatorname{Hom}_A(T, -)$ of the regular components from $\mathcal{R}(A)$ form a family $\mathcal{R}(B)$ of regular components in $\Gamma(\operatorname{mod} B)$.*

(e) *The images under the functor $\operatorname{Hom}_A(T, -)$ of the postprojective torsion A-modules form a postprojective component $\mathcal{P}(B)$ containing all indecomposable projective B-modules but no injective modules.*

(f) *$\Gamma(\operatorname{mod} B)$ is the disjoint union of $\mathcal{P}(B)$, $\mathcal{R}(B)$, and $\mathcal{Q}(B)$, and we have*

$$\operatorname{Hom}_B(\mathcal{R}(B), \mathcal{P}(B)) = 0, \quad \operatorname{Hom}_B(\mathcal{Q}(B), \mathcal{P}(B)) = 0, \quad \operatorname{Hom}_B(\mathcal{Q}(B), \mathcal{R}(B)) = 0.$$

(g) *$\operatorname{pd} Z \leq 1$ and $\operatorname{id} Z \leq 1$ for all regular modules Z and all but finitely many nonisomorphic indecomposable B-modules Z in $\mathcal{P}(B) \cup \mathcal{Q}(B)$.*

Proof. (a) and (b). Because the postprojective component $\mathcal{P}(A)$ of the quiver $\Gamma(\operatorname{mod} A)$ is isomorphic to $(-\mathbb{N})Q_A^{\operatorname{op}}$, it contains infinitely many sections, all isomorphic to Q_A^{op}. Because, on the other hand, T has finitely many nonisomorphic indecomposable direct summands, $\mathcal{P}(A)$ contains a section Δ such that the full translation subquiver $\mathcal{P}_\Delta$ of $\mathcal{P}(A)$ consisting of all successors of Δ contains no indecomposable direct summand of T. Because T is postprojective, it follows from (2.5) and (2.13) that $\mathcal{T}(T) = \{M_A \mid \operatorname{Ext}_A^1(T, M) = 0\} = \{M_A \mid \operatorname{Hom}_A(M, \tau T) = 0\}$ contains all the modules from $\mathcal{P}_\Delta$, as well as all the regular and preinjective modules. Moreover, all nontorsion, and in particular all torsion-free, modules must precede Δ and hence are postprojective.

(c)–(f). Let Σ be the section in $\mathcal{C}_T$ constructed as in (3.5). Because T has no preinjective direct summand, we know from (4.1) that $\mathcal{C}_T$ contains no projective module. Further, by (3.5), any proper successor of Σ in $\mathcal{C}_T$ lies in $\mathcal{X}(T)$. It follows from (b) and the equivalence $\mathcal{X}(T) \cong \mathcal{F}(T)$ that Σ has only finitely many successors.

On the other hand, the translation subquiver $\mathcal{P}_\Delta$ of $\mathcal{P}(A)$ lies in $\mathcal{T}(T)$ and, by (VI.5.3), its image under the functor $\operatorname{Hom}_A(T, -)$ is a full translation quiver closed under successors lying in some component $\mathcal{P}(B)$ of

$\Gamma(\operatorname{mod} B)$. For the same reason, the image of $\mathcal{R}(A)$ under the functor $\operatorname{Hom}_A(T, -)$ is a family of regular components of $\Gamma(\operatorname{mod} B)$.

Observe that $\Gamma(\operatorname{mod} B)$ is infinite, hence it has no finite component. Because $(\mathcal{X}(T), \mathcal{Y}(T))$ is a splitting torsion pair in $\operatorname{mod} B$, we get that all the indecomposable modules from $\mathcal{X}(T)$ belong to $\mathcal{C}_T$, and $\mathcal{P}(B)$ is the image under the functor $\operatorname{Hom}_A(T, -)$ of $\mathcal{P}(A) \cap \mathcal{T}(T)$. Clearly, $\mathcal{P}(B)$ is a postprojective component containing all the indecomposable projective modules (because $\mathcal{P}(A) \cap \mathcal{T}(T)$ contains all the indecomposable direct summands of T). Also, $\mathcal{Q}(B) = \mathcal{C}_T$ is a preinjective component containing all the indecomposable injective B-modules, and $\Gamma(\operatorname{mod} B)$ is the disjoint union of $\mathcal{P}(B)$, $\mathcal{Q}(B)$, and the family $\mathcal{R}(B)$ of regular components. Finally, applying (2.5), (2.13) and using that $(\mathcal{X}(T), \mathcal{Y}(T))$ is a torsion pair, we obtain (f).

(g) Because all the indecomposable projective B-modules belong to $\mathcal{P}(B)$ (thus have only finitely many nonisomorphic predecessors), whereas all the indecomposable injective B-modules belong to $\mathcal{Q}(B)$ (thus have only finitely many nonisomorphic successors), we have

$$\operatorname{Hom}_B(DB, \tau Z) = 0 \quad \text{and} \quad \operatorname{Hom}_B(\tau^{-1}Z, B) = 0$$

for all but finitely many nonisomorphic indecomposable B-modules Z in $\mathcal{P}(B) \cup \mathcal{Q}(B)$ and for all regular modules Z. We then apply (IV.2.7). □

Under the assumptions and with the notation of Theorem 4.5, we may visualise the situation in the following picture:

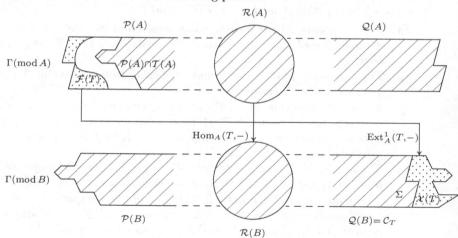

Here, and as usual, we denote by the classes $\mathcal{T}(T)$ and $\mathcal{Y}(T)$ and by the classes $\mathcal{F}(T)$ and $\mathcal{X}(T)$.

As can be seen, if B is not hereditary itself, its module category is very close to that of a hereditary algebra. Indeed, with the preceding notation, the functor $\mathrm{Hom}_A(T, -)$ induces an equivalence between the additive full subcategories of $\mathrm{mod}\,A$ generated by the indecomposables from $\mathcal{P}(A) \cap \mathcal{T}(T)$, $\mathcal{R}(A)$, and $\mathcal{Q}(A)$, and the additive full subcategories of $\mathrm{mod}\,B$ generated by the indecomposables from $\mathcal{P}(B)$, $\mathcal{R}(B)$, and $\mathcal{Q}(B) \cap \mathcal{Y}(T)$, respectively, and all but finitely many nonisomorphic indecomposable A-modules or B-modules, respectively, belong to one of these subcategories. Also, $\mathrm{gl.dim}\,B \leq 2$ and $\mathrm{pd}\,Z \leq 1$, $\mathrm{id}\,Z \leq 1$ for all but finitely many nonisomorphic indecomposable B-modules Z in $\mathcal{P}(B) \cup \mathcal{Q}(B)$ and for all regular modules Z. One may then think of $\Gamma(\mathrm{mod}\,B)$ as "concealing" some hereditary full subcategory involving all but finitely many nonisomorphic indecomposable B-modules. This explains the following terminology.

4.6. Definition. Let Q be a finite, connected, and acyclic quiver that is not a Dynkin quiver. An algebra B is called **concealed of type** Q if there exists a postprojective tilting module T over the path algebra $A = KQ$ such that $B = \mathrm{End}\,T_A$.

Clearly, the statement that Q is not a Dynkin quiver just means that A is representation–infinite.

We quote the analogue of (4.5) for the tilted algebras arising from preinjective tilting modules. Its proof is similar to that of (4.5) and therefore is omitted.

4.7. Theorem. *Let A be a representation–infinite hereditary algebra, T be a preinjective tilting A-module, and $B = \mathrm{End}\,T_A$.*

(a) *$\mathcal{F}(T)$ contains all but finitely many nonisomorphic indecomposable A-modules and any indecomposable A-module not in $\mathcal{F}(T)$ is preinjective.*

(b) *$\mathcal{T}(T)$ contains finitely many nonisomorphic indecomposable A-modules and all of them are preinjective.*

(c) *The connecting component $\mathcal{C}_T$ of $\Gamma(\mathrm{mod}\,B)$ determined by T is a postprojective component $\mathcal{P}(B)$ containing all indecomposable projective modules and all indecomposable modules from $\mathcal{Y}(T)$ but no injective module.*

(d) *The images under the functor $\mathrm{Ext}_A^1(T, -)$ of the regular components from $\mathcal{R}(A)$ form a family $\mathcal{R}(B)$ of regular components in $\Gamma(\mathrm{mod}\,B)$.*

(e) *The images under the functor $\mathrm{Ext}_A^1(T, -)$ of the preinjective torsionfree A-modules form a preinjective component $\mathcal{Q}(B)$ containing all indecomposable injective B-modules but no projective modules.*

(f) *$\Gamma(\mathrm{mod}\,B)$ is the disjoint union of $\mathcal{P}(B)$, $\mathcal{R}(B)$, and $\mathcal{Q}(B)$ and*

$$\mathrm{Hom}_B(\mathcal{R}(B), \mathcal{P}(B)) = 0, \ \mathrm{Hom}_B(\mathcal{Q}(B), \mathcal{P}(B)) = 0, \ \mathrm{Hom}_B(\mathcal{Q}(B), \mathcal{R}(B)) = 0.$$

(g) *$\mathrm{pd}\,Z \leq 1$ and $\mathrm{id}\,Z \leq 1$, for all regular modules Z and all but finitely many nonisomorphic indecomposable modules Z in $\mathcal{P}(B) \cup \mathcal{Q}(B)$.* $\square$

Under the assumptions and with the notation of Theorem 4.7, we may visualise the situation as in the following picture:

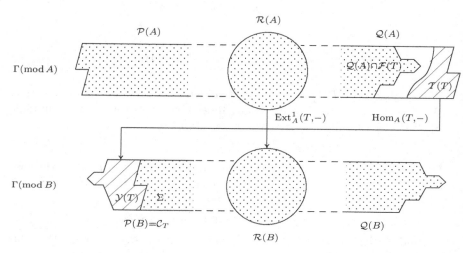

The functor $\mathrm{Ext}_A^1(T, -)$ induces an equivalence between the additive full subcategories of $\mathrm{mod}\,A$ generated by the indecomposable modules from $\mathcal{P}(A)$, $\mathcal{R}(A)$, and $\mathcal{Q}(A) \cap \mathcal{F}(T)$ and the additive full subcategories of $\mathrm{mod}\,B$ generated by the indecomposables from $\mathcal{P}(B) \cap \mathcal{X}(T)$, $\mathcal{R}(B)$, and $\mathcal{Q}(B)$, respectively. Thus, as before, one may think of $\mathrm{mod}\,B$ as "concealing" a hereditary full subcategory involving all but finitely many nonisomorphic indecomposable modules. In fact, one can prove (see Exercise 6.9) that, for a representation–infinite hereditary algebra A, an algebra B is of the form $\mathrm{End}\,T_A$ for some postprojective tilting A-module T if and only if

$$B \cong \mathrm{End}\,T'_A,$$

for some preinjective tilting A-module T'. Thus the class of concealed algebras coincides with the class obtained from representation–infinite hereditary algebras by preinjective tilting modules.

4.8. Examples. (a) Let A be the path algebra of the Euclidean quiver Q:

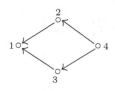

of type $\widetilde{\mathbb{A}}_3$. Consider the indecomposable A-modules:

$$T_1 = P(1) = K \begin{smallmatrix} & 0 & \\ \nearrow & & \searrow \\ & & 0 \\ \searrow & & \nearrow \\ & 0 & \end{smallmatrix} 0 \qquad T_2 = K \begin{smallmatrix} & 0 & \\ \nearrow & & \searrow \\ & & K \\ \searrow & & \nearrow \\ & K & \end{smallmatrix} K$$

$$T_3 = K \begin{smallmatrix} & K & \\ \nearrow 1 & & \searrow 1 \\ & & K \\ \searrow & & \nearrow \\ & 0 & \end{smallmatrix} K \qquad T_4 = I(4) = 0 \begin{smallmatrix} & 0 & \\ \nearrow & & \searrow \\ & & K \\ \searrow & & \nearrow \\ & 0 & \end{smallmatrix} K$$

We see that $T_1 = P(1)$ is postprojective, whereas $T_4 = I(4)$ is preinjective. We claim that T_2 and T_3 are regular. Indeed, consider the simple A-module $S(2)$; it has a minimal projective presentation

$$0 \longrightarrow P(1) \xrightarrow{\ p\ } P(2) \longrightarrow S(2) \longrightarrow 0.$$

Hence, by (IV.2.4), $\tau S(2)$ is the kernel of $\nu p : \nu P(1) \to \nu P(2)$. Because $\nu P(1) \cong I(1)$ and $\nu P(2) \cong I(2)$, we get $\tau S(2) \cong T_2$. Similarly, $\tau^{-1} S(2) \cong T_2$ and $\tau^{-1} S(3) \cong T_3 \cong \tau S(3)$. Thus there exist cycles

$$T_2 \to * \to S(2) \to * \to T_2 \text{ and } T_3 \to * \to S(3) \to * \to T_2$$

in $\text{mod}\,A$. In particular, T_2 and T_3 lie in neither $\mathcal{P}(A)$ nor $\mathcal{Q}(A)$. This shows our claim. Moreover, it is easy to check that $\tau I(4) = 0 \begin{smallmatrix} 1 \\ 1 \end{smallmatrix} 1$.

Let $T_A = \bigoplus_{i=1}^{4} T_i$. Then (IV.2.14) yields the isomorphisms

$$\text{Ext}^1_A(T,T) \cong D\text{Hom}_A(T, \tau T) \cong D\text{Hom}_A(T, S(2) \oplus S(3) \oplus \tau I(4)) = 0,$$

and consequently T is a tilting module. Because Q is Euclidean and T contains both a postprojective and a preinjective direct summand, it follows from (4.4) that $B = \text{End}\,T_A$ is representation–finite. In fact, B is given by the quiver

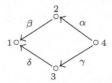

bound by $\alpha\beta = 0$, $\gamma\delta = 0$. The Auslander–Reiten quiver $\Gamma(\text{mod}\,B)$ is given by

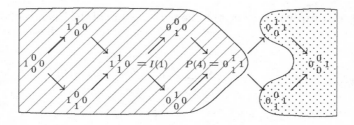

We note that the indecomposable modules $1\,{}^1_1\,0$, $0\,{}^0_1\,0$, $0\,{}^1_0\,0$, $0\,{}^1_1\,1$ form a section Σ in $\Gamma(\mathrm{mod}\,B)$ isomorphic to Q^{op}.

(b) Let A be the path algebra of the quiver $Q : \ \overset{1}{\circ}\longleftarrow\overset{2}{\circ}\longleftarrow\overset{3}{\circ}$.
Then the beginning of the postprojective component $\mathcal{P}(A)$ of $\Gamma(\mathrm{mod}\,B)$ is of the form

$$
\begin{array}{ccccccc}
 & & 221 & & & 463 & \cdots \\
 & \nearrow\!\!\!\nearrow & & \searrow & \nearrow\!\!\!\nearrow & & \searrow \\
 & 110 & & & 342 & & \cdots \\
 \nearrow & & \searrow & \nearrow & & \searrow & \nearrow \\
100 & & 010 & & 332 & \cdots
\end{array}
$$

and the end of the preinjective component $\mathcal{Q}(A)$ of $\Gamma(\mathrm{mod}\,A)$ is of the form

$$
\begin{array}{ccccccc}
\cdots & 269 & & & 023 & & 001 \\
 & \nearrow\!\!\!\nearrow & & \searrow & \nearrow\!\!\!\nearrow & & \searrow & \nearrow \\
\cdots & & 146 & & & 012 & \\
 & \searrow & & \nearrow & & \searrow & \nearrow \\
\cdots & 034 & & & 112
\end{array}
$$

Consider the module $T_A = S(1) \oplus I(1) \oplus I(3)$. Then (IV.2.14) yields

$$\mathrm{Ext}^1_A(T,T) \cong D\mathrm{Hom}_A(T,\tau T) \cong D\mathrm{Hom}_A(T, \tau I(1) \oplus \tau I(3)) = 0,$$

and hence T is a tilting A-module. The tilted algebra $B = \mathrm{End}\,T_A$ is given by the quiver

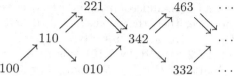

bound by $\alpha\gamma = 0$, $\beta\gamma = 0$. Because the hereditary algebra given by the full subquiver with points 2 and 3 equals the quotient of B by the two-sided ideal generated by the idempotent e_1 corresponding to the point 1, and is representation–infinite (it is indeed isomorphic to the Kronecker algebra), we conclude from (VII.2.2) that B is also representation–infinite. This shows that in (4.4) the restriction that A be the path algebra of a Euclidean quiver is essential.

(c) Let A be the path algebra of the Kronecker quiver $\overset{1}{\circ}\rightleftarrows\overset{2}{\circ}$. Then $\mathcal{P}(A)$ and $\mathcal{Q}(A)$ are respectively of the forms

$$
\mathcal{P}(A): \quad
\begin{array}{ccccccc}
 & P(2) & & \tau^{-1}P(2) & & \tau^{-2}P(2) & \\
 \nearrow\!\!\!\nearrow & & \searrow\!\!\!\searrow & \nearrow\!\!\!\nearrow & & \searrow\!\!\!\searrow & \nearrow\!\!\!\nearrow & & \searrow\!\!\!\searrow \\
P(1) & & \tau^{-1}P(1) & & \tau^{-2}P(1) & & \cdots
\end{array}
$$

$$
\mathcal{Q}(A): \quad
\begin{array}{ccccccc}
\cdots & \tau^2 I(1) & & \tau I(1) & & I(1) & \\
 \nearrow\!\!\!\nearrow & & \searrow\!\!\!\searrow & \nearrow\!\!\!\nearrow & & \searrow\!\!\!\searrow & \nearrow\!\!\!\nearrow & & \searrow\!\!\!\searrow \\
\cdots & & \tau^2 I(2) & & \tau I(2) & & I(2) & \cdots
\end{array}
$$

If M is an indecomposable A-module not isomorphic to the simple injective module $S(2) = I(2)$, then $\operatorname{Hom}_A(P(1), M) \neq 0$. Similarly, if N is an indecomposable A-module not isomorphic to the simple projective module $S(1) = P(1)$, then $\operatorname{Hom}_A(P(2), N) \neq 0$. This first implies that there exists no tilting module $T = T_1 \oplus T_2$ such that T_1 is indecomposable postprojective and T_2 is indecomposable preinjective. Indeed, assuming that this is the case, then there exist $t, s \geq 0$ and two indecomposable modules: P projective and I injective, such that $T_1 \cong \tau^{-t}P$ and $T_2 \cong \tau^s I$. In view of (IV.2.14) and (IV.2.15), this gives $0 = D\operatorname{Ext}^1_A(T_2, T_1) \cong \operatorname{Hom}_A(T_1, \tau T_2) \cong \operatorname{Hom}_A(\tau^{-t}P, \tau^{s+1}I) \cong \operatorname{Hom}_A(P, \tau^{t+s+1}I)$, which contradicts the preceding remarks. Consequently, any tilted algebra obtained from A is representation–infinite (by (4.3)(b)).

The same remarks also show that if $a \in \{1, 2\}$ and $s \geq 0$, $t \geq 1$, then

$$\operatorname{Hom}_A(\tau^{-s}P(a), \tau^{-s-t}P(a)) \cong \operatorname{Hom}_A(P(a), \tau^{-t}P(a)) \neq 0;$$

therefore, if $T = T_1 \oplus T_2$ is a postprojective tilting module, with T_1 and T_2 indecomposable, then T_1 and T_2 belong to distinct τ-orbits. Assume thus that $a \neq b$ and $s, t \geq 0$ are such that $T_1 = \tau^{-s}P(a)$ and $T_2 = \tau^{-s-t}P(b)$. Then (IV.2.14) and (IV.2.15) yield the isomorphisms

$$D\operatorname{Ext}^1_A(T_2, T_1) \cong \operatorname{Hom}_A(T_1, \tau T_2) \cong \operatorname{Hom}_A(\tau^{-s}P(a), \tau^{-s-t+1}P(b))$$
$$\cong \operatorname{Hom}_A(P(a), \tau^{-t+1}P(b))$$

and this vanishes if and only if $a = 2$, $b = 1$, and $t \leq 1$, that is, if and only if $T_A \cong \tau^{-s}P(1) \oplus \tau^{-s}P(2)$, or $T_A \cong \tau^{-s}P(2) \oplus \tau^{-s-1}P(1)$ for some $s \geq 0$.

Similarly, if T is a preinjective tilting module, then $T \cong \tau^s I(1) \oplus \tau^s I(2)$ or $T \cong \tau^{s+1}I(1) \oplus \tau^s I(2)$, for some $s \geq 0$. Finally, we prove in the second volume of this book that, for any regular indecomposable A-module R, we have $\operatorname{Ext}^1_A(R, R) \neq 0$ and consequently R cannot be a summand of any tilting module. This shows that we have obtained all the possible tilting modules. As a consequence, any tilted algebra from A is concealed and isomorphic to A.

(d) Let A be the path algebra of the Euclidean quiver

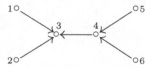

of type $\widetilde{\mathbb{D}}_5$. The beginning of the postprojective component $\mathcal{P}(A)$ is of the form

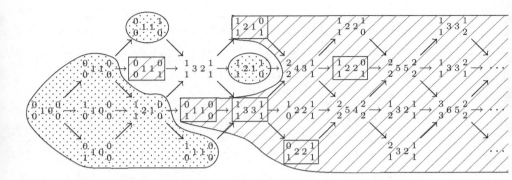

It is easily verified that the postprojective A-module

$$T_A = \begin{smallmatrix} & 0 & \\ 0 & 1 & 1 & 0 \\ & 1 & \end{smallmatrix} \oplus \begin{smallmatrix} & 0 & \\ 1 & 1 & 1 & 0 \\ & 0 & \end{smallmatrix} \oplus \begin{smallmatrix} & 1 & \\ 1 & 2 & 1 & 0 \\ & 1 & \end{smallmatrix} \oplus \begin{smallmatrix} & 1 & \\ 1 & 3 & 3 & 1 \\ & 1 & \end{smallmatrix} \oplus \begin{smallmatrix} & 1 & \\ 1 & 2 & 2 & 0 \\ & 1 & \end{smallmatrix} \oplus \begin{smallmatrix} & 0 & \\ 1 & 2 & 2 & 1 \\ & 1 & \end{smallmatrix}$$

is a tilting A-module. Therefore $B = \operatorname{End} T_A$ is a concealed algebra of type $\widetilde{\mathbb{D}}_5$. It is given by the quiver

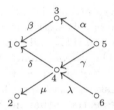

bound by $\alpha\beta = \gamma\delta$. The postprojective component $\mathcal{P}(B)$ of $\Gamma(\operatorname{mod} B)$ is the image of $\mathcal{P}(A) \cap \mathcal{T}(T)$ under the action of the functor $\operatorname{Hom}_A(T, -)$ and is of the form

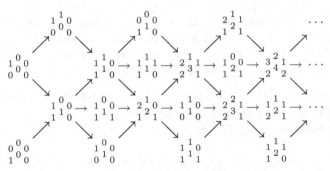

whereas the preinjective component $\mathcal{Q}(B)$ of $\Gamma(\operatorname{mod} B)$ equals the connecting component $\mathcal{C}_T$ determined by T and is of the form

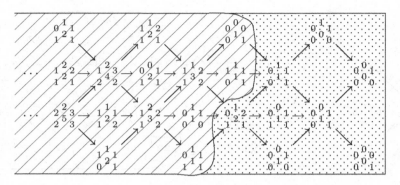

We note that the indecomposable modules $0\,{}^0_1\,0\atop 0\ \ \ 1$, $1\,{}^1_1\,1\atop 0\ \ \ 1$, $1\,{}^1_3\,2\atop 1\ \ \ 2$, $1\,{}^2_3\,2\atop 1\ \ \ 2$, $0\,{}^1_1\,1\atop 0\ \ \ 0$, and $0\,{}^1_1\,1\atop 1\ \ \ 1$ form a section $\Sigma \cong Q^{\mathrm{op}}$ in $\Gamma(\mathrm{mod}\,B)$.

VIII.5. The criterion of Liu and Skowroński

To decide whether a given algebra is tilted, we need some intrinsic characterisation. The objective of this section is to give such a characterisation, obtained independently by Liu [111] and Skowroński [156]. This result uses the concept of section. There exist many other characterisations, using related concepts such as that of slice (see, for instance, [145]). But the criterion of Liu and Skowroński is very useful for practical applications. Our presentation here follows essentially that in [158].

Let A be an algebra. We recall that an A-module M is said to be faithful if its right annihilator $\mathcal{I}_M = \{a \in A \mid Ma = 0\}$ vanishes. We showed in (VI.2.2) that an A-module M is faithful if and only if A_A is cogenerated by M_A, or equivalently, if and only if $D(A)_A$ is generated by M_A.

Let A be an algebra. We recall from (VI.2.2) that any tilting A-module is faithful and from (VI.6.3) that any Gen–minimal faithful A-module is a partial tilting module. We now give an alternate sufficient condition for a faithful A-module to be a partial tilting module.

5.1. Lemma. *Let A be an algebra and M be a faithful A-module.*

(a) *If $\mathrm{Hom}_A(M, \tau M) = 0$, then $\mathrm{pd}\,M \leq 1$.*
(b) *If $\mathrm{Hom}_A(\tau^{-1}M, M) = 0$, then $\mathrm{id}\,M \leq 1$.*

Proof. We only prove (a); the proof of (b) is similar. Because the module M is faithful, there exist $t \geq 1$ and an epimorphism $M^t \to DA$, by (VI.2.2). Applying the functor $\mathrm{Hom}_A(-, \tau M)$ yields a monomorphism

$\text{Hom}_A(DA, \tau M) \to \text{Hom}_A(M, \tau M)^t$. Hence $\text{Hom}_A(DA, \tau M) = 0$ so, by (IV.2.7), we get $\text{pd}\, M \le 1$. $\qquad\square$

Thus, if M is a faithful module such that $\text{Hom}_A(M, \tau M) = 0$, it is a partial tilting module (because $\text{pd}\, M \le 1$ and there are isomorphisms $\text{Ext}_A^1(M, M) \cong D\text{Hom}_A(M, \tau M) = 0$, by (IV.2.14)).

We now need the following lemma, relating the Auslander–Reiten translates of the same module in two module categories.

5.2. Lemma. *Let A be an algebra, $\mathcal{I}$ be a two-sided ideal of A, and $B = A/\mathcal{I}$. If M is a B-module, then the Auslander–Reiten translate $\tau_B M$ of M in $\text{mod}\, B$ is a submodule of the Auslander–Reiten translate $\tau_A M$ of M in $\text{mod}\, A$.*

Proof. For any module N_A, we set

$$\mathbf{t}_{\mathcal{I}}(N) = \{n \in N; \; n\mathcal{I} = 0\}.$$

It is easy to see that $\mathbf{t}_{\mathcal{I}}(N) \subseteq N$ is a B-module and, for each homomorphism $f \in \text{Hom}_A(N, L)$, the restriction $\mathbf{t}_{\mathcal{I}}(f) : \mathbf{t}_{\mathcal{I}}(N) \to \mathbf{t}_{\mathcal{I}}(L)$ of f to $\mathbf{t}_{\mathcal{I}}(N)$ is a homomorphism of B-modules. Obviously, we have defined a covariant functor $\mathbf{t}_{\mathcal{I}} : \text{mod}\, A \longrightarrow \text{mod}\, B$.

Assume now that M_B is a B-module. Without loss of generality, we may assume that M_A is indecomposable. First we note that if M is projective when viewed as an A-module, then M_B is projective. Indeed, if $g : X \to Y$ is an epimorphism of B-modules, then it is an A-module epimorphism and $\text{Hom}_B(M, g) : \text{Hom}_B(M, X) \to \text{Hom}_B(M, Y)$ is surjective, because $\text{Hom}_B(M, Z) = \text{Hom}_A(M, Z)$ for any B-module Z.

Assume now that M_B is not projective. Then M_A is not projective, and there exists an almost split sequence $0 \longrightarrow \tau_A M \overset{f}{\longrightarrow} E \overset{g}{\longrightarrow} M \longrightarrow 0$ in $\text{mod}\, A$. Applying the functor $\mathbf{t}_{\mathcal{I}}$ yields an exact sequence in $\text{mod}\, B$

$$0 \longrightarrow \mathbf{t}_{\mathcal{I}}(\tau_A M) \overset{\mathbf{t}_{\mathcal{I}}(f)}{\longrightarrow} \mathbf{t}_{\mathcal{I}}(E) \overset{\mathbf{t}_{\mathcal{I}}(g)}{\longrightarrow} M \longrightarrow 0,$$

where $\mathbf{t}_{\mathcal{I}}(M) = M$, because M is a B-module. The homomorphism $\mathbf{t}_{\mathcal{I}}(g)$ is right almost split in $\text{mod}\, B$. Indeed, it is clearly not a retraction and, if X_B is a B-module and $u : X_B \to M_B$ is not a retraction, then $u : X_A \to M_A$ viewed as a homomorphism of A-modules is not a retraction. Because g is right almost split in $\text{mod}\, A$, u lifts to a homomorphism $v : X_A \to E_A$ in $\text{mod}\, A$ such that $u = gv$. It is clear that $\text{Im}\, v \subseteq \mathbf{t}_{\mathcal{I}}(E)$, because $X\mathcal{I} = 0$. Consequently, u lifts to a homomorphism $v : X_B \to \mathbf{t}_{\mathcal{I}}(E)$ in $\text{mod}\, B$ such that $u = \mathbf{t}_{\mathcal{I}}(g)v$, and we are done.

Because M_B is not projective, there exists an almost split sequence

$$0 \longrightarrow \tau_B M \overset{f'}{\longrightarrow} E' \overset{g'}{\longrightarrow} M \longrightarrow 0$$

in mod B. Because $t_{\mathcal{I}}(g)$ is right almost split in mod B and g' is not a retraction, there exists a homomorphism $h : E' \to t_{\mathcal{I}}(E)$ of B-modules such that $t_{\mathcal{I}}(g)h = g'$. It follows that h is a section because g' is minimal right almost split. Consequently, we get a commutative diagram with exact rows

$$
\begin{array}{ccccccccc}
0 & \longrightarrow & \tau_B M & \xrightarrow{\ \ f'\ \ } & E' & \xrightarrow{\ \ g'\ \ } & M & \longrightarrow & 0 \\
 & & \downarrow & & h\downarrow & & 1_M\downarrow & & \\
0 & \longrightarrow & t_{\mathcal{I}}(\tau_A M) & \xrightarrow{\ t_{\mathcal{I}}(f)\ } & t_{\mathcal{I}}(E) & \xrightarrow{\ t_{\mathcal{I}}(g)\ } & t_{\mathcal{I}}(M) & \longrightarrow & 0 \\
 & & \downarrow & & \downarrow & & 1_M\downarrow & & \\
0 & \longrightarrow & \tau_A M & \xrightarrow{\ \ f\ \ } & E & \xrightarrow{\ \ g\ \ } & M & \longrightarrow & 0
\end{array}
$$

where the vertical homomorphisms are injective. As a consequence, $\tau_B M$ is isomorphic to a submodule of $t_{\mathcal{I}}(\tau_A M)$ and thus to a submodule of $\tau_A M$. $\square$

The following lemma, obtained in [157], is crucial in the sequel.

5.3. Lemma. *Let A be an algebra and n be the rank of the group $K_0(A)$. Assume that an A-module M is a direct sum of m pairwise nonisomorphic indecomposable modules and $\mathrm{Hom}_A(M, \tau M) = 0$. Then $m \le n$.*

Proof. Let $\mathcal{I}_M$ be the right annihilator of M, that is, $\mathcal{I}_M = \{a \in A \mid Ma = 0\}$. Then $\mathcal{I}_M$ is a two-sided ideal of A. Thus, if $B = A/\mathcal{I}_M$, we have, by (5.2), that $\tau_B M$ is a submodule of $\tau_A M = \tau M$. The assumption that $\mathrm{Hom}_A(M, \tau M) = 0$ implies that $\mathrm{Hom}_B(M, \tau_B M) = 0$. Because M is a faithful B-module, we deduce from (5.1)(a) that M is a partial tilting B-module. By Bongartz's lemma (VI.2.4), there exists a B-module N such that $M \oplus N$ is a tilting B-module. By (VI.4.4), $m \le \mathrm{rk}\,K_0(B)$. On the other hand, clearly, $\mathrm{rk}\,K_0(B) \le n$. $\square$

To motivate the assumptions of the following lemma, we recall that, if Σ is a section in a component of the Auslander–Reiten quiver $\Gamma(\mathrm{mod}\,B)$ of an algebra B, say, then, by (1.4), if U_B belongs to Σ and there exists an irreducible homomorphism $V \to U$, then V_B belongs to either Σ or $\tau\Sigma = \{\tau W \mid W \text{ on } \Sigma\}$; similarly, if there exists an irreducible morphism $U \to V$, then V belongs to either Σ or $\tau^{-1}\Sigma = \{\tau^{-1}W \mid W \text{ on } \Sigma\}$.

5.4. Lemma. *Let B be an algebra, $\mathcal{C}$ be a component of $\Gamma(\mathrm{mod}\,B)$, and Σ be a finite and acyclic connected full subquiver of $\mathcal{C}$.*

(a) *Assume that if U belongs to Σ and there exists an irreducible morphism $V \to U$, then V belongs to either Σ or $\tau\Sigma$. Then any homomorphism $f : Y \to U$ between indecomposables U on Σ and Y not on Σ must factor through a direct sum of modules from $\tau\Sigma$.*

(b) *Assume that if U belongs to Σ and there exists an irreducible morphism $U \to V$, then V belongs to either Σ or $\tau^{-1}\Sigma$. Then any homomorphism $g : U \to X$ between indecomposables U on Σ and X not on Σ must factor through a direct sum of modules from $\tau^{-1}\Sigma$.*

Proof. We only prove (a); the proof of (b) is similar. Assume that $f : Y \to U$ is a homomorphism between indecomposables U on Σ and Y not on Σ. Because Σ is finite and acyclic, we prove the statement by induction on an admissible sequence of sources in Σ (see (VII.5)). Assume first that U is a source in Σ, and consider the right minimal almost split morphism $u : E \to U$. Then every indecomposable summand of E belongs to $\tau\Sigma$. Because f factors through u, we are done. Assume that U is not a source, and consider the right minimal almost split morphism $u : E \to U$. Then $E = E' \oplus E''$, where all the indecomposable summands of E' belong to $\tau\Sigma$, whereas all the indecomposable direct summands of E'' belong to Σ and are predecessors of U in the admissible sequence. Then f factors through

$$u = [u'\ u''] : E' \oplus E'' \longrightarrow U.$$

Because the homomorphism $Y \longrightarrow E''$ thus obtained factors through a direct sum of modules from $\tau\Sigma$, by the induction hypothesis, the proof is complete. $\square$

5.5. Lemma. *Let B be an algebra, $\mathcal{C}$ be a component of $\Gamma(\mathrm{mod}\,B)$ containing a finite section Σ, and T_B be the direct sum of all modules on Σ. Then $\mathrm{Hom}_B(T, \tau T) = 0$ if and only if $\mathrm{Hom}_B(\tau^{-1}T, T) = 0$.*

Proof. Let $p : P \to \tau^{-1}T$ be a projective cover. Applying (5.4)(a) to $\tau^{-1}\Sigma$, we get that p factors through a direct sum of modules from Σ. Consequently, there exist $t \geq 1$ and an epimorphism $f : T^t \to \tau^{-1}T$. Similarly, considering the injective envelope of τT, we find $s \geq 1$ and a monomorphism $g : \tau T \to T^s$.

Assume that $\mathrm{Hom}_B(T, \tau T) \neq 0$ and let $h : T \to \tau T$ be a nonzero homomorphism of B-modules. Applying (5.4)(b) to Σ, we get $r \geq 1$ and a factorisation $h = h_2 h_1$, where $h_1 : T \to (\tau^{-1}T)^r$ and $h_2 : (\tau^{-1}T)^r \to \tau T$. Then the composed homomorphism $g h_2 : (\tau^{-1}T)^r \to T^s$ is nonzero, and consequently $\mathrm{Hom}_B(\tau^{-1}T, T) \neq 0$. Similarly, $\mathrm{Hom}_B(\tau^{-1}T, T) \neq 0$ implies $\mathrm{Hom}_B(T, \tau T) \neq 0$. $\square$

Now we are able to prove an important criterion of Liu and Skowroński, which characterises the tilted algebras as being those algebras B having a faithful section Σ such that $\mathrm{Hom}_B(U, \tau V) = 0$ for all modules U and V from Σ.

342 CHAPTER VIII. TILTED ALGEBRAS

5.6. Theorem. *An algebra B is a tilted algebra if and only if the quiver $\Gamma(\mathrm{mod}\,B)$ contains a component $\mathcal{C}$ with a faithful section Σ such that $\mathrm{Hom}_B(U, \tau V) = 0$ for all modules U, V from Σ. Moreover, in this case, the direct sum T_B of all modules on Σ is a tilting B-module with $A = \mathrm{End}\,T_B$ hereditary, and $\mathcal{C}$ is the connecting component of $\Gamma(\mathrm{mod}\,B)$ determined by the tilting A-module $T_A^* = D(_AT)$.*

Proof. Let B be a tilted algebra; then there exist a hereditary algebra A and a tilting A-module T such that $B \cong \mathrm{End}\,T_A$. By (3.5), the class Σ of all modules of the form $\mathrm{Hom}_A(T, I)$, where I is indecomposable injective, forms a section in the connecting component $\mathcal{C}_T$ of $\Gamma(\mathrm{mod}\,B)$ determined by T.

By (VI.3.3), there is an isomorphism $\mathrm{Hom}_A(T, DA) \cong (DT)_B$ of B-modules. Moreover, the B-module

$$\mathrm{Hom}_A(T, DA) \cong D(_BT_A \otimes_A A) \cong (DT)_B$$

generates DB. Indeed, because $_BT$ is a tilting module, there exist $m \geq 1$ and a monomorphism $_BB \to {}_BT^m$. Hence we get an epimorphism $(DT)_B^m \to DB$.

Because the module $(DT)_B \cong \mathrm{Hom}_A(T, DA)$ is the direct sum of modules from Σ, we get from (VI.2.2) that Σ is faithful. Finally, by the connecting lemma (VI.4.9), the module $\tau^{-1}\mathrm{Hom}_A(T, DA) \cong \mathrm{Ext}_A^1(T, A)$ belongs to $\mathcal{X}(T)$, whereas $\mathrm{Hom}_A(T, DA) \in \mathcal{Y}(T)$. Thus, if U, V are two modules from Σ, we have $\mathrm{Hom}_B(\tau^{-1}U, V) = 0$. By (5.5), $\mathrm{Hom}_B(U, \tau V) = 0$. This shows the necessity.

For the sufficiency, let B be an algebra such that $\Gamma(\mathrm{mod}\,B)$ has a component $\mathcal{C}$ containing a faithful section Σ such that $\mathrm{Hom}_B(U, \tau V) = 0$ for all U, V from Σ. By (5.3) and our assumption, Σ is finite. Let T_B be the direct sum of all modules on Σ. We claim that T_B is a tilting module such that $A = \mathrm{End}\,T_B$ is hereditary. Then it follows from (VI.3.3) and (VI.4.4) that $T_A^* = D(_AT)$ is a tilting A-module such that the canonical homomorphism

$$\varphi : B \to \mathrm{End}\,T_A^*,$$

defined for $b \in B$, $t \in T$ and $f \in T^*$ by $\varphi(b)(f)(t) = f(tb)$, is an isomorphism. Moreover, there are isomorphisms

$$\mathrm{Hom}_A(T^*, DA) \cong \mathrm{Hom}_A(DT, DA) \cong \mathrm{Hom}_A(A, T) \cong T$$

of right B-modules, and hence $\mathcal{C}$ equals the component $\mathcal{C}_{T^*}$ of $\Gamma(\mathrm{mod}\,B)$ determined by T^* and Σ is the section constructed as in (3.5).

By hypothesis, T_B is a faithful module with $\mathrm{Hom}_B(T, \tau T) = 0$. By (5.5), $\mathrm{Hom}_B(\tau^{-1}T, T) = 0$. By (5.1), we have $\mathrm{pd}\,T_B \leq 1$ and $\mathrm{id}\,T_B \leq 1$, so that T_B

is a partial tilting B-module. Let $f_1, \ldots, f_d$ be a K-basis of $\operatorname{Hom}_B(B, T) \cong T_B$, then consider the monomorphism $f = [f_1, \ldots, f_d] : B \to T^d$. We have a short exact sequence

$$0 \longrightarrow B \xrightarrow{f} T^d \xrightarrow{g} U \longrightarrow 0,$$

where $U = \operatorname{Coker} f$. Because $\operatorname{pd} T_B \leq 1$ and B_B is projective, we have $\operatorname{pd} U_B \leq 1$, so that $\operatorname{pd}(T \oplus U) \leq 1$. We claim that

$$\operatorname{Ext}_B^1(T \oplus U, T \oplus U) = 0.$$

Applying the functor $\operatorname{Hom}_B(-, T)$ to the preceding short exact sequence yields an exact sequence

$$\operatorname{Hom}_B(T^d, T) \xrightarrow{\operatorname{Hom}_B(f, T)} \operatorname{Hom}_B(B, T) \longrightarrow \operatorname{Ext}_B^1(U, T) \longrightarrow \operatorname{Ext}_B^1(T^d, T) = 0,$$

because $\operatorname{Ext}_B^1(T, T) \cong D\operatorname{Hom}_B(T, \tau T) = 0$. Because, by definition of f, the homomorphism $\operatorname{Hom}_B(f, T)$ is surjective, $\operatorname{Ext}_B^1(U, T) = 0$. Applying $\operatorname{Hom}_B(U, -)$ to the same short exact sequence yields

$$0 = \operatorname{Ext}_B^1(U, T^d) \longrightarrow \operatorname{Ext}_B^1(U, U) \longrightarrow \operatorname{Ext}_B^2(U, B) = 0,$$

because $\operatorname{pd} U \leq 1$. Hence $\operatorname{Ext}_B^1(U, U) = 0$. Finally, applying $\operatorname{Hom}_B(T, -)$ yields

$$0 = \operatorname{Ext}_B^1(T, T^d) \to \operatorname{Ext}_B^1(T, U) \to \operatorname{Ext}_B^2(T, B) = 0,$$

because $\operatorname{pd} T \leq 1$. Hence $\operatorname{Ext}_B^1(T, U) = 0$. This completes the proof of our claim and shows that $T \oplus U$ is a tilting B-module.

We now show that $U \in \operatorname{add} T$. If this is not the case, let U' be an indecomposable direct summand of U that is not in $\operatorname{add} T$. Then there exists an epimorphism $T^d \to U \to U'$, and therefore $\operatorname{Hom}_B(T, U') \neq 0$. By (5.4)(b), we have $\operatorname{Hom}_B(\tau^{-1}T, U') \neq 0$. Because $\operatorname{id} T \leq 1$, we have, by (IV.2.14),

$$\operatorname{Ext}_B^1(U', T) \cong D\operatorname{Hom}_B(\tau^{-1}T, U') \neq 0,$$

a contradiction to $\operatorname{Ext}_B^1(U, T) = 0$.

This shows that T_B is a tilting module. It remains to show that $A = \operatorname{End} T_B$ is hereditary. Let P_A be indecomposable projective and $f : M \to P$ be a monomorphism with M indecomposable. It suffices to show that M_A is projective. The tilting module T_B determines a torsion pair $(\mathcal{T}(T), \mathcal{F}(T))$ in $\operatorname{mod} B$ and another $(\mathcal{X}(T), \mathcal{Y}(T))$ in $\operatorname{mod} A$. Because $P_A \in \mathcal{Y}(T)$, which is torsion-free, we have $M_A \in \mathcal{Y}(T)$. That is, there exists a homomorphism $g : U \to V$ in $\operatorname{mod} B$, with $U, V \in \mathcal{T}(T)$, $\operatorname{Hom}_B(T, g) = f$, $\operatorname{Hom}_B(T, U) =$

M_A, $\mathrm{Hom}_B(T, V) = P_A$, and V lying on Σ. Because $M \neq 0$, there exists an indecomposable projective A-module P'_A and a nonzero homomorphism $f' : P' \to M$. Then there exists a homomorphism $g' : V' \to U$ in $\mathrm{mod}\,B$, such that V' lies on Σ, $\mathrm{Hom}_B(T, V') = P'_A$ and $\mathrm{Hom}_B(T, g') = f'$. Because f is a monomorphism we have $ff' \neq 0$ and hence $gg' \neq 0$.

We prove that U belongs to Σ. Assume, to the contrary, that U does not belong to Σ. It then follows from (5.4)(a) that there exist homomorphisms of B-modules $t : W \to V$ and $h : U \to W$ such that $g = th$ and W is a direct sum of modules of $\tau\Sigma$. Because $thg' = gg' \neq 0$, there is a nonzero homomorphism $hg' : V' \to W$, and consequently $\mathrm{Hom}_B(T, \tau T) \neq 0$. This is a contradiction to our assumption on Σ. Consequently, U belongs to Σ and therefore the A-module $M_A = \mathrm{Hom}_B(T, U)$ is projective. This finishes the proof. $\square$

5.7. Examples. (a) Let B be the path K-algebra of the quiver

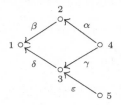

bound by two relations $\alpha\beta = \gamma\delta$ and $\varepsilon\delta = 0$ (see Example 1.3 (b)). Then the Auslander–Reiten quiver $\Gamma(\mathrm{mod}\,B)$ of B is given by

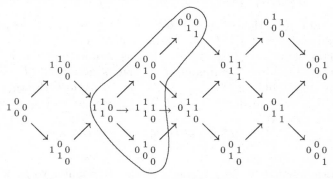

where the indecomposable modules are represented by their dimension vectors. We consider the illustrated section Σ of $\Gamma(\mathrm{mod}\,B)$. It is easily seen that any indecomposable projective B-module is a submodule of a module lying on Σ; hence, by (VII.2.2), Σ is a faithful section. Clearly, $\mathrm{Hom}_B(U, \tau V) = 0$ for all U, V on Σ. Therefore, applying (5.6), we get that B is a tilted algebra, and in fact that if T_B denotes the direct sum of the modules on Σ, then $A = \operatorname{End} T_B$ is hereditary and $T^* = D(_A T)$ is a tilting A-module such that $B = \operatorname{End} T_A^*$. A straightforward calculation shows that A is given by the Dynkin quiver:

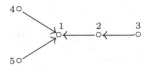

of type $\mathbb{D}_5$. We now compute the module T_A^* using the procedure explained in (VI.6.9). It is known that the points of Σ are of the form $\operatorname{Hom}_A(T^*, I(a))$, where a is a point in the quiver of A. Thus

$$\operatorname{Hom}_A(T^*, I(1)) = 1\ {}_1^1\ {}_0^0 \qquad \operatorname{Hom}_A(T^*, I(2)) = 0\ {}_1^0\ {}_0^0$$

$$\operatorname{Hom}_A(T^*, I(3)) = 0\ {}_1^0\ {}_1^0 \qquad \operatorname{Hom}_A(T^*, I(4)) = 1\ {}_1^1\ {}_0^1$$

$$\operatorname{Hom}_A(T^*, I(5)) = 0\ {}_0^1\ {}_0^0$$

Thus, if one writes

$$T_A^* = T_1^* \oplus T_2^* \oplus T_3^* \oplus T_4^* \oplus T_5^*,$$

with $T_1^*, T_2^*, T_3^*, T_4^*, T_5^*$ indecomposable, one gets

$$T_1^* = {}_0^1\,1\,0\,0, \qquad T_2^* = {}_1^1\,1\,0\,0, \qquad T_3^* = {}_0^1\,1\,1\,1,$$

$$T_4^* = {}_0^1\,0\,0\,0, \qquad T_5^* = {}_0^0\,0\,0\,1.$$

(b) Let B be given by the quiver

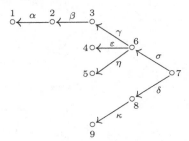

bound by two zero relations $\gamma\beta = 0$ and $\sigma\eta = 0$. Constructing the Auslander–Reiten quiver $\Gamma(\operatorname{mod} B)$ of B as usual yields

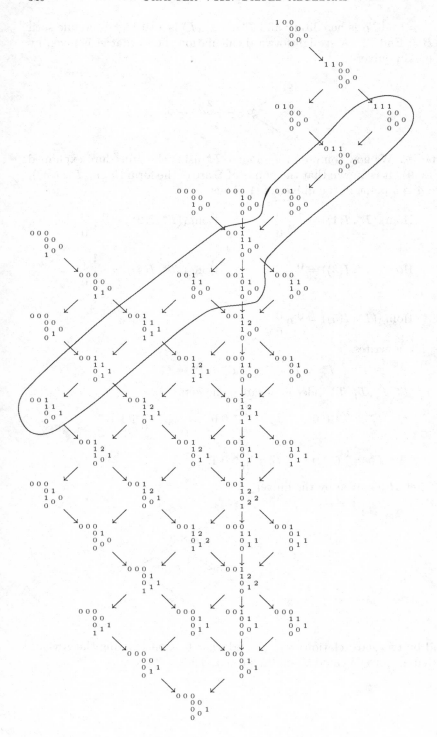

We consider the illustrated section Σ of $\Gamma(\mathrm{mod}\,B)$. Any indecomposable projective B-module is a submodule of a module on Σ, so Σ is faithful. Because $\mathrm{Hom}_B(U, \tau V) = 0$ for all modules U, V on Σ, we have, by (5.6), that B is a representation–finite tilted algebra of type Σ^{op}:

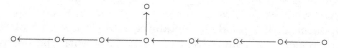

Observe that Σ^{op} is neither a Dynkin nor a Euclidean quiver.

(c) Let B' be given by the quiver

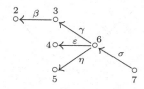

bound by $\gamma\beta = 0$, $\sigma\eta = 0$, and $\sigma\varepsilon = 0$. Thus B' is a quotient of the algebra B of Example (b). Then $\Gamma(\mathrm{mod}\,B')$ is the quiver

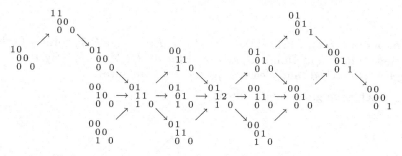

which contains no section. Therefore B' is not tilted. This shows that a quotient algebra of a tilted algebra is not necessarily tilted.

(d) Let B be given by the quiver

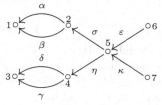

bound by $\sigma\alpha = 0$, $\sigma\beta = 0$, $\eta\gamma = 0$, and $\eta\delta = 0$. Then B is the gluing of three hereditary algebras: B_1 given by the full subquiver with points 1 and

2, B_2 given by the full subquiver with points 3 and 4, and B_3 given by the full subquiver with points 2, 4, 5, 6, and 7.

One can show that if M is an indecomposable B-module, then it is a module over one of the algebras B_1, B_2 and B_3 (see Exercise 14). Because the radical of $P(5)_B$ is equal to $S(2) \oplus S(4)$, $S(2)$ is also a simple injective B_1-module, whereas $S(4)$ is a simple injective B_2-module. We infer that the component $\mathcal{C}$ of $\Gamma(\operatorname{mod} B)$ containing $P(5)$ is a gluing of the preinjective components of $\Gamma(\operatorname{mod} B_1)$ and $\Gamma(\operatorname{mod} B_2)$ with the postprojective component of $\Gamma(\operatorname{mod} B_3)$, that is, $\mathcal{C}$ is of the form

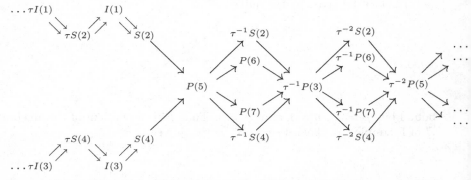

The modules $I(1)$, $S(2)$, $I(3)$, $S(4)$, $P(5)$, $P(6)$, and $P(7)$ form a faithful section Σ in $\mathcal{C}$ and $\operatorname{Hom}_B(U, \tau V) = 0$ for all U, V on Σ. By (5.6), the algebra B is tilted (and clearly representation–infinite).

(e) Let B be given by the quiver

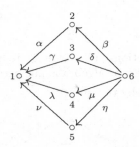

bound by the commutativity relations $\beta\alpha = \delta\gamma = \mu\lambda = \eta\nu$.

Denote by C the hereditary algebra given by the full subquiver with points 1, 2, 3, 4, and 5 and by D the hereditary algebra given by the full subquiver with points 2, 3, 4, 5, and 6.

Finally, let B' denote the algebra with the same quiver as B, bound by $\beta\alpha = \delta\gamma = \mu\lambda = \eta\nu = 0$.

Clearly, B' is a quotient of B and one can show that any indecomposable B'-module is a C-module or a D-module and that any indecomposable B-module not isomorphic to $P(6) \cong I(1)$ is a B'-module (see Exercise 15). By (IV.3.11), we have an almost split sequence of the form

$$0 \longrightarrow \operatorname{rad} P(6) \longrightarrow \operatorname{rad} P(6)/S(1) \oplus P(6) \longrightarrow P(6)/S(1) \longrightarrow 0$$

in the category $\operatorname{mod} B$. There is a decomposition

$$\operatorname{rad} P(6)/S(1) \cong S(2) \oplus S(3) \oplus S(4) \oplus S(5),$$

$\operatorname{rad} P(6)$ is the indecomposable injective C-module $I(1)_C$, whereas $P(6)/S(1)$ is the indecomposable projective D-module $P(6)_D$. Therefore, the component $\mathcal{C}$ of $\Gamma(\operatorname{mod} B)$ containing $P(6) = I(1)$ is the following gluing of the preinjective component of $\Gamma(\operatorname{mod} C)$ with the postprojective component of $\Gamma(\operatorname{mod} D)$:

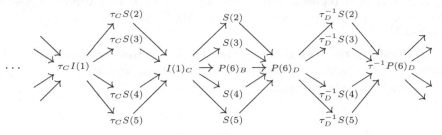

where τ_C and τ_D denote, respectively, the Auslander–Reiten translations in $\operatorname{mod} C$ and $\operatorname{mod} D$. The modules $S(2)$, $S(3)$, $P(6)_B$, $S(4)$, $S(5)$, $P(6)_D$ form a section Σ in $\mathcal{C}$. The indecomposable projective B-modules are submodules of $P(6)_B$ and so Σ is faithful. Because $\operatorname{Hom}_B(U, \tau V) = 0$ for all U, V on Σ, we deduce that B is tilted of type

We observe that $\Gamma(\operatorname{mod} B')$ is obtained from $\Gamma(\operatorname{mod} B)$ by removing $P(6)_B$ and all the arrows with source or target in $P(6)_B$. Thus, $\Gamma(\operatorname{mod} B')$ has a component $\mathcal{C}'$ obtained from $\mathcal{C}$ by removing $P(6)_B$. Moreover, the modules $I(1)_C \cong I(1)_{B'}$, $S(2)$, $S(3)$, $S(4)$, $S(5)$, and $P(6)_{B'} \cong P(6)_D$ form a faithful section Σ' in $\mathcal{C}'$ such that $\operatorname{Hom}_{B'}(U', \tau_{B'} V') = 0$ for all U', V' on Σ'. Therefore, B' is a tilted algebra of type $\Sigma'^{\operatorname{op}} \cong Q_B$.

VIII.6. Exercises

1. Construct $\mathbb{Z}\Sigma$ if Σ is one of the following quivers:

(a)

(b)

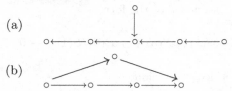

2. Let $\mathcal{C}$ be a component of the Auslander–Reiten quiver of an algebra B, having a faithful section Σ. Show that:

(a) if Σ has finitely many predecessors, then $\mathcal{C}$ is postprojective containing all projective modules and B is a tilted algebra.

(b) if Σ has finitely many successors, then $\mathcal{C}$ is preinjective containing all injective modules and B is a tilted algebra.

3. Let A be a representation–finite algebra and P be an indecomposable projective-injective A-module. Show that P belongs to any section in $\Gamma(\mathrm{mod}\,A)$.

4. Construct the postprojective and the preinjective component of the Auslander–Reiten quiver of each of the following algebras A:

(a) A is given by the quiver

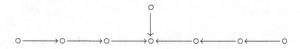

(b) A is given by the quiver

(c) A is given by the quiver

(d) A is given by the quiver

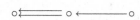

(e) A is given by the quiver

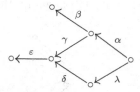

bound by three zero relations $\alpha\beta = 0$, $\gamma\varepsilon = 0$, $\delta\varepsilon = 0$.

(f) A is given by the quiver

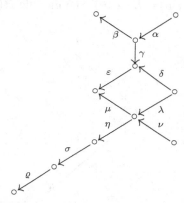

bound by five zero relations $\sigma\varrho = 0$, $\nu\eta\sigma = 0$, $\delta\varepsilon = \lambda\mu$, $\gamma\varepsilon = 0$, $\alpha\beta = 0$.

5. Let A be the Kronecker algebra, and for $\lambda \in K$, let $H_1(\lambda)$ be the indecomposable A-module given by

$$K \xrightarrow[\lambda]{\overset{1}{\longleftarrow}} K$$

where λ denotes the multiplication by λ (see Example 2.11).

(a) Compute a minimal projective presentation for $H_1(\lambda)$ and deduce that $\tau H_1(\lambda) \cong H_1(\lambda)$.

(b) Show that $\operatorname{Ext}^1_A(H_1(\lambda), H_1(\lambda)) \cong K$ and that the canonical short exact sequence

$$0 \longrightarrow H_1(\lambda) \longrightarrow H_2(\lambda) \longrightarrow H_1(\lambda) \longrightarrow 0$$

is almost split, where the module $H_2(\lambda)$ is given by

$$K^2 \xleftarrow[I_{2,\lambda}=\left[\begin{smallmatrix} \lambda & 0 \\ 1 & \lambda \end{smallmatrix}\right]]{\overset{1}{\longleftarrow}} K^2.$$

6. Let A be the hereditary algebra given by the quiver

Show that, for any pair $(\lambda, \mu) \in K^2 \setminus \{0\}$, the module $H(\lambda, \mu)$ given by

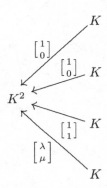

is regular and that $H(\lambda, \mu) \cong H(\lambda', \mu')$ if and only if the pairs (λ, μ) and (λ', μ') correspond to the same point on the projective line $\mathbb{P}_1(K)$.

7. Show that, up to isomorphism, there is only one multiplicity-free tilting module over the Nakayama algebra $A = K[t]/(t^m)$, where $m \geq 2$.

8. Let B be a tilted algebra and M be an indecomposable B-module. Show that $\operatorname{Ext}_B^2(M, M) = 0$.

9. Let B be a concealed algebra, that is, there exists a postprojective tilting module T over a hereditary algebra $A = KQ$ such that $B = \operatorname{End} T_A$. Show that the postprojective component $\mathcal{P}(B)$ of B contains a section isomorphic to Q^{op}. Deduce that there exists a preinjective tilting A-module T' such that $B \cong \operatorname{End} T'_A$.

10. Let A be a representation–finite algebra such that $\Gamma(\operatorname{mod} A)$ is acyclic. Show that A is tilted if and only if $\Gamma(\operatorname{mod} A)$ contains a section.

11. Show that each of the following algebras is a representation–finite tilted algebra.

(a) A given by the quiver

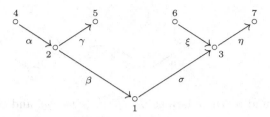

bound by three zero relations $\alpha\gamma = 0$, $\beta\sigma = 0$, $\xi\eta = 0$.

(b) A given by the quiver

(with $n \geq 3$), bound by $\alpha_{n-1}\ldots\alpha_1 = 0$.

(c) A given by the quiver

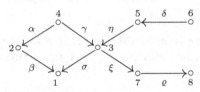

bound by two relations $\alpha\beta = \gamma\sigma$, $\eta\xi = 0$.

(d) A given by the quiver

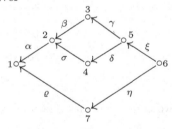

bound by two commutativity relations $\gamma\beta = \delta\sigma$, $\xi\gamma\beta\alpha = \eta\varrho$.

(e) A given by the quiver

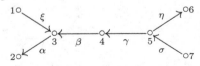

bound by the zero relation $\sigma\gamma\beta\alpha = 0$.

(f) A given by the quiver

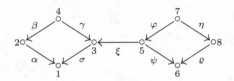

bound by the relations $\beta\alpha = \gamma\sigma$, $\varphi\psi = \eta\varrho$, and $\varphi\xi\sigma = 0$.

(g) A given by the quiver

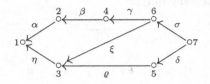

bound by two commutativity relations $\gamma\beta\alpha = \xi\eta$ and $\sigma\xi = \delta\varrho$.

12. Show that each of the following K-algebras B is a tilted algebra. Then compute a hereditary algebra A and a tilting A-module T such that $B \cong \operatorname{End} T_A$.

(a) B given by the quiver

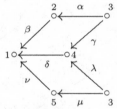

bound by two commutativity relations $\alpha\beta = \gamma\delta$ and $\lambda\delta = \mu\nu$.

(b) B given by the quiver

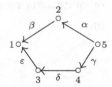

bound by the commutativity relation $\alpha\beta = \gamma\delta\varepsilon$.

(c) B given by the quiver

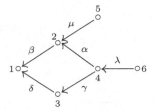

bound by two relations $\alpha\beta = \gamma\delta$ and $\lambda\gamma = 0$.

13. Show that each of the following algebras B is a concealed algebra:

(a) B given by the quiver

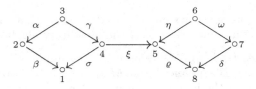

bound by two commutativity relations $\alpha\beta = \gamma\sigma$ and $\eta\varrho = \omega\delta$.

(b) B given by the quiver

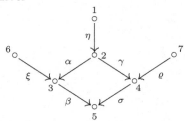

bound by the commutativity relation $\alpha\beta = \gamma\sigma$.

(c) B given by the quiver

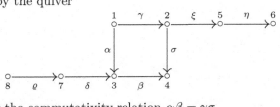

bound by the commutativity relation $\alpha\beta = \gamma\sigma$.

(d) B given by the quiver

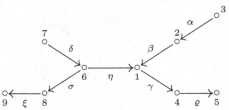

bound by two zero relations $\delta\sigma = 0$ and $\alpha\beta\gamma\varrho = 0$.

14. (a) Let A be given by the quiver $1 \circ \underset{\beta}{\overset{\alpha}{\rightleftarrows}} \overset{2}{\circ} \overset{\gamma}{\leftarrow} \circ 3$ bound by $\gamma\alpha = 0$, $\gamma\beta = 0$. Show that any indecomposable A-module $M = (M_i, \varphi_\alpha)$ with $M_1 \neq 0$ is such that $(\operatorname{Ker}\varphi_\alpha) \cap (\operatorname{Ker}\varphi_\beta) = 0$; deduce that if $M \not\cong S(3)$ and $M \not\cong P(3)$, then M is an indecomposable module over the Kronecker algebra.

(b) Let B be as in Example 5.7 (d). Show that any indecomposable B-module is a module over one of the hereditary algebras B_1, B_2 or B_3.

15. (a) Let B, B', C, D be as in the Example 5.7 (e). Show that any indecomposable B'-module $M = (M_i, \varphi_\alpha)$ such that $M_1 \neq 0$ and $M_6 \neq 0$ must have one of the homomorphisms φ_α, φ_γ, φ_λ, or φ_ν a monomorphism. Deduce that any indecomposable B'-module is a C-module or a D-module.

(b) Show that any indecomposable B-module not isomorphic to $P(6) \cong I(1)$ is a B'-module.

Chapter IX

Directing modules and postprojective components

Let A be an algebra. We studied in Chapter VIII some types of components of the Auslander–Reiten quiver $\Gamma(\mathrm{mod}\,A)$ of A that are acyclic, that is, that contain no cyclic paths, such as the postprojective, the preinjective, and the connecting component of a tilted algebra. We now study more generally those indecomposable modules that lie on no cycle of nonzero nonisomorphisms in the module category. These modules are called directing modules. Although their properties generalise those of modules lying in one of the aforementioned components, they also enjoy some properties of their own. For instance, we show that any algebra having a sincere and directing indecomposable module is a tilted algebra. We next study the class of representation–directed algebras, which are those algebras having the property that each indecomposable module is directing, and we show in particular that these algebras are representation–finite. It is usually difficult to predict whether a given algebra is representation–directed; we give here an easily verified sufficient condition — the so-called separation condition — for an algebra to have a postprojective component and so to be representation–directed whenever it is representation–finite. The last two sections are devoted, respectively, to algebras having the property that all their indecomposable projective modules belong to postprojective components and to the classification of the tilted algebras of type $\mathbb{A}_n$.

IX.1. Directing modules

We recall from (VIII.2) the definitions of path and cycles in a module category. Let A be an algebra. A **path** in $\mathrm{mod}\,A$ is a sequence

$$M_0 \xrightarrow{f_1} M_1 \xrightarrow{f_2} M_2 \longrightarrow \cdots \longrightarrow M_{t-1} \xrightarrow{f_t} M_t$$

of nonzero nonisomorphisms $f_1, \ldots, f_t$ between indecomposable A-modules $M_0, M_1, \ldots M_t$ with $t \geq 1$. We then say that M_0 is a **predecessor** of M_t or that M_t is a **successor** of M_0. A path in $\mathrm{mod}\,A$ is called a **cycle** if its source module M_0 is isomorphic with its target M_t. An indecomposable A-module that lies on no cycle in $\mathrm{mod}\,A$ is called a **directing module**.

Clearly, the requirement that the $f_1, \ldots, f_t$ are nonzero nonisomorphisms amounts to say that they belong to

$$\mathrm{rad}_A = \mathrm{rad}_{\mathrm{mod}\,A},$$

the radical of the category $\operatorname{mod} A$ (see Section A.3 of the Appendix). Because the arrows of $\Gamma(\operatorname{mod} A)$ represent irreducible morphisms, any path between points in $\Gamma(\operatorname{mod} A)$ induces a path in $\operatorname{mod} A$. The converse, however, is generally not true; indeed, the f_i may map between indecomposables lying in distinct components of $\Gamma(\operatorname{mod} A)$.

Our first lemma provides examples of directing modules.

1.1. Lemma. (a) *Let A be an algebra and C be a postprojective or preinjective component of $\Gamma(\operatorname{mod} A)$. Then every indecomposable A-module in C is directing.*

(b) *Let H be a hereditary algebra, T be a tilting H-module, $A = \operatorname{End} T_H$, and C_T be the connecting component of $\Gamma(\operatorname{mod} A)$ determined by T. Then every indecomposable A-module in C_T is directing.*

(c) *Let A be a representation–finite hereditary or tilted algebra. Then every indecomposable A-module is directing.*

Proof. (a) This is just (VIII.2.6).

(b) Let M_A be an indecomposable in C_T and suppose, to the contrary, that there exists a cycle

$$M = M_0 \xrightarrow{f_1} M_1 \xrightarrow{f_2} M_2 \longrightarrow \cdots \longrightarrow M_{t-1} \xrightarrow{f_t} M_t = M,$$

where $t \geq 1$, the homomorphisms $f_1, \ldots, f_t$ are nonzero nonisomorphisms, and the modules M_i are indecomposable. By (VIII.3.5), C_T contains a finite section Σ such that all predecessors of Σ belong to the torsion-free part $\mathcal{Y}(T)$, and all its proper successors belong to the torsion part $\mathcal{X}(T)$. Moreover, C_T is acyclic. Then there exists i such that $1 \leq i \leq t$ and there is no path of irreducible morphisms from M_{i-1} to M_i.

Let r be the least integer such that $1 \leq r \leq t$ and there is no path of irreducible morphisms from M_{r-1} to M_r. Then $M = M_0, \ldots, M_{r-1}$ belong to C_T. Now (IV.5.1) yields a chain of irreducible morphisms

$$M_{r-1} = U_0 \to U_1 \to \cdots \to U_p$$

such that U_p is a proper successor of Σ in C_T and $\operatorname{rad}_A(U_p, M_r) \neq 0$. In particular, $U_p \in \mathcal{X}(T)$. Because $\mathcal{X}(T)$ is closed under successors, we have $M_r \in \mathcal{X}(T)$ and consequently $M = M_t \in \mathcal{X}(T)$. Similarly, let s be the maximal integer such that $1 \leq s \leq t$ and there is no path of irreducible morphisms from M_{s-1} to M_s in $\operatorname{mod} A$. Then the modules $M_s, \ldots, M_t = M$ belong to C_T and there is a chain of irreducible morphisms $V_q \to \cdots \to V_1 \to V_0 = M_s$ such that V_q is a predecessor of Σ in C_T and $\operatorname{rad}_A(M_{s-1}, V_q) \neq 0$. In particular, $V_q \in \mathcal{Y}(T)$. Because $\mathcal{Y}(T)$ is closed under predecessors, we have $M_{s-1} \in \mathcal{Y}(T)$ and consequently $M = M_0 \in \mathcal{Y}(T)$. Therefore $M \in \mathcal{X}(T) \cap \mathcal{Y}(T)$, a contradiction. Hence M is directing.

(c) This follows easily from (b). $\square$

It is important to observe that, although in Lemma 1.1, all Auslander–Reiten components considered are acyclic, there exist examples of directing modules lying in components containing cyclic paths.

1.2. Example. Consider the algebra A given by the quiver

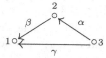

bound by $\alpha\beta = 0$ (see Example IV.4.14). Then $\Gamma(\mathrm{mod}\,A)$ is given by

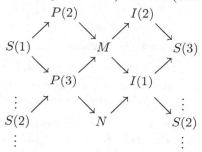

where $M = (P(2) \oplus P(3))/S(1)$, $N = P(3)/S(2)$, and we identify the two copies of $S(2)$ along the dotted lines. Clearly, $S(1)$, $P(2)$, $I(2)$, and $S(3)$ are directing, but none of the other indecomposable modules is.

We now look at the support of a directing module. Let $A = KQ_A/\mathcal{I}$ be a bound quiver presentation of an algebra A and M be an A-module. The **support** of M is the full subquiver supp M of Q_A generated by all the points $i \in (Q_A)_0$ such that $(\mathbf{dim}\,M)_i \neq 0$ (equivalently, such that $\mathrm{Hom}_A(P(i), M) \neq 0$). An indecomposable A-module M is called **sincere** whenever its support equals Q_A (thus, for instance, any faithful A-module is clearly sincere).

Observe that if e_j denotes the primitive idempotent corresponding to $j \in (Q_A)_0$ and $e = \sum_{j \notin (\mathrm{supp}\,M)_0} e_j$, then M is sincere viewed as a module over the algebra A/AeA, called the **support algebra** of M.

We recall from (VIII.1) that to say that supp M is a convex subquiver of Q_A means that any path in Q_A having its source and its target in supp M lies entirely in supp M.

1.3. Proposition. *Let $A = KQ_A/\mathcal{I}$ and M_A be a directing indecomposable A-module. Then the support supp M of M is a convex subquiver of Q_A.*

Proof. Assume to the contrary that supp M is not convex. Then there exists a path $a_0 \xrightarrow{\alpha_1} a_1 \xrightarrow{\alpha_2} \cdots \xrightarrow{\alpha_m} a_m$ in Q_A such that $m \geq 2$, $a_0, a_m \in$

$(\operatorname{supp} M)_0$ but $a_1, \ldots, a_{m-1} \notin (\operatorname{supp} M)_0$. Let $\alpha_1 = \beta_1, \ldots, \beta_s$ be all the arrows in Q_A from a_0 to a_1 and $\alpha_m = \gamma_1, \ldots, \gamma_t$ be all the arrows in Q_A from a_{m-1} to a_m. Let J be the two-sided ideal of KQ_A generated by all paths of the form $\beta_i \delta$ or $\delta \gamma_j$, with $\delta \in (Q_A)_1$, $1 \le i \le s$, $1 \le j \le t$. Consider the algebra $A' = KQ_A/(I + J)$. Because M is annihilated by J, it is an A'-module. Moreover, $\operatorname{Hom}_{A'}(P(a_0)_{A'}, M) \ne 0$ and $\operatorname{Hom}_{A'}(M, I(a_m)_{A'}) \ne 0$. For any r with $1 \le r \le m - 1$, let U_r denote a uniserial A'-module of length two having $S(a_r)$ as top and $S(a_{r+1})$ as socle. Then there exists a path in the category $\operatorname{mod} A'$

$$I(a_m)_{A'} \longrightarrow S(a_{m-1}) \longrightarrow U_{m-2} \longrightarrow S(a_{m-2}) \longrightarrow$$

$$\cdots \longrightarrow S(a_2) \longrightarrow U_1 \longrightarrow S(a_1) \longrightarrow P(a_0)_{A'},$$

where the homomorphisms are the obvious ones. Therefore we get a cycle

$$M \longrightarrow I(a_m)_{A'} \longrightarrow S(a_{m-1}) \longrightarrow \cdots \longrightarrow S(a_1) \longrightarrow P(a_0)_{A'} \longrightarrow M$$

in $\operatorname{mod} A'$, hence also in $\operatorname{mod} A$, because A' is a quotient of A. This contradicts the hypothesis that M is directing and finishes the proof. $\square$

1.4. Proposition. *Let A be an algebra and M be a directing indecomposable A-module. Then $\operatorname{End} M \cong K$ and $\operatorname{Ext}_A^j(M, M) = 0$ for all $j \ge 1$.*

Proof. Because M is directing and indecomposable, $\operatorname{mod} A$ contains no cycle of the form $M \to M$; hence $\operatorname{rad} \operatorname{End}_A M = \operatorname{rad}_A(M, M) = 0$, and so $\operatorname{End}_A M \cong K$. Denote by $\mathcal{U}$ the class of all predecessors of M in $\operatorname{mod} A$. We show by induction on $j \ge 1$ that $\operatorname{Ext}_A^j(U, M) = 0$ for all U in $\mathcal{U}$. This will clearly imply our claim. Assume $j = 1$. If $0 \ne \operatorname{Ext}_A^1(U, M) \cong D\overline{\operatorname{Hom}}_A(M, \tau U)$ for some U in $\mathcal{U}$, there exists a nonzero homomorphism $M \to \tau U$, hence a cycle $M \to \tau U \to * \to U \to \ldots \to M$, and we get a contradiction. Therefore $\operatorname{Ext}_A^1(U, M) = 0$ for all U in $\mathcal{U}$. Assume that $\operatorname{Ext}_A^j(U, M) = 0$ for some $j \ge 1$ and all U in $\mathcal{U}$. Take U in $\mathcal{U}$ and a short exact sequence $0 \longrightarrow V \longrightarrow P \longrightarrow U \longrightarrow 0$ with P projective. By (A.4.5) of the Appendix, we have $\operatorname{Ext}_A^{j+1}(U, M) \cong \operatorname{Ext}_A^j(V, M)$, for $j \ge 1$, and the latter vanishes, because all indecomposable summands of V belong to $\mathcal{U}$. This finishes the proof. $\square$

1.5. Corollary. *Let A be an algebra of finite global dimension and M be a directing indecomposable A-module. Then $\dim M$ is a positive root of the Euler quadratic form q_A of A.*

Proof. It follows from (1.4) and (III.3.13) that

$$q_A(\dim M) = \sum_{i \ge 0} (-1)^i \dim_K \operatorname{Ext}_A^i(M, M) = 1. \qquad \square$$

IX.2. Sincere directing modules

In this section, we show that any algebra having a sincere directing module is tilted. Further, we show how to construct a faithful section in the Auslander–Reiten quiver of such an algebra. For this purpose, we need a definition: A path $M_0 \to \cdots \to M_t$ in the Auslander–Reiten quiver of an algebra A is called **sectional** if, for all i with $1 < i \le t$, we have $\tau M_i \not\cong M_{i-2}$. Clearly, if all the M_i belong to a section in $\Gamma(\mathrm{mod}\,A)$, then each such path between the M_i is sectional. Our first proposition due to Bautista and Smalø [28] (see also [35]) says that if the composition of the irreducible morphisms corresponding to a path in $\Gamma(\mathrm{mod}\,A)$ vanishes, then this path cannot be sectional.

2.1. Proposition. *Let A be an algebra, $M_1, \ldots, M_{n+1}$ be indecomposable A-modules, and $f_i : M_i \to M_{i+1}$, $1 \le i \le n$, be irreducible morphisms. If the composition $f_n \ldots f_1$ either equals zero or there is a commutative diagram*

$$M_1 \xrightarrow{\ f_1\ } M_2 \xrightarrow{\ f_2\ } \cdots \longrightarrow M_n \xrightarrow{\ f_n\ } M_{n+1}$$

$$\searrow{\scriptstyle h} \qquad\qquad\qquad \nearrow{\scriptstyle g}$$

$$N$$

where N is an indecomposable module not isomorphic to M_n, $h : M_1 \to N$ is a homomorphism of A-modules and $g : N \to M_{n+1}$ is an irreducible morphism, then there exists l such that $3 \le l \le n + 1$ and $\tau M_l \cong M_{l-2}$.

Proof. We use induction on n. Assume $n = 1$. Because f_1 is irreducible, then $f_1 \ne 0$ and $f_1 = gh$, with $g : N \to M_2$ irreducible. It follows that h is a section. Because N and M_2 are indecomposable, h is an isomorphism. This contradicts our hypothesis that $N \not\cong M_1$.

Assume $n > 1$ and let $f = f_{n-1} \ldots f_1$. Consider first the case where $f_n f = 0$. If $f = 0$, the result follows from the induction hypothesis. If $f \ne 0$, then f_n is not a monomorphism, so it is an epimorphism. Hence the module M_{n+1} is not projective and there exists an almost split sequence of the form

$$0 \longrightarrow \tau M_{n+1} \xrightarrow{\ \left[\begin{smallmatrix} f'_n \\ l' \end{smallmatrix}\right]\ } M_n \oplus L \xrightarrow{\ [f_n\ l]\ } M_{n+1} \longrightarrow 0$$

with f'_n and l' irreducible. Applying $\mathrm{Hom}_A(M_1, -)$ yields a left exact sequence

$$0 \longrightarrow \mathrm{Hom}_A(M_1, \tau M_{n+1}) \xrightarrow{\ \mathrm{Hom}_A\left(M_1, \left[\begin{smallmatrix} f'_n \\ l' \end{smallmatrix}\right]\right)\ } \mathrm{Hom}_A(M_1, M_n \oplus L)$$

$$\xrightarrow{\ \mathrm{Hom}_A(M_1, [f_n\ l])\ } \mathrm{Hom}_A(M_1, M_{n+1}).$$

Because $f_n f = 0$, we have $\begin{bmatrix} f \\ 0 \end{bmatrix} \in \operatorname{Ker} \operatorname{Hom}_A(M_1, [f_n \ l])$; hence there exists $k : M_1 \to \tau M_{n+1}$ such that $f = f_n' k$. If $M_{n-1} \cong \tau M_{n+1}$,and we are done. Otherwise, the irreducibility of $f_n' : \tau M_{n+1} \to M_n$ yields the result by the induction hypothesis applied to $N = \tau M_{n+1}$, $g = f_n'$ and to

$$M_1 \xrightarrow{f_1} M_2 \xrightarrow{f_2} \cdots \longrightarrow M_{n-1} \xrightarrow{f_{n-1}} M_n \xrightarrow{f_n} M_{n+1}$$

with arrows k and f_n' into τM_{n+1}.

This finishes the proof in case $f_n f = 0$. Assume now $f_n f = gh \neq 0$, with $g : N \to M_{n+1}$ irreducible and N indecomposable not isomorphic to M_n. We claim that M_{n+1} is not projective. Assume to the contrary that M_{n+1} is projective. Then the irreducible morphisms f_n and g are not epimorphisms; hence they are monomorphisms. Because $N \not\cong M_n$ and $\operatorname{rad} M_{n+1}$ is the unique maximal submodule of M_{n+1}, then the modules $\operatorname{Im} f_n$ and $\operatorname{Im} g$ are distinct direct summands of $\operatorname{rad} M_{n+1}$ and therefore $\operatorname{Im} f_n \cap \operatorname{Im} g = 0$. On the other hand, the relation $f_n f = gh$ implies $\operatorname{Im} f_n \cap \operatorname{Im} g \neq 0$, and we get a contradiction. Consequently, M_{n+1} is not projective.

Because f_n and g are irreducible, $N \not\cong M_n$ and M_{n+1} is not projective, then there exists an almost split sequence of the form

$$0 \longrightarrow \tau M_{n+1} \xrightarrow{\begin{bmatrix} f_n' \\ g' \\ l' \end{bmatrix}} M_n \oplus N \oplus L \xrightarrow{[f_n \ g \ l]} M_{n+1} \longrightarrow 0.$$

Applying $\operatorname{Hom}_A(M_1, -)$ yields a left exact sequence

$$0 \longrightarrow \operatorname{Hom}_A(M_1, \tau M_{n+1}) \xrightarrow{\operatorname{Hom}_A\left(M_1, \begin{bmatrix} f_n' \\ g' \\ l' \end{bmatrix}\right)} \operatorname{Hom}_A(M_1, M_n \oplus N \oplus L)$$

$$\xrightarrow{\operatorname{Hom}_A(M_1, [f_n \ g \ l])} \operatorname{Hom}_A(M_1, M_{n+1}).$$

Because $f_n f = gh$, we have $\begin{bmatrix} f \\ -h \\ 0 \end{bmatrix} \in \operatorname{Ker} \operatorname{Hom}_A(M_1, [f_n \ g \ l])$. Hence there exists $k : M_1 \to \tau M_{n+1}$ such that $f = f_n' k$. If $M_{n-1} \cong \tau M_{n+1}$, we are done. Otherwise, the irreducibility of $f_n' : \tau M_{n+1} \to M_n$ yields the result by the induction hypothesis applied to $N = \tau M_{n+1}$ and $g = f_n'$. □

A first, easy, important consequence of (2.1) is the following fact, mentioned earlier.

2.2. Corollary. *Let A be an algebra. If $M_1 \xrightarrow{f_1} M_2 \xrightarrow{f_2} \cdots \xrightarrow{f_{t-1}} M_t$ is a path of irreducible morphisms corresponding to a sectional path in $\Gamma(\operatorname{mod} A)$, then $f_{t-1} \ldots f_1 \neq 0$.* □

A second consequence of (2.1) is that no sectional path is a cycle.

2.3. Corollary. *Let A be an algebra. If $M_1 \xrightarrow{f_1} M_2 \xrightarrow{f_2} \cdots \xrightarrow{f_{t-1}} M_t$ is a sectional path in $\Gamma(\text{mod } A)$, then $M_1 \not\cong M_t$.*

Proof. Assume to the contrary that $M_1 \cong M_t$. By (2.2), $f = f_{t-1} \cdots f_1$ is a nonzero endomorphism of M_1, which is not an isomorphism (because the homomorphisms $f_1, \ldots, f_{t-1}$ are irreducible). Because $\text{End } M_1$ is local, f is nilpotent. But then the given sectional cycle induces a longer one in which the composition of the homomorphisms is zero, a contradiction to (2.2). □

We now proceed to the proof of our main result. We need two lemmas.

2.4. Lemma. *Let A be an algebra and M be a directing indecomposable A-module. Let $f : P \to P'$ be a nonzero homomorphism between indecomposable projective A-modules. Then the induced homomorphism*

$$\text{Hom}_A(f, M) : \text{Hom}_A(P', M) \to \text{Hom}_A(P, M)$$

is either a monomorphism or an epimorphism.

Proof. Assume to the contrary that $\text{Hom}_A(f, M)$ is neither a monomorphism nor an epimorphism, and set $U = \text{Coker } f$. Because $\text{Hom}_A(f, M)$ is not a monomorphism, $\text{Hom}_A(U, M) \cong \text{Ker } \text{Hom}_A(f, M) \neq 0$. For an indecomposable projective A-module eA, we have functorial isomorphisms $\text{Hom}_A((eA)^t, DM) \cong \text{Hom}_A(Ae, DM) \cong eDM \cong D(Me) \cong D\text{Hom}_A(eA, M)$, where, as usual, $(-)^t = \text{Hom}_A(-, A)$. It follows that the diagram

$$
\begin{array}{ccc}
D\text{Hom}_A(P, M) & \xrightarrow{\ D\text{Hom}_A(f,M)\ } & D\text{Hom}_A(P', M) \\
\cong \downarrow & & \cong \downarrow \\
\text{Hom}_A(P^t, DM) & \xrightarrow{\ \text{Hom}_A(f^t,DM)\ } & \text{Hom}_A(P'^t, DM)
\end{array}
$$

is commutative. Because the linear map $\text{Hom}_A(f, M)$ is not an epimorphism, $D\text{Hom}_A(f, M)$ and $\text{Hom}_A(f^t, DM)$ are not monomorphisms. Consequently, $\text{Hom}_A(\text{Coker } f^t, DM) \cong \text{Ker } \text{Hom}_A(f^t, DM) \neq 0$. However, P and P' are indecomposable projective A-modules; hence $P \xrightarrow{f} P' \to U \to 0$ is a minimal projective presentation, so that $\text{Coker } f^t = \text{Tr } U$. Hence we get $\text{Hom}_A(\text{Tr } U, DM) \neq 0$ and therefore $\text{Hom}_A(M, \tau U) \neq 0$. We know that $\text{Hom}_A(U, M) \neq 0$. Also U, being a quotient of P', has a simple top and hence is indecomposable. We deduce the existence of a cycle $M \to \tau U \to * \to U \to M$ in $\text{mod } A$, contrary to the assumed directedness of M. □

As we observed before, any faithful module is sincere (for example, any tilting module is sincere). We next show a partial converse of this statement.

2.5. Lemma. *Let A be an algebra. Then any sincere and directing indecomposable A-module is faithful.*

Proof. Let M be a sincere and directing indecomposable A-module, and let $e_1, \ldots, e_n$ be a complete set of primitive orthogonal idempotents of A. Suppose to the contrary that the right annihilator $R = \{a \in A \mid Ma = 0\}$ of M is nonzero. Then there exist i, j such that $1 \le i, j \le n$ and $e_i R e_j \ne 0$. Let $x \in R$ be such that $e_i x e_j \ne 0$. Because $e_i A e_j \cong \operatorname{Hom}_A(e_j A, e_i A)$, the element $e_i x e_j$ induces a nonzero homomorphism $f_x : e_j A \to e_i A$, $e_j a \mapsto (e_i x e_j) e_j a$, for $a \in A$. Because M is sincere, $M e_i \cong \operatorname{Hom}_A(e_i A, M) \ne 0$ and $M e_j \cong \operatorname{Hom}_A(e_j A, M) \ne 0$. But our choice of x guarantees that $\operatorname{Hom}_A(f_x, M) = 0$, so that $\operatorname{Hom}_A(f_x, M)$ is neither a monomorphism nor an epimorphism, a contradiction to (2.4). Consequently, M is faithful. $\square$

We are now able to prove the main result of this section due to Ringel [145].

2.6. Theorem. *Let A be an algebra having a sincere and directing indecomposable module M.*

   (a) *If $\mathcal{C}$ is a component of $\Gamma(\operatorname{mod} A)$ containing M, then $\mathcal{C}$ contains a faithful section Σ containing M.*

   (b) *A is a tilted algebra.*

Proof. Let M be a sincere and directing indecomposable A-module and $\mathcal{C}$ be the component of $\Gamma(\operatorname{mod} A)$ containing M. Let Σ denote the full subquiver of $\mathcal{C}$ consisting of all the successors U of M in $\mathcal{C}$ having the property that every path from M to U in $\operatorname{mod} A$ is sectional (that is, there exists no path of the form $M \to \cdots \to \tau W \to * \to W \to \cdots \to U$, in $\operatorname{mod} A$ with W indecomposable). Because M is directing, M itself belongs to Σ. Further, for any $U, V \in \Sigma_0$ we have $\operatorname{Hom}_A(U, \tau V) = 0$; indeed, a nonzero homomorphism from U to τV yields a path $M \to \cdots \to U \to \tau V \to * \to V$, a contradiction to the assumption that $V \in \Sigma_0$. Next, by (2.5), M is faithful. We prove that Σ is a section of $\mathcal{C}$; then applying (VIII.5.6) will complete the proof.

We notice that Σ has the following property: If there exists a path

$$M \to \cdots \to N \to \cdots \to U$$

with $N \in \mathcal{C}_0$ and $U \in \Sigma_0$, then $N \in \Sigma_0$. If this is not the case, then there exists a nonsectional path $M \to \cdots \to \tau W \to * \to W \to \cdots \to N$; hence, by composition, a nonsectional path from M to U, which is a contradiction. This implies that Σ is convex: If $U_0 \to \cdots \to U_t$ is a path with $U_0, U_t \in \Sigma_0$, then there exists a path $M \to \cdots \to U_0 \to \cdots \to U_t$ so that all U_i belong to Σ.

We claim that Σ is acyclic. If not, then there exists a cycle

$$U_0 \to U_1 \to \cdots \to U_r = U_0$$

in Σ. Because all these modules lie in Σ, we have $\tau U_i \ncong U_j$ for all i, j. Consequently, this is a sectional cycle, a contradiction to (2.3).

It clearly follows from the definition of Σ that it contains at most one module from each τ-orbit in $\mathcal{C}$. We claim that Σ intersects each τ-orbit in $\mathcal{C}$. Suppose that this is not the case. Because $\mathcal{C}$ is a connected translation quiver, there exist modules $U, V \in \mathcal{C}_0$ such that $U \in \Sigma_0$, the τ-orbit of V does not intersect Σ but U and V have neighbouring orbits; that is, there exist $p, q \in \mathbb{Z}$ and an arrow $\tau^p U \to \tau^q V$ or an arrow $\tau^q V \to \tau^p U$. If $p \le 0$, then $\tau^p U$ cannot precede any indecomposable projective module $P \in \mathcal{C}_0$; indeed, if this is the case, then U itself precedes P, and the sincerity of M yields a cycle $M \to \cdots \to U \to \cdots \to P \to M$, contrary to the assumption that M is directing. Similarly, if $p \ge 1$, then $\tau^p U$ cannot succeed any indecomposable injective module $I \in \mathcal{C}_0$. Indeed, if this is the case, then U itself succedes I, and the sincerity of M yields a path $M \to I \to \cdots \to \tau^p U \to * \to \tau^{p-1} U \to \cdots \to U$, a contradiction, because $U \in \Sigma_0$.

It is easily shown that these two remarks imply the existence of an arrow $N = \tau^l V \to U$. It follows from our assumption that $N \notin \Sigma_0$. Because N precedes $U \in \Sigma_0$, we have no path $M \to \cdots \to N$. In particular, N is not injective, because M is sincere. Now the arrow $U \to \tau^{-1} N$ induces a path $M \to \cdots \to U \to \tau^{-1} N$. Our assumption implies that $\tau^{-1} N \notin \Sigma_0$; hence there exist an indecomposable L_A and a path

$$M \longrightarrow \cdots \longrightarrow \tau L \longrightarrow * \longrightarrow L = L_0 \longrightarrow L_1 \longrightarrow \cdots \longrightarrow L_t = \tau^{-1} N.$$

We have $\operatorname{Hom}_A(L_i, A) = 0$ for all i with $0 \le i \le t$. Indeed, if this is not the case, then there exist an indecomposable projective A-module P' and an i such that $0 \le i \le t$ and $\operatorname{Hom}_A(L_i, P') \ne 0$, and the sincerity of M yields $\operatorname{Hom}_A(P', M) \ne 0$, hence a cycle

$$M \longrightarrow \cdots \longrightarrow \tau L \longrightarrow * \longrightarrow L = L_0 \longrightarrow L_1 \longrightarrow \cdots \longrightarrow L_i \longrightarrow P' \longrightarrow M,$$

which is a contradiction. This implies that, for any i such that $1 \le i \le t$, we have $0 \ne \operatorname{Hom}_A(L_{i-1}, L_i) = \underline{\operatorname{Hom}}_A(L_{i-1}, L_i)$ and so

$$0 \ne D\operatorname{Hom}_A(L_{i-1}, L_i) = D\underline{\operatorname{Hom}}_A(L_{i-1}, L_i) \cong \operatorname{Ext}_A^1(L_i, \tau L_{i-1})$$

$$\cong D\overline{\operatorname{Hom}}_A(\tau L_{i-1}, \tau L_i) \subseteq D\operatorname{Hom}_A(\tau L_{i-1}, \tau L_i).$$

We thus deduce the existence of a sequence of nonzero homomorphisms

$$\tau L = \tau L_0 \longrightarrow \tau L_1 \longrightarrow \cdots \longrightarrow \tau L_t = N;$$

hence a path $M \to \cdots \to \tau L = \tau L_0 \to \tau L_1 \to \cdots \to \tau L_t = N$, which is a contradiction. This shows that Σ intersects any τ-orbit in $\mathcal{C}$ and thus is a section. Because Σ is faithful, according to (VIII.5.6), A is a tilted algebra. $\square$

Dually, one can show that (with the same hypothesis and notation) the full subquiver of $\mathcal{C}$ consisting of all the predecessors V of M having the property that every path from V to M is sectional, is a faithful section in $\mathcal{C}$, to which we can apply (VIII.5.6).

The converse of (2.6) is clearly not true. For instance, the algebra given by the quiver $\underset{\circ}{1} \xleftarrow{\quad\beta\quad} \underset{\circ}{2} \xleftarrow{\quad\alpha\quad} \underset{\circ}{3}$ bound by $\alpha\beta = 0$ is tilted but has no sincere indecomposable module.

2.7. Corollary. *Let A be an algebra and M be a sincere and directing indecomposable A-module. Then* $\mathrm{gl.dim}\, A \leq 2$, $\mathrm{pd}\, M \leq 1$ *and* $\mathrm{id}\, M \leq 1$.

Proof. Because A is a tilted algebra, (VIII.3.2) yields $\mathrm{gl.dim}\, A \leq 2$. Moreover, we have $\mathrm{Hom}_A(I, \tau M) = 0$ for every indecomposable injective A-module I. Indeed, $\mathrm{Hom}_A(M, I) \neq 0$ and $\mathrm{Hom}_A(I, \tau M) \neq 0$ yield a cycle $M \to I \to \tau M \to * \to M$, a contradiction. Consequently, $\mathrm{pd}\, M \leq 1$ by (IV.2.7). Dually, $\mathrm{id}\, M \leq 1$. □

The next corollary asserts that if M is a directing indecomposable A-module, then there exists a tilted algebra B (which is a quotient algebra of A) such that M is a sincere and directing B-module. Thus the structure of the directing modules over any algebra A is completely determined by those over the tilted quotients of A.

2.8. Corollary. *Let A be an algebra and M be a directing indecomposable A-module. Then the support algebra B of M is tilted.*

Proof. Clearly, M is a sincere and indecomposable B-module. Also, Because B is a quotient of A, a cycle in $\mathrm{mod}\, B$ induces a cycle in $\mathrm{mod}\, A$, so M is a directing B-module. Applying (2.6) yields that B is tilted. □

IX.3. Representation–directed algebras

In this section we study the algebras having the property that every indecomposable module is directing. However, we start with a more general result asserting that directing modules (over an arbitrary algebra) are uniquely determined by their composition factors.

3.1. Proposition. *Let A be an algebra and M, N be indecomposable A-modules. If M is directing and $\mathbf{dim}\, M = \mathbf{dim}\, N$, then $M \cong N$.*

Proof. Let B be the support algebra of M. It follows from (2.7) that $\mathrm{gl.dim}\, B \leq 2$. In particular, the Euler characteristic $\langle -, - \rangle_B$ of B is defined (see III.3.11). Moreover, because M is sincere when viewed as a B-module, $\mathrm{pd}\, M_B \leq 1$ and $\mathrm{id}\, M_B \leq 1$, again by (2.7). Finally, by (1.5), $\mathbf{dim}\, M$ is a root of the quadratic form q_B, because M is indecomposable and directing (when viewed as a B-module).

Assume, to the contrary, that $M \not\cong N$ and $\dim M = \dim N$. Clearly, B is also the support algebra of N. Because $\text{pd}\, M_B \leq 1$, $\text{Ext}^2_B(N, M) = 0$ and, according to (III.3.13) and (1.5), we have

$$1 = q_B(\dim M) = \langle \dim M, \dim M \rangle_B = \langle \dim M, \dim N \rangle_B$$
$$= \dim_K \text{Hom}_B(M, N) - \dim_K \text{Ext}^1_B(M, N);$$

hence $\text{Hom}_A(M, N) = \text{Hom}_B(M, N) \neq 0$. Similarly, $\text{id}\, M_B \leq 1$ implies that $\text{Ext}^2_B(N, M) = 0$. It follows that

$$1 = q_B(\dim M) = \langle \dim M, \dim M \rangle_B = \langle \dim N, \dim M \rangle_B$$
$$= \dim_K \text{Hom}_B(N, M) - \dim_K \text{Ext}^1_B(N, M);$$

hence $\text{Hom}_A(N, M) = \text{Hom}_B(N, M) \neq 0$. This gives a cycle $M \to N \to M$ in $\text{mod}\, A$, contrary to the assumption that M is directing. Consequently, there is an isomorphism $M \cong N$ of A-modules. $\square$

The hypothesis in (3.1) that M is directing is essential; as is shown by Example (1.2), the indecomposable modules $P(3)$ and $I(1)$ have the same composition factors but are clearly not isomorphic.

Proposition 3.1 and Lemma 1.1 imply that all postprojective and all preinjective indecomposable modules as well as all indecomposables that belong to the connecting component of a tilted algebra are uniquely determined by their composition factors.

We saw in (VIII.4.3) that the Auslander–Reiten quiver of any representation–finite tilted algebra is acyclic and, consequently, any indecomposable module is directing. On the other hand, the Example VIII.5.7 (c) shows that there exist representation–finite algebras with acyclic Auslander–Reiten quivers that are not tilted. This motivates the following definition.

3.2. Definition. An algebra is called **representation–directed** if every indecomposable A-module is directing.

We recall that Gabriel's theorem (VII.5.10) provides a bijection between the indecomposable modules over a representation–finite hereditary algebra and the roots of the corresponding quadratic form. The same result holds more generally for a representation–directed algebra with global dimension at most two.

3.3. Theorem. *Let A be a representation–directed K-algebra with* gl.dim $A \leq 2$. *The Euler quadratic form q_A of A is weakly positive, and the correspondence $M \mapsto \dim M$ defines a bijection between the isomorphism classes of indecomposable A-modules and the positive roots of q_A.*

Proof. Because A is representation–directed, every indecomposable A-module M is directing and, according to (1.5), the dimension vector $\dim M$ of M is a positive root of q_A.

Let $\mathbf{x}$ be a positive vector in $K_0(A)$. Then there exists a nonzero A-module M such that $\mathbf{x} = \dim M$. Choose such a module M with $\dim_K(\operatorname{End} M)$ as small as possible. Let $M = \bigoplus_{i=1}^m M_i$ be a decomposition of M into indecomposable summands. We claim that $\operatorname{Ext}^1_A(M_j, M_i) = 0$ for any pair (i, j) with $i \neq j$. Suppose that this is not the case. Then $\operatorname{Ext}^1_A(\bigoplus_{j \neq i} M_j, M_i) \neq 0$ for some i and therefore there exists a nonsplit exact sequence

$$0 \longrightarrow M_i \longrightarrow N \longrightarrow \bigoplus_{j \neq i} M_j \longrightarrow 0.$$

It follows that $\dim N = \dim(M_i \oplus (\bigoplus_{j \neq i} M_j)) = \dim M$. By (VIII.2.8), we get $\dim_K \operatorname{End}_A N < \dim_K \operatorname{End}_A(M_i \oplus \bigoplus_{j \neq i} M_j) = \dim_K \operatorname{End}_A M$, which contradicts the minimality of M. Consequently, $\operatorname{Ext}^1_A(M_j, M_i) = 0$ whenever $i \neq j$.

Because each M_i is directing, we also have $\operatorname{Ext}^1_A(M_i, M_i) = 0$ for any i, by (1.4). Therefore $\operatorname{Ext}^1_A(M, M) = 0$ and, because gl.dim $A \leq 2$, we have

$$q_A(\mathbf{x}) = q_A(\dim M) = \dim_K \operatorname{End} M + \dim_K \operatorname{Ext}^2_A(M, M) > 0.$$

Thus, q_A is weakly positive. Moreover, if $\mathbf{x} = \dim M$ is a positive root of q_A, then $1 = \dim_K \operatorname{End} M + \dim_K \operatorname{Ext}^2_A(M, M)$. It follows that $\operatorname{End} M \cong K$ and M is indecomposable.

Also, if M, N are indecomposable A-modules such that $\dim M = \dim N$, then (3.1) implies $M \cong N$. Hence, in view of (1.5), $M \mapsto \dim M$ establishes a bijection between the set of isomorphism classes of indecomposable A-modules and the set of positive roots of q_A. $\square$

3.4. Corollary. *Any representation–directed algebra is representation–finite.*

Proof. Assume that A is a representation–directed algebra. Let $A = KQ_A/\mathcal{I}$ be a bound quiver presentation of A, and let M be an indecomposable A-module. By our assumption, M is directing and, according to (2.8), the support algebra B of M is a tilted algebra, whose quiver supp M is, by (1.3), a convex full subquiver of Q_A. It follows from (3.3) that the quadratic form q_B of B is weakly positive and that $M \mapsto \dim M$ defines a bijection between the isomorphism classes of indecomposable B-modules and the positive roots of q_B. But, by (VII.3.4), a weakly positive quadratic form has only finitely many positive roots. Therefore B is representation–finite. Because the finite quiver Q_A has only finitely many convex full subquivers, A is also representation–finite. $\square$

Note that (3.3) and (3.4) apply in particular to all representation–finite tilted algebras.

3.5. Lemma. *Let A be a connected representation–finite algebra. Then A is representation–directed if and only if it admits a postprojective component.*

Proof. Because A is representation–finite, it follows from (IV.5.4) that the Auslander–Reiten quiver $\Gamma(\mathrm{mod}\,A)$ of A is connected. Assume that A is representation–directed. Because $\Gamma(\mathrm{mod}\,A)$ is connected, obviously $\Gamma(\mathrm{mod}\,A)$ is a postprojective component. Conversely, assume that A admits a postprojective component. Because $\Gamma(\mathrm{mod}\,A)$ is connected, it coincides with its postprojective component. In particular, all indecomposable A-modules are directing, by (1.1). □

Because the support algebra of any directing module is a tilted algebra, we may look at a representation–directed algebra as being a gluing of finitely many representation–finite tilted algebras given by the supports of the indecomposable modules.

3.6. Examples. (a) Let A be given by the quiver

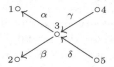

bound by four zero relations $\gamma\alpha = 0$, $\delta\alpha = 0$, $\gamma\beta = 0$, and $\delta\beta = 0$. Then $\Gamma(\mathrm{mod}\,A)$ is the quiver

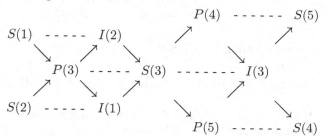

In particular, A is representation–directed. Also, we have in $\Gamma(\mathrm{mod}\,A)$ a section given by the modules $I(1)$, $I(2)$, $S(3)$, $P(4)$, $P(5)$. Hence A is tilted of type $\widetilde{\mathbb{D}}_4$, and so gl.dim $A \leq 2$. Applying (3.3), we get that q_A is weakly positive, and the dimension vectors $\begin{smallmatrix}1&&0\\&0&\\0&&0\end{smallmatrix}$, $\begin{smallmatrix}0&&0\\&0&\\1&&0\end{smallmatrix}$, $\begin{smallmatrix}1&&0\\&1&\\1&&0\end{smallmatrix}$, $\begin{smallmatrix}1&&0\\&0&\\0&&0\end{smallmatrix}$, $\begin{smallmatrix}0&&0\\&1&\\1&&0\end{smallmatrix}$, $\begin{smallmatrix}0&&0\\&1&\\1&&0\end{smallmatrix}$, $\begin{smallmatrix}0&&1\\&0&\\0&&0\end{smallmatrix}$, $\begin{smallmatrix}0&&0\\&1&\\1&&1\end{smallmatrix}$, $\begin{smallmatrix}0&&1\\&0&\\0&&0\end{smallmatrix}$, $\begin{smallmatrix}0&&0\\&0&\\0&&1\end{smallmatrix}$ of the indecomposable A-modules form a complete list of the positive roots of q_A. On the other hand, q_A is not positive definite, because it is $\mathbb{Z}$-congruent to the Euler form of hereditary algebra of Euclidean type $\widetilde{\mathbb{D}}_4$ (see (VI.4.7), (VII.4.2)).

   (b) Let $n \geq 2$ be a positive integer. Consider the algebra given by the

quiver

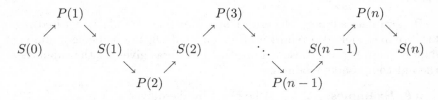

bound by $\alpha_{i+1}\alpha_i = 0$ for all i, $1 \leq i \leq n-1$. The simple module $S(i)$ has a minimal projective resolution

$$0 \longrightarrow P(0) \xrightarrow{f_1} P(1) \xrightarrow{f_2} \cdots \longrightarrow P(i-1) \xrightarrow{f_i} P(i) \longrightarrow S(i) \longrightarrow 0$$

and, for each j, $S(j) = \operatorname{Coker} f_j$. Therefore, $\operatorname{pd} S(i) = i$ for any i, and gl.dim $A = n$. On the other hand, $\Gamma(\operatorname{mod} A)$ is of the form

$$
\begin{array}{ccccccc}
P(1) & & & P(3) & & & P(n) \\
\nearrow \quad \searrow & & \nearrow \quad \searrow & & & \nearrow \quad \searrow \\
S(0) \qquad S(1) & & S(2) & & \ddots & S(n-1) \qquad S(n) \\
& \searrow \quad \nearrow & & & \searrow \quad \nearrow \\
& P(2) & & & P(n-1) \\
\end{array}
$$

and so A is representation–directed. Therefore there exist representation–directed algebras of arbitrary finite global dimension. Let $n = 3$ (thus gl.dim $A = 3$) and consider the module $M = S(0) \oplus S(3)$. Using the projective resolution, we get $\operatorname{Ext}^1_A(S(3), S(0)) = 0$, $\operatorname{Ext}^2_A(S(3), S(0)) = 0$ while $\operatorname{Ext}^3_A(S(3), S(0)) \cong K$. Because the module $S(0)$ is projective, we have $\operatorname{Ext}^i_A(S(0), S(3)) = 0$ for all $i \geq 1$. Similarly, $\operatorname{Ext}^i_A(S(0), S(0)) = 0$ for all $i \geq 1$. Finally, $\operatorname{Ext}^i_A(S(3), S(3)) = 0$ for all $i \geq 1$, because $S(3)$ is directing and (1.4) applies. Hence, according to (III.3.13),

$$q_A(\operatorname{\mathbf{dim}} M) = \sum_{i \geq 0} (-1)^i \dim_K \operatorname{Ext}^i_A(S(0) \oplus S(3), S(0) \oplus S(3))$$

$$= \dim_K \operatorname{End} S(0) + \dim_K \operatorname{End} S(3) - \dim_K \operatorname{Ext}^3_A(S(3), S(0)) = 1.$$

Therefore, $\operatorname{\mathbf{dim}} M$ is a positive root of q_A, but it is not the dimension vector of an indecomposable A-module. This shows that the assumption on the global dimension of A in (3.3) is essential.

(c) Let A be given by the quiver

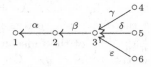

bound by $\beta\alpha = 0$, $\gamma\beta = 0$, $\delta\beta = 0$, $\varepsilon\beta = 0$. It follows from the imposed relations that any indecomposable A-module is an indecomposable module over one of the hereditary algebras: H_1 given by the points 1 and 2, H_2

given by the points 2 and 3, or H_3 given by the points 3, 4, 5, and 6. Hence $\Gamma(\mathrm{mod}\,A)$ is of the form

$$
\begin{array}{ccccccccc}
& \begin{smallmatrix}0\\1\,1\,0\,0\\0\end{smallmatrix} & & & \begin{smallmatrix}1\\0\,0\,1\,0\\0\end{smallmatrix} & & \begin{smallmatrix}0\\0\,0\,1\,1\\1\end{smallmatrix} & & \begin{smallmatrix}1\\0\,0\,0\,0\\0\end{smallmatrix} \\
\nearrow & & \searrow & \nearrow & & \searrow & \nearrow & & \searrow \\
\begin{smallmatrix}0\\1\,0\,0\,0\\0\end{smallmatrix} & & \begin{smallmatrix}0\\0\,1\,0\,0\\0\end{smallmatrix} & & \begin{smallmatrix}0\\0\,0\,1\,0\\0\end{smallmatrix}\!\to\!\begin{smallmatrix}0\\0\,0\,1\,1\\0\end{smallmatrix}\!\to\!\begin{smallmatrix}0\\0\,0\,2\,1\\1\end{smallmatrix}\!\to\!\begin{smallmatrix}1\\0\,0\,1\,0\\1\end{smallmatrix}\!\to\!\begin{smallmatrix}1\\0\,0\,1\,1\\1\end{smallmatrix}\!\to\!\begin{smallmatrix}0\\0\,0\,0\,1\\0\end{smallmatrix} \\
& \searrow & & \nearrow & \searrow & & \nearrow & \searrow & \\
& \begin{smallmatrix}0\\0\,1\,1\,0\\0\end{smallmatrix} & & & \begin{smallmatrix}0\\0\,0\,1\,0\\1\end{smallmatrix} & & \begin{smallmatrix}1\\0\,0\,1\,1\\0\end{smallmatrix} & & \begin{smallmatrix}0\\0\,0\,0\,0\\1\end{smallmatrix}
\end{array}
$$

Thus A is representation–directed. The simple modules $S(i)$, with $i = 4, 5, 6$ have minimal projective resolutions of the form

$$0 \longrightarrow P(1) \longrightarrow P(2) \longrightarrow P(3) \longrightarrow P(i) \longrightarrow S(i) \longrightarrow 0.$$

Hence $\mathrm{pd}\,S(i) = 3$ for $i = 4, 5, 6$. Clearly, $\mathrm{pd}\,S(3) = 2$, $\mathrm{pd}\,S(2) = 1$, and $\mathrm{pd}\,S(1) = 0$. Then $\mathrm{gl.dim}\,A = 3$. Calculating the extension spaces $\mathrm{Ext}^s_A(S(i), S(j))$ for $s \geq 1$ and $1 \leq i, j \leq 6$, we find that each of the spaces

$$\mathrm{Ext}^1_A(S(2), S(1)),\ \mathrm{Ext}^1_A(S(3), S(2)),\ \mathrm{Ext}^1_A(S(4), S(3)),\ \mathrm{Ext}^1_A(S(5), S(3)),$$
$$\mathrm{Ext}^1_A(S(6), S(3)),\ \mathrm{Ext}^2_A(S(3), S(1)),\ \mathrm{Ext}^2_A(S(4), S(2)),\ \mathrm{Ext}^2_A(S(5), S(2)),$$
$$\mathrm{Ext}^2_A(S(6), S(2)),\ \mathrm{Ext}^3_A(S(4), S(1)),\ \mathrm{Ext}^3_A(S(5), S(1)),\ \mathrm{Ext}^3_A(S(6), S(1))$$

is isomorphic to K, whereas the remaining spaces vanish. Thus, for any vector $\mathbf{x} = \left(\begin{smallmatrix} & x_4 & \\ x_1\, x_2\, x_3\, x_5 \\ & x_6 & \end{smallmatrix}\right) \in K_0(A)$, the Euler form $q_A(\mathbf{x})$ of A is defined by the formula

$$q_A(\mathbf{x}) = \sum_{i=1}^{6} x_i^2 - \sum_{i,j=1}^{6} a_{ij}^{(1)} x_i x_j + \sum_{i,j=1}^{6} a_{ij}^{(2)} x_i x_j - \sum_{i,j=1}^{6} a_{ij}^{(3)} x_i x_j,$$

where $a_{ij}^{(s)} = \dim_K \mathrm{Ext}^s_A(S(i), S(j))$ for $s = 1, 2, 3$. It follows that

$$\begin{aligned}
q_A(\mathbf{x}) =\ & x_1^2 + x_2^2 + x_3^2 + x_4^2 + x_5^2 + x_6^2 - x_1 x_2 - x_2 x_3 - x_3 x_4 - x_3 x_5 - x_3 x_6 \\
& + x_1 x_3 + x_2 x_4 + x_2 x_5 + x_2 x_6 - x_1 x_4 - x_1 x_5 - x_1 x_6.
\end{aligned}$$

In particular, for $\mathbf{x} = \left(\begin{smallmatrix} & 1 & \\ 1\, 0\, 1\, 1 \\ & 1 & \end{smallmatrix}\right)$, we have $q_A(\mathbf{x}) = 0$. Hence q_A is not weakly positive. Moreover, $\mathbf{y} = \left(\begin{smallmatrix} & 1 & \\ 1\, 0\, 2\, 1 \\ & 1 & \end{smallmatrix}\right)$ satisfies $q_A(\mathbf{y}) = 1$, and $\mathbf{y}$ is clearly not the dimension vector of an indecomposable A-module. Also, for $\mathbf{z} = \left(\begin{smallmatrix} & 1 & \\ 1\, 1\, 1\, 1 \\ & 1 & \end{smallmatrix}\right)$, we have $q_A(\mathbf{z}) = 2$. On the other hand, for any indecomposable A-module M, we have, by (1.5), $q_A(\mathbf{dim}\,M) = 1$.

IX.4. The separation condition

The aim of this section is to give an easily verified sufficient (though by no means necessary) combinatorial criterion for an algebra to have a post-projective component and thus to be representation–directed whenever it is representation–finite (see (3.5)). Because representation–directed algebras have an acyclic ordinary quiver, we assume throughout this section that all algebras we deal with have an acyclic ordinary quiver.

4.1. Definition. Let A be an algebra with an acyclic quiver Q_A.

(a) An indecomposable projective module $P(a)_A$ is said to have a **separated radical** if, for any distinct indecomposable summands M and N of rad $P(a)$, the supports supp M and supp N lie in distinct connected components of the full subquiver $Q_A(\overrightarrow{a})$ of Q_A generated by the nonpredecessors of a. The algebra A is said to satisfy the **separation condition** if each indecomposable projective A-module has a separated radical.

(b) An indecomposable injective module $I(a)_A$ is said to have a **separated socle factor** if, for any distinct indecomposable summands M and N of $I(a)/\operatorname{soc} I(a)$, the supports supp M and supp N lie in distinct connected components of the full subquiver $Q_A(\overleftarrow{a})$ of Q_A generated by the nonsuccessors of a. The algebra A is said to satisfy the **coseparation condition** if each indecomposable injective A-module has a separated socle factor.

Thus, A satisfies the separation condition if and only if the opposite algebra A^{op} satisfies the coseparation condition.

Clearly, if an indecomposable projective module $P(a)_A$ has a separated radical, then two distinct indecomposable summands of rad $P(a)$ are necessarily nonisomorphic. On the other hand, if $P(a)_A$ has an indecomposable radical, then it has a separated radical. Trivially, any simple projective has a separated radical.

4.2. Examples. (a) Let A be given by the quiver

bound by $\alpha\beta = \gamma\delta$. The radical of each indecomposable projective is indecomposable or zero. Hence A satisfies the separation condition.

(b) Let A be given by the same quiver as in (a), bound by $\alpha\beta = 0$, $\gamma\delta = 0$. Here, rad $P(a) = S(b) \oplus S(c)$ and $Q_A(\overrightarrow{a})_0 = \{b, c, d\}$, thus $Q_A(\overrightarrow{a})$ is connected. Hence $P(a)$ does not have a separated radical. Thus A does not satisfy the separation condition, even though $P(b)$, $P(c)$, and $P(d)$ have separated radicals. One shows that the algebra A is representation–finite and has a postprojective component.

(c) Let A be given by the same quiver as in (a), bound by $\gamma\delta = 0$. Here, $\operatorname{rad} P(a) = P(b) \oplus S(c)$, and so $P(a)$ does not have a separated radical.

(d) Let A be given by the quiver

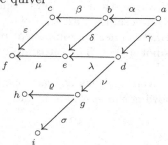

bound by $\alpha\delta = \gamma\lambda$, $\beta\varepsilon = \delta\mu$, $\lambda\mu = 0$, $\nu\sigma = 0$, $\alpha\beta = 0$. Then A satisfies the separation condition.

(e) There exist algebras satisfying the separation condition, but not the coseparation condition. Let A be given by the quiver

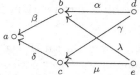

bound by $\alpha\beta = \gamma\delta$, $\lambda\beta = \mu\delta$. Each indecomposable projective has indecomposable (or zero) radical, hence A satisfies the separation condition. On the other hand, neither $I(b)$ nor $I(c)$ has a separated socle factor.

The examples should inspire the reader for the following picture. The algebra A satisfies the separation condition if and only if, for any $a \in (Q_A)_0$, the full subquiver of Q_A generated by a and $Q_A(\overrightarrow{a})$ has the following shape

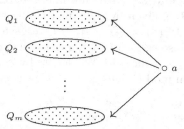

with no walk not passing through a between two distinct connected components Q_i and Q_j of $Q_A(\overrightarrow{a})$.

The following lemma, used in Section 6, is also strongly suggested by the preceding examples.

4.3. Lemma. *Let A be an algebra such that Q_A is a tree. Then A satisfies the separation condition. Conversely, if A is bound only by zero relations, satisfies the separation condition, and is representation–finite, then Q_A is a tree.*

Proof. If Q_A is a tree, then it follows from (III.2.2) that $\operatorname{rad} P(a)$ is a direct sum of indecomposables with simple top, the support of each being contained in a distinct connected component of $Q_A(\overrightarrow{a})$. For the converse, assume that Q_A is not a tree. Then it contains a full subquiver Q' that is a (nonoriented) cycle.

Because Q_A is acyclic, Q' has at least one source a and one sink b, so that it has the following shape, where w and w' are walks

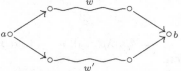

Because A is representation–finite, so is the algebra B given by the full subquiver Q' with the inherited relations. Hence Q' is bound by at least one relation, which is necessarily a zero relation. But then $\operatorname{rad} P(a)_A$ is not separated, a contradiction. □

We want to show that an algebra satisfying the separation condition admits a postprojective component. Clearly, this sufficient condition is not necessary. Indeed, the algebra of Example 4.2 (b) is representation–directed (as one verifies easily by direct computation of its Auslander–Reiten quiver) and thus admits a postprojective component, but it does not satisfy the separation condition.

We need some notation. Assume that A satisfies the separation condition and that $a \in (Q_A)_0$ is a source of Q_A. Letting B denote the algebra given by the quiver $Q_A(\overrightarrow{a})$ with the inherited relations, we get $B = \prod_{i=1}^m B_i$, where each B_i is given by a distinct connected component of $Q_A(\overrightarrow{a})$. We may write $\operatorname{rad} P(a) = \bigoplus_{i=1}^m R_i$, where each R_i is an indecomposable B_i-module. Because each B_i is a quotient algebra of A, any B_i-module can be considered as an A-module. We denote by τ_{B_i} and τ_A the Auslander–Reiten translations in $\operatorname{mod} B_i$ and in $\operatorname{mod} A$, respectively.

4.4. Lemma. *Assume that A is an algebra satisfying the separation condition. Let $a \in (Q_A)_0$ and let B, B_i, R_i be as earlier. Assume that $\Gamma(\operatorname{mod} B_i)$ has a postprojective component $\mathcal{P}_i$ and let $M \in (\mathcal{P}_i)_0$ be such that R_i is not a proper predecessor of M.*

(a) *Every predecessor of M in $\Gamma(\operatorname{mod} A)$ is a predecessor of M in $\mathcal{P}_i$.*

(b) *If $M \not\cong R_i$, then $\tau_{B_i}^{-1} M \cong \tau_A^{-1} M$.*

Proof. We use induction on the number $n(M)$ of predecessors of M in $\mathcal{P}_i$. If $n(M) = 1$, then M is a simple projective B_i-module, hence a simple projective A-module, so that (a) follows trivially. On the other hand, any irreducible morphism in $\operatorname{mod} A$ of source M has a projective target $P(b)$, so that if $M \not\cong R_i$, then $b \neq a$, because R_i is the unique indecomposable B_i-module that is a radical summand of $P(a)$ and so $P(b)$ is a projective B_i-module. But then the cokernel term in the almost split sequence

$$0 \longrightarrow M \longrightarrow \bigoplus P(b) \longrightarrow \tau_A^{-1} M \longrightarrow 0$$

is a B_i-module. This implies (b).

For the induction step, we first claim that, for each irreducible morphism $L \to M$ in $\operatorname{mod} A$, with L indecomposable, L is a B_i-module. Indeed, if M is a projective A-module, then it is a projective B_i-module; hence L, being a submodule of M, is also a B_i-module. If M is not a projective A-module, then it is not a projective B_i-module, and $n(\tau_{B_i} M) < n(M)$. The induction hypothesis gives $M \cong \tau_{B_i}^{-1} \tau_{B_i} M \cong \tau_A^{-1} \tau_{B_i} M$ so that $\tau_A M \cong \tau_{B_i} M$ is a B_i-module. The almost split sequence $0 \longrightarrow \tau_{B_i} M \longrightarrow L \oplus L' \longrightarrow M \longrightarrow 0$ in $\operatorname{mod} A$ guarantees that L is a B_i-module. This implies (a).

We now show (b). Assume $M \not\cong R_i$. To prove that $\tau_{B_i}^{-1} M \cong \tau_A^{-1} M$, it suffices to show that $\tau_A^{-1} M$ is a B_i-module. Let $M \to N$ be an irreducible morphism in $\operatorname{mod} A$, with N indecomposable. We claim that N is a B_i-module. If N is not projective, then there exists an irreducible morphism $\tau_A N \to M$ and the claim implies that $\tau_A N$ is a B_i-module. Hence so is $N \cong \tau_A^{-1} \tau_A N \cong \tau_{B_i}^{-1} \tau_A N \cong \tau_{B_i}^{-1} \tau_A N$. If N is projective, then because $M \not\cong R_i$, we have $N \cong P(b)$ for some $b \neq a$. We claim that $P(b)_A$ is actually a projective B_i-module. Indeed, if this is not the case, then b is a predecessor of a so that we have a path $b = b_0 \to b_1 \to \cdots \to b_t = a$, with $t \geq 1$. Because M is a radical summand of $P(b)$, there exists a direct successor b' of b lying in $\operatorname{supp} M$, which is a convex full subquiver of Q_{B_i}, by (1.1)(a) and (1.3). On the other hand, because $\operatorname{supp} R_i$ is also a convex full subquiver of Q_{B_i}, there is an arrow $a \to a'$, with $a' \in (Q_{B_i})_0$. Because B_i is connected, there exists a walk $b' - \cdots - a' \longleftarrow a \longleftarrow \cdots \longleftarrow b_1$ in Q_A with b' and b_1 both direct successors of b. By hypothesis, $P(b)$ has a separated radical, and M is a summand of $\operatorname{rad} P(b)$. Therefore b_1 must lie in the support of M, a contradiction to the fact that M is a B_i-module. This proves our claim.

We are now able to show that $\tau_A^{-1} M$ is a B_i-module. If M is an injective A-module, then it is certainly injective as a B_i-module. If M is not an injective A-module, then, in the almost split sequence

$$0 \longrightarrow M \longrightarrow E \longrightarrow \tau_A^{-1} M \longrightarrow 0$$

in $\operatorname{mod} A$, we have just shown that E is a B_i-module. Hence so is $\tau_A^{-1} M$. $\square$

4.5. Theorem. *Let A be an algebra.*

(a) *If A satisfies the separation condition, then A admits a postprojective component.*

(b) *If A satisfies the coseparation condition, then A admits a preinjective component.*

Proof. We only prove (a); (b) follows from (a) and the standard duality $D : \operatorname{mod} A \to \operatorname{mod} A^{\mathrm{op}}$.

We use induction on $|(Q_A)_0|$; we have two cases to consider. Assume first that there exists a source $a \in (Q_A)_0$ and a radical summand R_i of $P(a)$ that does not belong to a postprojective component of $\Gamma(\operatorname{mod} B_i)$ (with the preceding notation). By induction, $\Gamma(\operatorname{mod} B_i)$ admits a postprojective component $\mathcal{P}$. By (4.4)(a), $\mathcal{P}$ is a postprojective component of $\Gamma(\operatorname{mod} A)$.

If this is not the case, then we construct a postprojective component of $\Gamma(\operatorname{mod} A)$ by constructing a sequence $(\mathcal{P}_n)$ of full subquivers of $\Gamma(\operatorname{mod} A)$ such that:

(a) Each $\mathcal{P}_n$ is finite, connected, acyclic, and closed under predecessors.

(b) $\tau_A^{-1}\mathcal{P}_n \cup \mathcal{P}_n \subseteq \mathcal{P}_{n+1}$.

Then $\mathcal{P} = \bigcup_{n \geq 0} \mathcal{P}_n$ is the wanted postprojective component.

We start by setting $\mathcal{P}_0 = \{S\}$, where S is a simple projective. To obtain $\mathcal{P}_{n+1}$ from $\mathcal{P}_n$, we consider the (finite) set $\mathcal{S}$ of indecomposable modules M in $\mathcal{P}_n$ having the property that $\tau_A^{-1}M$ is not in $\mathcal{P}_n$. We let $\mathcal{P}_{n+1}$ be the full subquiver of $\Gamma(\operatorname{mod} A)$ generated by $\mathcal{P}_n$ and , for each M in $\mathcal{S}$, all the predecessors of $\tau_A^{-1}M$ in $\Gamma(\operatorname{mod} A)$. If $\mathcal{S}$ is empty, we let $\mathcal{P}_{n+1} = \mathcal{P}_n$. Clearly, $\mathcal{P}_{n+1}$ satisfies (b). We must show that it satisfies (a).

For this purpose, we start by numbering the modules $M_1, \ldots, M_t$ in $\mathcal{S}$ in such a way that if M_i precedes M_j, then $i < j$ (this is possible because $\mathcal{P}_n$ is acyclic). We use induction on i. We show the induction step. Consider the almost split sequence $0 \longrightarrow M_{i+1} \longrightarrow E \longrightarrow \tau_A^{-1}M_{i+1} \longrightarrow 0$ in $\operatorname{mod} A$. We must show that if L is an indecomposable summand of E, then L has only finitely many predecessors and is directing. If L is projective, say $L = P(a)$, then by assumption, each of the radical summands R_i of $P(a)$ lies in a postprojective component of $\Gamma(\operatorname{mod} B_i)$ and the statement follows from (4.4)(a). If L is not projective, then either L is in $\mathcal{P}_n$ and we are done, or L is not in $\mathcal{P}_n$ and then the existence of an irreducible morphism $\tau_A L \to M_{i+1}$, together with the fact that M_{i+1} is in $\mathcal{P}_n$, which is closed under predecessors, implies that $\tau_A L$ is in $\mathcal{P}_n$. Consequently, $\tau_A L \cong M_j$ for some $j \leq i$; then $L \cong \tau_A^{-1}M_j$ satisfies our assumption by the induction hypothesis. The case $i = 1$ is shown likewise. $\qquad\square$

We now consider the situation from another point of view. As we have seen, a representation–finite algebra satisfying the separation condition is representation–directed. We wish to characterise, among the representation–directed algebras, which ones satisfy the separation condition. For this purpose, we need a new combinatorial invariant introduced in [40].

4.6. Definition. Let (Γ, τ) be a postprojective component of an Auslander–Reiten quiver $\Gamma(\operatorname{mod} A))$, viewed as a translation subquiver of $(\Gamma(\operatorname{mod} A)), \tau)$. The **orbit quiver** $\operatorname{Orb}(\Gamma)$ of Γ is defined as follows. The points of $\operatorname{Orb}(\Gamma)$ are the τ-orbits ω_x of the points $x \in \Gamma_0$ (and thus are in a bijective correspondence with the projectives in Γ). For a projective $p \in \Gamma_0$, let $x_1, \ldots, x_s$ be all its direct predecessors and for each i with $1 \leq i \leq s$, let n_i be the number of arrows from x_i to p, and let p_i be the unique projective in the τ-orbit of x_i; then put n_i arrows from ω_{p_i} to ω_p in $\operatorname{Orb}(\Gamma)$.

One may thus speak of the orbit quiver of $\Gamma(\operatorname{mod} A)$, where A is a representation–directed algebra.

Let (Γ, τ) be a postprojective component of an Auslander–Reiten quiver. There exists an arrow $\omega_x \to \omega_y$ in $\operatorname{Orb}(\Gamma)$ if and only if the τ-orbit of x contains a direct predecessor of the unique projective in the τ-orbit of y. If this is the case, then there exists a path in Γ from the projective in the τ-orbit of x to the projective in the τ-orbit of y. Also, because Γ is acyclic, so is the orbit quiver $\operatorname{Orb}(\Gamma)$.

4.7. Examples. (a) Let A be as in (4.2)(a). Then $\Gamma(\operatorname{mod} A)$ is given by

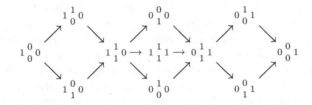

and obviously the orbit quiver $\operatorname{Orb}(\Gamma(\operatorname{mod} A))$ is given by

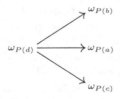

(b) Let B be as in (4.2)(b). Then $\Gamma(\operatorname{mod} B)$ is given by

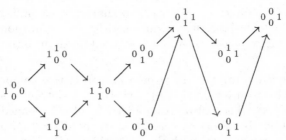

It is clear that the orbit quiver $\mathrm{Orb}(\Gamma(\operatorname{mod} B))$ is given by

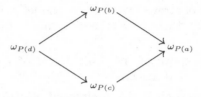

In these examples, both algebras A and B are representation–directed. The first satisfies the separation condition (and its orbit quiver is a tree), whereas the second does not (and its orbit quiver is not a tree).

4.8. Theorem. *Let A be a connected and representation–directed algebra. Then A satisfies the separation condition if and only if the orbit quiver $\mathrm{Orb}(\Gamma(\operatorname{mod} A))$ is a tree.*

Proof. (a) The necessity is shown by induction on $|(Q_A)_0|$. Assume that A satisfies the separation condition. Because $\Gamma(\operatorname{mod} A)$ is acyclic and has only finitely many projective points, there exist an indecomposable projective A-module $P(a)$ having no other indecomposable projective as a successor. This choice guarantees that a is a source in Q_A. Let B be the (not necessarily connected) algebra whose quiver is the full subquiver of Q_A generated by all points except a, with the inherited relations. Let $B = B_1 \times \ldots \times B_m$, where $B_1, \ldots, B_m$ are connected algebras, and

$$\operatorname{rad} P(a) = R_1 \oplus \ldots \oplus R_m,$$

where each R_i is an indecomposable B_i-module. Because a is a source, each B_i satisfies the separation condition and the induction hypothesis implies that $\mathrm{Orb}(\Gamma(\operatorname{mod} B_i))$ is a tree. We notice that if an indecomposable A-module M is not a proper successor of $P(a)$, then $\operatorname{Hom}_A(P(a), M) = 0$; hence M has its support entirely contained in B, so that it is a B_i-module for some i. For each $i \in \{1, \ldots, m\}$, let $P(b_i)$ be the unique indecomposable projective B_i-module in the τ-orbit of R_i. Then $\mathrm{Orb}(\Gamma(\operatorname{mod} A))$ is constructed from the disjoint union of the trees $\mathrm{Orb}(\Gamma(\operatorname{mod} B_i))$ by adding one extra point $\omega_{P(a)}$ and an arrow $\omega_{P(i)} \to \omega_{P(a)}$ for each i. Hence it is a tree.

(b) For sufficiency, we suppose that $\mathrm{Orb}(\Gamma(\mathrm{mod}\,A))$ is a tree but that A does not satisfy the separation condition. There exists $a \in (Q_A)_0$ such that $P(a)$ does not have a separated radical. We may choose a so that, for each proper successor a' of a in Q_A, the module $P(a')$ has a separated radical. Because A is representation–directed, it is representation–finite, hence, by (IV.4.9), distinct direct summands of $\mathrm{rad}\,P(a)$ are not isomorphic. Then there exist two nonisomorphic indecomposable summands M, N of $\mathrm{rad}\,P(a)$ and two points b_1 in $\mathrm{supp}\,M$ and b_t in $\mathrm{supp}\,N$ that are connected by a walk

$$b_1 \,\text{---}\, b_2 \,\text{---}\, \cdots \,\text{---}\, b_t$$

in $Q_A(\overrightarrow{a})$. Let $c, d \in (Q_A)_0$ and $r, s \geq 0$ be such that $M \cong \tau^{-r} P(c)$ and $N \cong \tau^{-s} P(d)$. Because b_1 is in $\mathrm{supp}\,M$, we have $\mathrm{Hom}_A(P(b_1), M) \neq 0$; hence we have a path from $P(b_1)$ to M and similarly a path from $P(b_t)$ to N in $\Gamma(\mathrm{mod}\,A)$. Consequently, the points $c, b_1, \ldots, b_t, d$ all belong to the same connected component Q of $Q_A(\overrightarrow{a})$. Let B be the algebra given by the quiver Q with the inherited relations. By our assumption on a, the algebra B satisfies the separation condition. Because it is a quotient of A, it is representation–finite. Hence B is representation–directed, because $\Gamma(\mathrm{mod}\,B)$ is postprojective, by (4.5)(a), and the necessity part yields that $\mathrm{Orb}(\Gamma(\mathrm{mod}\,B))$ is a tree. On the other hand, the hypothesis that $\mathrm{Orb}(\Gamma(\mathrm{mod}\,A))$ is a tree implies that $c \neq d$ (otherwise, we would have two arrows from $\omega_{P(c)}$ to $\omega_{P(a)}$). Consequently, $\mathrm{Orb}(\Gamma(\mathrm{mod}\,A))$ contains two distinct arrows, $\omega_{P(c)} \to \omega_{P(a)}$ and $\omega_{P(d)} \to \omega_{P(a)}$, and hence a cycle

$$\omega_{P(a)} \xleftarrow{\hspace{1em}} \omega_{P(c)} \,\text{---}\cdots\text{---}\, \omega_{P(b_1)} \,\text{---}\, \cdots \,\text{---}\, \omega_{P(b_t)} \,\text{---}\cdots\text{---}\, \omega_{P(d)} \xrightarrow{\hspace{1em}} \omega_{P(a)},$$

contrary to the hypothesis that $\mathrm{Orb}(\Gamma(\mathrm{mod}\,A))$ is a tree. $\qquad\square$

IX.5. Algebras such that all projectives are postprojective

We know that if A is a representation–directed or concealed algebra, then $\Gamma(\mathrm{mod}\,A)$ has a postprojective component containing all indecomposable projective A-modules [see (VIII.4.5)]. This is not true in general. For instance, the algebra A given by the quiver

$$\underset{1}{\circ} \overset{\beta}{\underset{\gamma}{\rightleftarrows}} \underset{2}{\circ} \overset{\alpha}{\longleftarrow} \underset{3}{\circ}$$

bound by $\alpha\beta = 0$ is such that the module $P(3)$ is not postprojective. Indeed, the algebra has a unique postprojective component equal to that of the path algebra of the full subquiver generated by points 1 and 2, and it is easily seen that $\mathrm{rad}\,P(3)$ (which is indecomposable) does not lie in this component.

In general, we have the following characterisation of algebras having the property that all indecomposable projectives are postprojective.

5.1. Proposition. *Let A be an algebra and $\Gamma(\mathrm{mod}\, A)$ the Auslander–Reiten quiver of A. The following three conditions are equivalent:*

(a) *The quiver $\Gamma(\mathrm{mod}\, A)$ admits postprojective components the union of which contains all indecomposable projective A-modules.*

(b) *There is a common bound on the length of paths in $\mathrm{mod}\, A$ the targets of which are indecomposable projective A-modules.*

(c) *The number of paths in $\Gamma(\mathrm{mod}\, A)$ the targets of which are indecomposable projective A-modules is finite.*

Proof. Assume (a). It follows from (VIII.2.5) that each path in $\mathrm{mod}\, A$ with target that is an indecomposable projective A-module is of finite length. Then (b) follows at once.

Because the quiver $\Gamma(\mathrm{mod}\, A)$ is locally finite, (b) implies (c) trivially. Now we assume (c) and prove (a). Let $\mathcal{C}$ be a component in $\Gamma(\mathrm{mod}\, A)$ that contains an indecomposable projective A-module. We claim that $\mathcal{C}$ is postprojective. Let $\mathcal{D}$ denote the full translation subquiver of $\mathcal{C}$ generated by all modules in $\mathcal{C}$ that are predecessors of a projective module in $\mathcal{C}$. Clearly, by our assumption, $\mathcal{D}$ is finite, acyclic and closed under predecessors. In particular, for any M in $\mathcal{D}$, there exist $r \geq 0$ and an indecomposable projective module P in $\mathcal{D}$ such that $\tau^r M \cong P$. We now prove that, for any N in $\mathcal{C}$, there exist $s \geq 0$ and a module M in $\mathcal{D}$ such that $N \cong \tau^{-s} M$. Clearly, this will imply that $N \cong \tau^{-t} P$, for some $t \geq 0$ and some indecomposable projective P.

Let N be a module in $\mathcal{C}$, and assume it is not in $\mathcal{D}$. Because $\mathcal{C}$ is connected, there exists a walk

$$M = M_0 \text{ --- } M_1 \text{ --- } \cdots \text{ --- } M_m \text{ --- } M_{m+1} = N$$

in $\mathcal{C}$, for some M in $\mathcal{D}$. We may assume that none of the modules $M_1, \dots, M_m$ belongs to $\mathcal{D}$. Then the modules $M_1, \dots, M_{m+1}$ are not projective; hence there is a walk

$$\tau M_1 \text{ --- } \cdots \text{ --- } \tau M_m \text{ --- } \tau M_{m+1}.$$

By induction, we conclude that the module $\tau M_{m+1} = \tau N$ is of the form $\tau^{-s} L$ for some $s \geq 0$ and some L in $\mathcal{D}$, and consequently $N \cong \tau^{-s-1} L$.

We complete the proof by showing that $\mathcal{C}$ is acyclic. Assume that

$$L = L_1 \longrightarrow L_2 \longrightarrow \cdots \longrightarrow L_t = L$$

is a cycle in $\mathcal{C}$. There is an integer $r \geq 0$ such that $\tau^r L_i$ is projective for some i and $\tau^r L_j \neq 0$ for all $j \neq i$, where i and j are such that $1 \leq i, j \leq t$.

Hence, there is a cycle

$$\tau^r L = \tau^r L_1 \longrightarrow \cdots \longrightarrow \tau^r L_i \longrightarrow \cdots \longrightarrow \tau^r L_t = \tau^r L$$

in $\mathcal{C}$ passing through the projective module $\tau^r L_i$. Thus there are paths in $\mathcal{C}$ of arbitrarily large length with a target that is the projective module $\tau^r L_i$, a contradiction. □

We now aim to prove the following theorem, which will play an important rôle later.

5.2. Theorem. *Let A be an algebra and assume that $\Gamma(\mathrm{mod}\, A)$ admits postprojective components the union of which contains all indecomposable projective A-modules. Then, for any idempotent $e \in A$, $\Gamma(\mathrm{mod}\,(A/AeA))$ admits postprojective components the union of which contains all indecomposable projective A/AeA-modules.*

Proof. It follows from (5.1) that there is a common bound, say m, on the length of paths in $\mathrm{mod}\, A$ with targets that are indecomposable projective A-modules. We prove that any path in $\mathrm{mod}\,(A/AeA)$ with a target that is an indecomposable projective A/AeA-module is of length at most m. The result will follow from (5.1). Let

$$M_r \xrightarrow{f_r} M_{r-1} \longrightarrow \cdots \longrightarrow M_1 \xrightarrow{f_1} M_0 = P'$$

be a path in $\mathrm{mod}\,(A/AeA)$, with P' projective. There exists an indecomposable projective A-module P such that $P' = P/PeA$, and we have an exact sequence

$$0 \longrightarrow PeA \xrightarrow{u_0} P \xrightarrow{v_0} P' \longrightarrow 0$$

in $\mathrm{mod}\, A$. Constructing successively fibered products along the f_i yields a commutative diagram in $\mathrm{mod}\, A$ with exact rows:

$$
\begin{array}{ccccccccc}
0 & \longrightarrow & PeA & \xrightarrow{u_r} & N_r & \xrightarrow{v_r} & M_r & \longrightarrow & 0 \\
& & \| & & \downarrow & & \downarrow{\scriptstyle f_r} & & \\
0 & \longrightarrow & PeA & \xrightarrow{u_{r-1}} & N_{r-1} & \xrightarrow{v_{r-1}} & M_{r-1} & \longrightarrow & 0 \\
& & \| & & \downarrow & & \downarrow & & \\
& & \vdots & & \vdots & & \vdots & & \\
& & \| & & \downarrow & & \downarrow & & \\
0 & \longrightarrow & PeA & \xrightarrow{u_1} & N_1 & \xrightarrow{v_1} & M_1 & \longrightarrow & 0 \\
& & \| & & \downarrow & & \downarrow{\scriptstyle f_1} & & \\
0 & \longrightarrow & PeA & \xrightarrow{u_0} & P & \xrightarrow{v_0} & P' & \longrightarrow & 0
\end{array}
$$

We note that $\operatorname{Im} u_i = N_i eA$, and $M_i \cong N_i/N_i eA$ for each i such that $1 \le i \le r$. Hence

$$N_i = L_i \oplus L_i',$$

where L_i is an indecomposable A-module, and L_i' is an A-module such that $L_i' = L_i' eA$. Moreover, v_i induces an isomorphism $L_i/L_i eA \cong M_i$ for each i. Hence we get a commutative diagram in $\operatorname{mod} A$ with exact rows

$$
\begin{array}{ccccccccc}
0 & \longrightarrow & L_r eA & \xrightarrow{u_r'} & L_r & \xrightarrow{v_r'} & M_r & \longrightarrow & 0 \\
& & \downarrow{\scriptstyle f_r''} & & \downarrow{\scriptstyle f_r'} & & \downarrow{\scriptstyle f_r} & & \\
0 & \longrightarrow & L_{r-1} eA & \xrightarrow{u_{r-1}'} & L_{r-1} & \xrightarrow{v_{r-1}'} & M_{r-1} & \longrightarrow & 0 \\
& & \downarrow & & \downarrow & & \downarrow & & \\
& & \vdots & & \vdots & & \vdots & & \\
& & \downarrow & & \downarrow & & \downarrow & & \\
0 & \longrightarrow & L_1 eA & \xrightarrow{u_1'} & L_1 & \xrightarrow{v_1'} & M_1 & \longrightarrow & 0 \\
& & \downarrow{\scriptstyle f_1''} & & \downarrow{\scriptstyle f_1'} & & \downarrow{\scriptstyle f_1} & & \\
0 & \longrightarrow & PeA & \xrightarrow{u_0} & P & \xrightarrow{v_0} & P' & \longrightarrow & 0
\end{array}
$$

where all the homomorphisms are the obvious ones. Beacause f_i belongs to $\operatorname{rad}_A(M_i, M_{i-1})$ for each i, we infer that

$$f_i' \in \operatorname{rad}_A(L_i, L_{i-1})$$

for each i. Hence, we deduce the existence of a path

$$L_r \xrightarrow{f_r'} L_{r-1} \longrightarrow \cdots \longrightarrow L_1 \xrightarrow{f_1'} P$$

in $\operatorname{mod} A$ with target in the projective module P, so that $r \le m$. This finishes the proof. $\qquad\square$

Our next question is whether a postprojective component containing all projectives also contains enough sincere indecomposable modules. To motivate our result, we start with the following two examples.

5.3. Examples. (a) Let A be given by the quiver

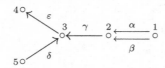

bound by $\delta\varepsilon = 0$, $\alpha\gamma = 0$. Then $\Gamma(\mathrm{mod}\,A)$ has a unique postprojective component $\mathcal{P}(A)$ of the form

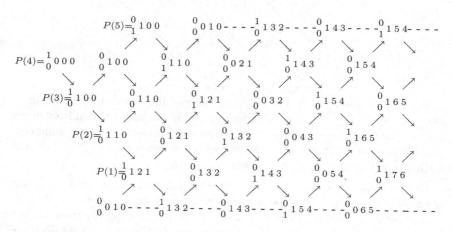

where the modules along the horizontal dotted lines have to be identified. One sees that $\mathcal{P}(A)$ contains all the indecomposable projective modules and that the dimension vector of any module in $\mathcal{P}(A)$ is zero at either point 4 or point 5. Hence $\mathcal{P}(A)$ does not contain sincere indecomposables. We note that the modules $P(1) = {}^1_0 1\,2\,1$, $\tau^{-1}P(2) = {}^0_0 1\,2\,1$, $\tau^{-2}P(3) = {}^0_1 1\,2\,1$, $\tau^{-3}P(4) = {}^0_0 0\,2\,1$, and $\tau^{-2}P(5) = {}^1_0 1\,3\,2$ form a section Σ of underlying graph $\tilde{\mathbb{A}}_4$. It is easily seen that any indecomposable projective A-module is a submodule of a module on Σ, hence by (VI.2.2), Σ is a faithful section. Clearly, $\mathrm{Hom}_A(U, \tau V) = 0$ for all U, V on Σ. Applying (VIII.5.6) yields that A is a tilted algebra of type $\tilde{\mathbb{A}}_4$. It is not concealed. Indeed, $\Gamma(\mathrm{mod}\,A)$ has a preinjective component of the form

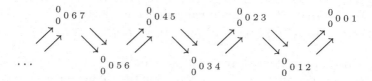

and hence A cannot be concealed by (VIII.4.5)(c).

(b) Let A be given by the quiver

$$\overset{1}{\circ} \xleftarrow{\quad\gamma\quad} \overset{2}{\circ} \overset{\alpha}{\underset{\beta}{\rightleftarrows}} \overset{3}{\circ}$$

bound by $\alpha\gamma = 0$. Then $\Gamma(\mathrm{mod}\,A)$ has a unique postprojective component

$\mathcal{P}(A)$ of the form

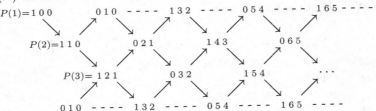

where the modules along the horizontal dotted lines have to be identified. One sees that $\mathcal{P}(A)$ contains all indecomposable projectives, infinitely many sincere indecomposable modules, and infinitely many nonsincere indecomposable modules. On the other hand, one shows easily, as in (a), that A is a tilted algebra of type $\widetilde{A}_2$ but is not concealed.

Our present objective is to show that this situation does not occur for concealed algebras. Let Q be a finite, connected, and acyclic quiver that is not Dynkin. We prove that if B is concealed of type Q, then all but finitely many modules from the unique postprojective component $\mathcal{P}(B)$ of $\Gamma(\mathrm{mod}\, B)$ are sincere. We start by proving that this is the case for the path algebra $A = KQ$ of Q. We need two lemmas.

5.4. Lemma. *Assume that $A = KQ$, where Q is a finite, connected, and acyclic quiver that is not Dynkin. Let P and P' be two indecomposable projective A-modules. Then the sequence $\dim_K \mathrm{Hom}_A(P, \tau^{-m}P')$, with $m \geq 1$, is not bounded.*

Proof. We recall from (VIII.2.1) that the unique postprojective component $\mathcal{P}(A)$ of $\Gamma(\mathrm{mod}\, A)$ consists of all modules $\tau^{-m}P(j)$, where $j \in Q_0$ and $m \geq 0$. For $i, j \in Q_0$, let

$$d_{ij} = \overline{\lim_{m \to \infty}}\ \dim_K \mathrm{Hom}_A(P(i), \tau^{-m}P(j)).$$

Because Q is not Dynkin, the algebra A is representation–infinite. It follows from (IV.5.4) that $\mathcal{P}(A)$ is infinite and the dimensions

$$\dim_K \tau^{-m}P(j) = \sum_{i \in Q_0} \dim_K \mathrm{Hom}_A(P(i), \tau^{-m}P(j))$$

of the indecomposable postprojective A-modules $\tau^{-m}P(j)$ are unbounded. Consequently, not all d_{ij} are finite. We claim that in fact all d_{ij} are infinite.

Let $b \to a$ be an arrow in Q. Because, according to (VII.1.6), there exist isomorphisms $e_b(\mathrm{rad}\, A/\mathrm{rad}^2 A)e_a \cong \mathrm{Irr}(P(a), P(b)) \cong \mathrm{Irr}(I(a), I(b))$, there exist irreducible morphisms $P(a) \to P(b)$ and $I(a) \to I(b)$. It follows that there exist almost split sequences of the form

$$0 \longrightarrow P(a) \longrightarrow P(b) \oplus E \longrightarrow \tau^{-1}P(a) \longrightarrow 0,$$

$$0 \longrightarrow P(b) \longrightarrow \tau^{-1}P(a) \oplus F \longrightarrow \tau^{-1}P(b) \longrightarrow 0,$$

and all their nonzero terms are postprojective. Because the component $\mathcal{P}(A)$ is infinite, according to (VIII.2.1), for each m, the modules $\tau^{-m}P(a)$ and $\tau^{-m}P(b)$ are nonzero. Hence, there exist almost split sequences of the form

$$0 \longrightarrow \tau^{-m}P(a) \longrightarrow \tau^{-m}P(b) \oplus \tau^{-m}E \longrightarrow \tau^{-m-1}P(a) \longrightarrow 0,$$
$$0 \longrightarrow \tau^{-m}P(b) \longrightarrow \tau^{-m-1}P(a) \oplus \tau^{-m}F \longrightarrow \tau^{-m-1}P(b) \longrightarrow 0$$

for each $m \geq 1$. Applying the exact functor $\operatorname{Hom}_A(P(i), -)$, we get exact sequences, and we easily conclude that

$$d_{ib} \leq 2d_{ia} \quad \text{and} \quad d_{ia} \leq 2d_{ib}$$

for any $i \in Q_0$. Consequently, d_{ib} is infinite if and only if d_{ia} is infinite. Further, (III.2.11) and (IV.2.15) yield

$$\begin{aligned}
\dim_K \operatorname{Hom}_A(P(i), \tau^{-m}P(j)) &= \dim_K \operatorname{Hom}_A(\tau^{-m}P(j), I(i)) \\
&= \dim_K \operatorname{Hom}_A(P(j), \tau^m I(i)) \\
&= \dim_K \operatorname{Hom}_A(\tau^m I(i), I(j)).
\end{aligned}$$

Analogously, there exist almost split sequences of the form

$$0 \longrightarrow \tau I(b) \longrightarrow I(a) \oplus E' \longrightarrow I(b) \longrightarrow 0,$$
$$0 \longrightarrow \tau I(a) \longrightarrow \tau I(b) \oplus F' \longrightarrow I(a) \longrightarrow 0,$$

and all their nonzero terms are preinjective. By (VIII.2.1), the preinjective component $\mathcal{Q}(A)$ of $\Gamma(\operatorname{mod} A)$ is infinite and the modules $\tau^m I(a)$ and $\tau^m I(b)$ are nonzero for all $m \geq 0$. Hence, there exist almost split sequences of the form

$$0 \longrightarrow \tau^{m+1} I(b) \longrightarrow \tau^m I(a) \oplus \tau^m E' \longrightarrow \tau^m I(b) \longrightarrow 0,$$
$$0 \longrightarrow \tau^{m+1} I(a) \longrightarrow \tau^{m+1} I(b) \oplus \tau^m F' \longrightarrow \tau^m I(a) \longrightarrow 0$$

for each $m \geq 1$. Applying the exact functor $\operatorname{Hom}_A(-, I(j))$, we get exact sequences and we easily conclude that

$$d_{aj} \leq 2d_{bj} \quad \text{and} \quad d_{bj} \leq 2d_{aj}$$

for any $j \in Q_0$ and, consequently, d_{bj} is infinite if and only if d_{aj} is infinite. Our claim then follows from the connectedness of Q. $\qquad\square$

In the following lemma and proposition, we need the notions of reflection of a quiver and associated reflection functors, as defined in (VII.5).

5.5. Lemma. *Assume that $A = KQ$, where Q is a finite, connected, and acyclic quiver that is not Dynkin. Let a be a sink in Q and A' be the path algebra of the quiver $\sigma_a Q$. Then all but finitely many indecomposable postprojective A'-modules are sincere if and only if all but finitely many indecomposable postprojective A-modules are sincere.*

Proof. Consider the APR-tilting module $T[a]_A = \tau^{-1}S(a) \oplus (\bigoplus_{b \neq a} P(b))$ and the reflection functors $S_a^+ = \operatorname{Hom}_A(T[a], -) : \operatorname{mod} A \to \operatorname{mod} A'$ and $S_a^- = - \otimes_{A'} T[a] : \operatorname{mod} A' \to \operatorname{mod} A$ as defined in (VII.5). Because $S(a)_A = P(a)_A$ is a simple projective A-module whereas $S(a)_{A'}$ is a simple injective A'-module, it follows from (VII.5.3) that S_a^+ and S_a^- induce an equivalence between the full subcategory of $\operatorname{mod} A$ consisting of all indecomposable postprojective A-modules except $S(a)_A$ and the full subcategory of $\operatorname{mod} A'$ consisting of all indecomposable postprojective A'-modules. Moreover, a is a source in $\sigma_a Q$ with corresponding projective module

$$P(a)_{A'} = \operatorname{Hom}_A(T[a], \tau^{-1}S(a)) = S_a^+ \tau^{-1}S(a).$$

Let $M \ncong S(a)$ be an indecomposable postprojective A-module. In view of (IV.2.15) and (VII.5.3), we have

$$\operatorname{Hom}_A(P(a)_A, M) \cong \operatorname{Hom}_A(S(a), M) \cong \operatorname{Hom}_A(\tau^{-1}S(a), \tau^{-1}M)$$
$$\cong \operatorname{Hom}_{A'}(S_a^+ \tau^{-1}S(a), S_a^+ \tau^{-1}M)$$
$$\cong \operatorname{Hom}_{A'}(P(a)_{A'}, \tau^{-1}S_a^+ M)$$

and, for any $b \neq a$,

$$\operatorname{Hom}_A(P(b), M) \cong \operatorname{Hom}_{A'}(S_a^+ P(b), S_a^+ M) \cong \operatorname{Hom}_{A'}(P(b)_{A'}, S_a^+ M).$$

This establishes the lemma. $\square$

5.6. Proposition. *Assume that $A = KQ$, where Q is a finite, connected, and acyclic quiver that is not Dynkin.*

  (a) *All but finitely many indecomposable postprojective A-modules are sincere.*

  (b) *All but finitely many indecomposable preinjective A-modules are sincere.*

Proof. (a) Because Q is not Dynkin, according to (VIII.2.1) the postprojective component $\mathcal{P}(A)$ of $\Gamma(\operatorname{mod} A)$ is infinite. Suppose, to the contrary, that $\mathcal{P}(A)$ contains infinitely many nonsincere indecomposable modules. Then there exists $a \in Q_0$ such that $\operatorname{Hom}_A(P(a), M) = 0$ for infinitely many modules M in $\mathcal{P}(A)$. We claim that we may assume a to be a source in Q. Indeed, if this is not the case, then by (VII.5.1), there exists an admissible sequence of sources $a_1, \ldots, a_t$ such that a is a source of $\sigma_{a_t} \ldots \sigma_{a_1} Q$.

Invoking (5.5) completes the proof of our claim. Therefore, assume that a is a source in Q. Letting Q_a denote the full subquiver or Q generated by all points except a, and $H = KQ_a$, it follows from our assumption that $\mathcal{P}(A)$ contains infinitely many indecomposable H-modules. We may write $H = B \times C$, with B connected and such that $\mathcal{P}(A)$ contains an infinite sequence $(M_i)_{i \geq 1}$ of indecomposable B-modules. We recall that any indecomposable module from $\mathcal{P}(A)$ has only finitely many indecomposable predecessors in $\operatorname{mod} A$. Because B-modules are A-modules, each of the M_i has only finitely many indecomposable predecessors in $\operatorname{mod} B$. But B is a representation–infinite hereditary algebra, so we infer that all M_i are postprojective B-modules (indeed, it follows from the definition of a preinjective component that preinjective modules have infinitely many preinjective predecessors whereas, if R is a regular indecomposable B-module, there exists an indecomposable projective B-module P such that $\operatorname{Hom}_B(P, R) \neq 0$ and (IV.5.1) shows that R has infinitely many postprojective predecessors). Further, because the postprojective component of $\Gamma(\operatorname{mod} B)$ has finitely many τ-orbits, each indecomposable postprojective B-module is a predecessor of some M_i, and hence all indecomposable postprojective B-modules lie in $\mathcal{P}(A)$. Let $\operatorname{rad} P(a) = N \oplus N'$, where N is a B-module and N' is a C-module. Clearly, N is nonzero (because A is connected) and projective. By (5.4), there exists an indecomposable nonprojective postprojective B-module U such that $\dim_K \operatorname{Hom}_B(N, U) \geq 3$. Applying the functor $\operatorname{Hom}_A(\tau_B^{-1}U, -)$ to the short exact sequence $0 \to N \oplus N' \to P(a) \to S(a) \to 0$ yields an exact sequence

$$0 = \operatorname{Hom}_A(\tau_B^{-1}U, S(a)) \longrightarrow \operatorname{Ext}_A^1(\tau_B^{-1}U, N \oplus N')$$
$$\longrightarrow \operatorname{Ext}_A^1(\tau_B^{-1}U, P(a)) \longrightarrow \operatorname{Ext}_A^1(\tau_B^{-1}U, S(a)) = 0,$$

because $\tau_B^{-1}U$ is a B-module, and $S(a)$ is an injective A-module. Moreover, because $A = KQ$ is hereditary, so is B; hence the projective dimension of the B-module $\tau_B^{-1}U$ is at most 1, and we have

$$\operatorname{Ext}_A^1(\tau_B^{-1}U, N \oplus N') = \operatorname{Ext}_B^1(\tau_B^{-1}U, N) \cong D\operatorname{Hom}_B(N, U).$$

Consequently, $\dim_K \operatorname{Ext}_A^1(\tau_B^{-1}U, P(a)) \geq 3$. Let

$$0 \to P(a) \to V \to \tau_B^{-1}U \to 0$$

be a nonsplit short exact sequence in $\operatorname{mod} A$. It follows from (VIII.2.8) that

$$\dim_K \operatorname{End}_A V < \dim_K \operatorname{End}_A(P(a) \oplus \tau_B^{-1}U)$$
$$= \dim_K \operatorname{End}_A P(a) + \dim_K \operatorname{End}_A(\tau_B^{-1}U) = 2$$

because $\operatorname{Hom}_A(P(a), \tau_B^{-1}U) = 0$ and

$$\operatorname{Hom}_A(\tau_B^{-1}U, P(a)) = \operatorname{Hom}_B(\tau_B^{-1}U, N) = 0$$

(because N is projective in $\bmod B$). Therefore, $\dim_K \operatorname{End}_A V = 1$ and V_A is indecomposable. Moreover, V belongs to $\mathcal{P}(A)$, because it is a predecessor of $\tau_B^{-1} U$. On the other hand, we have

$$
\begin{aligned}
q_A(\dim V) &= \langle \dim V, \dim V \rangle_A \\
&= \langle \dim P(a) + \dim \tau_B^{-1} U, \dim P(a) + \dim \tau_B^{-1} U \rangle_A \\
&= q_A(\dim P(a)) + q_A(\dim \tau_B^{-1} U) + \langle \dim P(a), \dim \tau_B^{-1} U \rangle_A \\
&\quad + \langle \dim \tau_B^{-1} U, \dim P(a) \rangle_A \\
&\leq 1 + 1 + 0 - 3 = -1.
\end{aligned}
$$

Therefore, $1 - \dim_K \operatorname{Ext}_A^1(V, V) = q_A(\dim V) < 0$, and so $\operatorname{Ext}_A^1(V, V) \neq 0$, which contradicts the fact that V lies in $\mathcal{P}(A)$ and finishes the proof of (a). Because (b) follows from (a) and from the duality $D : \bmod A \to \bmod A^{\mathrm{op}}$, the proposition is proved. □

We finally prove the announced result.

5.7. Theorem. *Let Q be a finite, connected, and acyclic quiver that is not Dynkin, and let B be a concealed algebra of type Q. Then all but finitely many indecomposable postprojective B-modules are sincere.*

Proof. Let $A = KQ$ and $B = \operatorname{End} T_A$ for some postprojective tilting module T_A. We know from (VIII.4.5) that the unique postprojective component $\mathcal{P}(B)$ of $\Gamma(\bmod B)$ consists of modules of the form $\operatorname{Hom}_A(T, M)$, where M ranges over all but finitely many isomorphism classes of indecomposable postprojective A-modules. Moreover, in view of (VI.3.10), if $T = T_1 \oplus \ldots \oplus T_n$ is a decomposition of T into indecomposable A-modules, then the modules $\operatorname{Hom}_A(T, T_i)$ form a complete set of representatives of the indecomposable projective B-modules, and these modules lie in $\mathcal{P}(B)$. Fix an index $i \in \{1, \ldots, n\}$. Because T_i lies in the postprojective component $\mathcal{P}(A)$ of $\Gamma(\bmod A)$, there exist $a_i \in Q_0$ and $m_i \geq 0$ such that $T_i \cong \tau^{-m_i} P(a_i)$. Further, in view of (VI.3.8) and (IV.2.15), for any indecomposable module M from $\mathcal{P}(A)$ with $\operatorname{Hom}_A(T, M) \neq 0$, there are isomorphisms

$$
\operatorname{Hom}_B(\operatorname{Hom}_A(T, T_i), \operatorname{Hom}_A(T, M)) \cong \operatorname{Hom}_A(T_i, M) \cong \operatorname{Hom}_A(P(a_i), \tau^{m_i} M).
$$

Because, by (5.6), $\operatorname{Hom}_A(P(a_i), N) \neq 0$ for all but finitely many modules N in $\mathcal{P}(A)$, we deduce that $\operatorname{Hom}_B(\operatorname{Hom}_A(T, T_i), X) \neq 0$ for all but finitely many modules X in $\mathcal{P}(B)$, as required. □

IX.6. Gentle algebras and tilted algebras of type $\mathbb{A}_n$

In this section, we consider a class of algebras, the gentle algebras, because they offer a particularly interesting example and because we need in the sequel a subclass, that of the tilted algebras of type $\mathbb{A}_n$. We give here a complete classification of the latter.

6.1. Definition. Let A be an algebra with acyclic quiver Q_A. The algebra $A \cong KQ_A/\mathcal{I}$ is called **gentle** if the bound quiver $(Q_A, \mathcal{I})$ has the following properties:

(G1) Each point of Q_A is the source and the target of at most two arrows.

(G2) For each arrow $\alpha \in (Q_A)_1$, there is at most one arrow β and one arrow γ such that $\alpha\beta \notin \mathcal{I}$ and $\gamma\alpha \notin \mathcal{I}$.

(G3) For each arrow $\alpha \in (Q_A)_1$, there is at most one arrow ξ and one arrow ζ such that $\alpha\xi \in I$ and $\zeta\alpha \in I$.

(G4) The ideal $\mathcal{I}$ is generated by paths of length two.

If Q_A is a tree, the gentle algebra $A \cong KQ_A/\mathcal{I}$ is called an algebra given by a gentle tree, or simply, a **gentle tree algebra**.

6.2. Examples. The following three bound quiver algebras are gentle:

(a) the algebra A given by the quiver

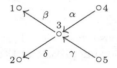

bound by $\alpha\beta = 0$, $\gamma\delta = 0$;

(b) the algebra B given by the quiver

$$\underset{1}{\circ} \xleftarrow{\varepsilon} \underset{2}{\circ} \xleftarrow{\delta} \underset{3}{\circ} \xrightarrow{\gamma} \underset{4}{\circ} \xleftarrow{\beta} \underset{5}{\circ} \xleftarrow{\alpha} \underset{6}{\circ}$$

bound by $\alpha\beta = 0$, $\delta\varepsilon = 0$; and

(c) the algebra C given by the quiver

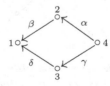

bound by $\alpha\beta = 0$, $\gamma\delta = 0$.

We now show that the tilted algebras of type $\mathbb{A}_n$ are gentle. To do so, we start by proving a lemma measuring the Hom-spaces in a hereditary algebra of type $\mathbb{A}_n$. We notice first that, over a hereditary algebra of type $\mathbb{A}_n$, the middle term of any almost split sequences is a direct sum of at most two indecomposable modules [this indeed follows from (IV.3.9), (VII.1.6), and (VII.5.13)]. Consequently, every point in the Auslander–Reiten quiver is the source or target of at most two sectional paths. We need the following notation.

Let A be a representation–directed algebra satisfying the separation condition, and assume that the middle term of any almost split sequence in $\mathrm{mod}\,A$ is a direct sum of at most two indecomposable modules. Let M be an indecomposable A-module. Draw the two maximal sectional paths starting at M (that is, sectional paths, that are not properly contained in other sectional paths). They have respective targets M_1 and M_2, and they determine a full subquiver Σ of $\Gamma(\mathrm{mod}\,A)$ with underlying graph $\mathbb{A}_n$. We construct $\mathbb{Z}\Sigma$ in which there is a unique maximal sectional path starting at each of M_1 and M_2. These two sectional paths intersect at a point X in $\mathbb{Z}\Sigma$ (which may not correspond to an indecomposable A-module). We then let $\mathcal{R}(M)$ denote the set of all indecomposable A-modules N such that there is a path

$$M \longrightarrow \cdots \longrightarrow N \longrightarrow \cdots \longrightarrow X$$

in $\mathbb{Z}\Sigma$. For example, let A be the path algebra of the quiver

$$\underset{6}{\circ}\longrightarrow\underset{5}{\circ}\longrightarrow\underset{1}{\circ}\longleftarrow\underset{2}{\circ}\longleftarrow\underset{3}{\circ}\longleftarrow\underset{4}{\circ}$$

and M_A be the indecomposable A-module such that $\mathbf{dim}\,M = 011110$. We have indicated in the following picture of $\Gamma(\mathrm{mod}\,A)$ the points of $\mathcal{R}(M)$ by black dots:

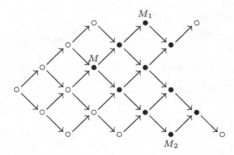

6.3. Lemma. *Let A be a representation–directed algebra satisfying the separation condition, and assume that the middle term of any almost split sequences in $\mathrm{mod}\,A$ is a direct sum of at most two indecomposable modules. Let M and N be indecomposable A-modules. Then $\dim_K \mathrm{Hom}_A(M,N) = 1$ if and only if $N \in \mathcal{R}(M)$, and $\mathrm{Hom}_A(M,N) = 0$ otherwise.*

Proof. Let $0 \to N' \to N \to N'' \to 0$ be a short exact sequence in $\operatorname{mod} A$ such that the modules N' and N'' are indecomposable and $\operatorname{Hom}_A(M, N') \neq 0$. Applying the functor $\operatorname{Hom}_A(M, -)$ yields an exact sequence

$$0 \longrightarrow \operatorname{Hom}_A(M, N') \longrightarrow \operatorname{Hom}_A(M, N) \longrightarrow \operatorname{Hom}_A(M, N'') \longrightarrow \operatorname{Ext}_A^1(M, N').$$

Assume $\operatorname{Ext}_A^1(M, N') \neq 0$. By (IV.2.13), there exists a homomorphism $N' \to \tau M$ that induces a cycle

$$M \to N' \to \tau M \to * \to M,$$

contrary to the assumption that A is representation–directed. This shows that $\operatorname{Ext}_A^1(M, N') = 0$, and we get

$$\dim_K \operatorname{Hom}_A(M, N) = \dim_K \operatorname{Hom}_A(M, N') + \dim_K \operatorname{Hom}_A(M, N''),$$

that is, the function $f_M = \dim_K \operatorname{Hom}_A(M, -)$ is additive on short exact sequences with indecomposable end terms, provided it is nonzero on the first term. Clearly, $f_M(M) = 1$. Also, by (IV.5.6), if $f_M(N) \neq 0$, then N is a successor of M. The result follows from an easy induction. $\qquad\square$

6.4. Corollary. *Let A be a hereditary algebra of type $\mathbb{A}_n$ and T_A be a tilting module. Then $B = \operatorname{End} T_A$ is a gentle algebra.*

Proof. Let $T(a)$ be an indecomposable summand of T and $T(b)$ be another indecomposable summand such that $\operatorname{Hom}_A(T(a), T(b)) \neq 0$. Assume first that $T(a)$ is not injective. Because

$$\operatorname{Hom}_A(\tau^{-1} T(a), T(b)) \cong D\operatorname{Ext}_A^1(T(b), T(a)) = 0,$$

$T(b)$ is a successor of $T(a)$ but not of $\tau^{-1} T(a)$; hence it lies on one of the (at most two) maximal sectional paths starting with $T(a)$. This is also (trivially) the case if $T(a)$ is injective, for then $\mathcal{R}(T(a))$ is reduced to these two paths. Because

$$\dim_K \operatorname{Hom}_A(T(a), T(b)) \leq 1,$$

in view of (VI.3.10) there is exactly one nonzero path from b to a in Q_B. Similarly, if $T(c)$ is another summand of T such that $\operatorname{Hom}_A(T(c), T(a)) \neq 0$, then $T(c)$ lies on one of the (at most two) sectional paths ending with $T(a)$, and there is exactly one nonzero path from a to c in Q_B. This shows (G1).

If $T(c)$, $T(a)$, and $T(b)$ are as described earlier and they lie on the same sectional path, then $\operatorname{Hom}_A(T(c), T(b)) \neq 0$ (by (2.2)). If, on the other hand, they do not lie on the same sectional path, then in particular $T(c)$ is not injective and $\operatorname{Hom}_A(\tau^{-1}T(c), T(b)) = 0$ implies $\operatorname{Hom}_A(T(c), T(b)) = 0$ (see the following picture). This shows (G2) and (G3).

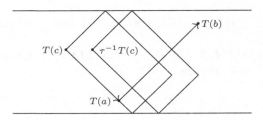

Because there are at most two sectional paths starting or ending at each indecomposable summand of T, the argument also proves (G4).   □

We now show that tilted algebras of type $\mathbb{A}_n$ are given by gentle trees. For this purpose, it suffices, by (4.3), to show that they satisfy the separation condition.

6.5. Proposition. *Let A be a representation–finite hereditary algebra, T_A be a tilting module, and $B = \operatorname{End} T_A$. Then B satisfies the separation condition.*

Proof. Assume to the contrary that B does not satisfy the separation condition. Then there exists $a \in (Q_B)_0$ such that $\operatorname{rad} P(a)_B$ has two indecomposable summands M_1 and M_2, having two points b_1 and b_2 in their respective supports, which are connected by a walk in $Q_B(\overrightarrow{a})$. By (VI.3.8), this walk induces a walk linking $P(b_1)$ and $P(b_2)$ in $\Gamma(\operatorname{mod} B)$ (on which no module has a in its support). Because there exist paths from $P(b_1)$ to M_1 and $P(b_2)$ to M_2, this yields a closed walk w in $\Gamma(\operatorname{mod} B)$

$$P(a) \longleftarrow M_1 \longleftarrow \cdots \longleftarrow P(b_1) \text{——} \cdots \text{——} P(b_2) \longrightarrow \cdots \longrightarrow M_2 \longrightarrow P(a).$$

We can, of course, assume w to be of minimal length. By (VIII.3.5), $\Gamma(\operatorname{mod} B)$ is acyclic, hence w contains sources and sinks. One can suppose that every sink corresponds to a projective B-module. Indeed, if the sink U is such that U_B is not projective, then we can replace U by τU and each arrow $V \to U$ by the corresponding arrow $\tau U \to V$ (note that this process does not affect the length of w). On the other hand, the minimality of w implies that, repeatedly applying this process, we cannot reach another point of the original walk w.

Let Y be a point of w that is not a sink. There exists a path from Y to some sink P, but P_B is projective and hence belongs to the torsion-free class $\mathcal{Y}(T_A)$. Because T_A is splitting, $\mathcal{Y}(T_A)$ is closed under predecessors. Thus $Y \in \mathcal{Y}(T_A)$, and all modules on w belong to $\mathcal{Y}(T_A)$.

Let Z be a source on w and put $N_A = Z \otimes_B T$. If N is not injective, then by (VI.5.2), the almost split sequence starting with $Z \cong \operatorname{Hom}_A(T, N)$ lies entirely in $\mathcal{Y}(T_A)$; hence we may replace Z by $\tau^{-1}Z$ and each arrow $Z \to Y$ by the corresponding arrow $Y \to Z$, thus obtaining a new walk w' in $\mathcal{Y}(T_A)$ of the same length as w and such that, for each source Z on w' the A-module N is injective. We note that w' may have sinks that do not correspond to projectives: what matters to us is that it still lies entirely inside $\mathcal{Y}(T_A)$.

Applying the functor $- \otimes_B T$, we obtain a closed walk $\overline{w}'$ in $\Gamma(\operatorname{mod} A)$ having all its sources injective. This however is impossible, because A is a hereditary algebra of Dynkin type and hence satisfies the coseparation condition.   □

6.6. Corollary. *Let B be a tilted algebra of type $\mathbb{A}_n$. Then B is a gentle tree algebra.*

Proof. This follows from (6.4), (6.5), and (4.3).   □

Our present objective is to characterise among the gentle tree algebras which ones are tilted of type $\mathbb{A}_n$. That they are not all so is shown by Example 6.2 (b). In fact, we show in (6.11) that this is essentially the only "bad" example.

6.7. Proposition. *Let $A = KQ_A/\mathcal{I}$ be a gentle tree K-algebra and $n = |(Q_A)_0|$. Then there exists a sequence of algebras $A = A_0, A_1, \ldots, A_m$ and a sequence of separating tilting modules $T_{A_0}^{(0)}, \ldots, T_{A_{m-1}}^{(m-1)}$ such that*

$$A_{j+1} = \operatorname{End} T_{A_j}^{(j)}, \quad or \quad A_{j+1} = (\operatorname{End} T_{A_j}^{(j)})^{\mathrm{op}}$$

for all $j \in \{1, \ldots, m\}$, and A_m is hereditary of type $\mathbb{A}_n$. In particular, the algebra A is representation–finite.

Proof. We show that we can tilt A to another algebra given by a gentle tree, having one fewer relation. The statement follows from an obvious induction on the number of relations on Q_A. Up to duality, we can assume that Q_A has a sink with exactly one neighbour so that the bound quiver of A has the form

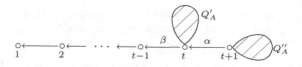

with $\alpha\beta = 0$ and $t \geq 2$. It follows that the beginning of the Auslander–Reiten quiver $\Gamma(\operatorname{mod} A)$ has the form

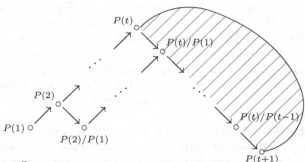

We define $T_A = \bigoplus_{i=1}^n T(i)$ by

$$T(i) = \begin{cases} P(t)/P(t-i) & 1 \leq i < t, \\ P(i) & i \geq t. \end{cases}$$

It is easy to see that T_A is a tilting module. We now show that T_A is separating. Let $\mathcal{A}$ denote the additive full subcategory of $\operatorname{mod} A$ consisting of direct sums of the indecomposable modules the support of which lies completely inside $\{1, \ldots, t-1\}$.

We claim that $\mathcal{F}(T) = \mathcal{A}$, whereas $\mathcal{T}(T)$ consists of direct sums of the remaining indecomposable modules. Indeed, because for each $i < t$ and each indecomposable A-module M, $\operatorname{Hom}_A(P(t)/P(i), M)$ is a subspace of $\operatorname{Hom}_A(P(t), M)$, we have $\operatorname{Hom}_A(T, M) = 0$ if and only if $\operatorname{Hom}_A(P(j), M) = 0$ for all $j \geq t$. This shows that $\mathcal{F}(T) = \mathcal{A}$.

To show that $\mathcal{T}(T)$ consists of direct sums of the remaining indecomposable modules, it suffices, by maximality of the torsion class, to show that if $M \notin \mathcal{A}$ is indecomposable, then $\operatorname{Hom}_A(M, -)|_{\mathcal{A}} = 0$. So, let $N \in \mathcal{A}$ be an indecomposable A-module such that $\operatorname{Hom}_A(M, N) \neq 0$. Applying (IV.5.1) repeatedly, our assumptions that $\operatorname{Hom}_A(M, N) \neq 0$ and $M \notin \mathcal{A}$ imply that $\operatorname{Hom}_A(M, P(1)) \neq 0$, but this is impossible, because $P(1)$ is simple projective.

We claim that the bound quiver of $B = \operatorname{End} T_A$ has the following form

$$Q'_B \underset{t}{\overset{}{\bigcirc}} \circ \overset{\alpha_t}{\underset{t-1}{\longleftarrow}} \circ \longleftarrow \cdots \longleftarrow \circ \overset{\alpha_1}{\underset{1}{\longleftarrow}} \underset{t+1}{\circ} \bigcirc Q''_B$$

where $Q'_B = Q'_A$ and is bound by the same relations as Q'_A, whereas $Q''_B = Q''_A$ and is bound by the same relations as Q''_A. Moreover, the path

$$t+1 \to 1 \to \cdots \to t-1 \to t$$

is not bound; there exists a relation of the form $\xi\alpha_1$ in B if and only if there exists a corresponding relation $\xi\alpha$ in A; and there are no other relations in B involving the arrows $\alpha_1, \ldots, \alpha_t$.

Because the points of Q_B lying inside Q'_B or Q''_B correspond to those summands of T_A that are the indecomposable projective A-modules corresponding to the points of Q'_A or Q''_A, respectively, we deduce that $Q'_B = Q'_A$, $Q''_B = Q''_A$, and they are bound by the same relations as Q'_A or Q''_A, respectively.

In view of (VI.3.10), the existence of the irreducible morphisms

$$P(t) \longrightarrow P(t)/P(1) \longrightarrow \cdots \longrightarrow P(t)/P(t-1) \longrightarrow P(t+1)$$

in $\operatorname{mod} A$ implies the existence of the arrows

$$t+1 \xrightarrow{\alpha_1} 1 \xrightarrow{\alpha_2} \cdots \longrightarrow t-1 \xrightarrow{\alpha_t} t$$

in Q_B. Clearly, for $i \in \{1, \ldots, t-1\}$, there is no homomorphism from $P(t)/P(i)$ to a projective corresponding to a point in Q'_A and no homomorphism from a projective corresponding to a point in Q''_A to $P(t)/P(i)$. On the other hand, all the homomorphisms from projectives corresponding to points of Q'_A to $P(t)/P(i)$ must factor through $P(t)$, and all the homomorphisms from $P(t)/P(i)$ to projectives corresponding to points of Q''_A must factor through $P(t+1)$. Thus Q_B has the required form.

Next, if there exists a relation starting in $a \in (Q''_A)_0)$ and ending in t, it must be of the form $\xi\alpha = 0$, where $\xi : a \to t+1$. It is replaced in B by a relation of the form $\xi\alpha_1 = 0$, because $\operatorname{Hom}_A(P(t), P(a)) = 0$ implies $\operatorname{Hom}_A(P(t)/P(1), P(a)) = 0$.

We claim that there are no new relations. Indeed, a new relation can either start at Q''_B and end at some $i \in \{1, \ldots, t-1\}$ or start at some $i \in \{1, \ldots, t-1\}$ and end in Q'_B. Suppose $a \in (Q'_A)_0$ is such that $\operatorname{Hom}_A(P(a), P(t)/P(i)) = 0$ but $\operatorname{Hom}_A(P(a), P(t)) \neq 0$. Then there exists a nonzero homomorphism $P(a) \to P(t)$ having its image in $P(i) \subset P(t)$, which is a contradiction.

Finally, if $b \in (Q''_A)_0$ is such that $\operatorname{Hom}_A(P(t)/P(i), P(b)) = 0$ but $\operatorname{Hom}_A(P(t+1), P(b))$ is nonzero, we again have $\operatorname{Hom}_A(P(t), P(b)) = 0$ and hence one of the zero relations discussed earlier. Thus, in particular, B is given by a gentle tree with one fewer zero relation.

To finish the proof, assume that A is representation–infinite. Because T is separating, B is also representation–infinite. But applying this process inductively, we end with a hereditary algebra of type $\mathbb{A}_n$, which is representation–infinite and thus we have a contradiction. $\square$

6.8. Corollary. *Let* $A = KQ_A/\mathcal{I}$ *be a gentle tree algebra.*

(a) *If* $n = |(Q_A)_0|$, *there exists a hereditary algebra* H *of type* $\mathbb{A}_n$; *a sequence of algebras* $H = A_0, A_1, \ldots, A_m = A$; *and a sequence of splitting tilting modules* $T_{A_0}^{(0)}, T_{A_1}^{(1)}, \ldots, T_{A_{m-1}}^{(m-1)}$ *such that*

$$A_{i+1} = \operatorname{End} T_{A_i}^{(i)} \quad \text{or} \quad A_{i+1} = (\operatorname{End} T_{A_i}^{(i)})^{\operatorname{op}}$$

for all $i \in \{1, \ldots, m-1\}$.

(b) *The algebra* A *satisfies the separation condition.*

(c) *The middle term of any almost split sequences is a direct sum of at most two indecomposable modules.*

(d) *Assume that* M *and* N *are indecomposable modules in* mod A. *Then* $\dim_K \mathrm{Hom}_A(M, N) = 1$ *if and only if* $N \in \mathcal{R}(M)$, *and* $\dim_K \mathrm{Hom}_A(M, N) = 0$ *otherwise.*

Proof. (a) This follows from the fact that $B \cong \mathrm{End}\, T_A$, where T_A is a separating tilting module, if and only if $A \cong (\mathrm{End}\,_B T)^{\mathrm{op}}$, where $_B T$ is a splitting tilting module.

(b) Because Q_A is a tree, we just apply (4.3).

(c) We apply the description of the almost split sequences in (VI.5.2).

(d) We apply (c) and (6.3). $\qquad\qquad\qquad\qquad\qquad\qquad\qquad$ $\square$

6.9. Lemma. *Let* B *be a tilted algebra of type* $\mathbb{A}_n$. *Then the bound quiver of* B *contains no full bound subquiver of the form*

$$\underset{1}{\circ} \xleftarrow{\;\;\delta\;\;} \underset{2}{\circ} \xleftarrow{\;\;\gamma\;\;} \underset{3}{\circ} \!-\!\! \cdots \!-\!\! \underset{t-2}{\circ} \xleftarrow{\;\;\beta\;\;} \underset{t-1}{\circ} \xleftarrow{\;\;\alpha\;\;} \underset{t}{\circ}$$

with $t \geq 4$, $\alpha\beta = 0$, $\gamma\delta = 0$; *all unoriented edges may be oriented arbitrarily; and there are no other zero relations between* 2 *and* $t-1$.

Proof. Assume first that the bound quiver of B contains such a subquiver with $t \geq 5$; then consider the indecomposable B-module M having as support the subquiver $3 \circ \!-\!\!\!-\!\! \cdots \!-\!\!\!-\! \circ\, t-2$ (that is, $M_i = K$ for $3 \leq i \leq t-2$ and $M_i = 0$ otherwise).

We claim that $\mathrm{pd}\, M > 1$. To construct the projective cover of M, we take all the sources $s_1, \ldots, s_k$ in $\mathrm{supp}\, M$, then $\mathrm{top}\, M \cong \bigoplus_{i=1}^{k} S(s_i)$ and the projective cover of M is $P = \bigoplus_{i=1}^{k} P(s_i)$. It remains to show that the kernel of the canonical surjection $p : P \to M$ is not projective. But there exists a source s_i and a path $s_i \to \cdots \to 3$, and $P(s_i)$ contains a submodule L, which is a direct summand of $\mathrm{Ker}\, p$, has simple top $S(2)$ but no simple composition factors isomorphic to $S(1)$. Now L is not projective; if it were, it would have $S(1)$ as a composition factor. Then $\mathrm{pd}\, M > 1$. Similarly, $\mathrm{id}\, M > 1$. Therefore, by (VIII.3.2)(e), B is not tilted.

It remains to consider the case where $t = 4$. Here, the bound quiver of B has the form

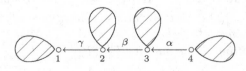

bound by $\alpha\beta = 0$, $\beta\gamma = 0$. We write the beginning of a minimal projective resolution for $S(4)$. Clearly, the canonical surjection $P(4) \to S(4)$ has in its kernel a summand Z having simple top $S(3)$. The projective cover of Z being $P(3)$, the canonical surjection $P(3) \to Z$ has in its kernel a summand Y having simple top $S(2)$. The kernel of the canonical surjection $P(2) \to Y$ has a summand X having simple top $S(1)$. Thus, the beginning of a minimal projective resolution for $S(4)$ is

$$P(1) \oplus P_1 \longrightarrow P(2) \oplus P_2 \longrightarrow P(3) \oplus P_3 \longrightarrow P(4) \longrightarrow S(4) \longrightarrow 0.$$

hence, $\operatorname{pd} S(4) \geq 3$ and so $\operatorname{gl.dim} B \geq 3$. Consequently, by (VIII.3.2)(e), the algebra B is not tilted.   □

6.10. Lemma. *Let A' be a gentle tree algebra with bound quiver*

$$\underset{a_0}{\circ}\!\!-\!\!-\!\!\circ-\cdots-\circ\overset{\beta_1}{\longleftarrow}\underset{a_1}{\circ}\overset{\alpha_1}{-\!\!-}\circ-\cdots-\circ\overset{\alpha_2}{\longrightarrow}\underset{a_2}{\circ}\overset{\beta_2}{\longrightarrow}\circ\cdots\circ-\!\!-\underset{a_r}{\circ}-\!\!-\circ\cdots\circ\overset{}{\longleftarrow}\underset{a_{r+1}}{\circ}$$

such that there is no zero relation having its midpoint between a_j and a_{j+1}; there is a zero relation of midpoint a_r pointing left or right according to whether r is odd or even; and no two consecutive zero relations point in the same direction. Assume that there exists a path $I(a_0) \to \cdots \to P(a_{r+1})$ in $\Gamma(\operatorname{mod} A')$. Then $r \leq 1$ and $\operatorname{Hom}_{A'}(I(a_0), P(a_{r+1})) \neq 0$.

Proof. Let, for each j such that $0 \leq j \leq r$, A'_j denote the (hereditary) algebra given by the full subquiver of $Q_{A'}$:

$$\underset{a_j}{\circ}\ \overline{}\ \circ\ \overline{}\ \cdots\ \overline{}\ \circ\ \overline{}\ \circ\ a_{j+1}$$

Then it is easily seen that $\Gamma(\operatorname{mod} A')$ has the following shape

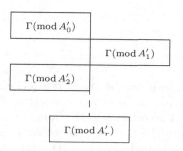

and $\Gamma(\operatorname{mod} A'_j) \cap \Gamma(\operatorname{mod} A'_{j+1}) = \{S(a_{j+1})\}$. In particular, the existence of a path from $I(a_0)$ to $P(a_{r+1})$ implies that the path must factor over $S(a_1)$, $r \leq 1$, and the quiver of A' is of the form

$$\underset{a_0}{\circ}\!\overset{}{\longleftarrow}\!\!-\!\!\circ\!\overset{}{\longleftarrow}\!\overline{}\ \cdots\ \overset{\beta_1}{\longleftarrow}\underset{a_1}{\circ}\!\overset{\alpha_1}{\longleftarrow}\ \cdots\ \overset{}{\longleftarrow}\underset{a_2}{\circ}$$

bound by $\alpha_1\beta_1 = 0$. Clearly, we then have $\operatorname{Hom}_{A'}(I(a_0), P(a_2))) \neq 0$.   □

6.11. Theorem. *A gentle tree algebra A is tilted of type $\mathbb{A}_n$ if and only if its bound quiver contains no full bound subquiver of the form*

$$
\underset{1}{\circ}\overset{\delta}{\longleftarrow}\underset{2}{\circ}\overset{\gamma}{\longleftarrow}\underset{3}{\circ}\!\!-\!\!\!-\circ\!\!-\!\!\!- \cdots -\!\!\!-\circ\!\!-\!\!\!-\underset{t-2}{\circ}\overset{\beta}{\longleftarrow}\underset{t-1}{\circ}\overset{\alpha}{\longleftarrow}\underset{t}{\circ}
$$

with $t \geq 4$, $\alpha\beta = 0$, $\gamma\delta = 0$; all unoriented edges may be oriented arbitrarily; and there are no other zero relations between 2 and $t - 1$.

Proof. Thanks to (6.9), we only need to show the sufficiency. We construct a section in $\Gamma(\mathrm{mod}\,A)$. We consider the connected full subquiver of $\Gamma(\mathrm{mod}\,A)$ consisting of those M such that there is a path $M \to \cdots \to P(s)$ for some source s in Q_A, and Σ is the right border of this subquiver, that is, Σ is the connected full subquiver of $\Gamma(\mathrm{mod}\,A)$ consisting of those M such that there is a path from M to $P(s)$ for some source s in Q_A, and every such path is sectional.

It follows from the definition of Σ that it is convex and that it intersects each τ-orbit at most once. We now show that Σ intersects each τ-orbit.

First, we notice that no indecomposable projective A-module is a proper successor of Σ. Indeed, let $P(a)$ be an indecomposable projective. There is a source s_a in Q_A and a path $s_a \to \cdots \to a$ that induces a path $P(a) \to \cdots \to P(s_a)$ in $\Gamma(\mathrm{mod}\,A)$. This establishes our claim.

Next, we show that no indecomposable injective A-module is a proper predecessor of Σ. Assume that there exists a path from $I(a)$ to Σ. We claim that $I(a)$ in fact lies on Σ. There exists a source s_a in Q_A and a path $I(a) \to \cdots \to P(s_a)$ in $\Gamma(\mathrm{mod}\,A)$. The hypothesis shows that the walk linking a to s_a in Q_A is of one of the forms

$$
\text{(I)} \quad \underset{a=a_0}{\circ}\!\!-\!\!\!-\cdots-\!\!\!-\circ\overset{\beta_1}{\longleftarrow}\underset{a_1}{\circ}\overset{\alpha_1}{\longleftarrow}\circ\cdots-\!\!\!-\circ\overset{\alpha_2}{\longrightarrow}\underset{a_2}{\circ}\overset{\beta_2}{\longrightarrow}\circ\!\!-\!\!\!-\cdots-\!\!\!-\circ\!\!-\!\!\!-\underset{a_r}{\circ}\!\!-\!\!\!-\cdots\longleftarrow-\!\!\!-\underset{a_{r+1}=s_a}{\circ}
$$

$$
\text{(II)} \quad \underset{a=a_0}{\circ}\!\!-\!\!\!-\cdots-\!\!\!-\circ\overset{\alpha_1}{\longrightarrow}\underset{a_1}{\circ}\overset{\beta_1}{\longrightarrow}\circ\cdots-\!\!\!-\circ\overset{\beta_2}{\longleftarrow}\underset{a_2}{\circ}\overset{\alpha_2}{\longleftarrow}\circ\!\!-\!\!\!-\cdots-\!\!\!-\circ\!\!-\!\!\!-\underset{a_r}{\circ}\!\!-\!\!\!-\cdots\longleftarrow-\!\!\!-\underset{a_{r+1}=s_a}{\circ}
$$

where unoriented edges may be oriented arbitrarily; there are zero relations with midpoints $a_1, \ldots, a_r$, no two consecutive of which are oriented in the same direction; and no other zero relations.

We consider only the case (I); the other is similar. Let A' be the algebra given by the bound quiver of (I), and let $I'(a)$ and $P'(s_a)$ denote, respectively, the indecomposable injective A'-module corresponding to a, and the indecomposable projective A'-module corresponding to s_a. Let $E : \mathrm{mod}\,A' \to \mathrm{mod}\,A$ be the full, faithful, and exact embedding defined by $E(M)_i = M_i$ if $i \in (Q_{A'})_0$; $E(M)_i = 0$ if $i \notin (Q_{A'})_0$; $E(M)_\alpha = M_\alpha$ if $\alpha \in (Q_{A'})_1$; and $E(M)_\alpha = 0$ if $\alpha \notin (Q_{A'})_1$ (under the identification of modules over A and A' with representations of corresponding bound quivers). Then if $R : \mathrm{mod} \to \mathrm{mod}\,A'$ denotes the restriction functor, we have

$RE = 1_{\operatorname{mod} A'}$. Thus $EI'(a)_a \neq 0$, and hence there exists a nonzero homomorphism $EI'(a) \to I(a)$. Similarly, we have a nonzero homomorphism $P(s_a) \to EP'(s_a)$. Thus, the existence of a path $I(a) \to \cdots \to P(s_a)$ implies the existence of a path

$$EI'(a) \longrightarrow I(a) \longrightarrow M_1 \longrightarrow \cdots \longrightarrow M_m \longrightarrow P(s_a) \longrightarrow EP'(s_a).$$

We claim that, by applying the functor R, this yields a path in $\Gamma(\operatorname{mod} A')$ from $I'(a)$ to $P'(s_a)$. Indeed, because $RE = 1_{\operatorname{mod} A'}$, this occurs if $\operatorname{supp} M_j \cap Q_{A'} \neq \emptyset$ for all j with $1 \leq j \leq m$.

Let Q_b denote the branch

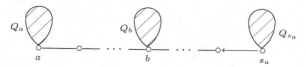

of the tree $Q_{A'}$ attached at the point b of $Q_{A'}$. Suppose that $\operatorname{supp} M_j \cap Q_{A'} = \emptyset$ for some j. Because $\operatorname{supp} I(a) \cap Q_{A'} \neq \emptyset$ and $\operatorname{supp} P(s_a) \cap Q_{A'} \neq \emptyset$, there exist t_1 and t_2 with $t_1 < t_2$ such that all M_t (with $t_1 \leq t < t_2$) have their supports not intersecting $Q_{A'}$, whereas $M_{t_1 - 1}$ and M_{t_2} have their supports intersecting $Q_{A'}$. Because M_{t_1} is indecomposable and $\operatorname{supp} M_{t_1} \cap Q_{A'} = \emptyset$, there exists $b \in (Q_{A'})_0$ such that $\operatorname{supp} M_{t_1} \subseteq Q_b$. For the same reason, all the M_t, with $t_1 \leq t < t_2$, have their supports inside the same Q_b. However, $\operatorname{Hom}_A(M_{t_1 - 1}, M_{t_1}) \neq 0$ and $\operatorname{Hom}_A(M_{t_2 - 1}, M_{t_2}) \neq 0$ imply that

$$\operatorname{supp} M_{t_1 - 1} \cap \operatorname{supp} M_{t_1} \neq \emptyset \quad \text{and} \quad \operatorname{supp} M_{t_2 - 1} \cap \operatorname{supp} M_{t_2} \neq \emptyset.$$

Therefore, $b \in \operatorname{supp} M_{t_1 - 1}$ and $b \in \operatorname{supp} M_{t_2}$; hence there exist nonzero homomorphisms $f_1 : P(b) \to M_{t_1 - 1}$ and $f_2 : P(b) \to M_{t_2}$. Let g denote the composition $M_{t_1 - 1} \longrightarrow M_{t_1} \longrightarrow \cdots \longrightarrow M_{t_2}$. Because $b \notin \operatorname{supp} M_{t_1}$, we have $\operatorname{Hom}_A(P(b), M_{t_1}) = 0$; hence $gf_1 = 0$. But, by (6.8)(d), any two paths from $P(b)$ to M_{t_2} give rise to the same homomorphism, up to scalar multiplication, hence $\operatorname{Hom}_A(P(b), M_{t_2}) = 0$, which is a contradiction.

We thus have the required path in $\Gamma(\operatorname{mod} A')$. Then (6.10) yields $\operatorname{Hom}_{A'}(I'(a), P'(s_a)) \neq 0$. Hence $\operatorname{Hom}_A(EI'(a), EP'(s_a)) \neq 0$ implies that $\operatorname{Hom}_A(I(a), P(s_a)) \neq 0$. Because $I(a)$ is injective and $P(s_a) \neq 0$ is projective, according to (6.3) and (6.8), there is a sectional path from $I(a)$ to $P(s_a)$, and so $I(a)$ lies on Σ.

This completes the proof that Σ is a section. Clearly, $\operatorname{Hom}_A(U, \tau V) = 0$ for all U, V on Σ. To apply (VIII.5.6), it suffices to observe that the direct sum $\bigoplus_{M \in \Sigma_0} M$ is a tilting module (and therefore is faithful): indeed, we have just seen that the number of points on Σ equals the rank of the group $K_0(A)$, and that $\operatorname{Ext}_A^1(U, V) = 0$ for all U, V on Σ; on the other hand,

$\text{Hom}_A(DA, \tau U) = 0$ for all $U \in \Sigma_0$, because no injective lies on the left of Σ, and thus $\text{pd}\, U \le 1$. □

To sum up, we have proved the following useful fact.

6.12. Corollary. *An algebra is tilted of type* $\mathbb{A}_n$ *if and only if its bound quiver is a finite connected full bound subquiver of the infinite tree*

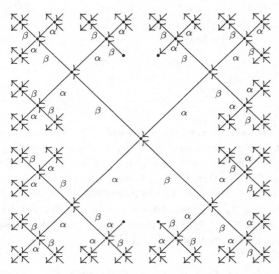

bound by all possible relations of the forms $\alpha\beta = 0$ *and* $\beta\alpha = 0$ *and contains no full bound subquiver of the form*

$$\underset{1}{\circ} \xleftarrow{\delta} \underset{2}{\circ} \xleftarrow{\gamma} \underset{3}{\circ} \text{---} \circ \text{---} \cdots \text{---} \circ \text{---} \underset{t-2}{\circ} \xleftarrow{\beta} \underset{t-1}{\circ} \xleftarrow{\alpha} \underset{t}{\circ}$$

with $t \ge 4$, $\alpha\beta = 0$, $\gamma\delta = 0$, *all unoriented edges may be oriented arbitrarily; and there are no other zero relations between* 2 *and* $t - 1$.

Observe that an algebra is a gentle tree algebra if and only if its bound quiver is a finite full bound subquiver of the infinite bound tree presented in (6.12). Moreover, it follows from (6.8) that any gentle tree algebra may be obtained from a hereditary algebra of type $\mathbb{A}_n$ by a finite sequence of tilts and cotilts.

IX.7. Exercises

1. Let A be a representation–finite algebra. Show that every path in $\text{mod}\, A$ gives rise to a path in $\Gamma(\text{mod}\, A)$, and conversely.

2. Let A be the path algebra of the Kronecker quiver o⇄o. Give an example of a sincere A-module M that is not faithful, and exhibit a cycle containing M.

3. Show that each of the following algebras is representation–directed but not tilted:

(a) A given by the quiver

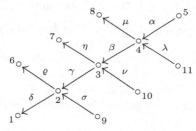

bound by $\alpha\beta\gamma = 0$, $\beta\gamma\delta = 0$, and $\gamma\delta\varepsilon = 0$;

(b) A given by the quiver

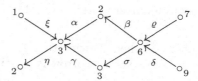

bound by $\alpha\mu = 0$, $\lambda\beta = 0$, $\beta\eta = 0$, $\nu\gamma = 0$, $\gamma\varrho = 0$, and $\sigma\delta = 0$;

(c) A given by the quiver

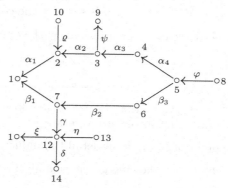

bound by $\beta\alpha = \sigma\gamma$, $\varrho\beta = 0$, $\varrho\sigma = 0$, $\xi\eta = 0$, and $\alpha\eta = 0$; and

(d) A given by the quiver

bound by $\alpha_2\alpha_1 = 0$, $\alpha_4\alpha_3 = 0$, $\beta_2\beta_1 = 0$, $\beta_3\beta_2 = 0$, $\varphi\alpha_4 = 0$, $\varphi\beta_3 = 0$, $\alpha_3\psi = 0$, $\varrho\alpha_1 = 0$, $\beta_2\gamma = 0$, $\gamma\xi = 0$, and $\eta\delta = 0$.

4. Show that representation–directed algebras have finite global dimension.

5. Let A be a representation–directed algebra and $a \in (Q_A)_0$. Show that $P(a)$ does not have a separated radical if and only if there is a closed walk in $\Gamma(\operatorname{mod} A)$ of the form

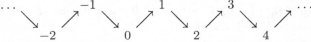

$$P(a) \xleftarrow{\ \alpha_0\ } M_1 \xrightarrow{\ \alpha_1\ } M_2 \xrightarrow{\ \alpha_2\ } \cdots \xrightarrow{\ \alpha_{t-1}\ } M_t \xrightarrow{\ \alpha_t\ } P(a),$$

where, for each i with $1 \leq i \leq t$, we have $\operatorname{supp} M_i \subseteq Q_A(\overrightarrow{a})$.

6. Let A be a representation–finite algebra satisfying the separation condition.

(a) Let M and N be two indecomposable A-modules such that there exists a path in $\Gamma(\operatorname{mod} A)$ from M to N. Show that any two such paths contain exactly the same number of arrows.

(b) Let M be an indecomposable A-module and let $\mathbb{Z}\mathbb{A}_2$ be the infinite quiver

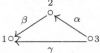

Find a unique translation quiver morphism $\pi : \Gamma(\operatorname{mod} A) \to \mathbb{Z}\mathbb{A}_2$ such that $\pi(M) = 0$.

7. In the proof of (2.6), when showing that Σ intersects each τ-orbit in $\mathcal{C}$, show in detail that there exists an arrow $N = \tau^l V \to U$.

8. In the proof of (4.5), do the case where $i = 1$.

9. Let A be a representation–finite algebra. Show that A satisfies the separation condition if and only if it satisfies the coseparation condition.

10. Let A be the gentle algebra given by the quiver

$$
\begin{array}{ccc}
 & 2 & \\
 & \circ & \\
\beta \nearrow & & \nwarrow \alpha \\
1 \circ & \xleftarrow{\quad} & \circ 3 \\
 & \gamma &
\end{array}
$$

bound by $\alpha\beta = 0$. Show that for every nonprojective separating tilting module T, the ordinary quiver of $\operatorname{End} T$ has a cycle.

11. In the proof of (6.3), do in detail the induction step.

12. For each of the following gentle tree algebras, construct a sequence as in (6.7):

(a) A given by the quiver

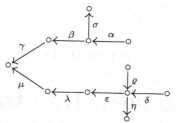

bound by $\alpha\beta = 0$, $\beta\gamma = 0$, $\lambda\mu = 0$, $\delta\varepsilon = 0$, $\varrho\eta = 0$ and
(b) A given by the quiver

$$\underset{n}{\circ} \xleftarrow{\alpha_{n-1}} \underset{n-1}{\circ} \xleftarrow{\alpha_{n-2}} \underset{n-2}{\circ} \longleftarrow \quad \cdots \quad \xleftarrow{\alpha_2} \underset{2}{\circ} \xleftarrow{\alpha_1} \underset{1}{\circ}$$

bound by $\alpha_i\alpha_{i+1} = 0$ for all i such that $1 \le i < n - 1$.

13. Show that an algebra B is tilted of the form $B = \operatorname{End} T_A$, where A is the path algebra $K\Delta$ of the quiver $\Delta : 1 \longleftarrow 2 \longleftarrow \cdots \longleftarrow n - 1 \longleftarrow n$ and T_A is a tilting A-module if and only if B is given by a finite connected full bound subquiver of the infinite tree

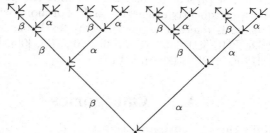

bound by all possible relations of the form $\alpha\beta = 0$.

14. Let A be a K-algebra given by the quiver

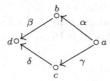

bound by $\alpha\beta = 0$ and $\gamma\delta = 0$. Show that the algebra A is representation–finite, representation–directed, and has a postprojective component. Describe the Auslander–Reiten quiver $\Gamma(\operatorname{mod} A)$ of A.

Appendix A

Categories, functors, and homology

For the convenience of the reader, we collect here the notations and terminology we use on categories, functors, and homology, and we recall some of the basic facts from category theory and homological algebra needed in the book.

We introduce the notions of category, additive category, K-category, abelian category, and the (Jacobson) radical of an additive category. We also collect basic facts from category theory and homological algebra. In this appendix we do not present proofs of the results, except for a few classical theorems that we frequently use in the book. For more details and complete proofs, the reader is referred to the following textbooks and papers on this subject [1], [2], [24], [41], [46], [47], [66], [70], [77], [95], [111], [112], [114], [115], [125], [129], [133], [148], and [149].

A.1. Categories

1.1. Definition. A **category** is a triple $\mathcal{C} = (\mathrm{Ob}\,\mathcal{C},\ \mathrm{Hom}\,\mathcal{C},\ \circ)$, where $\mathrm{Ob}\,\mathcal{C}$ is called the **class of objects** of $\mathcal{C}$, $\mathrm{Hom}\,\mathcal{C}$ is called the **class of morphisms** of $\mathcal{C}$, and $\circ$ is a partial binary operation on morphisms of $\mathcal{C}$ satisfying the following conditions:

(a) to each pair of objects X, Y of $\mathcal{C}$, we associate a set $\mathrm{Hom}_{\mathcal{C}}(X,Y)$, called the **set of morphisms** from X to Y, such that if $(X,Y) \neq (Z,U)$ then the intersection of the sets $\mathrm{Hom}_{\mathcal{C}}(X,Y)$ and $\mathrm{Hom}_{\mathcal{C}}(Z,U)$ is empty; and

(b) for each triple of objects X, Y, Z of $\mathcal{C}$, the operation

$$\circ : \mathrm{Hom}_{\mathcal{C}}(Y,Z) \times \mathrm{Hom}_{\mathcal{C}}(X,Y) \longrightarrow \mathrm{Hom}_{\mathcal{C}}(X,Z), \quad (g,f) \mapsto g \circ f$$

(called the **composition** of f and g), is defined and has the following two properties:

(i) $h \circ (g \circ f) = (h \circ g) \circ f$, for every triple $f \in \mathrm{Hom}_{\mathcal{C}}(X,Y)$, $g \in \mathrm{Hom}_{\mathcal{C}}(Y,Z)$, $h \in \mathrm{Hom}_{\mathcal{C}}(Z,U)$ of morphisms; and

(ii) for each object X of $\mathcal{C}$, there exists an element $1_X \in \operatorname{Hom}_{\mathcal{C}}(X, X)$, called the identity morphism on X, such that if $f \in \operatorname{Hom}_{\mathcal{C}}(X, Y)$ and $g \in \operatorname{Hom}_{\mathcal{C}}(Z, X)$ then $f \circ 1_X = f$ and $1_X \circ g = g$.

We often write $f : X \longrightarrow Y$ or $X \overset{f}{\longrightarrow} Y$ instead of $f \in \operatorname{Hom}_{\mathcal{C}}(X, Y)$, and we say that f is a morphism from X to Y. We also write $X \in \operatorname{Ob} \mathcal{C}$ to mean that X is an object of $\mathcal{C}$.

We say that a diagram in the category $\mathcal{C}$ is commutative whenever the composition of morphisms along any two paths with the same source and target are equal. For instance, we say that the diagram

$$
\begin{array}{ccc}
X & \overset{f}{\longrightarrow} & Y \\
{\scriptstyle h}\downarrow & & \downarrow{\scriptstyle g} \\
V & \overset{i}{\longrightarrow} & Z
\end{array}
$$

is **commutative** if $g \circ f = i \circ h$.

1.2. Definition. Let $\mathcal{C}$ be a category. A category $\mathcal{C}'$ is a subcategory of $\mathcal{C}$ if the following four conditions are satisfied:

(a) the class $\operatorname{Ob} \mathcal{C}'$ of objects of $\mathcal{C}'$ is a subclass of the class $\operatorname{Ob} \mathcal{C}$ of objects of $\mathcal{C}$;

(b) if X, Y are objects of $\mathcal{C}'$, then $\operatorname{Hom}_{\mathcal{C}'}(X, Y) \subseteq \operatorname{Hom}_{\mathcal{C}}(X, Y)$;

(c) the composition of morphisms in $\mathcal{C}'$ is the same as in $\mathcal{C}$; and

(d) for each object X of $\mathcal{C}'$, the identity morphism $1'_X$ in $\operatorname{Hom}_{\mathcal{C}'}(X, X)$ coincides with the identity morphism 1_X in $\operatorname{Hom}_{\mathcal{C}}(X, X)$.

A subcategory $\mathcal{C}'$ of $\mathcal{C}$ is said to be **full** if $\operatorname{Hom}_{\mathcal{C}'}(X, Y) = \operatorname{Hom}_{\mathcal{C}}(X, Y)$ for all objects X, Y of $\mathcal{C}'$.

Let X and Y be objects of a category $\mathcal{C}$. Any morphism $h : X \longrightarrow X$ in $\mathcal{C}$ is called an **endomorphism** of X. A morphism $u : X \longrightarrow Y$ in $\mathcal{C}$ is a **monomorphism** if for each object Z in $\operatorname{Ob} \mathcal{C}$ and each pair of morphisms $f, g \in \operatorname{Hom}_{\mathcal{C}}(Z, X)$ such that $u \circ f = u \circ g$, we have $f = g$. A morphism $p : X \longrightarrow Y$ in $\mathcal{C}$ is an **epimorphism** if for each object Z in $\operatorname{Ob} \mathcal{C}$ and each pair of morphisms $f, g \in \operatorname{Hom}_{\mathcal{C}}(Y, Z)$ such that $f \circ p = g \circ p$ we have $f = g$. A morphism $u : X \longrightarrow Y$ in $\mathcal{C}$ is an **isomorphism** if there exists a morphism $v : Y \longrightarrow X$ in $\mathcal{C}$ such that $uv = 1_Y$ and $vu = 1_X$. In this case, the morphism v is uniquely determined by u, it is called the **inverse** of u, and it is denoted by u^{-1}.

If there exists an isomorphism $u : X \longrightarrow Y$ in $\mathcal{C}$, we say that the objects X and Y are **isomorphic** in $\mathcal{C}$, and we write $X \cong Y$. It is easy to see that any isomorphism is both a monomorphism and an epimorphism. The converse implication does not hold in general (see Exercise 6.4).

A **direct sum** (or a **coproduct**) of the objects $X_1, \ldots, X_n$ of $\mathcal{C}$ is an object $X_1 \oplus \ldots \oplus X_n$ of $\mathcal{C}$ together with morphisms

$$u_j : X_j \longrightarrow X_1 \oplus \ldots \oplus X_n$$

for $j = 1, \ldots, n$, such that for each object Z in $\mathrm{Ob}\,\mathcal{C}$ and for each set of morphisms $f_1 : X_1 \longrightarrow Z, \ldots, f_n : X_n \longrightarrow Z$ in $\mathcal{C}$, there exists a unique morphism $f : X_1 \oplus \ldots \oplus X_n \longrightarrow Z$ such that $f_j = f \circ u_j$ for all $j = 1, \ldots, n$.

If such an object $X_1 \oplus \ldots \oplus X_n$ exists, it is unique, up to isomorphism. We often write $\bigoplus_{j=1}^{n} X_j$ instead of $X_1 \oplus \ldots \oplus X_n$.

For each $j \in \{1, \ldots, n\}$, the morphism $u_j : X_j \longrightarrow X_1 \oplus \ldots \oplus X_n$ is called the jth **summand embedding** (or summand injection).

1.3. Definition. A category $\mathcal{C}$ is an **additive category** if the following conditions are satisfied:

(a) for any finite set of objects $X_1, \ldots, X_n$ of $\mathcal{C}$ there exists a direct sum $X_1 \oplus \ldots \oplus X_n$ in $\mathcal{C}$;

(b) for each pair $X, Y \in \mathrm{Ob}\,\mathcal{C}$, the set $\mathrm{Hom}_{\mathcal{C}}(X, Y)$ of all morphisms from X to Y in $\mathcal{C}$ is equipped with an abelian group structure;

(c) for each triple of objects X, Y, Z of $\mathcal{C}$, the composition of morphisms in $\mathcal{C}$

$$\circ : \mathrm{Hom}_{\mathcal{C}}(Y, Z) \times \mathrm{Hom}_{\mathcal{C}}(X, Y) \longrightarrow \mathrm{Hom}_{\mathcal{C}}(X, Z)$$

is **bilinear**, that is, $(f + f') \circ g = f \circ g + f' \circ g$ and $f \circ (g + g') = f \circ g + f \circ g'$, for all morphisms $f, f' \in \mathrm{Hom}_{\mathcal{C}}(Y, Z)$ and all morphisms $g, g' \in \mathrm{Hom}_{\mathcal{C}}(X, Y)$; and

(d) there exists an object $0 \in \mathrm{Ob}\,\mathcal{C}$ (called the **zero object** of $\mathcal{C}$) such that the identity morphism 1_0 is the element zero of the abelian group $\mathrm{Hom}_{\mathcal{C}}(0, 0)$.

It is easy to see that the zero object of an additive category $\mathcal{C}$ is uniquely determined, up to isomorphism.

For any additive category $\mathcal{C}$, we define the **opposite category** $\mathcal{C}^{\mathrm{op}}$ of $\mathcal{C}$ to be the additive category the objects of which are the objects of $\mathcal{C}$, $\mathrm{Hom}_{\mathcal{C}^{\mathrm{op}}}(X, Y) = \mathrm{Hom}_{\mathcal{C}}(Y, X)$ for all objects X and Y in $\mathrm{Ob}\,\mathcal{C}$; the addition in $\mathrm{Hom}_{\mathcal{C}^{\mathrm{op}}}(X, Y)$ is the addition in $\mathrm{Hom}_{\mathcal{C}}(Y, X)$; and the composition $\circ'$ in $\mathrm{Hom}\,\mathcal{C}^{\mathrm{op}}$ is given by the formula $g \circ' f = f \circ g$, where $\circ$ is the composition in $\mathrm{Hom}\,\mathcal{C}$. It is clear that $(\mathcal{C}^{\mathrm{op}})^{\mathrm{op}} = \mathcal{C}$.

Assume that $\mathcal{C}$ is an additive category and let $X_1 \oplus \ldots \oplus X_n \in \mathrm{Ob}\,\mathcal{C}$ be the direct sum of objects $X_1, \ldots, X_n$ of $\mathcal{C}$. Let $u_j : X_j \longrightarrow X_1 \oplus \ldots \oplus X_n$ be the jth summand embedding. One can show that, for each $j \in \{1, \ldots, n\}$, there exists a morphism $p_j : X_1 \oplus \ldots \oplus X_n \longrightarrow X_j$ (called the jth **summand projection**) such that $p_j \circ u_j = 1_{X_j}$, $p_j \circ u_i = 0$ for all $i \neq j$ and

$u_1 \circ p_1 + \ldots + u_n \circ p_n = 1_{X_1 \oplus \ldots \oplus X_n}$. Moreover, given a set of morphisms $g_1 : X \longrightarrow X_1, \ldots, g_m : X \longrightarrow X_n$ in $\mathcal{C}$, there exists a unique morphism $g : X \longrightarrow X_1 \oplus \cdots \oplus X_n$ such that $p_j \circ g = g_j$ for $j = 1, \ldots, n$.

In presenting morphisms between direct sums of objects in an additive category $\mathcal{C}$, we use the following **matrix notation**. Given a set of morphisms

$$f_1 : X_1 \longrightarrow Y, \ldots, f_n : X_n \longrightarrow Y \text{ and } g_1 : Y \longrightarrow Z_1, \ldots, g_m : Y \longrightarrow Z_m$$

in $\mathcal{C}$ we denote by

$$f = [f_1 \ \ldots \ f_n] : X_1 \oplus \cdots \oplus X_n \longrightarrow Y, \quad g = \begin{bmatrix} g_1 \\ \vdots \\ g_m \end{bmatrix} : Y \longrightarrow Z_1 \oplus \cdots \oplus Z_m$$

the unique morphisms f and g in $\mathcal{C}$ such that $f \circ u_j = f_j$ for $j = 1, \ldots, n$ and $p_i \circ g = g_i$ for $i = 1, \ldots, m$, where $u_j : X_j \longrightarrow X_1 \oplus \cdots \oplus X_n$ is the jth summand embedding and $p_i : Z_1 \oplus \cdots \oplus Z_m \longrightarrow Z_i$ is the ith summand projection. If $X = X_1 \oplus \cdots \oplus X_n$ and $Z = Z_1 \oplus \cdots \oplus Z_m$, then any morphism $h : X \longrightarrow Z$ in $\mathcal{C}$ is identified with the $m \times n$ matrix

$$h = [h_{ij}] = \begin{bmatrix} h_{11} & h_{12} & \ldots & h_{1n} \\ h_{21} & h_{22} & \ldots & h_{2n} \\ \vdots & \vdots & \ddots & \vdots \\ h_{m1} & h_{m2} & \ldots & h_{mn} \end{bmatrix},$$

where $h_{ij} = p_i \circ h \circ u_j \in \mathrm{Hom}_{\mathcal{C}}(X_j, Z_i)$.

1.4. Definition. Let K be a field. A category $\mathcal{C}$ is a K-**category** if, for each pair $X, Y \in \mathrm{Ob}\, \mathcal{C}$, the set $\mathrm{Hom}_{\mathcal{C}}(X, Y)$ is equipped with a K-vector space structure such that the composition $\circ$ of morphisms in $\mathcal{C}$ is a K-bilinear map.

We note that for any object X of a K-category $\mathcal{C}$, the group

$$\mathrm{End}_{\mathcal{C}} X = \mathrm{Hom}_{\mathcal{C}}(X, X)$$

of all endomorphisms of X in $\mathcal{C}$, equipped with the multiplication $\circ$, is a K-algebra (not necessarily finite dimensional) with the identity 1_X. We call it the **endomorphism algebra** of X.

Throughout, we identify any object $X \in \mathcal{C}$ with the morphism $1_X \in \mathrm{Hom}_{\mathcal{C}}(X, X)$. This allows us to think about $\mathcal{C}$ as a class $\mathrm{Hom}\,\mathcal{C}$ of morphisms with the partial associative multiplication $\circ$ having "local" identities 1_X, where $X \in \mathrm{Ob}\, \mathcal{C}$. If, in addition, $\mathcal{C}$ is a K-category we think about $\mathcal{C}$ as a "partial" K-algebra $(\mathrm{Hom}\,\mathcal{C}, \circ, +)$ with "local" identities $1_X \in \mathrm{Hom}_{\mathcal{C}}(X, X)$ and local zeros $0_X \in \mathrm{Hom}_{\mathcal{C}}(X, X)$, where $X \in \mathrm{Ob}\, \mathcal{C}$; see [115].

Let $\mathcal{C}$ be an additive category and $f : X \longrightarrow Y$ be a morphism in $\mathcal{C}$. A **kernel** of f is an object $\mathrm{Ker}\, f$ together with a morphism $u : \mathrm{Ker}\, f \longrightarrow X$ satisfying the following two conditions: (1) $f \circ u = 0$, and (2) for any object Z of $\mathcal{C}$ and for any morphism $h : Z \longrightarrow X$ in $\mathcal{C}$ such that $f \circ h = 0$, there exists a unique morphism $h' : Z \longrightarrow \operatorname{Ker} f$ such that $h = u \circ h'$. A **cokernel** of f is an object $\operatorname{Coker} f$ together with a morphism $p : Y \longrightarrow \operatorname{Coker} f$ satisfying the following two conditions: (1) $p \circ f = 0$, and (2) for any object Z of $\mathcal{C}$ and for any morphism $g : Y \longrightarrow Z$ in $\mathcal{C}$ such that $h \circ f = 0$, there exists a unique morphism $g' : \operatorname{Coker} f \longrightarrow Z$ such that $g = g' \circ p$. It is clear that u is a monomorphism and p is an epimorphism.

Assume that every morphism in $\mathcal{C}$ admits a kernel and a cokernel. Then for each morphism $f : X \longrightarrow Y$ in $\mathcal{C}$, there exists a unique morphism $\overline{f}$ in $\mathcal{C}$ making the square in the following diagram

$$\operatorname{Ker} f \xrightarrow{\ u\ } X \xrightarrow{\ f\ } Y \xrightarrow{\ p\ } \operatorname{Coker} f$$
$$p' \downarrow \qquad \uparrow u'$$
$$\operatorname{Coker} u \xrightarrow{\ \overline{f}\ } \operatorname{Ker} p$$

commutative (that is, $f = u' \circ \overline{f} \circ p'$), where $u' : \operatorname{Ker} p \longrightarrow Y$ is the kernel of p and $p' : X \longrightarrow \operatorname{Coker} u$ is the cokernel of u. Indeed, because $p \circ f = 0$, there exists a unique morphism $f' : X \longrightarrow \operatorname{Ker} p$ such that $f = u' \circ f'$. Moreover, because $u' \circ f' \circ u = f \circ u = 0$ and u' is a monomorphism, $f' \circ u = 0$ and hence, by the definition of cokernel, there exists a unique morphism $\overline{f} : \operatorname{Coker} u \longrightarrow \operatorname{Ker} p$ such that $f' = \overline{f} \circ p'$. Consequently, the morphism $\overline{f}$ makes the preceding square commutative. One shows easily that $\overline{f}$ is unique. The object $\operatorname{Ker} p$ is called the **image** of f and is denoted by $\operatorname{Im} f$.

1.5. Definition. A category $\mathcal{C}$ is an **abelian category** if
(a) $\mathcal{C}$ is additive; and
(b) each morphism $f : X \longrightarrow Y$ in $\mathcal{C}$ admits a kernel $u : \operatorname{Ker} f \longrightarrow X$ of f and a cokernel $p : Y \longrightarrow \operatorname{Coker} f$ of f and the induced morphism $\overline{f} : \operatorname{Coker} u \longrightarrow \operatorname{Ker} p$ is an isomorphism.

Let $\mathcal{C}$ be an abelian category. A sequence (infinite or finite)

$$\cdots \longrightarrow X_{n+1} \xrightarrow{f_n} X_n \xrightarrow{f_{n-1}} X_{n-1} \longrightarrow \cdots$$

in $\mathcal{C}$ is said to be **exact** if $\operatorname{Ker} f_{n-1} = \operatorname{Im} f_n$, for all n. Any exact sequence of the form $0 \longrightarrow X \xrightarrow{f} Y \xrightarrow{g} Z \longrightarrow 0$ in $\mathcal{C}$ is called a **short exact sequence**.

Let K be a field and A be a K-algebra. In this book, we are mainly interested in the following two classes of abelian K-categories:
(1) the category $\operatorname{Mod} A$ of all right A-modules, and
(2) the full subcategory $\operatorname{mod} A$ of $\operatorname{Mod} A$ of finitely generated modules.
The objects of the module category $\operatorname{Mod} A$ (or $\operatorname{mod} A$) are the right A-modules (or the finitely generated A-modules). The set of morphisms between the modules M and N is the set $\operatorname{Hom}_A(M, N)$ of all A-module

homomorphisms $h : M \longrightarrow N$, endowed with the usual K-vector space structure. The composition of morphisms is just the composition of maps, and the direct sum $M \oplus N$ of two modules M and N is just the usual direct sum of K-vector spaces endowed with the A-module structure given by the formula $(m, n)a = (ma, na)$ for all $m \in M$, $n \in N$, and $a \in A$.

The kernel of a morphism $f : M \longrightarrow N$ in $\operatorname{Mod} A$ is the A-module $\operatorname{Ker} f = \{m \in M; \ f(m) = 0\}$, and the cokernel $\operatorname{Coker} f$ of f is the quotient A-module $N/\operatorname{Im} f$, where $\operatorname{Im} f = \{f(m); \ m \in M\}$ is the image of f.

It is clear that $\operatorname{Mod} A$ and $\operatorname{mod} A$ are abelian K-categories.

A.2. Functors

2.1. Definition. A **covariant functor** $T : \mathcal{C} \longrightarrow \mathcal{C}'$ from a category $\mathcal{C}$ to a category $\mathcal{C}'$ is defined by assigning to each object X of $\mathcal{C}$ an object $T(X)$ of $\mathcal{C}'$ and to each morphism $h : X \longrightarrow Y$ in $\mathcal{C}$ a morphism $T(h) : T(X) \longrightarrow T(Y)$ in $\mathcal{C}'$ such that the following conditions are satisfied:

(a) $T(1_X) = 1_{T(X)}$, for all objects X of $\mathcal{C}$; and

(b) for each pair of morphisms $X \xrightarrow{f} Y$ and $Y \xrightarrow{g} Z$ in $\mathcal{C}$, the equality $T(g \circ f) = T(g) \circ T(f)$ holds.

A **contravariant functor** $T : \mathcal{C} \longrightarrow \mathcal{C}'$ from a category $\mathcal{C}$ to a category $\mathcal{C}'$ is defined by assigning to each object X of $\mathcal{C}$ an object $T(X)$ of $\mathcal{C}'$, and to each morphism $h : X \longrightarrow Y$ in $\mathcal{C}$ a morphism $T(h) : T(Y) \longrightarrow T(X)$ in $\mathcal{C}'$ such that the following conditions are satisfied:

(a) $T(1_X) = 1_{T(Y)}$, for all objects X of $\mathcal{C}$; and

(b) for each pair of morphisms $X \xrightarrow{f} Y$ and $Y \xrightarrow{g} Z$ in $\mathcal{C}$, the equality $T(g \circ f) = T(f) \circ T(g)$ holds.

It is clear that any contravariant functor $T : \mathcal{C} \longrightarrow \mathcal{C}'$ can be viewed as a covariant functor $T : \mathcal{C} \longrightarrow \mathcal{C}'^{\mathrm{op}}$ or $T : \mathcal{C}^{\mathrm{op}} \longrightarrow \mathcal{C}'$ in an obvious way.

If $T : \mathcal{C} \longrightarrow \mathcal{C}'$ and $T' : \mathcal{C}' \longrightarrow \mathcal{C}''$ are functors, we define their composition $T'T : \mathcal{C} \longrightarrow \mathcal{C}''$ as follows. For each object X of $\mathcal{C}$, we set $T'T(X) = T'(T(X))$, and, for each morphism $X \xrightarrow{f} Y$ in $\mathcal{C}$, we set $T'T(f) = T'(T(f))$. It is easy to see that $T'T$ is a functor.

Given a pair of categories $\mathcal{C}$ and $\mathcal{D}$, we define their product $\mathcal{C} \times \mathcal{D}$ to be the category the objects of which are the pairs (C, D) with $C \in \operatorname{Ob} \mathcal{C}$, $D \in \operatorname{Ob} \mathcal{D}$, and morphisms $h : (C, D) \longrightarrow (C', D')$ are the pairs $h = (h_1, h_2)$, where $h_1 \in \operatorname{Hom}_{\mathcal{C}}(C, C')$ and $h_2 \in \operatorname{Hom}_{\mathcal{D}}(D, D')$. The composition $\circ$ in $\mathcal{C} \times \mathcal{D}$ is defined by $(g_1, g_2) \circ (h_1, h_2) = (g_1 \circ h_1, g_2 \circ h_2)$, for all $h_1 \in \operatorname{Hom}_{\mathcal{C}}(C, C')$, $g_1 \in \operatorname{Hom}_{\mathcal{C}}(C', C'')$, $h_2 \in \operatorname{Hom}_{\mathcal{D}}(D, D')$, and $g_2 \in \operatorname{Hom}_{\mathcal{D}}(D', D'')$. Any functor $F : \mathcal{C} \times \mathcal{D} \longrightarrow \mathcal{C}'$ is called a **bifunctor**.

Let $T, T' : \mathcal{C} \longrightarrow \mathcal{C}'$ be functors. A **functorial morphism** Ψ :

$T \longrightarrow T'$ (or a **natural transformation of functors**) is a family $\Psi = \{\Psi_X\}_{X \in \mathrm{Ob}\, \mathcal{C}}$ of morphisms $\Psi_X : T(X) \longrightarrow T'(X)$ such that, for any morphism $f : X \longrightarrow Y$ in $\mathcal{C}$, the diagram

$$
\begin{array}{ccc}
T(X) & \xrightarrow{\ \Psi_X\ } & T'(X) \\
{\scriptstyle T(f)}\downarrow & & \downarrow{\scriptstyle T'(f)} \\
T(Y) & \xrightarrow{\ \Psi_Y\ } & T'(Y)
\end{array}
$$

in $\mathcal{C}'$ is commutative. In this case, we write $\Psi : T \longrightarrow T'$. We call Ψ a **functorial isomorphism** (or a **natural equivalence of functors**) if, for any $X \in \mathrm{Ob}\, \mathcal{C}$, the morphism $\Psi_X : F(X) \longrightarrow F'(X)$ is an isomorphism in $\mathcal{C}'$.

A covariant functor $T : \mathcal{C} \longrightarrow \mathcal{C}'$ is called **an equivalence of categories** if there exist a functor $F : \mathcal{C}' \longrightarrow \mathcal{C}$ and functorial isomorphisms $\Psi : 1_{\mathcal{C}} \xrightarrow{\sim} FT$ and $\Phi : 1_{\mathcal{C}'} \xrightarrow{\sim} TF$, where $1_{\mathcal{C}'}$ and $1_{\mathcal{C}}$ are the identity functors on $\mathcal{C}'$ and $\mathcal{C}$, respectively. In this case, the functor F is called a **quasi-inverse** of T. If there exists an equivalence $\Psi : T \longrightarrow T'$ of categories $\mathcal{C}$ and $\mathcal{C}'$, then we say that $\mathcal{C}$ and $\mathcal{C}'$ are **equivalent categories**, and we write $\mathcal{C} \cong \mathcal{C}'$. A contravariant functor $D : \mathcal{C} \longrightarrow \mathcal{D}$ is an equivalence of categories if the induced covariant functor $D : \mathcal{C}^{\mathrm{op}} \longrightarrow \mathcal{D}$ is an equivalence of categories.

2.2. Definition. A contravariant functor $D : \mathcal{C} \longrightarrow \mathcal{D}$ that is an equivalence of categories is called a **duality**.

Let K be a field, A be a finite dimensional K-algebra, and A^{op} be the algebra opposite to A defined in Chapter I. An important example of a duality is the standard duality $D = \mathrm{Hom}_K(-, K) : \mathrm{mod}\, A \longrightarrow \mathrm{mod}\, A^{\mathrm{op}}$, defined in (I.2.9), between the category $\mathrm{mod}\, A$ of finite dimensional right A-modules and the category $\mathrm{mod}\, A^{\mathrm{op}}$ of finite dimensional left A-modules.

A covariant functor $T : \mathcal{C} \longrightarrow \mathcal{C}'$ is called **dense** if, for any object A of $\mathcal{C}'$, there exists an object C in $\mathcal{C}$ and an isomorphism $T(C) \cong A$. We say that T is **full** if the map

$$T_{XY} : \mathrm{Hom}_{\mathcal{C}}(X, Y) \longrightarrow \mathrm{Hom}_{\mathcal{C}'}(T(X), T(Y)),$$

given by $f \mapsto T(f)$, is surjective for all objects X and Y of $\mathcal{C}$. If T_{XY} is an injective map, for all $X, Y \in \mathrm{Ob}\, \mathcal{C}$, the functor T is called **faithful**.

Assume that $T : \mathcal{C} \longrightarrow \mathcal{C}'$ is a covariant functor between additive categories $\mathcal{C}$ and $\mathcal{C}'$. We say that T **preserves direct sums** if, for any objects $X_1, X_2 \in \mathrm{Ob}\, \mathcal{C}$, the morphisms $T(X_1) \xrightarrow{\ T(u_1)\ } T(X_1 \oplus X_2) \xleftarrow{\ T(u_2)\ } T(X_2)$

induced by the direct summand embeddings $X_1 \xrightarrow{u_1} X_1 \oplus X_2 \xleftarrow{u_2} X_2$ yield an isomorphism $T(X_1) \oplus T(X_2) \xrightarrow{\simeq} T(X_1 \oplus X_2)$. The functor T is **additive** if T preserves direct sums, and, for all $X, Y \in \mathrm{Ob}\,\mathcal{C}$, the map $T_{XY} : \mathrm{Hom}_{\mathcal{C}}(X, Y) \longrightarrow \mathrm{Hom}_{\mathcal{C}'}(T(X), T(Y))$, given by $h \mapsto T(h)$, satisfies $T(f + g) = T(f) + T(g)$, for all $f, g \in \mathrm{Hom}_{\mathcal{C}}(X, Y)$.

If $\mathcal{C}$ and $\mathcal{C}'$ are K-categories, then $T : \mathcal{C} \longrightarrow \mathcal{C}'$ is called K-**linear** if T is additive and T_{XY} is a K-linear map for all $X, Y \in \mathrm{Ob}\,\mathcal{C}$.

A full, faithful, and K-linear covariant functor $T : \mathcal{C} \longrightarrow \mathcal{C}'$ between additive K-categories $\mathcal{C}$ and $\mathcal{C}'$ is called a **fully faithful embedding**. In other words, a K-linear functor T is a fully faithful embedding if, for each pair X and Y of objects of $\mathcal{C}$, the map $T_{XY} : \mathrm{Hom}_{\mathcal{C}}(X, Y) \longrightarrow \mathrm{Hom}_{\mathcal{C}'}(T(X), T(Y))$ is an isomorphism of K-vector spaces.

Throughout the text, we agree that the unqualified term "functor" always means a covariant functor. Moreover, by a functor between additive categories (or K-categories), we always mean an additive functor (or a K-linear functor, respectively).

Assume that $\mathcal{C}$ and $\mathcal{C}'$ are abelian categories. A covariant additive functor $T : \mathcal{C} \longrightarrow \mathcal{C}'$ is **right exact** (or **left exact**) if, for any exact sequence $X \xrightarrow{f} Y \xrightarrow{g} Z \longrightarrow 0$ (or exact sequence $0 \longrightarrow X \xrightarrow{f} Y \xrightarrow{g} Z$) in $\mathcal{C}$, the induced sequence

$$T(X) \xrightarrow{T(f)} T(Y) \xrightarrow{T(g)} T(Z) \longrightarrow 0$$

(or $0 \longrightarrow T(X) \xrightarrow{T(f)} T(Y) \xrightarrow{T(g)} T(Z)$, respectively) in $\mathcal{C}'$ is exact. The functor T is **exact** if it is both left and right exact.

It is obvious that the corresponding definitions for contravariant functors are analogous to the ones for covariant functors. In particular, a contravariant additive functor $F : \mathcal{C} \longrightarrow \mathcal{C}'$ between abelian categories $\mathcal{C}$ and $\mathcal{C}'$ is **left exact** (or **right exact**) if, for any exact sequence $X \xrightarrow{f} Y \xrightarrow{g} Z \longrightarrow 0$ (or exact sequence $0 \longrightarrow X \xrightarrow{f} Y \xrightarrow{g} Z$) in $\mathcal{C}$, the induced sequence

$$0 \longrightarrow F(Z) \xrightarrow{F(g)} F(Y) \xrightarrow{F(f)} F(X)$$

(or $F(Z) \xrightarrow{F(g)} F(Y) \xrightarrow{F(f)} F(X) \longrightarrow 0$, respectively) in $\mathcal{C}'$ is exact.

2.3. Definition. *Let* $\mathcal{A} \xrightleftharpoons[R]{L} \mathcal{B}$ *be a pair of additive covariant functors between abelian categories* $\mathcal{A}$ *and* $\mathcal{B}$. *The functor* L *is* **left adjoint** *to* R *and* R *is* **right adjoint** *to* L *if there exists an isomorphism*

$$\mathrm{Hom}_{\mathcal{B}}(L(X), Y) \cong \mathrm{Hom}_{\mathcal{A}}(X, R(Y))$$

for any object X *of* $\mathcal{A}$ *and any object* Y *of* $\mathcal{B}$, *which is functorial at* X *and* Y.

It was shown in (I.2.11) that, given two K-algebras A, B and an A-B-bimodule $_AM_B$, the functor $L = - \otimes_A M_B : \operatorname{Mod} A \longrightarrow \operatorname{Mod} B$ is left adjoint to the Hom-functor $R = \operatorname{Hom}_B(_AM_B, -) : \operatorname{Mod} B \longrightarrow \operatorname{Mod} A$.

We state without proof the following useful lemma (see [6], [148]).

2.4. Lemma. *Let* $\mathcal{A}$ *and* $\mathcal{B}$ *be abelian categories and let* $\mathcal{A} \underset{R}{\overset{L}{\rightleftarrows}} \mathcal{B}$ *be a pair of additive covariant functors such that* L *is left adjoint to* R. *Then* L *is right exact and* R *is left exact.* $\square$

The following important fact is frequently used in the book.

2.5. Theorem. *A covariant functor* $T : \mathcal{C} \longrightarrow \mathcal{C}'$ *is an equivalence of categories if and only* T *is full, faithful, and dense.*

Proof. Assume that T is full, faithful, and dense. We define a quasi-inverse functor $F : \mathcal{C}' \longrightarrow \mathcal{C}$ of T as follows. For any $X' \in \operatorname{Ob} \mathcal{C}'$, we fix an object X of $\mathcal{C}$ and an isomorphism $\Phi_{X'} : X' \xrightarrow{\simeq} T(X)$ in $\mathcal{C}'$. We set $F(X') = X$. Given a morphism $f' \in \operatorname{Hom}_{\mathcal{C}'}(X', Y')$, we choose a morphism $f \in \operatorname{Hom}_{\mathcal{C}}(X, Y)$ making the following diagram

$$
\begin{array}{ccc}
X' & \xrightarrow{\Phi_{X'}} & T(X) \\
{\scriptstyle f'}\downarrow & & \downarrow{\scriptstyle T(f)} \\
Y' & \xrightarrow{\Phi_{Y'}} & T(Y)
\end{array}
$$

commutative. We set $F(f') = f$. It is easy to check that this procedure defines a covariant functor F. Moreover, for any $X' \in \operatorname{Ob} \mathcal{C}'$, the following diagram

$$
\begin{array}{ccc}
X' & \xrightarrow{\Phi_{X'}} & TF(X') \\
{\scriptstyle f'}\downarrow & & \downarrow{\scriptstyle TF(f')} \\
Y' & \xrightarrow{\Phi_{Y'}} & TF(Y')
\end{array}
$$

is commutative. This shows that the family $\{\Phi_{X'}\}_{X' \in \operatorname{Ob} \mathcal{C}'}$ of isomorphisms defines a functorial isomorphism $\Phi : 1_{\mathcal{C}'} \longrightarrow TF$.

Next, we define a functorial isomorphism $\Psi : 1_{\mathcal{C}} \longrightarrow FT$ as follows. For any $Z \in \operatorname{Ob} \mathcal{C}$, we set $X'_Z = T(Z)$. Then $\Phi_{T(Z)} = \Phi_{X'_Z}$ is the composed morphism

$$
T(Z) = X'_Z \xrightarrow{\Phi_{X'_Z}} TF(X'_Z) = T(FT(Z)).
$$

Because the functor T is full and faithful, there exists a unique isomorphism $\Psi_Z : Z \longrightarrow FT(Z)$ such that $T(\Psi_Z) = \Phi_{X'_Z} = \Phi_{T(Z)}$.

Let $g : Z \longrightarrow V$ be an arbitrary morphism in $\mathcal{C}$. We show that the following diagram

$$
\begin{array}{ccc}
Z & \xrightarrow{\ \Psi_Z\ } & FT(Z) \\
{\scriptstyle g}\downarrow & & \downarrow{\scriptstyle FT(g)} \\
V & \xrightarrow{\ \Psi_V\ } & FT(V).
\end{array}
$$

is commutative. Because $\Phi : 1_{\mathcal{C}'} \xrightarrow{\ \simeq\ } TF$ is a functorial isomorphism, the following diagram

$$
\begin{array}{ccc}
T(Z) & \xrightarrow{\ \Phi_{T(Z)}\ } & TF(T(Z)) \\
{\scriptstyle T(g)}\downarrow & & \downarrow{\scriptstyle TF(T(g))} \\
T(V) & \xrightarrow{\ \Phi_{T(V)}\ } & TF(T(V))
\end{array}
$$

is commutative. It follows from the choice of Ψ_Z and Ψ_V that $\Phi_{T(Z)} = T(\Psi_Z)$ and $\Phi_{T(V)} = T(\Psi_V)$. Hence, we get

$$
T(\Psi_V \circ g) = T(\Psi_V) \circ T(g) = TFT(g) \circ T(\Psi_Z) = T(FT(g) \circ \Psi_Z).
$$

Because T is faithful, the equality yields $\Psi_V \circ g = FT(g) \circ \Psi_Z$, as required. Consequently, the functorial morphism $\Psi : 1_{\mathcal{C}} \longrightarrow FT$ is a functorial isomorphism.

Now assume that $T : \mathcal{C} \longrightarrow \mathcal{C}'$ is an equivalence of categories and that $F : \mathcal{C}' \longrightarrow \mathcal{C}$ is a quasi-inverse of T. Let $\Psi : 1_{\mathcal{C}} \xrightarrow{\ \simeq\ } FT$ and $\Phi : 1_{\mathcal{C}'} \xrightarrow{\ \simeq\ } TF$ be functorial isomorphisms. Then, for any $X' \in \mathrm{Ob}\ \mathcal{C}'$, there is an isomorphism $X' \cong TF(X')$, and therefore T is dense. Moreover, for any morphism $f' : X' \longrightarrow Y'$ in $\mathcal{C}'$, the diagram

$$
\begin{array}{ccc}
X' & \xrightarrow{\ \Phi_{X'}\ } & TF(X') \\
{\scriptstyle f'}\downarrow & & \downarrow{\scriptstyle TF(f')} \\
Y' & \xrightarrow{\ \Phi_{Y'}\ } & TF(Y')
\end{array}
$$

is commutative. This implies that the functor F is faithful. Similarly, for any morphism $h : U \longrightarrow V$ in $\mathcal{C}$, the diagram

$$
\begin{array}{ccc}
U & \xrightarrow{\ \Psi_U\ } & FT(U) \\
{\scriptstyle h}\downarrow & & \downarrow{\scriptstyle FT(h)} \\
V & \xrightarrow{\ \Psi_V\ } & FT(V)
\end{array}
$$

is commutative. This implies that the functor T is faithful. To show that T is full, we take $f' \in \mathrm{Hom}_{\mathcal{C}'}(T(U), T(V))$, where $U, V \in \mathrm{Ob}\,\mathcal{C}$, and we set $h = \Psi_V^{-1} \circ F(f') \circ \Psi_U \in \mathrm{Hom}_{\mathcal{C}}(U, V)$. Then the commutativity of the diagram yields $F(f') = \Psi_V \circ \Psi_V^{-1} \circ F(f') \circ \Psi_U \circ \Psi_U^{-1} = \Psi_V \circ h \circ \Psi_U^{-1} = FT(h)$. It follows that $f' = T(h)$, because F is faithful. This finishes the proof. $\square$

2.6. Example. Let A be the lower triangular K-subalgebra

$$A = \begin{bmatrix} K & 0 \\ K & K \end{bmatrix} = \left\{ \left(\begin{smallmatrix} a & 0 \\ b & c \end{smallmatrix}\right);\ a, b, c \in K \right\}$$

of the matrix algebra $\mathbb{M}_2(K)$. We show that the category $\mathrm{mod}\,A$ of all finite dimensional right A-modules is equivalent with the category $\mathcal{M}ap_K$ of K-linear maps between K-vector spaces, which we will define.

We define $\mathcal{M}ap_K$ to be the category with objects that are triples (V, W, f), where V and W are finite dimensional K-vector spaces and $f : V \longrightarrow W$ is a K-linear map. A morphism from (V, W, f) to (V', W', f') in $\mathcal{M}ap_K$ is a pair (h_1, h_2) of K-linear maps such that the diagram

$$
\begin{array}{ccc}
V & \xrightarrow{\ f\ } & W \\
{\scriptstyle h_1}\downarrow & & \downarrow{\scriptstyle h_2} \\
V' & \xrightarrow{\ f'\ } & W'
\end{array}
$$

is commutative. If (h_1', h_2') is a morphism from (V', W', f') to (V'', W'', f'') in $\mathcal{M}ap_K$, we set $(h_1', h_2') \circ (h_1, h_2) = (h_1' h_1, h_2' h_2)$. It is easy to see that (h_1, h_2) is an isomorphism in $\mathcal{M}ap_K$ if and only if h_1 and h_2 are isomorphisms. The direct sum in $\mathcal{M}ap_K$ is defined by the formula

$$(V, W, f) \oplus (V', W', f') = (V \oplus V', W \oplus W', f \oplus f');$$

that is, it is the direct sum of the K-linear maps f and f'.

To construct an equivalence of categories

$$\rho : \mathrm{mod}\,A \longrightarrow \mathcal{M}ap_K,$$

we note that the matrices $e_1 = \left(\begin{smallmatrix} 1 & 0 \\ 0 & 0 \end{smallmatrix}\right)$, $e_2 = \left(\begin{smallmatrix} 0 & 0 \\ 0 & 1 \end{smallmatrix}\right)$, $e_{21} = \left(\begin{smallmatrix} 0 & 0 \\ 1 & 0 \end{smallmatrix}\right)$ form a basis of A over K, $1_A = e_1 + e_2$, $e_1 e_2 = e_2 e_1 = 0$, $e_{21} = e_2 e_{21} = e_{21} e_1$ and $e_1 e_{21} = e_{21} e_2 = 0$. It follows that every module X in $\mathrm{mod}\,A$, viewed as a K-vector space, has a direct sum decomposition $X = X e_1 \oplus X e_2$. Therefore, X uniquely determines the triple $\rho(X) = (V_X, W_X, f_X)$, where $V_X = X e_2$, $W_X = X e_1$, and $f_X : V_X \longrightarrow W_X$ is the K-linear map given by $f_X(v) = v e_{21} = v e_{21} e_1$, where $v = x e_2 \in V_X$. If $g : X \longrightarrow Y$ is a homomorphism of right A-modules, we define $\rho(g) : \rho(X) \longrightarrow \rho(Y)$ to be the pair $\rho(g) = (g_1, g_2)$, where $g_1 : V_X \longrightarrow V_Y$ and $g_2 : W_X \longrightarrow W_Y$ are the restrictions of g to V_X and to W_X, respectively. It is easily checked that ρ is a full, faithful, dense, and K-linear functor, and, according to

(2.5), the functor ρ is an equivalence of categories. The quasi-inverse $\rho_1 :$ $\mathcal{M}ap_K \longrightarrow \mathrm{mod}\,A$ of ρ is defined by attaching to any object (V, W, f) in $\mathcal{M}ap_K$ the K-vector space $X = W \oplus V$ with the right action $\cdot : X \times A \longrightarrow X$ of A on X defined by the formula $(w, v) \cdot \left(\begin{smallmatrix} a & 0 \\ b & c \end{smallmatrix}\right) = (wa + f(v)b, vc)$, where $v \in V$, $w \in W$ and $a, b, c \in K$.

2.7. Example. Let A and B be finite dimensional K-algebras, and let $_AM_B$ be a finite dimensional A-B-bimodule. We illustrate the notion of an equivalence of categories by showing that the category of modules over the lower triangular matrix K-algebra $C = \left(\begin{smallmatrix} B & 0 \\ {}_AM_B & A \end{smallmatrix}\right)$ is equivalent with a category rep($_AM_B$), called the category of **representations of the bimodule** $_AM_B$, and defined as follows.

The objects of rep($_AM_B$) are the triples $(X'_A, X''_B; \varphi)$, where X'_A is a module in $\mathrm{mod}\,A$, X''_B is an module in $\mathrm{mod}\,B$, and $\varphi : X' \otimes_A M_B \longrightarrow X''_B$ is a B-module homomorphism. A morphism from $(X'_A, X''_B; \varphi)$ to $(Y'_A, Y''_B; \psi)$ in rep($_AM_B$) is a pair $(f', f'') : (X'_A, X''_B; \varphi) \longrightarrow (Y'_A, Y''_B; \psi)$, where $f' : X'_A \longrightarrow Y'_A$ is an A-homomorphism and $f'' : X''_B \longrightarrow Y''_B$ is a B-homomorphism, making the diagram

$$(*) \qquad \begin{array}{ccc} X' \otimes_A M_B & \xrightarrow{\varphi} & X''_B \\ \downarrow{\scriptstyle f' \otimes M} & & \downarrow{\scriptstyle f''} \\ Y' \otimes_A M_B & \xrightarrow{\psi} & Y''_B \end{array}$$

commutative. The composition of morphisms and the direct sum in rep($_AM_B$) are defined componentwise. It is easy to check that rep($_AM_B$) is a K-category.

The set $C = \left(\begin{smallmatrix} B & 0 \\ {}_AM_B & A \end{smallmatrix}\right)$ of all matrices $\left(\begin{smallmatrix} b & 0 \\ m & a \end{smallmatrix}\right)$, where $a \in A$, $b \in B$, and $m \in M$, endowed with the multiplication given by the formula

$$\begin{pmatrix} b & 0 \\ m & a \end{pmatrix} \begin{pmatrix} f & 0 \\ v & e \end{pmatrix} = \begin{pmatrix} bf & 0 \\ mf+av & ae \end{pmatrix},$$

is a finite dimensional K-algebra with identity element $1 = e_B + e_A$, where $e_B = \left(\begin{smallmatrix} 1 & 0 \\ 0 & 0 \end{smallmatrix}\right)$, $e_A = \left(\begin{smallmatrix} 0 & 0 \\ 0 & 1 \end{smallmatrix}\right)$. We define a functor

$$F : \mathrm{mod}\,C \longrightarrow \mathrm{rep}(_AM_B)$$

as follows. For each module X_C in $\mathrm{mod}\,C$, we set $F(X_C) = (X'_A, X''_B; \varphi)$, where $X'_A = Xe_A$, $X''_B = Xe_B$, and $\varphi : X' \otimes_A M_B \longrightarrow X''_B$ is a B-module homomorphism defined by $\varphi(x' \otimes m) = x' \cdot \left(\begin{smallmatrix} 0 & 0 \\ m & 0 \end{smallmatrix}\right) = x' \cdot \left(\begin{smallmatrix} 0 & 0 \\ m & 0 \end{smallmatrix}\right)e_B$. If $f : X_C \longrightarrow Y_C$ is a C-module homomorphism, we define $F(f) : F(X) \longrightarrow F(Y)$ to be the pair (f', f''), where $f' : Xe_A \longrightarrow Ye_A$ is the A-homomorphism given by $xe_A \mapsto f(xe_A) = f(xe_A)e_A$, and $f'' : Xe_B \longrightarrow Ye_B$ is the B-homomorphism $xe_B \mapsto f(xe_B) = f(xe_B)e_B$. A straightforward calculation shows that the diagram $(*)$ is commutative and therefore $F(f)$ is a morphism in $\mathrm{rep}(_AM_B)$. It is easy to check that F is a covariant K-linear functor.

To show that F is faithful, we note that if $F(f) = 0$ then $f' = 0$, $f'' = 0$, and, in view of the equality $1 = e_B + e_A$, we get $f(x) = f(xe_A) + f(xe_B) = f'(xe_A) + f''(xe_B) = 0$, for all $x \in X$. Hence $f = 0$ and it follows that the K-linear map $f \mapsto F(f)$ is injective and therefore the functor F is faithful. In view of (2.5), to prove that F is an equivalence of categories, it remains to show that F is dense and full. For this purpose, take an object $(X'_A, X''_B; \varphi)$ in $\mathrm{rep}(_AM_B)$. The K-vector space $X = X''_B \oplus X'_A$ endowed with the right C-action $\cdot : X \times C \longrightarrow X$ defined by the formula

$$(x'', x') \cdot \left(\begin{smallmatrix} b & 0 \\ m & a \end{smallmatrix}\right) = (x''b + \varphi(x' \otimes m), x'a)$$

for $x' \in X'_A$, $x'' \in X''_B$, $a \in A$, $b \in B$ and $m \in M$, is a right C-module. It is immediate that $F(X) \cong (X'_A, X''_B; \varphi)$, so F is dense. Now let (f', f'') : $(X'_A, X''_B; \varphi) \longrightarrow (Y'_A, Y''_B; \psi)$ be a morphism in $\mathrm{rep}(_AM_B)$. A simple calculation shows that the K-linear map $f = \begin{bmatrix} f'' & 0 \\ 0 & f' \end{bmatrix} : X''_B \oplus X'_A \longrightarrow Y''_B \oplus Y'_A$ is a homomorphism of right C-modules $X = X''_B \oplus X'_A$ and $Y = Y''_B \oplus Y'_A$ such that $F(f) = (f', f'')$. This shows that F is full. Consequently, the functor F is an equivalence of categories.

Usually we identify right C-module X with $F(X)$. In other words, we view a module X in $\mathrm{mod}\, C$ as a triple $X_C = (X'_A, X''_B; \varphi)$, where X'_A is a module in $\mathrm{mod}\, A$, X''_B is a module in $\mathrm{mod}\, B$, and $\varphi : X' \otimes_A M_B \longrightarrow X''_B$ is a B-module homomorphism. Any C-module homomorphism $f : X \longrightarrow Y$ is identified with the pair $f = (f', f'')$, where $f' : X'_A \longrightarrow Y'_A$ is an A-homomorphism and $f'' : X''_B \longrightarrow Y''_B$ is a B-homomorphism such that the diagram $(*)$ is commutative.

In view of the adjunction isomorphism (I.2.11), the C-module X can be also identified with the triple $X_C = (X'_A, X''_B; \overline{\varphi})$, where X'_A and X''_B are as given earlier, and

$$\overline{\varphi} : X'_A \longrightarrow \mathrm{Hom}_B(_AM_B, X''_B)$$

is the A-homomorphism adjoint to φ defined by $\overline{\varphi}(x')(m) = \varphi(x' \otimes m)$.

The preceding discussion can be summarised as follows. If A and B are finite dimensional K-algebras and $_AM_B$ is a finite dimensional A-B-bimodule, then there exist equivalences of categories

$$\mathrm{mod}\left(\begin{smallmatrix} B & 0 \\ _AM_B & A \end{smallmatrix}\right) \cong \mathrm{rep}(_AM_B) \cong \mathrm{mod}\left(\begin{smallmatrix} A & _AM_B \\ 0 & B \end{smallmatrix}\right). \tag{2.8}$$

The left-hand equivalence is given by the functor F in (2.7). Its quasi-inverse is defined by associating to any object $(X'_A, X''_B; \varphi)$ in $\mathrm{rep}(_AM_B)$

the K-vector space $X_B'' \oplus X_A'$ endowed with the right action
$$\cdot : (X_B'' \oplus X_A') \times \left(\begin{smallmatrix} B & 0 \\ {}_A M_B & A \end{smallmatrix} \right) \longrightarrow X_B'' \oplus X_A'$$
defined by the formula $(x'', x') \cdot \left(\begin{smallmatrix} b & 0 \\ m & a \end{smallmatrix} \right) = (x''b + \varphi(x' \otimes m), x'a)$ for $x' \in X_A'$, $x'' \in X_B''$, $a \in A$, $b \in B$, $m \in M$, and to any morphism (f', f'') : $(X_A', X_B''; \varphi) \longrightarrow (Y_A', Y_B''; \psi)$ the homomorphism $f'' \oplus f' : X_B'' \oplus X_A' \longrightarrow Y_B'' \oplus Y_A'$ of right $\left(\begin{smallmatrix} B & 0 \\ {}_A M_B & A \end{smallmatrix} \right)$-modules. The right-hand equivalence in (2.8) can be proved similarly. One can deduce from (2.8) an equivalence mod $\left(\begin{smallmatrix} K & 0 \\ K & K \end{smallmatrix} \right) \cong$ Map_K constructed in (2.6).

We finish this section with basic properties of the categories of functors from module categories over K-algebras to the category of K-vector spaces.

Let A be a finite dimensional K-algebra. An important rôle in Auslander-Reiten theory is played by the categories $Fun^{op}A$ and $Fun\,A$ of the contravariant, and covariant, respectively, K-linear functors from the category $\mathrm{mod}\,A$ of finitely generated right A-modules into the category $\mathrm{mod}\,K$ of finite dimensional K-vector spaces, which we now define as follows.

We define $Fun^{op}A$ (and $Fun\,A$) to be the category the objects of which are all contravariant (and covariant) K-linear functors $T : \mathrm{mod}\,A \longrightarrow \mathrm{mod}\,K$, respectively, from the category $\mathrm{mod}\,A$ of finite dimensional right A-modules to the category $\mathrm{mod}\,K$ of finite dimensional K-vector spaces. Given a pair of K-linear functors $T, S : \mathrm{mod}\,A \longrightarrow \mathrm{mod}\,K$, we define the set $\mathrm{Hom}(S, T)$ of morphisms from S to T to be the set of all functorial morphisms $\Phi : S \longrightarrow T$. If T, T', T'' are functors in $Fun^{op}A$ (or in $Fun\,A$) and $T \xrightarrow{\Psi} T' \xrightarrow{\Phi} T''$ are functorial morphisms given by $\Psi = \{\Psi_X\}_X$ and $\Phi = \{\Phi_X\}_X$, we define the composite functorial morphism $\Phi \circ \Psi : T \longrightarrow T''$ of Ψ and Φ by the formula $\Phi \circ \Psi = \{\Phi_X \Psi_X\}_X$, where X runs through all modules in $\mathrm{mod}\,A$. A routine calculation shows that $Fun^{op}A$ and $Fun\,A$ are categories.

Assume that S and T is a pair of functors in $Fun^{op}A$ (or in $Fun\,A$). We say that S is a **subfunctor** of T if there is a functorial morphism $\mathbf{u} = \{\mathbf{u}_X\}_X : S \longrightarrow T$ such that, for each module X in $\mathrm{mod}\,A$, we have $S(X) \subseteq T(X)$ and the K-linear homomorphism $\mathbf{u}_X : S(X) \longrightarrow T(X)$ is the inclusion.

We are now able to prove that the functor categories $Fun^{op}A$ and $Fun\,A$ are abelian.

2.9. Theorem. *For any finite dimensional K-algebra A, the categories $Fun^{op}A$ and $Fun\,A$ are abelian K-categories.*

Proof. First, we prove that $Fun^{op}A$ and $Fun\,A$ are additive K-categories. Let T, T' be a pair of functors in $Fun^{op}A$ (or in $Fun\,A$). Let $\lambda, \lambda' \in K$ and $\Psi, \Psi' : T \longrightarrow T'$ be functorial morphisms given by $\Psi = \{\Psi_X\}_X$ and

$\Psi' = \{\Psi'_X\}_X$, where X runs through all modules in $\bmod A$. We define the functorial morphism $\Psi\lambda + \Psi'\lambda' : T \longrightarrow T'$ by the formula $\Psi\lambda + \Psi'\lambda' = \{\Psi_X\lambda + \Psi'_X\lambda'\}_X$. A routine calculation shows that we have defined a K-vector space structure on the set of morphisms $\mathrm{Hom}(T, T')$. Moreover, the composition

$$\circ : \mathrm{Hom}(T', T'') \times \mathrm{Hom}(T, T') \longrightarrow \mathrm{Hom}(T, T'')$$

is a K-bilinear map. Further, we define the direct sum of a finite set of functors $T_1, \ldots, T_n$ in $\mathcal{F}un^{\mathrm{op}}A$ (or in $\mathcal{F}un\, A$) to be the functor $T_1 \oplus \ldots \oplus T_n$ together with direct summand embeddings

$$\mathbf{u}_j = \{\mathbf{u}_{j,X}\}_X : T_j \longrightarrow T_1 \oplus \ldots \oplus T_n,$$

for $j = 1, \ldots, n$, defined as follows. For each module X in $\bmod A$, we set $(T_1 \oplus \ldots \oplus T_n)(X) = T_1(X) \oplus \ldots \oplus T_n(X)$ and

$$\mathbf{u}_{j,X} : T_j(X) \longrightarrow T_1(X) \oplus \ldots \oplus T_n(X)$$

is the jth direct summand embedding. For each A-homomorphism $f : X \to Y$ in $\bmod A$, we set $(T_1 \oplus \ldots \oplus T_n)(f) = T_1(f) \oplus \ldots \oplus T_n(f)$. A direct calculation shows that $T_1 \oplus \ldots \oplus T_n$ is the direct sum of functors $T_1, \ldots, T_n$ in the categories $\mathcal{F}un^{\mathrm{op}}A$ and $\mathcal{F}un\, A$, respectively.

Finally, we define the zero functor by associating to each X in $\bmod A$ the zero vector space, and to each A-homomorphism $f : X \to Y$ in $\bmod A$ the zero map. It is clear that the zero functor is the unique zero object in $\mathcal{F}un^{\mathrm{op}}A$ and $\mathcal{F}un\, A$. Consequently, $\mathcal{F}un^{\mathrm{op}}A$ and $\mathcal{F}un\, A$ are additive K-categories.

It remains to prove that the categories $\mathcal{F}un^{\mathrm{op}}A$ and $\mathcal{F}un\, A$ are abelian.

Let $\Psi = \{\Psi_X\}_X : T \longrightarrow T'$ be a functorial morphism in $\mathcal{F}un^{\mathrm{op}}A$ (or in $\mathcal{F}un\, A$), where X runs through all modules in $\bmod A$. We define the kernel $\mathrm{Ker}\,\Psi$ of Ψ and the image $\mathrm{Im}\,\Psi$ of Ψ to be the subfunctor of T and the subfunctor of T' given by the formulas $(\mathrm{Ker}\,\Psi)(X) = \mathrm{Ker}\,\Psi_X$ and $(\mathrm{Im}\,\Psi)(X) = \mathrm{Im}\,\Psi_X$, for each module X in $\bmod A$. Further, we define the cokernel $\mathrm{Coker}\,\Psi$ of Ψ by associating to each module X in $\bmod A$ the quotient vector space $\mathrm{Coker}\,\Psi_X = T'(X)/\mathrm{Im}\,\Psi_X$, and to each A-homomorphism $f : X \to Y$ in $\bmod A$ the unique K-linear map $(\mathrm{Coker}\,\Psi)(f)$ such that the diagram

$$\begin{array}{ccc} T'(X) & \xrightarrow{\ \mathbf{p}_X\ } & \mathrm{Coker}\,\Psi_X \\ {\scriptstyle T'(f)}\uparrow & & \uparrow{\scriptstyle (\mathrm{Coker}\,\Psi)(f)} \\ T'(Y) & \xrightarrow{\ \mathbf{p}_Y\ } & \mathrm{Coker}\,\Psi_Y \end{array} \quad (\text{or } {\scriptstyle T'(f)}\downarrow \quad \begin{array}{ccc} T'(X) & \xrightarrow{\ \mathbf{p}_X\ } & \mathrm{Coker}\,\Psi_X \\ & & \downarrow{\scriptstyle (\mathrm{Coker}\,\Psi)(f)} \\ T'(Y) & \xrightarrow{\ \mathbf{p}_Y\ } & \mathrm{Coker}\,\Psi_Y \end{array})$$

is commutative, where $\mathbf{p}_X$ and $\mathbf{p}_Y$ are the canonical projections. A routine calculation shows that $\mathrm{Coker}\,\Psi$ is a functor and, together with the functorial morphism $\mathbf{p} = \{\mathbf{p}_X\} : T' \longrightarrow \mathrm{Coker}\,\Psi$, it is the cokernel of the morphism Ψ in the category $\mathcal{F}un^{\mathrm{op}}A$ (or in $\mathcal{F}un\,A$, respectively). By applying the previous definitions, it is easy to see that the categories $\mathcal{F}un^{\mathrm{op}}A$ and $\mathcal{F}un\,A$ are abelian. In particular, it follows from the proof that a short sequence in $\mathcal{F}un^{\mathrm{op}}A$ (or in $\mathcal{F}un\,A$)

$$0 \longrightarrow T' \xrightarrow{\ \Phi\ } T \xrightarrow{\ \Psi\ } T'' \longrightarrow 0$$

is exact if and only if, for each module X in $\mathrm{mod}\,A$, the induced sequence of K-vector spaces

$$0 \longrightarrow T'(X) \xrightarrow{\ \Phi_X\ } T(X) \xrightarrow{\ \Psi_X\ } T''(X) \longrightarrow 0$$

is exact.   □

The categories $\mathcal{F}un^{\mathrm{op}}A$ and $\mathcal{F}un A$ are studied in detail in Section IV.6. We now give an example showing that the category $\mathcal{F}un^{\mathrm{op}}A$ is equivalent to the category $\mathrm{Mod}\,B$ of right modules over a finite dimensional algebra B if the algebra A is representation–finite, that is, if the number of the isomorphism classes of the indecomposable modules in $\mathrm{mod}\,A$ is finite.

2.10. Example. Assume that A is a representation–finite K-algebra and let $M_1, \ldots, M_n$ be a complete set of the isomorphism classes of the indecomposable modules in $\mathrm{mod}\,A$. Let $M = M_1 \oplus \ldots \oplus M_n$. The finite dimensional K-algebra

$$B = \mathrm{End}\,M$$

is known as the **Auslander algebra** (see [21], [31], [151], [164]) of the representation–finite algebra A. Consider the K-linear functor

$$\mathbb{H} : \mathcal{F}un^{\mathrm{op}}A \longrightarrow \mathrm{Mod}\,B$$

defined as follows. If $T : \mathrm{mod}\,A \longrightarrow \mathrm{mod}\,K$ is a contravariant functor, we denote by $\mathbb{H}(T)$ the vector space $T(M)$ endowed with the structure of right B-module given by $xf = T(f)(x)$, for all $x \in T(M)$ and $f \in B$. If $\Psi = \{\Psi_X\}_X : T \longrightarrow T'$ is a functorial morphism in $\mathcal{F}un^{\mathrm{op}}A$, where X runs through all modules in $\mathrm{mod}\,A$, then we take for $\mathbb{H}(\Psi) : \mathbb{H}(T) \longrightarrow \mathbb{H}(T')$ the B-module homomorphism $\Psi_M : T(M) \longrightarrow T'(M)$. One shows that $\mathbb{H}$ is a K-linear functor which establishes an equivalence of categories

$$\mathcal{F}un^{\mathrm{op}}A \cong \mathrm{Mod}\,B.$$

This follows from the fact that every functor $T : \mathrm{mod}\,A \longrightarrow \mathrm{mod}\,K$ is uniquely determined by its restriction to M, that is, by the B-module $T(M)$, because the algebra A is representation–finite (see [12], [13], [115], [146], and [150] for details).

Hence, it follows from (IV.6.8) and (IV.6.11) that the projective dimension (defined in Section A.4) of any simple right B-module is at most 2. This implies [as will be seen in (4.8)] that the global dimension of the algebra B is at most 2.

A.3. The radical of a category

Following Kelly [103], we introduce here the notion of a radical $\mathrm{rad}_{\mathcal{C}}$ of any additive category $\mathcal{C}$ (see also Mitchell [115]). We collect elementary properties of the radical $\mathrm{rad}_{\mathcal{C}}$, mainly in case $\mathcal{C}$ is the category $\mathrm{mod}\, A$ of finite dimensional modules over a finite dimensional K-algebra A. More information on $\mathrm{rad}_A := \mathrm{rad}_{\mathrm{mod}\, A}$ can be found in [21]. We try in this book to show that the radical rad_A of $\mathrm{mod}\, A$ and its powers rad_A^m, where $m \geq 2$, are very efficient tools for describing the structure of the module category $\mathrm{mod}\, A$.

3.1. Definition. Let $\mathcal{C}$ be an additive K-category. A class $\mathcal{I}$ of morphisms of $\mathcal{C}$ is a **two-sided ideal** in $\mathcal{C}$ if $\mathcal{I}$ has the following properties:

(a) for each $X \in \mathrm{Ob}\,\mathcal{C}$, the zero morphism $0_X \in \mathrm{Hom}_{\mathcal{C}}(X, X)$ belongs to $\mathcal{I}$;

(b) if $f, g : X \longrightarrow Y$ are morphisms in $\mathcal{I}$ and $\lambda, \mu \in K$, then $f\lambda + g\mu \in \mathcal{I}$;

(c) if $f \in \mathcal{I}$ and g is a morphism in $\mathcal{C}$ that is left-composable with f, then $g \circ f \in \mathcal{I}$; and

(d) if $f \in \mathcal{I}$ and h is a morphism in $\mathcal{C}$ that is right-composable with f, then $f \circ h \in \mathcal{I}$.

Equivalently, a two-sided ideal $\mathcal{I}$ of $\mathcal{C}$ can be thought as a subfunctor

$$\mathcal{I}(-, -) \subseteq \mathrm{Hom}_{\mathcal{C}}(-, -) : \mathcal{C}^{\mathrm{op}} \times \mathcal{C} \longrightarrow \mathrm{Mod}\, K$$

of the bifunctor $\mathrm{Hom}_{\mathcal{C}}(-, -)$, defined by assigning to each pair (X, Y) of objects X, Y of $\mathcal{C}$ a K-subspace $\mathcal{I}(X, Y)$ of $\mathrm{Hom}_{\mathcal{C}}(X, Y)$ such that:

(i)   if $f \in \mathcal{I}(X, Y)$ and $g \in \mathrm{Hom}_{\mathcal{C}}(Y, Z)$, then $gf \in \mathcal{I}(X, Z)$; and

(ii) if $f \in \mathcal{I}(X, Y)$ and $h \in \mathrm{Hom}_{\mathcal{C}}(U, X)$, then $fh \in \mathcal{I}(U, Z)$.

Given a two-sided ideal $\mathcal{I}$ in an additive K-category $\mathcal{C}$, we define the **quotient category** $\mathcal{C}/\mathcal{I}$ to be the category the objects of which are the objects of $\mathcal{C}$ and the space of morphisms from X to Y in $\mathcal{C}/\mathcal{I}$ is the quotient space

$$\mathrm{Hom}_{\mathcal{C}/\mathcal{I}}(X, Y) = \mathrm{Hom}_{\mathcal{C}}(X, Y)/\mathcal{I}(X, Y)$$

of $\mathrm{Hom}_{\mathcal{C}}(X, Y)$ modulo the subspace $\mathcal{I}(X, Y)$. In particular, if $\mathcal{X}$ is a class of objects of $\mathcal{C}$, then $\mathcal{C}/[\mathcal{X}]$ denotes the quotient category of $\mathcal{C}$ modulo the two-sided ideal $[\mathcal{X}]$ in $\mathcal{C}$ consisting of all morphisms having a factorisation through a direct sum of objects from $\mathcal{X}$.

It is easy to see that the quotient category $\mathcal{C}/\mathcal{I}$ is an additive K-category and the projection functor $\pi : \mathcal{C} \longrightarrow \mathcal{C}/\mathcal{I}$ assigning to each $f : X \to Y$ in $\mathcal{C}$ the coset $f + \mathcal{I} \in \mathrm{Hom}_{\mathcal{C}/\mathcal{I}}(X, Y)$ is a K-linear functor. Moreover π is full and dense and $\mathrm{Ker}\,\pi = \mathcal{I}$.

By the **kernel of a K-linear functor** $T : \mathcal{C} \longrightarrow \mathcal{C}'$ we mean the class $\mathrm{Ker}\,T$ of all morphisms $h : A \longrightarrow B$ in $\mathcal{C}$ such that $T(h) = 0$. It is easy to check that $\mathrm{Ker}\,T$ is a two-sided ideal in $\mathcal{C}$ and the isomorphism theorem for algebras generalises to additive K-categories as follows.

3.2. Lemma. *Let $T : \mathcal{C} \longrightarrow \mathcal{C}'$ be a full, dense, and K-linear functor between additive K-categories $\mathcal{C}$ and $\mathcal{C}'$. Then T induces a K-linear equivalence of K-categories $\mathcal{C}/\mathrm{Ker}\,T \cong \mathcal{C}'$.* $\square$

3.3. Definition. (a) The (Jacobson) **radical** of an additive K-category $\mathcal{C}$ is the two-sided ideal $\mathrm{rad}_{\mathcal{C}}$ in $\mathcal{C}$ defined by the formula

$$\mathrm{rad}_{\mathcal{C}}(X, Y) = \big\{ h \in \mathcal{C}(X, Y);\ 1_X - g \circ h \text{ is invertible for any } g \in \mathcal{C}(Y, X) \big\}$$

for all objects X and Y of $\mathcal{C}$.

(b) Given $m \geq 1$, we define the mth power $\mathrm{rad}_{\mathcal{C}}^{m} \subseteq \mathrm{rad}_{\mathcal{C}}$ of $\mathrm{rad}_{\mathcal{C}}$ by taking for $\mathrm{rad}_{\mathcal{C}}^{m}(X, Y)$ the subspace of $\mathrm{rad}_{\mathcal{C}}(X, Y)$ consisting of all finite sums of morphisms of the form

$$X = X_0 \xrightarrow{h_1} X_1 \xrightarrow{h_2} X_2 \longrightarrow \cdots \longrightarrow X_{m-1} \xrightarrow{h_m} X_m = Y,$$

where $h_j \in \mathrm{rad}_{\mathcal{C}}(X_{j-1}, X_j)$. In case $\mathcal{C} = \mathrm{mod}\,A$ is the category of finitely generated right A-modules, we set

$$\mathrm{rad}_A = \mathrm{rad}_{\mathrm{mod}\,A}.$$

It is clear that the intersection

$$\mathrm{rad}_A^{\infty} = \bigcap_{m=1}^{\infty} \mathrm{rad}_A^{m}$$

of all powers rad_A^{m} of rad_A is a two-sided ideal of $\mathrm{mod}\,A$, known as the **infinite radical** of $\mathrm{mod}\,A$.

3.4. Lemma. *Let $\mathcal{C}$ be an additive K-category.*
(a) *For each $m \geq 1$, $\mathrm{rad}_{\mathcal{C}}^{m}$ is a two-sided ideal of $\mathcal{C}$.*
(b) *Let $X_1, \ldots X_n, Y_1 \ldots, Y_m$ be objects in $\mathcal{C}$. A morphism*

$$f = \begin{bmatrix} f_{11} & f_{12} & \cdots & f_{1n} \\ f_{21} & f_{22} & \cdots & f_{2n} \\ \vdots & \vdots & \ddots & \vdots \\ f_{m1} & f_{m2} & \cdots & f_{mn} \end{bmatrix} : \bigoplus_{i=1}^{n} X_i \longrightarrow \bigoplus_{j=1}^{m} Y_j$$

in C belongs to $\mathrm{rad}_C(\oplus_{i=1}^n X_i,\ \oplus_{j=1}^m Y_j)$ *if and only if the morphism* f_{ji} :
$X_i \longrightarrow Y_j$ *belongs to* $\mathrm{rad}_C(X_i,\ Y_j)$ *for* $i=1,\ldots,n$ *and* $j=1,\ldots,m$.

Proof. (a) We only prove the statement for $m=1$; because the proof is similar for $m \geq 2$.

Assume that $f \in \mathrm{rad}_C(X,Y)$ and let $h' : Y \longrightarrow Z'$ be a morphism in C. It follows that, for any $g' : Z' \longrightarrow X$, the morphism $1_X - g' \circ h' \circ f$ is invertible and therefore $h' \circ f \in \mathrm{rad}_C(X,Z')$ for any morphism h'. Let $h : Z \longrightarrow X$ be a morphism in C. We prove that $f \circ h \in \mathrm{rad}_C(Z,Y)$ by showing that $1_Z - g \circ f \circ h$ is invertible for any morphism $g : Y \longrightarrow Z$. By the assumption, there exists $\varphi : X \longrightarrow X$ such that $(1_X - h \circ g \circ f) \circ \varphi = 1_X$ and $\varphi \circ (1_X - h \circ g \circ f) = 1_X$. It follows that $(1_Z - g \circ f \circ h) \circ (1_Z + g \circ f \circ \varphi \circ h) = 1_Z$ and $(1_Z + g \circ f \circ \varphi \circ h) \circ (1_Z - g \circ f \circ h) = 1_Z$, and we are done.

Now we prove that if $f, f' \in \mathrm{rad}_C(X,Y)$, then $f - f' \in \mathrm{rad}_C(X,Y)$ by showing that the morphism $1_X - g \circ (f - f')$ is invertible, for any morphism $g : Y \longrightarrow X$ in the category C. Because $f \in \mathrm{rad}_C(X,Y)$, $t(1_X - g \circ f) = 1_X$ and $(1_X - g \circ f)t = 1_X$, for some morphism $t : X \longrightarrow X$. Because $f' \in \mathrm{rad}_C(X,Y)$, $t'(1_X - (-t \circ g) \circ f') = 1_X$ for some morphism $t' : X \longrightarrow X$. Thus $t' \circ t(1_X - g \circ (f - f')) = 1_X$. Further, by the first part of the proof, we get $f' \circ t \in \mathrm{rad}_C(X,Y)$, and therefore $(1_X - (-g) \circ (f' \circ t))t'' = 1_X$ for some $t'' : X \longrightarrow X$. It follows that $(1_X - g \circ (f - f')) \circ t \circ t'' = 1_X$ and therefore $1_X - g \circ (f - f')$ is invertible for any morphism $g : Y \longrightarrow X$ in C as required.

(b) If $f = [f_{ji}] : \bigoplus_{i=1}^n X_i \longrightarrow \bigoplus_{j=1}^m Y_j$ is a morphism in C then

$$f_{ji} = p_j \circ f \circ u_i \in C(X_i, Y_j) \quad \text{and} \quad f = \sum_{i=1}^n \sum_{j=1}^m f_{ji},$$

where $u_i : X_i \longrightarrow X_1 \oplus X_2 \oplus \cdots \oplus X_n$ is the ith summand embedding and $p_j : Z_1 \oplus \cdots \oplus Z_m \longrightarrow Z_j$ is the jth summand projection. Thus (b) is a consequence of (a). □

3.5. Proposition. *Let C be an additive K-category.*

(a) *For any object Z in C,* $\mathrm{rad}_C(Z,Z)$ *is the Jacobson radical of the endomorphism algebra* $\mathrm{End}_C Z = \mathrm{Hom}_C(Z,Z)$ *of Z.*

(b) *Assume that X and Y are objects of C such that the K-algebras* $\mathrm{Hom}_C(X,X)$ *and* $\mathrm{Hom}_C(Y,Y)$ *are local; that is, each of them has a unique maximal ideal. Then* $\mathrm{rad}_C(X,Y)$ *is the vector space of all nonisomorphisms from X to Y in C. In particular, if $X \not\cong Y$ then* $\mathrm{rad}_C(X,Y) = \mathrm{Hom}_C(X,Y)$.

Proof. The statement (a) follows from the definition of the radical and (I.1.3).

(b) If $f \in \mathrm{rad}_{\mathcal{C}}(X,Y)$ then f is not invertible because, otherwise, in view of (I.1.3), the element $0 = 1 - f^{-1} \circ f$ would be invertible, which is a contradiction.

Assume that $f : X \longrightarrow Y$ is a nonzero nonisomorphism in $\mathcal{C}$. We show that f belongs to $\mathrm{rad}_{\mathcal{C}}(X,Y)$.

First, we prove that for any morphism $g : Y \longrightarrow X$ in $\mathcal{C}$, the endomorphism $g \circ f : X \longrightarrow X$ is not invertible. Assume to the contrary that $g \circ f$ is invertible. Let $s : X \longrightarrow X$ be such that $s \circ g \circ f = 1_X$. It follows that the element $e = f \circ s \circ g \in \mathrm{Hom}_{\mathcal{C}}(Y,Y)$ is nonzero and the equality $(1_Y - e) \circ e = 0$ holds. Then, in view of (I.1.3), $e \notin \mathrm{rad}(\mathrm{Hom}_{\mathcal{C}}(Y,Y))$, because otherwise $1_Y - e$ is invertible and the equality $(1_Y - e) \circ e = 0$ yields $e = 0$, which is a contradiction. Because, by our assumption, $\mathrm{rad}(\mathrm{Hom}_{\mathcal{C}}(Y,Y))$ is the unique maximal right ideal, there exist $r \in \mathrm{rad}(\mathrm{Hom}_{\mathcal{C}}(Y,Y))$ and $h \in \mathrm{Hom}_{\mathcal{C}}(Y,Y)$ such that $1_Y = r + e \circ h$. It follows from (I.1.3) that the element $e \circ h = 1_Y - r \in \mathrm{Hom}_{\mathcal{C}}(Y,Y)$ is invertible. If $t \in \mathrm{Hom}_{\mathcal{C}}(Y,Y)$ is such that $e \circ h \circ t = 1_Y$, then the equality $(1_Y - e) \circ e = 0$ yields $1_Y - e = (1_Y - e) \circ e \circ h \circ t = 0$. It follows that f is invertible and $f^{-1} = s \circ g$, contrary to our assumption that f is not an isomorphism.

Because $g \circ f : X \longrightarrow X$ has no left inverse and $\mathrm{rad}_{\mathcal{C}}(X,X)$ is the unique maximal left ideal of $\mathrm{Hom}_{\mathcal{C}}(X,X)$, $g \circ f \in \mathrm{rad}_{\mathcal{C}}(X,X)$ and, by (I.1.3), the element $1_X - g \circ f$ is invertible for any $g : Y \longrightarrow X$. This shows that $f \in \mathrm{rad}_{\mathcal{C}}(X,Y)$ and finishes the proof. $\square$

The description of the radical of morphism spaces given in (3.5) is very useful in applications for $\mathcal{C} = \mathrm{mod}\, A$, because we proved in Chapter I that finite dimensional indecomposable modules satisfy the hypothesis of the proposition.

The following corollary indicates that indecomposable objects with local endomorphism algebras are somewhat akin to indecomposable finitely generated modules over finite dimensional algebras.

3.6. Corollary. *Let X be an object of an additive K-category $\mathcal{C}$.*

(a) *If the endomorphism algebra $\mathrm{End}_{\mathcal{C}} X = \mathrm{Hom}_{\mathcal{C}}(X,X)$ of X is local, then X is indecomposable.*

(b) *Assume that $\mathcal{C}$ is abelian. If X is indecomposable and $\dim_K \mathrm{End}_{\mathcal{C}} X$ is finite, then the K-algebra $\mathrm{End}_{\mathcal{C}} X$ is local.*

Proof. (a) Assume to the contrary that X decomposes as $X = X_1 \oplus X_2$ with both X_1 and X_2 nonzero. Then there exist projections $p_i : X \longrightarrow X_i$ and injections $u_i : X_i \longrightarrow X$ (for $i = 1, 2$) such that $u_1 \circ p_1 + u_2 \circ p_2 = 1_X$, but neither $u_1 \circ p_1$ nor $u_2 \circ p_2$ is invertible. This is a contradiction because of (I.4.6).

(b) By (I.4.6), it is sufficient to prove that any idempotent $e \in \mathrm{End}_{\mathcal{C}} X$

equals zero or the identity 1_X. However, for such an idempotent e, a simple calculation shows that $X \cong \operatorname{Ker} e \oplus \operatorname{Ker}(1-e)$. Our claim follows from the indecomposability of X.   □

A.4. Homological algebra

We collect in this section basic notions and elementary facts from homological algebra needed in the book. In particular, we define the functors Ext_A^n and Tor_n^A, the projective and injective dimensions of a module, and the global dimension of an algebra and we give several characterisations of them. For more detailed information on homological algebra, the reader is referred to [41], [47], [77], [95], [111], [125], [148], and [168].

Throughout this section, K is a field and A is a K-algebra (not necessarily finite dimensional).

A **chain complex** in the category $\operatorname{Mod} A$ is a sequence

$$C_\bullet : \ \ldots \longrightarrow C_{n+2} \xrightarrow{d_{n+1}} C_{n+1} \xrightarrow{d_n} C_n \xrightarrow{d_{n-1}} C_{n-1} \longrightarrow \ \ldots \xrightarrow{d_2} C_1 \xrightarrow{d_1} C_0 \xrightarrow{d_0} 0$$

of right A-modules connected by A-homomorphisms such that $d_n d_{n+1} = 0$ for all $n \geq 0$. A **cochain complex** in the category $\operatorname{Mod} A$ is a sequence

$$C^\bullet : \ 0 \xrightarrow{d_0} C^0 \xrightarrow{d^1} C^1 \xrightarrow{d^2} \ \ldots \ \longrightarrow C^{n-1} \xrightarrow{d^{n-1}} C^n \xrightarrow{d^n} C^{n+1} \xrightarrow{d^{n+1}} C^{n+2} \longrightarrow \ \ldots$$

of right A-modules connected by A-homomorphisms such that $d^{n+1} d^n = 0$ for all $n \geq 0$. For each $n \geq 0$, the nth homology A-module of the chain complex $C_\bullet$ and the nth cohomology A-module of the cochain complex $C^\bullet$ are the quotient A-modules

$$H_n(C_\bullet) = \operatorname{Ker} d_n / \operatorname{Im} d_{n+1} \quad \text{and} \quad H^n(C^\bullet) = \operatorname{Ker} d^n / \operatorname{Im} d^{n-1},$$

respectively.

We start with two simple lemmas.

4.1. Lemma. *Let e be an idempotent of a finite dimensional K-algebra A, and let*

$$C^\bullet : \ 0 \xrightarrow{d_0} C^0 \xrightarrow{d^1} C^1 \xrightarrow{d^2} \ \ldots \ \longrightarrow C^{n-1} \xrightarrow{d^{n-1}} C^n \xrightarrow{d^n} C^{n+1} \xrightarrow{d^{n+1}} C^{n+2} \longrightarrow \ \ldots$$

be a cochain complex in $\operatorname{mod} A$. For every $n \geq 0$, there exists a functorial isomorphism $H^n(C^\bullet e) \cong H^n(C^\bullet)e$.

Proof. For each $n \geq 0$, we denote by $d_e^{n-1} : C^{n-1}e \longrightarrow C^n e$ and $d_{1-e}^{n-1} : C^{n-1}(1-e) \longrightarrow C^n(1-e)$ the restriction of d^{n-1} to $C^{n-1}e$ and $C^{n-1}(1-e)$, respectively. Because e is an idempotent, $C^\bullet e$ and $C^\bullet(1-e)$ are subcomplexes of $C^\bullet$ such that $C^\bullet = C^\bullet e \oplus C^\bullet(1-e)$. Moreover, for each $n \geq 0$, we have direct sum decompositions

$\operatorname{Ker} d^{n+1} = (\operatorname{Ker} d^{n+1})e \oplus (\operatorname{Ker} d^{n+1})(1-e) = \operatorname{Ker} d_e^{n+1} \oplus \operatorname{Ker} d_{1-e}^{n+1}$ and
$\operatorname{Im} d^n = (\operatorname{Im} d^n)e \oplus (\operatorname{Im} d^n)(1-e) = \operatorname{Im} d_e^n \oplus \operatorname{Im} d_{1-e}^n$.
Hence we get

$$H^n(C^\bullet) = \frac{\operatorname{Ker} d^{n+1}}{\operatorname{Im} d^n} \cong \frac{\operatorname{Ker} d_e^{n+1}}{\operatorname{Im} d_e^n} \oplus \frac{\operatorname{Ker} d_{1-e}^{n+1}}{\operatorname{Im} d_{1-e}^n} \cong H^n(C^\bullet e) \oplus H^n(C^\bullet(1-e)).$$

Because obviously $H^n(C^\bullet e)e = H^n(C^\bullet e)$ and $H^n(C^\bullet(1-e))e = 0$, we get
$H^n(C^\bullet)e \cong H^n(C^\bullet e)e = H^n(C^\bullet e)$. $\square$

4.2. Lemma. *Assume that A is a finite dimensional K-algebra. Let*
$D = \operatorname{Hom}_K(-, K) : \operatorname{mod} A \longrightarrow \operatorname{mod} A^{op}$ *be the standard duality and let*

$$C^\bullet : 0 \xrightarrow{d_0} C^0 \xrightarrow{d^1} C^1 \xrightarrow{d^2} \dots \longrightarrow C^{n-1} \xrightarrow{d^{n-1}} C^n \xrightarrow{d^n} C^{n+1} \xrightarrow{d^{n+1}} C^{n+2} \longrightarrow \dots$$

be a cochain complex in $\operatorname{mod} A$. Then $DC^\bullet$ is a chain complex in $\operatorname{mod} A^{op}$, and there exists a functorial isomorphism $H_n(DC^\bullet) \cong DH^n(C^\bullet)$ for every $n \geq 0$.

Proof. For each $n \geq 0$, there is a short exact sequence

$$0 \longrightarrow \operatorname{Im} d^n \longrightarrow \operatorname{Ker} d^{n+1} \longrightarrow H^n(C^\bullet) \longrightarrow 0.$$

By applying the duality D, we get the exact sequence

$$0 \longrightarrow DH^n(C^\bullet) \longrightarrow D(\operatorname{Ker} d^{n+1}) \longrightarrow D(\operatorname{Im} d^n) \longrightarrow 0$$

of left A-modules. On the other hand, because D is a duality, we get

$$\begin{aligned} D(\operatorname{Ker} d^{n+1}) &\cong & \operatorname{Coker} Dd^{n+1} &= & DC^n/\operatorname{Im} Dd^{n+1}, \\ D(\operatorname{Im} d^n) &\cong & DC^n/\operatorname{Ker} Dd^n, \end{aligned}$$

see (I.5.13). It then follows that the exact sequence

$$0 \longrightarrow DH^n(C^\bullet) \longrightarrow DC^n/\operatorname{Im} Dd^{n+1} \longrightarrow DC^n/\operatorname{Ker} Dd^n \longrightarrow 0$$

yields an isomorphism $DH^n(C^\bullet) \cong \operatorname{Ker} Dd^n/\operatorname{Im} Dd^{n+1} = H_n(DC^\bullet)$, which is obviously functorial. $\square$

Let K be a field and A be a K-algebra. We recall that any right A-module has a projective resolution and an injective resolution in $\operatorname{Mod} A$. If, in addition, A is finite dimensional over K, then any module in $\operatorname{mod} A$ has a minimal projective resolution and a minimal injective resolution in $\operatorname{mod} A$ (see Chapter I).

4.3. Definition. Let K be a field and A be an arbitrary K-algebra.

(a) The **projective dimension** of a right A-module M is the non-negative integer $\operatorname{pd} M = m$ such that there exists a projective resolution

$$0 \longrightarrow P_m \xrightarrow{h_m} P_{m-1} \longrightarrow \cdots \longrightarrow P_1 \xrightarrow{h_1} P_0 \xrightarrow{h_0} M \longrightarrow 0$$

of M of length m and M has no projective resolution of length $m-1$, if such a number m exists. If M admits no projective resolution of finite length, we define the projective dimension $\operatorname{pd} M$ of M to be infinity.

(b) An **injective dimension** of an A-module N is the nonnegative integer $\operatorname{id} N = m$ such that there exists an injective resolution

$$0 \longrightarrow N \xrightarrow{h^0} I^0 \xrightarrow{h^1} I^1 \longrightarrow \cdots \longrightarrow I^{m-1} \xrightarrow{h^m} I^m \longrightarrow 0$$

of N of length m and N has no injective resolution of length $m - 1$, if such a number m exists. If N admits no injective resolution of finite length, we define the injective dimension $\operatorname{id} N$ of N to be infinity.

One can show that the projective dimension of a module M is the length of a minimal projective resolution of M. Similarly, the injective dimension of a module N is the length of a minimal injective resolution of N.

The **right global dimension** and the **left global dimension** of a K-algebra A are defined to be the numbers

$$\text{r.gl.dim} A \;=\; \max \left\{ \operatorname{pd} M; \quad M \text{ is a right } A\text{-module} \right\} \text{ and}$$

$$\text{l.gl.dim} A \;=\; \max \left\{ \operatorname{pd} L; \quad L \text{ is a left } A\text{-module} \right\},$$

respectively, if these numbers exist; otherwise, we say that the right global dimension of A (or the left global dimension of A, respectively) is infinity.

It follows from the previous definitions that $\operatorname{pd} M = 0$ if and only if M is projective and $\operatorname{id} M = 0$ if and only if M is injective. One can prove that $\operatorname{gl.dim} K[t] = 1$ and, clearly, the global dimension of any finite dimensional semisimple K-algebra is zero.

4.4. Example. Let B be the algebra $K[t]/(t^2)$. Then the map $h : B \longrightarrow B$ given by $b \mapsto tb$ is a homomorphism of B-modules, $\operatorname{Ker} h = \operatorname{rad} B$, $B/\operatorname{rad} B \cong \operatorname{rad} B$, and the sequence

$$\cdots \longrightarrow B \xrightarrow{h} B \xrightarrow{h} B \longrightarrow \cdots \xrightarrow{h} B \xrightarrow{h} B,$$

together with the canonical epimorphism $h_0 : B \longrightarrow B/\operatorname{rad} B$, is a minimal projective resolution of the B-module $B/\operatorname{rad} B \cong K$. It follows that $\operatorname{pd}(B/\operatorname{rad} B) = \infty$ and $\text{r.gl.dim} B = \infty$.

Let A be a K-algebra. For each $m \geq 0$, the mth **extension bifunctor**

$$\operatorname{Ext}_A^m : (\operatorname{Mod} A)^{\mathrm{op}} \times \operatorname{Mod} A \longrightarrow \operatorname{Mod} K$$

is defined as follows. Given two modules M and N in $\operatorname{Mod} A$, we take a projective resolution $P_\bullet$ of M and construct the induced cochain complex

$$\operatorname{Hom}_A(P_\bullet, N) : 0 \longrightarrow \operatorname{Hom}_A(P_0, N) \xrightarrow{\operatorname{Hom}_A(h_1, N)} \operatorname{Hom}_A(P_1, N) \longrightarrow \cdots$$

$$\cdots \longrightarrow \operatorname{Hom}_A(P_m, N) \xrightarrow{\operatorname{Hom}_A(h_{m+1}, N)} \operatorname{Hom}_A(P_{m+1}, N) \longrightarrow \cdots$$

of K-vector spaces. We define $\mathrm{Ext}_A^m(M, N)$ to be the mth cohomology K-vector space $H^m(\mathrm{Hom}_A(P_\bullet, N))$ of the cochain complex $\mathrm{Hom}_A(P_\bullet, N)$, that is,

$$\mathrm{Ext}_A^m(M, N) = H^m(\mathrm{Hom}_A(P_\bullet, N)) = \mathrm{Ker}\,\mathrm{Hom}_A(h_{m+1}, N)/\mathrm{Im}\,\mathrm{Hom}_A(h_m, N),$$

where we set $h_0 = 0$. One shows that, up to isomorphism, the definition does not depend on the choice of the projective resolution of M. If $f : M \longrightarrow M'$ is a homomorphism of A-modules and $P'_\bullet$ is a projective resolution of M', then one can easily show that there is a commutative diagram

$$
\begin{array}{ccccccccccccc}
\cdots \longrightarrow & P_m & \xrightarrow{h_m} & P_{m-1} & \longrightarrow \cdots \longrightarrow & P_1 & \xrightarrow{h_1} & P_0 & \xrightarrow{h_0} & M & \to 0 \\
& \downarrow{\scriptstyle f_m} & & \downarrow{\scriptstyle f_{m-1}} & & \downarrow{\scriptstyle f_1} & & \downarrow{\scriptstyle f_0} & & \downarrow{\scriptstyle f} & \\
\cdots \longrightarrow & P'_m & \xrightarrow{h'_m} & P'_{m-1} & \longrightarrow \cdots \longrightarrow & P'_1 & \xrightarrow{h'_1} & P'_0 & \xrightarrow{h'_0} & M' & \to 0
\end{array}
$$

The system $f_\bullet = \{f_m\}_{m \in \mathbb{N}}$ (called a resolution of the homomorphism f) induces the commutative diagram

$$
\begin{array}{ccccc}
0 \longrightarrow & \mathrm{Hom}_A(P'_0, N) & \xrightarrow{\mathrm{Hom}_A(h'_1, N)} & \mathrm{Hom}_A(P'_1, N) & \longrightarrow \cdots \\
& \downarrow{\scriptstyle \mathrm{Hom}_A(f_0, N)} & & \downarrow{\scriptstyle \mathrm{Hom}_A(f_1, N)} & \\
0 \longrightarrow & \mathrm{Hom}_A(P_0, N) & \xrightarrow{\mathrm{Hom}_A(h_1, N)} & \mathrm{Hom}_A(P_1, N) & \longrightarrow \cdots
\end{array}
$$

$$
\begin{array}{ccccc}
\cdots \longrightarrow & \mathrm{Hom}_A(P'_m, N) & \xrightarrow{\mathrm{Hom}_A(h'_{m+1}, N)} & \mathrm{Hom}_A(P'_{m+1}, N) & \longrightarrow \cdots \\
& \downarrow{\scriptstyle \mathrm{Hom}_A(f_m, N)} & & \downarrow{\scriptstyle \mathrm{Hom}_A(f_{m+1}, N)} & \\
\cdots \longrightarrow & \mathrm{Hom}_A(P_m, N) & \xrightarrow{\mathrm{Hom}_A(h_{m+1}, N)} & \mathrm{Hom}_A(P_{m+1}, N) & \longrightarrow \cdots
\end{array}
$$

It follows that $\mathrm{Hom}_A(f_m, N)(\mathrm{Ker}\,\mathrm{Hom}_A(h'_{m+1}, N)) \subseteq \mathrm{Ker}\,\mathrm{Hom}_A(h_{m+1}, N)$ and $\mathrm{Hom}_A(f_m, N)(\mathrm{Im}\,\mathrm{Hom}_A(h'_m, N)) \subseteq \mathrm{Im}\,\mathrm{Hom}_A(h_m, N)$.

Therefore, the homomorphism $\mathrm{Hom}_A(f_m, N)$ induces a K-linear map $\mathrm{Ext}_A^m(f, N) : \mathrm{Ext}_A^m(M', N) \longrightarrow \mathrm{Ext}_A^m(M, N)$. One shows that $\mathrm{Ext}_A^m(f, N)$ does not depend on the choice of the resolution $f_\bullet$ of f and that

$$\mathrm{Ext}_A^m(-, N) : \mathrm{Mod}\,A \longrightarrow \mathrm{Mod}\,K$$

is a contravariant additive functor.

Let $g : N \longrightarrow N'$ be a homomorphism of right A-modules. It is clear that the family $\mathrm{Hom}_A(P_\bullet, g) = \{\mathrm{Hom}_A(P_m, g)\}_{m \in \mathbb{N}}$ defines a morphism $\mathrm{Hom}_A(P_\bullet, g) : \mathrm{Hom}_A(P_\bullet, N) \longrightarrow \mathrm{Hom}_A(P_\bullet, N')$ of cochain complexes, that is, the diagram

$$0 \to \operatorname{Hom}_A(P_0, N) \xrightarrow{\operatorname{Hom}_A(h_1, N)} \operatorname{Hom}_A(P_1, N) \longrightarrow \cdots$$

$$\Big\downarrow \operatorname{Hom}_A(P_0, g) \qquad\qquad\qquad \Big\downarrow \operatorname{Hom}_A(P_1, g)$$

$$0 \to \operatorname{Hom}_A(P_0, N') \xrightarrow{\operatorname{Hom}_A(h_1, N')} \operatorname{Hom}_A(P_1, N') \longrightarrow \cdots$$

$$\cdots \longrightarrow \operatorname{Hom}_A(P_m, N) \xrightarrow{\operatorname{Hom}_A(h_{m+1}, N)} \operatorname{Hom}_A(P_{m+1}, N) \longrightarrow \cdots$$

$$\Big\downarrow \operatorname{Hom}_A(P_m, g) \qquad\qquad\qquad\qquad \Big\downarrow \operatorname{Hom}_A(P_{m+1}, g)$$

$$\cdots \longrightarrow \operatorname{Hom}_A(P_m, N') \xrightarrow{\operatorname{Hom}_A(h_{m+1}, N')} \operatorname{Hom}_A(P_{m+1}, N') \longrightarrow \cdots$$

is commutative. It follows that

$$\operatorname{Hom}_A(P_m, g)(\operatorname{Ker}\operatorname{Hom}_A(h_{m+1}, N)) \subseteq \operatorname{Ker}\operatorname{Hom}_A(h_{m+1}, N') \text{ and}$$
$$\operatorname{Hom}_A(P_m, g)(\operatorname{Im}\operatorname{Hom}_A(h_m, N)) \subseteq \operatorname{Im}\operatorname{Hom}_A(h_m, N'),$$

and therefore $\operatorname{Hom}_A(P_m, g)$ induces a K-linear map

$$\operatorname{Ext}_A^m(M, g) : \operatorname{Ext}_A^m(M, N) \longrightarrow \operatorname{Ext}_A^m(M, N').$$

One shows that $\operatorname{Ext}_A^m(M, g)$ does not depend on the choice of the resolution $P_\bullet$ of M and that $\operatorname{Ext}_A^m(M, -) : \operatorname{Mod} A \longrightarrow \operatorname{Mod} K$ is a covariant additive functor. Consequently, we have defined an additive bifunctor $\operatorname{Ext}_A^m(-, -)$ for any $m \geq 0$. One can show that the K-vector space $\operatorname{Ext}_A^m(M, N)$ is isomorphic to the mth cohomology K-vector space of the cochain complex $\operatorname{Hom}_A(M, I^\bullet)$, where $I^\bullet$ is an injective resolution of the module N.

4.5. Theorem. (a) *For any right A-modules M and N, there is a functorial isomorphism $\operatorname{Ext}_A^0(M, N) \cong \operatorname{Hom}_A(M, N)$.*

(b) *Let M and N be right A-modules. Then any short exact sequence $0 \longrightarrow X \longrightarrow Y \longrightarrow Z \longrightarrow 0$ in $\operatorname{Mod} A$ induces two long exact sequences*

$$0 \longrightarrow \operatorname{Hom}_A(Z, N) \longrightarrow \operatorname{Hom}_A(Y, N) \longrightarrow \operatorname{Hom}_A(X, N)$$

$$\xrightarrow{\delta_0} \operatorname{Ext}_A^1(Z, N) \longrightarrow \operatorname{Ext}_A^1(Y, N) \longrightarrow \operatorname{Ext}_A^1(X, N)$$

$$\vdots \qquad\qquad \vdots \qquad\qquad \vdots$$

$$\cdots \xrightarrow{\delta_{m-1}} \operatorname{Ext}_A^m(Z, N) \longrightarrow \operatorname{Ext}_A^m(Y, N) \longrightarrow \operatorname{Ext}_A^m(X, N)$$

$$\xrightarrow{\delta_m} \operatorname{Ext}_A^{m+1}(Z, N) \longrightarrow \quad \cdots \qquad , \qquad \text{and}$$

$$0 \longrightarrow \operatorname{Hom}_A(M, X) \longrightarrow \operatorname{Hom}_A(M, Y) \longrightarrow \operatorname{Hom}_A(M, Z)$$

$$\xrightarrow{\delta_0} \operatorname{Ext}_A^1(M, X) \longrightarrow \operatorname{Ext}_A^1(M, Y) \longrightarrow \operatorname{Ext}_A^1(M, Z)$$

$$\vdots \qquad\qquad \vdots \qquad\qquad \vdots$$

$$\cdots \xrightarrow{\delta_{m-1}} \operatorname{Ext}_A^m(M, X) \longrightarrow \operatorname{Ext}_A^m(M, Y) \longrightarrow \operatorname{Ext}_A^m(M, Z)$$

$$\xrightarrow{\delta_m} \operatorname{Ext}_A^{m+1}(M, X) \longrightarrow \quad \cdots$$

$$\square$$

By applying (4.5), one proves the following useful results.

4.6. Corollary. (a) $\operatorname{pd} M = m$ *if and only if* $\operatorname{Ext}_A^{m+1}(M, -) = 0$ *and* $\operatorname{Ext}_A^m(M, -) \neq 0$.
   (b) $\operatorname{id} N = m$ *if and only if* $\operatorname{Ext}_A^{m+1}(-, N) = 0$ *and* $\operatorname{Ext}_A^m(-, N) \neq 0$.
   (c) $\operatorname{r.gl.dim} A = \max \{ \operatorname{id} N; \quad N \text{ is a right } A\text{-module} \}$.   □

4.7. Proposition. *Let* $0 \longrightarrow L \longrightarrow M \longrightarrow N \longrightarrow 0$ *be a short exact sequence in* $\operatorname{Mod} A$.
   (a) $\operatorname{pd} N \leq \max(\operatorname{pd} M, 1 + \operatorname{pd} L)$, *and the equality holds if* $\operatorname{pd} M \neq \operatorname{pd} L$.
   (b) $\operatorname{pd} L \leq \max(\operatorname{pd} M, -1 + \operatorname{pd} N)$, *and the equality holds if* $\operatorname{pd} M \neq \operatorname{pd} N$.
   (c) $\operatorname{pd} M \leq \max(\operatorname{pd} L, \operatorname{pd} N)$, *and the equality holds if* $\operatorname{pd} N \neq 1 + \operatorname{pd} L$.
                                                                              □

In computing the global dimension of an algebra, the following result due to Auslander [10] is very useful.

4.8. Theorem. *If* A *is a finite dimensional* K-algebra, then
$$\operatorname{r.gl.dim} A = \max \{ \operatorname{pd} S; \quad S \text{ is a simple right } A\text{-module} \}$$
$$= 1 + \max \{ \operatorname{pd}(\operatorname{rad} eA); \quad e \in A \text{ is a primitive idempotent} \}.$$
                                                                              □

Assume that A is a finite dimensional K-algebra. It follows from (4.8) that $\operatorname{r.gl.dim} A$ is the minimal number m such that, for each simple right A-module S, the functor $\operatorname{Ext}_A^{m+1}(S, -) : \operatorname{Mod} A \longrightarrow \operatorname{Mod} K$ is zero. Hence, one concludes that $\operatorname{r.gl.dim} A$ is the minimal number m such that, for each pair of modules M and N in $\operatorname{mod} A$, we have $\operatorname{Ext}_A^{m+1}(M, N) = 0$. In view of (4.6), this yields
$$\operatorname{r.gl.dim} A = \max \{ \operatorname{id} N; \quad N \text{ is in } \operatorname{mod} A \}$$
$$= \max \{ \operatorname{pd} M; \quad M \text{ is in } \operatorname{mod} A \}.$$
Obviously, a similar formula holds for the left global dimension of A. Hence, by applying the standard duality $D : \operatorname{mod} A \longrightarrow \operatorname{mod} A^{\operatorname{op}}$, we get the following result.

4.9. Corollary. *If* A *is a finite dimensional* K-algebra, then $\operatorname{r.gl.dim} A = \operatorname{l.gl.dim} A$.
                                                                              □

The common number $\operatorname{r.gl.dim} A = \operatorname{l.gl.dim} A$ is denoted by $\operatorname{gl.dim} A$ and is called the **global dimension** of the finite dimensional K-algebra A.

For each $m \geq 0$, we define the m**th torsion bifunctor**
$$\operatorname{Tor}_m^A : \operatorname{Mod} A \times \operatorname{Mod} A^{\operatorname{op}} \longrightarrow \operatorname{Mod} K$$
as follows. Given a right A-module M and a left A-module N, we take a projective resolution $P_\bullet$ of M and denote by $P_\bullet \otimes_A N$ the induced chain complex

$$\cdots \longrightarrow P_m \otimes_A N \overset{h_m \otimes 1}{\longrightarrow} P_{m-1} \otimes_A N \longrightarrow \cdots \longrightarrow P_1 \otimes_A N \overset{h_1 \otimes 1}{\longrightarrow} P_0 \otimes_A N \to 0.$$

We define $\mathrm{Tor}_m^A(M, N)$ to be the mth homology vector space $H_m(P_\bullet \otimes_A N)$ of the chain complex $P_\bullet \otimes_A N$; that is,

$$\mathrm{Tor}_m^A(M, N) = H_m(P_\bullet \otimes_A N) = \mathrm{Ker}(h_m \otimes 1)/\mathrm{Im}(h_{m+1} \otimes 1).$$

One shows that the definition does not depend, up to isomorphism, on the choice of the projective resolution of M. If $f : M \longrightarrow M'$ is a homomorphism of right A-modules, $P_\bullet'$ a projective resolution of M', and $f_\bullet = \{f_m\}_{m \in \mathbb{N}}$ is a resolution of the homomorphism f, then $f_\bullet$ induces a morphism $f_\bullet \otimes_A 1_N : P_\bullet \otimes_A N \longrightarrow P_\bullet' \otimes_A N$ of chain complexes. The induced homomorphism of the mth homology K-vector spaces is denoted by $\mathrm{Tor}_m^A(f, N) : \mathrm{Tor}_m^A(M, N) \longrightarrow \mathrm{Tor}_m^A(M', N)$.

One shows that $\mathrm{Tor}_m^A(f, N)$ does not depend on the choice of the resolution $f_\bullet$ of f and that $\mathrm{Tor}_m^A(-, N) : \mathrm{Mod}\, A \longrightarrow \mathrm{Mod}\, K$ is a covariant additive functor. If $g : N \longrightarrow N'$ is a homomorphism of left A-modules, then, modifying the previous arguments, one defines a K-linear map $\mathrm{Tor}_m^A(M, g) : \mathrm{Tor}_m^A(M, N) \longrightarrow \mathrm{Tor}_m^A(M, N')$ and proves that $\mathrm{Tor}_m^A(M, -)$ is a covariant additive functor. One can show that the K-vector space $\mathrm{Tor}_m^A(M, N)$ is isomorphic to the mth homology vector space of the chain complex $M \otimes_A P_\bullet'$, where $P_\bullet'$ is a projective resolution of the left module N.

The following result is often used.

4.10. Theorem. *Let A be a K-algebra and M be a right A-module.*

(a) *For any left A-module N, there is a functorial isomorphism of K-vector spaces $\mathrm{Tor}_0^A(M, N) \cong M \otimes_A N$.*

(b) *Any short exact sequence $\mathbb{E} : 0 \longrightarrow X \longrightarrow Y \longrightarrow Z \longrightarrow 0$ of left A-modules induces a long exact sequence*

$$\cdots \qquad \longrightarrow \quad \mathrm{Tor}_{m+1}^A(M, Z)$$

$$\cdots \quad \longrightarrow \quad \mathrm{Tor}_m^A(M, X) \quad \longrightarrow \quad \mathrm{Tor}_m^A(M, Y) \quad \longrightarrow \quad \mathrm{Tor}_m^A(M, Z)$$

$$\vdots \qquad\qquad \vdots \qquad\qquad \vdots$$

$$\longrightarrow \quad \mathrm{Tor}_1^A(M, X) \quad \longrightarrow \quad \mathrm{Tor}_1^A(M, Y) \quad \longrightarrow \quad \mathrm{Tor}_1^A(M, Z)$$

$$\longrightarrow \quad M \otimes_A X \quad \longrightarrow \quad M \otimes_A Y \quad \longrightarrow \quad M \otimes_A Z \quad \to 0$$

depending functorially on M and $\mathbb{E}$.

(c) *Let N be a left A-module. Then any short exact sequence of right A-modules $\mathbb{E}' : 0 \longrightarrow X' \longrightarrow Y' \longrightarrow Z' \longrightarrow 0$ induces a long exact sequence*

$$\cdots \quad \longrightarrow \quad \operatorname{Tor}^A_{m+1}(Z', N)$$

$$\cdots \quad \longrightarrow \quad \operatorname{Tor}^A_m(X', N) \quad \longrightarrow \quad \operatorname{Tor}^A_m(Y', N) \quad \longrightarrow \quad \operatorname{Tor}^A_m(Z', N)$$

$$\vdots \qquad\qquad\qquad \vdots \qquad\qquad\qquad \vdots$$

$$\longrightarrow \quad \operatorname{Tor}^A_1(X', N) \quad \longrightarrow \quad \operatorname{Tor}^A_1(Y', N) \quad \longrightarrow \quad \operatorname{Tor}^A_1(Z', N)$$

$$\longrightarrow \quad X' \otimes_A N \quad \longrightarrow \quad Y' \otimes_A N \quad \longrightarrow \quad Z' \otimes_A N \quad \longrightarrow 0$$

depending functorially on N and $\mathbb{E}'$. $\qquad\qquad\qquad\qquad\qquad\qquad\qquad\square$

We finish this section with the following result.

Proposition. 4.11. *Let B be a finite dimensional K-algebra. For all modules Y and Z in $\operatorname{mod} B$, there exist functorial isomorphisms of K-vector spaces $\operatorname{Hom}_B(Y, DZ) \cong D(Y \otimes_B Z)$ and $D\operatorname{Ext}^1_B(Y, DZ) \cong \operatorname{Tor}^B_1(Y, Z)$.*

Proof. The first formula is just the adjoint isomorphism $D(X \otimes_B Z) = \operatorname{Hom}_K(X \otimes_B Z, K) \cong \operatorname{Hom}_B(X, DZ)$ for any module X in $\operatorname{mod} B$. To prove the second, take a projective resolution

$$\cdots \longrightarrow P_m \xrightarrow{d_m} P_{m-1} \xrightarrow{d_{m-1}} \cdots \longrightarrow P_2 \xrightarrow{d_2} P_1 \xrightarrow{d_1} P_0 \longrightarrow Y \longrightarrow 0$$

with each P_i finite dimensional projective. Applying the functorial isomorphism $Y \otimes_B Z \cong D\operatorname{Hom}_B(Y, DZ)$ proved in the first part, to each term of the complex

$$P_\bullet : \quad \cdots \longrightarrow P_m \xrightarrow{d_m} P_{m-1} \xrightarrow{d_{m-1}} \cdots \longrightarrow P_2 \xrightarrow{d_2} P_1 \xrightarrow{d_1} P_0 \longrightarrow 0$$

yields an isomorphism of complexes $P_\bullet \otimes_B Z \cong D\operatorname{Hom}_B(P_\bullet, DZ)$. Hence, by applying (4.2), we get the following functorial isomorphisms:

$$\begin{aligned} \operatorname{Tor}^B_1(Y, Z) &= H_1(P_\bullet \otimes_B Z) \cong H^1(D\operatorname{Hom}_B(P_\bullet, DZ)) \\ &\cong DH_1(\operatorname{Hom}_B(P_\bullet, DZ)) \cong D\operatorname{Ext}^1_B(Y, DZ). \qquad \square \end{aligned}$$

A.5. The group of extensions

We give an interpretation of the group $\operatorname{Ext}^1_A(N, L)$ in terms of the short exact sequences $0 \to L \to M \to N \to 0$ in $\operatorname{Mod} A$ by constructing a group $\mathcal{E}xt^1_A(N, L)$ of extensions of a right A-module L by a right A-module N and by establishing an isomorphism $\operatorname{Ext}^1_A(N, L) \cong \mathcal{E}xt^1_A(N, L)$. This interpretation of $\operatorname{Ext}^1_A(N, L)$ is frequently used throughout this book.

In the definition of $\mathcal{E}xt^1_A(N, L)$, we use the notions of fibered product and of amalgammed sum defined as follows.

5.1. Definition. (a) The **fibered product** (or **pull-back**) of a pair of homomorphisms $X \xrightarrow{f} Z \xleftarrow{g} Y$ of right A-modules is the submodule

$$P = \{(x, y) \in X \oplus Y; \ f(x) = g(y)\}$$

of $X \oplus Y$ together with two homomorphisms $X \xleftarrow{f'} P \xrightarrow{g'} Y$ defined by the formulas $f'(x,y) = x$ and $g'(x,y) = y$.

(b) The **amalgammed sum** (or **push-out**) of a pair of homomorphisms $X \xleftarrow{u} Z \xrightarrow{v} Y$ of right A-modules is the module

$$S = (X \oplus Y)/\{(u(z), -v(z)),\ z \in Z\}$$

together with two homomorphisms $X \xrightarrow{u'} S \xleftarrow{v'} Y$ defined by the formulas $u'(x) = \overline{(x,0)}$ and $v'(y) = \overline{(0,y)}$, where $\overline{(x,y)}$ is the image of $(x,y) \in X \oplus Y$ under the canonical epimorphism $X \oplus Y \longrightarrow S$.

The following result is easily verified.

5.2. Lemma. (a) *If* (P, f', g') *is the fibered product of* $X \xrightarrow{f} Z \xleftarrow{g} Y$, *then* $ff' = gg'$ *and, for any pair of homomorphisms* $X \xleftarrow{f''} P' \xrightarrow{g''} Y$ *such that* $ff'' = gg''$, *there exists a unique homomorphism* $t : P' \longrightarrow P$ *such that the diagram*

$$
\begin{array}{ccc}
P' & & f'' \\
& \searrow^{t} & \\
& P & \xrightarrow{f'} \quad X \\
g'' & \downarrow^{g'} & \quad\downarrow^{f} \\
& Y & \xrightarrow{g} \quad Z
\end{array}
$$

is commutative.

(b) *If* (S, u', v') *is the amalgammed sum of* $X \xleftarrow{u} Z \xrightarrow{v} Y$, *then* $u'u = v'v$ *and, for any pair of homomorphisms* $X \xrightarrow{u''} S' \xleftarrow{v''} Y$ *such that* $u''u = v''v$, *there exists a unique homomorphism* $r : S \longrightarrow S'$ *such that the diagram*

$$
\begin{array}{ccc}
Z & \xrightarrow{u} & X \\
v\downarrow & & \downarrow u' \\
Y & \xrightarrow{v'} & S \quad u'' \\
& & \searrow^{r} \\
v'' & & S'
\end{array}
$$

is commutative.

$\square$

The following result will be frequently used.

5.3. Proposition. *Let* $0 \longrightarrow L \xrightarrow{f} M \xrightarrow{g} N \longrightarrow 0$ *be a short exact sequence in* $\mod A$.

(a) *If* $v : V \longrightarrow N$ *is an A-module homomorphism and* (V', v', g') *is the* *fibered product of* $V \xrightarrow{v} N \xleftarrow{g} M$, *then there exists a commutative diagram*

$$
\begin{array}{ccccccccc}
0 & \longrightarrow & L & \xrightarrow{r} & V' & \xrightarrow{v'} & V & \longrightarrow & 0 \\
& & \downarrow{\scriptstyle 1_L} & & \downarrow{\scriptstyle g'} & & \downarrow{\scriptstyle v} & & \\
0 & \longrightarrow & L & \xrightarrow{f} & M & \xrightarrow{g} & N & \longrightarrow & 0
\end{array}
\tag{5.4}
$$

with exact rows.

(b) *If* $u : L \longrightarrow U$ *is an A-module homomorphism and* (U', f', u') *is the amalgammed sum of* $M \xleftarrow{f} L \xrightarrow{u} U$, *then there exists a commutative diagram*

$$
\begin{array}{ccccccccc}
0 & \longrightarrow & L & \xrightarrow{f} & M & \xrightarrow{g} & N & \longrightarrow & 0 \\
& & \downarrow{\scriptstyle u} & & \downarrow{\scriptstyle f'} & & \downarrow{\scriptstyle 1_N} & & \\
0 & \longrightarrow & U & \xrightarrow{u'} & U' & \xrightarrow{t} & N & \longrightarrow & 0
\end{array}
\tag{5.5}
$$

with exact rows.

(c) *If there exist commutative diagrams* (5.4) *and* (5.5) *with exact rows then* V' *is isomorphic to the fibered product of* $V \xrightarrow{v} N \xleftarrow{g} M$ *and* U' *is isomorphic to the amalgammed sum of* $U \xleftarrow{u} L \xrightarrow{f} M$.

The proof can be found in [6], [41], and [148]. □

Any short exact sequence $0 \longrightarrow L \xrightarrow{f} M \xrightarrow{g} N \longrightarrow 0$ in mod A is called an **extension** of L by N. Two extensions

$$\mathbb{E} : 0 \longrightarrow L \xrightarrow{f} M \xrightarrow{g} N \longrightarrow 0 \quad \text{and} \quad \mathbb{E}' : 0 \longrightarrow L \xrightarrow{f'} M' \xrightarrow{g'} N \longrightarrow 0$$

are said to be **equivalent** if there exists a commutative diagram

$$
\begin{array}{ccccccccc}
\mathbb{E} : & 0 & \longrightarrow & L & \xrightarrow{f} & M & \xrightarrow{g} & N & \longrightarrow & 0 \\
& & & \downarrow{\scriptstyle 1_L} & & \downarrow{\scriptstyle h} & & \downarrow{\scriptstyle 1_N} & & \\
\mathbb{E}' : & 0 & \longrightarrow & L & \xrightarrow{f'} & M' & \xrightarrow{g'} & N & \longrightarrow & 0
\end{array}
$$

where h is an A-isomorphism. In this case, we write $\mathbb{E} \simeq \mathbb{E}'$. We denote by $\mathcal{E}(N, L)$ the set of all extensions of the A-module L by the A-module N. Given two extensions $\mathbb{E}$ and $\mathbb{E}'$ in $\mathcal{E}(N, L)$, we define their sum $\mathbb{E} + \mathbb{E}'$ to be the extension

$$\mathbb{E} + \mathbb{E}' : \quad 0 \longrightarrow L \xrightarrow{f''} M'' \xrightarrow{g''} N \longrightarrow 0,$$

where $M'' = W/V$ and $W = \{(m, m') \in M \oplus M'; \ g(m) = g'(m')\}$, $V = \{(f(x), -f'(x')) \in M \oplus M'; \ x \in L\}$. The homomorphisms f'' and g'' are induced by the homomorphisms $L \longrightarrow W$, $x \mapsto (f(x), 0)$, and $W \longrightarrow N$, $(m, m') \mapsto g(m)$, respectively.

Consider the set
$$\mathcal{E}xt^1_A(N, L) = \mathcal{E}(N, L)/ \simeq \qquad (5.6)$$

of the equivalence classes $[\mathbb{E}] = \mathbb{E}/\simeq$ of extensions E in $\mathcal{E}(N, L)$. The set $\mathcal{E}xt^1_A(N, L)$, equipped with the addition $[\mathbb{E}] + [\mathbb{E}'] = [\mathbb{E} + \mathbb{E}']$, is an abelian group. The class represented by the split extension is the zero element of $\mathcal{E}xt^1_A(N, L)$. We call $\mathcal{E}xt^1_A(N, L)$ the **group of extensions** of L by N.

If $\mathbb{E}$ is an extension and $v : V \longrightarrow N$, $u : L \longrightarrow U$ are A-homomorphisms then, in view of (5.3), there exist commutative diagrams (5.4) and (5.5) with exact rows and with the fibered product V' and the amalgammed sum U'. It follows from (5.3)(c) and (5.2) that $\mathbb{E}$, u, and the commutativity of (5.5) determine the lower exact row in (5.5) uniquely, up to equivalence of extensions. Similarly, $\mathbb{E}$, v, and the commutativity of (5.4) determine the upper exact row in (5.4) uniquely, up to equivalence of extensions.

We denote by $\mathcal{E}xt^1_A(N, u)[\mathbb{E}]$ the equivalence class in $\mathcal{E}xt^1_A(N, U)$ represented by the lower row in (5.5), and we call it the **extension induced by** u. Similarly, we denote by $\mathcal{E}xt^1_A(v, L)[\mathbb{E}]$ the equivalence class in $\mathcal{E}xt^1_A(V, L)$ represented by the upper row in (5.4), and we call it the **extension induced by** v. A straightforward calculation shows that, for any right A-modules N and L, we have defined two functors

$$\mathcal{E}xt^1_A(N, -) : \mathrm{mod}\, A \longrightarrow \mathcal{A}b \text{ and } \mathcal{E}xt^1_A(-, L) : (\mathrm{mod}\, A)^{\mathrm{op}} \longrightarrow \mathcal{A}b, \quad (5.7)$$

where $\mathcal{A}b$ is the category of abelian groups.

For each pair of A-modules L and N, the extension group $\mathcal{E}xt^1_A(N, L)$ is related with the first extension group $\mathrm{Ext}^1_A(N, L)$ by the group homomorphism

$$\chi : \mathcal{E}xt^1_A(N, L) \longrightarrow \mathrm{Ext}^1_A(N, L) \qquad (5.8)$$

defined as follows. Let $[\mathbb{E}]$ be an element of $\mathcal{E}xt^1_A(N, L)$ represented by the exact sequence $\mathbb{E} : 0 \longrightarrow L \overset{u}{\longrightarrow} M \longrightarrow N \longrightarrow 0$, and let

$$P_\bullet : \qquad \cdots \longrightarrow P_m \overset{h_m}{\longrightarrow} P_{m-1} \longrightarrow \cdots \longrightarrow P_1 \overset{h_1}{\longrightarrow} P_0$$

together with an epimorphism $h_0 : P_0 \longrightarrow N$ be a projective resolution of N. Because the module P_0 is projective, there exists a commutative diagram

$$
\begin{array}{ccccccccc}
P_2 & \overset{h_2}{\longrightarrow} & P_1 & \overset{h_1}{\longrightarrow} & P_0 & \overset{h_0}{\longrightarrow} & N & \longrightarrow & 0 \\
 & & \downarrow{\scriptstyle t_1} & & \downarrow{\scriptstyle t_0} & & \downarrow{\scriptstyle 1_N} & & \\
0 & \longrightarrow & L & \overset{u}{\longrightarrow} & M & \overset{v}{\longrightarrow} & N & \longrightarrow & 0
\end{array}
$$

It is easy to see that $\mathrm{Hom}_A(h_2, L)(t_1) = t_1 h_2 = 0$, and therefore the A-homomorphism t_1 belongs to $\mathrm{Ker}\,\mathrm{Hom}_A(h_2, L)$. If $t'_1 : P_1 \longrightarrow L$ and

$t'_0 : P_0 \longrightarrow M$ is another pair of A-homomorphisms making the diagram commutative, then $v(t_0 - t'_0) = h_0 - h_0 = 0$, and therefore there exists an A-homomorphism $s : P_0 \longrightarrow L$ such that $t_0 - t'_0 = us$. It follows that $u(t_1 - t'_1) = (t_0 - t'_0)h_1 = ush_1$, and the injectivity of u yields $t_1 - t'_1 = sh_1 = \operatorname{Hom}_A(h_1, L)(s) \in \operatorname{Im}\operatorname{Hom}_A(h_1, L)$. This shows that the coset

$$\chi[\mathbb{E}] = t_1 + \operatorname{Im}\operatorname{Hom}_A(h_1, L) \in \operatorname{Ext}^1_A(N, L)$$

of the A-homomorphism $t_1 \in \operatorname{Ker}\operatorname{Hom}_A(h_2, L)$ modulo $\operatorname{Im}\operatorname{Hom}_A(h_1, L)$ does not depend on the choice of t_1 and t_0, or on the choice of the extension $\mathbb{E}$ in the class $[\mathbb{E}]$. It is easy to check that χ is a group homomorphism.

The following important result is frequently used.

5.9. Theorem. *For any pair of A-modules M and N, the group homomorphism*

$$\chi : \mathcal{E}xt^1_A(N, L) \longrightarrow \operatorname{Ext}^1_A(N, L)$$

defined earlier is a functorial isomorphism.

For the proof the reader is referred to [6], [41], [111], and [148]. $\qquad\square$

A.6. Exercises

1. Let A, B be two K-algebras and $f : A \longrightarrow B$ be a surjective homomorphism. Let $\mathcal{A}_f$ denote the full subcategory of $\operatorname{Mod} A$ the objects of which are the modules M such that $M(\operatorname{Ker} f) = 0$.

(a) For any B-module X, we define $F(X)$ to be the vector space X equipped with the multiplication $\cdot : X \times A \to X$ given by $x \cdot a = xf(a)$, for all $x \in X$ and $a \in A$. Show that this multiplication is well-defined and induces a right A-module structure on X.

(b) Show that any homomorphism $\varphi : X \to Y$ of B-modules induces a homomorphism $F(\varphi) : F(X) \to F(Y)$ of A-modules, and deduce that $F : \operatorname{Mod} B \longrightarrow \operatorname{Mod} A$ is a functor.

(c) Show that the functor $F : \operatorname{Mod} B \longrightarrow \operatorname{Mod} A$ is additive, K-linear, full, faithful, and exact.

(d) Show that $F : \operatorname{Mod} B \longrightarrow \operatorname{Mod} A$ induces an equivalence of categories $\operatorname{Mod} B \xrightarrow{\simeq} \mathcal{A}_f$.

2. Prove that the upper row of the diagram (5.4) in Proposition 5.3 and the lower row of the diagram (5.5) in Proposition 5.3 are short exact sequences.

3. Prove that for each pair of A-modules M and N, the addition in $\mathcal{E}xt^1_A(M, N)$ (defined in Section 5) is associative and commutative.

4. Let $u : \mathbb{Z} \longrightarrow \mathbb{Q}$ be the embedding of the ring $\mathbb{Z}$ of integers in the field $\mathbb{Q}$ of rational numbers. Prove that u is a monomorphism and an epimorphism in the category of rings but that it is not an isomorphism in that category.

5. Let B be the algebra $K[t]/(t^2)$.
(a) Prove that the algebra B is self-injective, that is, the module B_B is an injective B-module.
(b) Show that the projective dimension of the simple one-dimensional B-module $S = B/\operatorname{rad} B$ is infinite and that the injective dimension of the simple B-module $B/\operatorname{rad} B \cong K$ is infinite, by applying the minimal projective resolution constructed in Example 4.4.
(c) For any B-module M and each $m \geq 0$, compute the extension groups $\operatorname{Ext}_B^m(S, M)$, $\operatorname{Ext}_B^m(M, S)$, and $\operatorname{Tor}_B^m(S, M)$.

6. Let A be a K-algebra and M be a right A-module.
(a) Show that the covariant functor $\operatorname{Hom}_A(M, -) : \operatorname{Mod} A \to \operatorname{Mod} K$ is left exact and that it is exact if and only if M is a projective module.
(b) Show that the functor $\operatorname{Hom}_A(-, M) : \operatorname{Mod} A \longrightarrow \operatorname{Mod} K$ is left exact and that it is exact if and only if M is an injective module.

7. Let A be a K-algebra and assume that the following diagram

$$
\begin{array}{ccccccccc}
0 & \longrightarrow & L & \xrightarrow{f} & M & \xrightarrow{g} & N & \longrightarrow & 0 \\
& & \downarrow{h'} & & \downarrow{h} & & \downarrow{h''} & & \\
0 & \longrightarrow & L' & \xrightarrow{f'} & M' & \xrightarrow{g'} & N' & \longrightarrow & 0
\end{array}
$$

in mod A is commutative and has exact rows. Prove that the following three conditions are equivalent:
(a) There exists a homomorphism $u : M \to L'$ of A-modules such that $uf = h'$.
(b) There exists a homomorphism $v : N \to M'$ of A-modules such that $g'v = h''$.
(c) There exist homomorphisms $u : M \to L'$ and $v : N \to M'$ of A-modules such that $f'u + vg = h$.

Bibliography

[1] I. T. Adamson, *Rings, Modules and Algebras*, Oliver and Boyd, Edinburgh, 1971.

[2] F. W. Anderson and K. R. Fuller, *Rings and Categories of Modules*, Graduate Texts in Mathematics 13, Springer-Verlag, New York, Heidelberg, Berlin, 1973 (new edition 1991).

[3] I. Assem, Tilted algebras of type $\mathbb{A}_n$, *Comm. Algebra*, 10(1982), 2121–39.

[4] I. Assem, Torsion theories induced by tilting modules, *Canad. J. Math.*, 36(1984), 899–913.

[5] I. Assem, Tilting theory - An introduction, in *Topics in Algebra, Part 1: Rings and Representations of Algebras*, Banach Center Publications, Vol. 26, PWN - Polish Scientific Publishers, Warszawa, 1990, pp. 127–80.

[6] I. Assem, *Algèbres et Modules*, Masson, Paris, 1997.

[7] I. Assem and D. Happel, Generalized tilted algebras of type $\mathbb{A}_n$, *Comm. Algebra*, 9(1981), 2101–25.

[8] I. Assem and A. Skowroński, Iterated tilted algebras of type $\widetilde{\mathbb{A}}_n$, *Math. Z.*, 195(1987), 269–90.

[9] I. Assem and A. Skowroński, Multicoil algebras, in Proc. the Sixth International Conference on Representations of Algebras, *Canadian Mathematical Society Conference Proceedings, AMS*, 14, 1993, pp. 29–68.

[10] M. Auslander, On the dimension of modules and algebras III, *Nagoya Math. J.*, 9(1955), 67–77.

[11] M. Auslander, Coherent functors, in Proc. Conf. on Categorical Algebra, La Jolla, Springer-Verlag, New York, 1966, pp. 189–231.

[12] M. Auslander, Representation theory of Artin algebras I, *Comm. Algebra,* 1(1974), 177–268.

[13] M. Auslander, Representation theory of Artin algebras II, *Comm. Algebra,* 1(1974), 269–310.

[14] M. Auslander, Large modules over Artin algebras, in *Algebra, Topology and Category Theory,* Academic Press, New York, 1976, pp. 3–17.

[15] M. Auslander, Functors and morphisms determined by objects, in Proc. Conf. on Representation Theory, Marcel Dekker, 1978, pp. 1–244.

[16] M. Auslander, Applications of morphisms determined by objects, in Proc. of the Philadelphia Conf., Lecture Notes in Pure and Applied Math., 37(1978), pp. 245–327.

[17] M. Auslander and M. Bridger, Stable module theory, *Memoirs Amer. Math. Soc.,* 94, 1969.

[18] M. Auslander, M. I. Platzeck, and I. Reiten, Coxeter functors without diagrams, *Trans. Amer. Math. Soc.,* 250(1979), 1–46.

[19] M. Auslander and I. Reiten, Representation theory of Artin algebras III, *Comm. Algebra,* 3(1975), 269–310.

[20] M. Auslander and I. Reiten, Representation theory of Artin algebras IV: Invariants given by almost split sequences, *Comm. Algebra,* 5(1977), 443–518.

[21] M. Auslander, I. Reiten, and S. Smalø, *Representation Theory of Artin Algebras,* Cambridge Studies in Advanced Mathematics 36, Cambridge University Press, Cambridge, New York, 1995.

[22] M. Auslander and S. O. Smalø, Almost split sequences in subcategories, *J. Algebra,* 69(1981), 426–54, Addendum, J. Algebra, 71(1981), 592–94.

[23] H. Bass, Finitistic dimension and a homological generalization of semi-primary rings, *Trans. Amer. Math. Soc.,* 95(1960), 466–88.

[24] H. Bass, *Algebraic K-theory,* W. A. Benjamin, Inc. New York, Amsterdam, 1968.

[25] R. Bautista, Irreducible morphisms and the radical of a category, *An. Inst. Mat. Nac. Autónoma México*, 22(1982), 83–135.

[26] R. Bautista, P. Gabriel, A. V. Roiter, and L. Salmerón, Representation-finite algebras and multiplicative bases, *Invent. Math.*, 81(1985), 217–85.

[27] R. Bautista, On algebras of strongly unbounded representation type, *Comment. Math. Helvetici*, 60(1985), 392–99.

[28] R. Bautista and S. Smalø, Nonexistent cycles, *Comm. Algebra*, 11(1983), 1755–67.

[29] R. Bautista and F. Larrión, Auslander–Reiten quivers for certain algebras of finite representation type, *J. London Math. Soc.*, 26(1982), 43–52.

[30] R. Bautista, F. Larrión, and L. Salmerón, On simply connected algebras *J. London Math. Soc.*, 27(1983), 212–20.

[31] D. J. Benson, *Representations and Cohomology, I: Basic Representation Theory of Finite Groups and Associative Algebras*, Cambridge Studies in Advanced Mathematics 30, Cambridge University Press, New York, 1991.

[32] I. N. Bernstein, I. M. Gelfand, and V. A. Ponomarev, Coxeter functors and Gabriel's theorem, *Uspiehi Mat. Nauk*, 28(1973), 19–33 (in Russian), English translation in *Russian Math. Surveys*, 28(1973), 17–32.

[33] K. Bongartz, Tilted algebras, in Proc. ICRA III (Puebla, 1980), Lecture Notes in Math. No. 903, Springer-Verlag, Berlin, Heidelberg, New York, 1981, pp. 26–38.

[34] K. Bongartz, True einfach zusammenhängende Algebren, *Comment. Math. Helvetici*, 57(1982), 282–330.

[35] K. Bongartz, On a result of Bautista and Smalø, *Comm. Algebra*, 11(1983), 2123–24.

[36] K. Bongartz, On omnipresent modules in preprojective components, *Comm. Algebra*, 11(1983), 2125–28.

[37] K. Bongartz, Algebras and quadratic forms, *J. London Math. Soc.*, 28(1983), 461–69.

[38] K. Bongartz, A criterion for finite representation type, *Math. Ann.*, 269(1984), 1–12.

[39] K. Bongartz, Critical simply connected algebras, *Manuscripta Math.*, 46(1984), 117–36.

[40] K. Bongartz and P. Gabriel, Covering spaces in representation theory, *Invent. Math.*, 65(1982), 331–78.

[41] N. Bourbaki, *Algèbres de Lie*, chap. IV, Masson, Paris, 1968.

[42] N. Bourbaki, *Algèbre homologique*, chap. X, Masson, Paris, 1980.

[43] S. Brenner, Decomposition properties of some small diagrams of modules, *Symposia Math. Inst. Naz. Alta Mat.* 13(1974), 127–41.

[44] S. Brenner, On four subspaces of a vector space, *J. Algebra*, 29(1974) 100–14.

[45] S. Brenner and M. C. R. Butler, The equivalence of certain functors occurring in the representation theory of algebras and species, *J. London Math. Soc.*, 14(1976) 183–87.

[46] S. Brenner and M. C. R. Butler, Generalisations of the Bernstein–Gelfand–Ponomarev reflection functors, in Proc. ICRA II (Ottawa, 1979), Lecture Notes in Math. No. 832, Springer-Verlag, Berlin, Heidelberg, New York, 1980, pp. 103–69.

[47] I. Bucur and A. Deleanu, *Introduction to the Theory of Categories and Functors*, Wiley-Interscience, London, New York, Sydney, 1969.

[48] E. Cartan and S. Eilenberg, *Homological Algebra*, Princeton University Press, 1956.

[49] P. M. Cohn, *Algebra*, Vols. 1, 2, and 3, Wiley, London, 1990.

[50] W. Crawley-Boevey, On tame algebras and bocses, *Proc. London Math. Soc.*, 56(1988), 451–83.

[51] W. Crawley-Boevey, Modules of finite length over their endomorphism rings, in *Representations of Algebras and Related Topics*, London Math. Soc. Lecture Notes 168(1992), 127–84.

[52] C. W. Curtis and I. Reiner, *Representation Theory of Finite Groups and Associative Algebras*, Wiley-Interscience, New York, 1962.

[53] S. Dickson, A torsion theory for abelian categories, *Trans. Amer. Math. Soc.*, 121(1966), 223–35.

[54] V. Dlab and C. M. Ringel, On algebras of finite representation type, *J. Algebra*, 33(1975), 306–94.

[55] V. Dlab and C. M. Ringel, Indecomposable representations of graphs and algebras, *Mem. Amer. Math. Soc.*, 173, 1976.

[56] P. Dowbor and A. Skowroński, On Galois coverings of tame algebras, *Archiv der Math.*, 44(1985), 522–29.

[57] P. Dowbor and A. Skowroński, Galois coverings of representation-infinite algebras, *Comment. Math. Helvetici*, 62(1987), 311–37.

[58] P. Dräxler, Auslander–Reiten quivers of algebras whose indecomposable modules are bricks, *Bull. London Math. Soc.* 23(1991), 141–45.

[59] Ju. A. Drozd, Coxeter transformations and representations of partially ordered sets, *Funkc. Anal. i Priložen.*, 8(1974), 34–42 (in Russian).

[60] Ju. A. Drozd, Tame and wild matrix problems, in *Representations and Quadratic Forms*, Akad. Nauk USSR, Inst. Matem., Kiev 1979, 39–74 (in Russian).

[61] J. A. Drozd and V. V. Kirichenko, *Finite Dimensional Algebras*, Springer-Verlag, Berlin, Heidelberg, New York, 1994.

[62] D. Eisenbud and P. Griffith, The structure of serial rings, *Pacific. J. Math.*, 366(1971), 109–21.

[63] D. Eisenbud and P. Griffith, Serial rings, *J. Algebra*, 17(1971), 389–400.

[64] K. Erdmann, *Blocks of Tame Representation Type and Related Algebras*, Lecture Notes in Math. No. 1428, Springer-Verlag, Berlin, Heidelberg, New York, 1990.

[65] C. Faith, *Algebra: Rings, Modules and Categories*, Springer-Verlag, Berlin, Heidelberg, New York, 1973.

[66] C. Faith, *Algebra II: Ring Theory*, Springer-Verlag, Berlin, Heidelberg, New York, 1976.

[67] P. Freyd, *Abelian Categories*, Harper and Row, New York, 1964.

[68] K. Fuller, Generalized uniserial rings and their Kupisch series, *Math. Z.*, 106(1968), 248–60.

[69] K. Fuller and I. Reiten, Note on rings of finite representation type and decompositions of modules, *Proc. Amer. Math. Soc.*, 50(1975), 92–4.

[70] P. Gabriel, Sur les catégories abéliennes localement noethériennes et leurs applications aux algèbres étudiées par Dieudonné, *Séminaire Serre*, Collège de France, Paris, 1960.

[71] P. Gabriel, Des catégories abéliennes, *Bull. Soc. Math. France*, 90(1962), 323–448.

[72] P. Gabriel, Unzerlegbare Darstellungen I, *Manuscripta Math.*, 6(1972), 71–103.

[73] P. Gabriel, Indecomposable representations II, *Symposia Mat. Inst. Naz. Alta Mat.*, 11(1973), 81–104.

[74] P. Gabriel, Représentations indécomposables, in *Séminaire Bourbaki* (1973-74), Lecture Notes in Math., No. 431, Springer-Verlag, Berlin, Heidelberg, New York, 1975, pp. 143–69.

[75] P. Gabriel, Auslander–Reiten sequences and representation–finite algebras, Proc. ICRA II (Ottawa, 1979), in Lecture Notes in Math. No. 903, Springer-Verlag, Berlin, Heidelberg, New York, Tokyo, 1981, pp. 1–71.

[76] P. Gabriel, The universal cover of a representation–finite algebra, in Lecture Notes in Math. No. 903, Springer-Verlag, Berlin, Heidelberg, New York, Tokyo, 1982, pp. 68–105.

[77] P. Gabriel and A. V. Roiter, *Representations of Finite Dimensional Algebras*, Algebra VIII, Encyclopaedia of Math. Sc., Vol. 73, Springer-Verlag, Berlin, Heidelberg, New York, 1992.

[78] S. I. Gelfand and Yu. I. Manin, *Methods of Homological Algebra*, Springer-Verlag, Berlin, Heidelberg, New York, 1996.

[79] I. M. Gelfand and V. A. Ponomarev, Indecomposable representations of the Lorentz group, *Uspechi Mat. Nauk* 2(1968), 1–60 (in Russian).

[80] I. M. Gelfand and V. A. Ponomarev, Problems of linear algebra and classification of quadruples of subspaces in a finite-dimensional vector space. *Coll. Math. Soc. Bolyai*, Tihany (Hungary), 5(1970), 163–237.

[81] K. R. Goodearl and R. B. Warfield, Jr., *An Introduction to Noncommutative Noetherian Rings*, London Mathematical Society Student Texts 16, Cambridge University Press, Cambridge, 1989.

[82] A. Grothendieck, Sur quelques points d'algèbre homologique, *Tohoku Math. J.*, 9(1957), 119–221.

[83] W.H. Gustafson, The history of algebras and their representations, in *Representations of Algebras*, Proc. (Workshop), (Puebla, Mexico, 1980), Lecture Notes in Math. No. 944, Springer-Verlag, Berlin, Heidelberg, New York, 1982, pp. 1–28.

[84] D. Happel, Composition factors of indecomposable modules, *Proc. Amer. Math. Soc.*, 86(1982), 29–31.

[85] D. Happel, *Triangulated Categories in the Representation Theory of Finite Dimensional Algebras*, London Math. Soc. Lecture Notes Series 119, Cambridge University Press, Cambridge, 1988.

[86] D. Happel, The converse of Drozd's theorem on quadratic forms, *Comm. Algebra*, 23(1995), 737–38.

[87] D. Happel, U. Preiser and C. M. Ringel, Vinsberg characterization of Dynkin diagrams using subadditive functions with applications to DTr-periodic modules, in *Representation Theory II*, Proc. ICRA II (Ottawa, 1979), Lecture Notes in Math. No. 832, Springer-Verlag, Berlin, Heidelberg, New York, 1980, pp. 280–94.

[88] D. Happel and C. M. Ringel, Construction of tilted algebras, in *Representations of Algebras*, Proc. ICRA III (Puebla, 1980), Lecture Notes in Math. No. 903, Springer-Verlag, Berlin, Heidelberg, New York, 1981, pp. 125–44.

[89] D. Happel and C. M. Ringel, Tilted algebras, *Trans. Amer. Math. Soc.*, 274(1982), 399–443.

[90] M. Harada and Y. Sai, On categories of indecomposable modules I, *Osaka J. Math.*, 8(1971), 309–21.

[91] I. Herzog, A test for finite representation type, *J. Pure Appl. Algebra*, 95(1994), 151–82.

[92] D. G. Higman, Indecomposable representations at characteristic p, *Duke Math. J.*, 21(1954), 377–81.

[93] M. Hoshino, Tilting modules and torsion theories, *Bull. London Math. Soc.*, 14(1982), 334–36.

[94] M. Hoshino, Splitting torsion theories induced by tilting modules, *Comm. Algebra*, 11(1983), 427–41.

[95] J. E. Humphreys, *Introduction to Lie Algebras and Representation Theory*, Springer-Verlag, New York, Heilelberg, Berlin, 1972.

[96] J. P. Jans, *Rings and Homology*, Holt, Rinehart and Winston, New York, Chicago, San Francisco, Toronto, London, 1964.

[97] N. Jacobson, The radical and semi-simplicity for arbitrary rings, *Amer. J. Math.*, 67(1945), 300–20.

[98] N. Janusz, Indecomposable representations of groups with cyclic Sylow subgroup, *Trans. Amer. Math. Soc.*,, 125(1966), 288–95.

[99] C. U. Jensen and H. Lenzing, *Model Theoretic Algebra With Particular Emphasis on Fields, Rings, Modules*, Algebra, Logic and Applications, Vol. 2, Gordon & Breach Science Publishers, New York, 1989.

[100] I. Kaplansky, Modules over Dedekind rings and valuation rings, *Trans. Amer. Math. Soc.*, 72(1952), 327–40.

[101] I. Kaplansky, On the dimension of modules and algebras X, A right hereditary ring which is not left hereditary, *Nagoya Math. J.*, 13(1958), 85–8.

[102] F. Kasch, *Modules and Rings*, Academic Press, London, New York, 1982.

[103] F. Kasch, M. Kneser, and H. Kupisch, Unzerlegbare modulare Darstellungen endlicher Gruppen mit zyklischer p-Sylow-Gruppe, *Archiv der Math.*, 8(1957), 320–1.

[104] G. M. Kelly, On the radical of a category, *J. Austral. Math. Soc.*, 4(1964), 299–307.

[105] O. Kerner, Tilting wild algebras, *J. London Math. Soc.*, 39(1989), 29–47.

[106] O. Kerner, Stable components of wild tilted algebras, *J. Algebra*, 142(1991), 37–57.

[107] G. Köthe, Verallgemeinerte Abelsche Gruppen mit hyperkomplexen Operatorenring, *Math. Z.*, 39(1934), 31–44.

[108] H. Kupish, Unzerlegbare Moduln endlicher Gruppen mit zyklischer p-Sylow Gruppe, *Math. Z.*, 108(1969), 77–104.

[109] S. Lang, *Algebra*, second edition, Addison-Wesley Publishing Company, Reading, Mass, 1984.

[110] H. Lenzing and J. A. de la Peña, Concealed-canonical algebras and separating tubular families, *Proc. London Math. Soc.*, 78(1999), 513–40.

[111] S. Liu, Tilted algebras and generalized standard Auslander–Reiten components, *Archiv der Math.*, 61(1993), 12–19.

[112] S. MacLane, *Homology*, Springer-Verlag, Berlin, Göttingen, Heidelberg, 1963.

[113] S. MacLane, *Categories for the Working Mathematicians*, Springer-Verlag, Berlin, 1972.

[114] R. Martinez and J. A. de la Peña, The universal cover of a quiver with relations, *J. Pure Appl. Algebra* 30(1983), 277–92.

[115] B. Mitchell, *Theory of Categories*, Academic Press, New York, 1966.

[116] B. Mitchell, Rings with several objects, *Advances in Math.*, 8(1972), 1–161.

[117] H. Meltzer and A. Skowroński, Group algebras of finite representation type, *Math. Z.*, 182(1983), 129–48; correction 187(1984), 563–9.

[118] K. Morita, Duality for modules and its applications to the theory of rings with minimum conditions, *Sci. Rep. Tokyo Kyoiku Daigaku*, A6(1958), 83–142.

[119] T. Nakayama, On Frobeniusean algebras, I, *Ann. Math.*, 40(1940), 611–33; II, ibid. 42(1941), 1–21.

[120] T. Nakayama, Note on uniserial and generalized uniserial rings, *Proc. Imp. Akad. Japan*, 16(1940), 285–9.

[121] L. A. Nazarova, Representations of quadruples, *Izv. Akad. Nauk SSSR,* 31(1967), 1361–78 (in Russian).

[122] L. A. Nazarova, Representations of quivers of infinite type, *Izv. Akad. Nauk SSSR*, 37(1973), 752–91 (in Russian).

[123] L. A. Nazarova and A. V. Roiter, On a problem of I. M. Gelfand, *Funk. Anal. i Priložen.*, 7(1973), 54–69 (in Russian).

[124] L. A. Nazarova and A.V. Roiter, Kategorielle Matrizen-Probleme und die Brauer-Thrall-Vermutung, *Mitt. Math. Sem. Giessen* 115(1975), 1–153.

[125] C. Nesbitt and W. M. Scott, Matrix algebras over algebraically closed field, *Trans. Amer. Math. Soc.*, 44(1943), 147–60.

[126] D. G. Northcott, *A First Course of Homological Algebra*, Cambridge University Press, Cambridge, 1973.

[127] S. Nowak and D. Simson, Locally Dynkin quivers and hereditary coalgebras whose left comodules are direct sums of finite dimensional comodules, *Comm. Algebra* 30(2002), 405–76.

[128] S. A. Ovsienko, Integral weakly positive forms, in *Schur Matrix Problems and Quadratic Forms*, Inst. Mat. Akad. Nauk USSR, Preprint 78.25, 1978, pp. 3–17 (in Russian)

[129] S. A. Ovsienko, A bound of roots of weakly positive forms, in *Representations and Quadratic Forms*, Acad. Nauk Ukr. S.S.R., Inst. Mat., Kiev, 1979, pp. 106–23 (in Russian).

[130] B. Pareigis, *Categories and Functors*, Academic Press, New York, London, 1970.

[131] R. S. Pierce, *Associative Algebras*, Springer-Verlag, New York, Heidelberg, Berlin, 1982.

[132] J. A. de la Peña, Algebras with hypercritical Tits form, in *Topics in Algebra, Part I: Rings and Representations of Algebras*, Banach Center Publications, Vol. 26, PWN Warszawa, 1990, pp. 353–69.

[133] J. A. de la Peña and A. Skowroński, Geometric and homological characterizations of polynomial growth strongly simply connected algebras, *Invent. math.*, 126(1996), 287–96.

[134] N. Popescu, *Abelian Categories with Applications to Rings and Modules*, Academic Press, London, New York, 1973.

[135] M. Prest, *Model Theory and Modules*, London Math. Soc. Lecture Notes Series 130, Cambridge University Press, Cambridge, 1988.

[136] I. Reiten, The use of almost split sequences in the representation theory of Artin algebras, in *Representations of Algebras*, Lecture Notes in Math. No. 944, Springer-Verlag, Berlin, Heidelberg, New York, 1981, pp. 29–104.

[137] Ch. Riedtmann, Algebren, Darstellungsköcher, Überlagerungen und zurück, *Comment. Math. Helvetici*, 55(1980), 199–224.

[138] C. M. Ringel, The indecomposable representations of the dihedral 2-groups, *Math. Ann.*, 214(1975), 19–34.

[139] C. M. Ringel, Representations of K-species and bimodules, *J. Algebra*, 41(1976), 269–302.

[140] C. M. Ringel, Report on the Brauer–Thrall conjectures: Rojter's theorem and the theorem of Nazarova and Rojter, in *Representation Theory II*, Proc. ICRA II (Ottawa, 1979), Lecture Notes in Math. No. 831, Springer-Verlag, Berlin, Heidelberg, New York, 1980, pp. 104–36.

[141] C. M. Ringel, Report on the Brauer–Thrall conjectures: Tame algebras, in *Representation Theory II*, Proc. ICRA II (Ottawa, 1979), Lecture Notes in Math. No. 831, Springer-Verlag, Berlin, Heidelberg, New York, 1980, pp. 137–287.

[142] C. M. Ringel, Kawada's theorem, in *Abelian Group Theory*, Lecture Notes in Math., No. 847, Springer-Verlag, Berlin, Heidelberg, New York, 1981, pp. 431–47.

[143] C. M. Ringel, Bricks in hereditary length categories, *Result. Math.*, 6(1983), 64–70.

[144] C. M. Ringel, Separating tubular series, in *Séminaire Bourbaki*, Lecture Notes in Math., No. 1029, Springer-Verlag, Berlin, Heidelberg, New York, 1983, pp. 134–58.

[145] C. M. Ringel, *Tame Algebras and Integral Quadratic Forms*, Lecture Notes in Math. No. 1099, Springer-Verlag, Berlin, Heidelberg, New York, Tokyo, 1984.

[146] C. M. Ringel, The regular components of Auslander–Reiten quiver of a tilted algebra, *Chinese Ann. Math.*, 9B(1988), 1–18.

[147] C. M. Ringel and H. Tachikawa, QF-3 rings, *J. Reine Angew. Math.*, 272(1975), 49–72.

[148] A. V. Roiter, The unboundedness of the dimension of the indecomposable representations of algebras that have an infinite number of indecomposable representations, *Izv. Acad. Nauk SSSR, Ser. Mat.*, 32(1968), 1275–82 (in Russian).

[149] J. J. Rotman, *An Introduction to Homological Algebra*, Academic Press, New York, 1979.

[150] M. Scott Osborne, *Basic Homological Algebra*, Springer-Verlag, New York, Berlin, Heidelberg, 2000.

[151] D. Simson, Functor categories in which every flat object is projective, *Bull. Polon. Acad. Sci., Ser. Math.*, 22(1974), 375–80.

[152] D. Simson, *Linear Representations of Partially Ordered Sets and Vector Space Categories*, Algebra, Logic and Applications, Vol. 4, Gordon & Breach Science Publishers, 1992.

[153] D. Simson, On representation types of module subcategories and orders, *Bull. Pol. Acad. Sci., Ser. Math.*, 41(1993), 77–93.

[154] D. Simson, On large indecomposable modules and right pure semisimple rings, *Algebra and Discrete Mathematics*, 2(2003), 93–117.

[155] A. Skowroński, Algebras of polynomial growth, in *Topics in Algebra, Part 1: Rings and Representations of Algebras*, Banach Center Publications, Vol. 26, PWN - Polish Scientific Publishers, Warszawa, 1990, pp. 535–68.

[156] A. Skowroński, Generalized standard Auslander–Reiten components without oriented cycles, *Osaka J. Math.*, 30(1993), 515–27.

[157] A. Skowroński, Regular Auslander–Reiten components containing directing modules, *Proc. Amer. Math. Soc.*, 120(1994), 19–26.

[158] A. Skowroński, Cycles in module categories, in *Finite Dimensional Algebras and Related Topics*, NATO ASI Series, Series C, Vol. 424, Kluwer Academic Publishers, 1994, 309–45.

[159] A. Skowroński, Generalized standard Auslander–Reiten components, *J. Math. Soc. Japan*, 46(1994), 517–43.

[160] A. Skowroński, Simply connected algebras and Hochschild cohomologies, *Canadian Mathematical Society Conference Proceedings*, AMS, 14, 1996, pp. 431–47.

[161] A. Skowroński, Module categories over tame algebras, *in* Workshop on Representations of Algebras, Mexico 1994, *Canadian Mathematical Society Conference Proceedings*, AMS, 19, 1996, pp. 281–313.

[162] S. O. Smalø, The inductive step of the second Brauer–Thrall conjecture, *Canad. J. Math.*, 2(1980), 342–9.

[163] S. O. Smalø, Torsion theories and tilting modules, *Bull. London Math. Soc.*, 16(1984), 518–22.

[164] B. Stenström, *Rings of Quotients*, Springer-Verlag, New York, Heidelberg, Berlin 1975.

[165] H. Tachikawa, *Quasi-Frobenius Rings and Generalizations*, Lecture Notes in Math. No. 351, Springer-Verlag, Berlin, Heidelberg, New York, Tokyo, 1973.

[166] H. Tachikawa and T. Wakamatsu, Tilting functors and stable equivalence for self-injective algebras, *J. Algebra*, 109(1987), 138–65.

[167] R. M. Thrall, On ahdir algebras, *Bull. Amer. Math. Soc.*, 53(1947), Abstract 22, 49–50.

[168] P. Webb, The Auslander–Reiten quiver of a finite group, *Math. Z.*, 179(1982), 79–121.

[169] C. A. Weibel, *An Introduction to Homological Algebra*, Cambridge Studies in Advanced Mathematics 38, Cambridge University Press, New York, 1997.

[170] K. Yamagata, On artinian rings of finite representation type, *J. Algebra*, 50(1978), 276–83.

[171] K. Yamagata, Frobenius algebras, in *Handbook of Algebra* (ed. M. Hazewinkel), Vol. 1, North-Holland Elsevier, Amsterdam, 1996, pp. 841–87.

[172] T. Yoshi, On algebras of bounded representation type, *Osaka Math. J.*, 8(1956), 51–105.

[173] W. Zimmermann, Einige Charakterisierung der Ringe über denen reine Untermoduln direkte Summanden sind, *Bayer. Akad. Wiss. Math.-Natur.*, Abt. II(1973), 77–9.

[174] B. Zimmermann–Huisgen, Rings whose right modules are direct sums of indecomposable modules, *Proc. Amer. Math. Soc.*, 77(1979), 191–7.

Index

451

List of symbols